Computational Intractability

Computational Intractability

A Guide to Algorithmic Lower Bounds

Erik D. Demaine, William Gasarch, and MohammadTaghi Hajiaghayi

The MIT Press
Cambridge, Massachusetts
London, England

The MIT Press
Massachusetts Institute of Technology
77 Massachusetts Avenue
Cambridge, MA 02139
mitpress.mit.edu

The MIT Press would like to thank the anonymous peer reviewers who provided comments on drafts of this book. The generous work of academic experts is essential for establishing the authority and quality of our publications. We acknowledge with gratitude the contributions of these otherwise uncredited readers.

This book was set in Times New Roman by Westchester Publishing Services. Printed and bound in the United States of America.

Library of Congress Cataloging-in-Publication Data

Names: Demaine, Erik D., 1981- author | Gasarch, William I. author |
Hajiaghayi, Mohammad Taghi, 1979- author
Title: Computational intractability : a guide to algorithmic lower bounds /
Erik D. Demaine, William Gasarch, MohammadTaghi Hajiaghayi.
Description: Cambridge, Massachusetts : The MIT Press, [2026] | Includes
bibliographical references and index.
Identifiers: LCCN 2026000530 (print) | LCCN 2026000531 (ebook) | ISBN
9780262550772 hardcover | ISBN 9780262381147 epub | ISBN 9780262381154
pdf
Subjects: LCSH: Algorithms | Computational complexity
Classification: LCC QA9.58 .D46 2026 (print) | LCC QA9.58 (ebook)
LC record available at https://lccn.loc.gov/2026000530
LC ebook record available at https://lccn.loc.gov/2026000531

EU Authorised Representative: Easy Access System Europe, Mustamäe tee 50, 10621 Tallinn, Estonia | Email: gpsr.requests@easproject.com

Contents

Part I NP and Its Variants

Preface

Some Problems Are Easy, Some Are Probably Hard

The following question permeates all of computer science and mathematics:

Given a problem, how hard is it to solve?

There are many aspects to hardness: time, space, communication, number-of-disk-calls, and others.

Why is it important to know that a problem is hard? If you know that obtaining (say) an exact solution to a problem is hard, then you may look for an approximation. If you know that the general case is hard, then you may want to look at special cases. There are many other scenarios.

Consider the following two problems:

1. Given a graph G, is there an ***Eulerian cycle***, which is a cycle that visits every *edge* exactly once? We call this problem EULERIAN CYCLE.
2. Given a graph G, is there a ***Hamiltonian cycle***, which is a cycle that visits every *vertex* exactly once? We call this problem HAMILTONIAN CYCLE.

For these two problems we will consider ***polynomial time*** (P) to be ***easy*** and NP-completeness to be ***hard***. We will define both of these terms in chapter 0.

In 1736 Euler showed (in modern terminology) that a graph G has an Eulerian cycle if and only if every vertex has even degree. Hence EULERIAN CYCLE is easy. Mathematicians later tried to find a characterization for HAMILTONIAN CYCLE. Note that the problem of finding a "characterization" was not well defined. In 1970 Cook [Coo71] and Levin [Lev73] developed the theory of NP-completeness, and in 1972 Karp showed that HAMILTONIAN CYCLE is NP-complete, meaning that it is unlikely to be in P. Thus the theory of NP-completeness not only showed that it is unlikely HAMILTONIAN CYCLE has a characterization but it also provided a way to state the questions rigorously.

P, NP, NP-Complete, and Our Book

In 1910 Pocklington [Poc10] analyzed two algorithms for solving quadratic congruences mod m (note that m takes $\lg m$ bits to represent) and noticed that

- one took time proportional to a power of $\lg m$, whereas
- the other took time proportional to $\sqrt{m}$.

In modern terms he was saying that one algorithm ran in polynomial (actually $O(\log m)$) time and the other algorithm took non-polynomial (actually $2^{O(\log m)}$) time. Unfortunately neither Pocklington, nor anyone else, pursued this distinction. Indeed, the notion that someone had earlier seen the distinction was not that well known until way after P and NP were defined (we did not know of Pocklington until we began working on this book). Pocklington's paper is the earliest reference to polynomial time that we know of.

In 1956 Gödel postal mailed (there was no email back then) a letter to von Neumann that, in modern terms, asked whether a problem that is NP-complete is actually in P. Unfortunately von Neumann never responded, and neither Gödel nor anyone else pursued the distinction. Indeed, this letter itself only came to light in 1989. Urquhart [Urq10] tells the entire story, plus why theoretical computer science did not emerge as a separate discipline until the 1960s.

In the early 1960s Cobham [Cob64] and Edmonds [Edm65] defined P and suggested it as a notion of efficient. Fortunately their definition was accepted and the stage was set for NP and NP-completeness.

Cook [Coo71] and Levin [Lev73] (see [Tra84] for a translation of Levin's article into English and some historical context) independently proved that SAT (Given a Boolean formula, is there a satisfying assignment?) is NP-complete. Shortly thereafter, Karp [Kar72] showed that 21 more natural problems are NP-complete problems. HAMILTONIAN CYCLE was one of them. Since then, literally thousands of problems have been proven NP-complete.

The initial papers used different definitions and notation. Garey and Johnson [GJ79] wrote a book that unified the definitions and notation, included most of what was known at the time, and became a bible for the field.

Garey and Johnson's book was published in 1979. There has been a lot of work since then on proving problems hard to solve. Hence a sequel is needed. While it is impossible for one book to encompass all of the work since then, this book will present some of that material. In particular we will cover, in broad terms, the following.

1. Since Garey and Johnson, several specific techniques have emerged for proving NP-hardness. We try to cover many of the major techniques, often illustrating them with examples that involve games and puzzles. In addition to personal preference, these examples make explicit the fun of finding reductions, and are useful for getting students excited about computer science.

2. Nuances on NP-completeness: hardness of approximation, counting (e.g., the number of satisfying assignments), truly exponential lower bounds (ETH), fixing parameters (FPT), and others.
3. Problems that are likely harder than NP (e.g., PSPACE, EXPTIME). For example, EXPTIME-complete problems are definitely not in P.
4. Problems that are in P but seem to require a certain polynomial amount of time (e.g., 3SUM-hard, APSP-hard).

Who Is the Audience for This Book

This book is based on a graduate course on the complexity of problems that Erik Demaine taught at MIT and Mohammad Hajiaghayi taught at the University of Maryland. The prerequisites for that course, and for the readers of this book, are (1) an undergraduate course in algorithms, (2) an undergraduate course in discrete mathematics, and, of course, (3) mathematical maturity. Exposure to computational complexity is not required, but it would not hurt.

Acknowledgments

This book evolved from scribe notes taken by students in a graduate course taught by Erik Demaine and Mohammad Hajiaghayi in the fall of 2014. We thank the scribes from both courses:

Scribes for Erik Demaine's course:
Aviv Adler, Hugo Akitaya, Jeffrey Bosboom, Aahlad Gogineni, Neil Gurram, Chennah Heroor, Adam Hesterberg, James Hirst, Justin Kopinsky, Jason Ku, Andrea Lincoln Quanquan Liu, Fermi Ma, Billy Moses, Asa Oines, Nathan Pinsker, William Qian, Mikhail Rudoy, Ariel Schvartzman, Jeffrey Shen, Rajan Udwani, Chelsea Voss, Erik Waingarten, Jonathan Weed, Kevin Wu, Patrick Yang, and Yun William Yu. Edited by TAs Sarah Eisenstat and Jayson Lynch.

Scribes for Mohammad Hajiaghayi's course:
Karthik Abinav, Melika Abolhassani, Aditya Acharya, Ahmed Abdelkader, Yingkang Cao, Joseph Carolan, Connor Clayton, Rushil Dandamudi, Philip Dasler, Sina Dehghani, Suchetan Dontha, Soheil Ehsani, Amin Shiraz Gilani, Huijing Gong, Thomas Pensyl, Manish Purohit, Saaed Seddighin, and Ateeq Sharfuddin.

We thank the following people for proofreading and helpful suggestions: Divesh Aggarwal, Victor Albert, Eric Allender, Ashwani Anand, Sricharan AR, Kiarash Banihashem, Huck Bennett, Gaëtan Berthe, Rob Brady, Gaston Brito, Josh Brunner, Yingkang Cao,

Joseph Carolan, Juan Chang, Nathan Cho, Lily Chung, Connor Clayton, Reiner Czerwinski, Rushil Dandamudi, Jenny Diomidova, Suchetan Dontha, Hossein Esfandiari, Yuval Filmus, Jon Forrest, Auguste Gezalyan, Amin Shiraz Gilani, Jacob Gilbert, Jacob Ginzburg, Evan Golub, Tom Hanson, Nathan Hayes, Della Hendrickson, Edmund Horsch, Hadi Jaffery, Gihan Jayatilaka, Matt Kovacs-Deak, Abrahim Ladha, Yi Lee, Harry Lewis, George Li, Vahid Liaghat, Issac Mammel, David Marcus, Erika Melder, Adam Melrod, Alexander Mendelson, Erik Metz, Grant Murray, Seyed Sajjad Nezhadi, Anthony Ostuni, Amadeo David De La Vega Parra, Grant Passmore, Jacob Prinz, Joel Rajakumar, Keivan Rezaei, Günter Rote, Rin Saito, Eashwar Sathyamurthy, Marcus Schaefer, Loren Schwiebert, Benjamin Sela, Jeffrey Shallit, Omri Shavit, Suho Shin, Morgan Sinclaire, Max Springer, Noah Stephens-Davidowitz, Samuel Tan, Georgios Tsimos, Ryan Williams, Patrick Yang, and Shaopeng Zhu.

We thank the following people for extra proofreading and scribe notes: Kiarash Banihashem, Samira Goudarzi, Jacob Gilbert, Nathan Hayes, Peyman Jabarzade, Erika Melder, and Jan Olkowski.

We thank Yaelle Goldschlag, Auguste Gezalyan, and Daniel Smolyak for many of the figures. We thank Auguste Gezalyan for technical support. We thank Costis Daskalakis, who was a guest lecturer in the course.

We thank Robert Hearn, who was an initial collaborator on this book, in particular for setting up the repository and the software used to write this book.

The first author thanks his father, Martin, for support throughout his life, in particular when creating this course and then book, and specifically for cinematography recording the lecture videos.

The second author thanks his wife, Carolyn, for her moral support and for helping him out when GitHub did something unexpected.

The third author thanks his wife, Arefeh, and his three children, Ilya, Sana, and Sarah, for their moral support and help. In particular thanks goes to Sana (7) and Ilya (9) for not proving $\mathsf{P} = \mathsf{NP}$, hence not making the book obsolete.

0

Computational Complexity Crash Course

0.1 Introduction

We begin with a brief introduction to the computational complexity needed to understand the rest of this book. In particular, we define some basic complexity classes, such as P, NP, coNP, PSPACE, EXPTIME, R, and how reductions define hardness and completeness. Figure 0.1 gives a visual overview of several of these classes and their relative difficulty. This chapter also serves as a general introduction to the entire book, with explicit pointers to the more advanced chapters.

0.2 Boolean Formulas and SAT

Boolean formulas and the problem SAT both motivate the definitions of P and NP, and are important in the study of NP.

Definition 0.1.

1. A ***Boolean variable*** is a variable that takes on the values TRUE and FALSE.
2. If x, y are Boolean variables, then the expression $x \wedge y$ is the AND of x and y. If x, y are both set to TRUE, then $x \wedge y$ is TRUE. All other settings set $x \wedge y$ to FALSE.
3. If x, y are Boolean variables, then the expression $x \vee y$ is the OR of x and y. If x, y are both set to FALSE, then $x \vee y$ is FALSE. All other settings set $x \wedge y$ to TRUE.
4. If x is a Boolean variable, then the expression $\neg x$ is the negation of x. If x is set to TRUE (FALSE), then $\neg x$ is set to FALSE (TRUE).

Definition 0.2. Boolean formula is defined inductively:

a. Any Boolean variable is a Boolean formula.
b. If A and B are Boolean formulas, then so are $(A \wedge B)$, $(A \vee B)$, and $\neg A$.

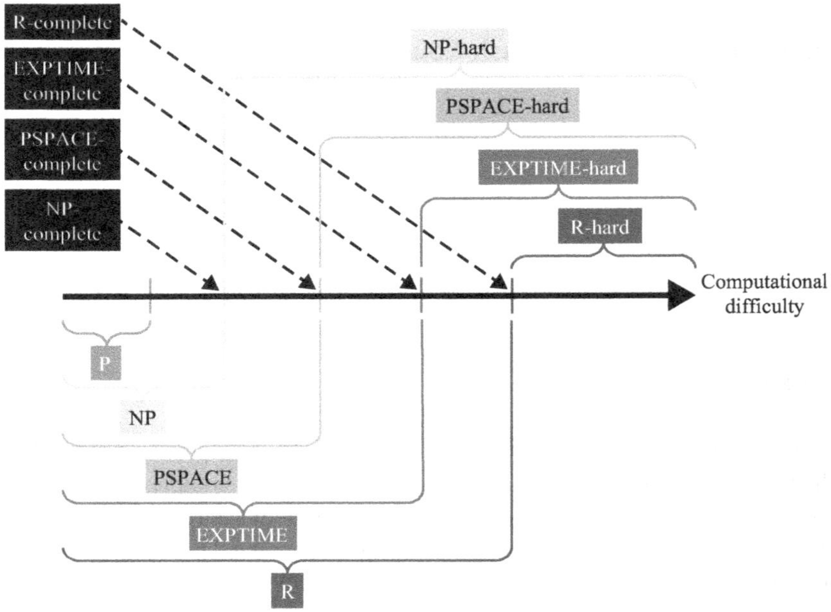

Figure 0.1
Hierarchy of classes.

Let $\varphi(x_1, \ldots, x_n)$ be a Boolean formula. The ***truth table*** for φ is a table of 2^n rows that gives, for each setting of the variables,

A ***literal*** is either a Boolean variable or its negation.

Convention 0.3. $((x_1 \wedge \neg x_2) \wedge x_3)$ is a Boolean formula. However, the parentheses are not needed. Hence we write this as $(x_1 \wedge \neg x_2 \wedge x_3)$. More generally we do not need parentheses when we have an $\wedge$ or an $\vee$ of many variables.

Example 0.4.

1. We give the truth table for

$$(x_1 \vee \neg x_2 \vee x_3) \wedge \neg x_3.$$

x_1	x_2	x_3	$(x_1 \vee \neg x_2 \vee x_3) \wedge \neg x_3$
TRUE	TRUE	TRUE	FALSE
TRUE	TRUE	FALSE	TRUE
TRUE	FALSE	TRUE	FALSE
TRUE	FALSE	FALSE	TRUE

x_1	x_2	x_3	$(x_1 \vee \neg x_2 \vee x_3) \wedge \neg x_3$
FALSE	TRUE	TRUE	FALSE
FALSE	TRUE	FALSE	FALSE
FALSE	FALSE	TRUE	FALSE
FALSE	FALSE	FALSE	TRUE

2. Assume you are given the formula $(x_1 \vee \neg x_2 \vee x_3) \wedge \neg x_3$, and you are asked whether there is some way to set the variables so that it returns TRUE. How would you do it? You could do the truth table in part 1. For three variables, that's $2^3 = 8$ rows, which isn't that many; however, you would not want to do this for (say) 20 variables. In this particular case, is there a shortcut? Yes. To make this formula TRUE, you need to make $\neg x_3$ TRUE, so you need to set x_3 to FALSE. Then, to make the first part true, set either x_1 to TRUE or x_2 to FALSE.

The example is a case of the following problem.

> SAT (short for SATISFIABILITY)
> *Instance:* A Boolean formula $\varphi(x_1, \ldots, x_n)$.
> *Question:* Is there a satisfying assignment for φ? That is, does there exist $(b_1, \ldots, b_n) \in \{\text{TRUE}, \text{FALSE}\}^n$ such that $\varphi(b_1, \ldots, b_n) = \text{TRUE}$?

The naïve algorithm is to try all 2^n $b_1, \ldots, b_n \in \{\text{TRUE}, \text{FALSE}\}^n$. This shows SAT can be solved in time roughly 2^n. Can we do better? Let's say that someone came up with an algorithm that worked in $2^{n/10}$. Would that be impressive? In practice it would work better than the naïve algorithm; however, it still looks like brute force with some tricks. How fast would an algorithm have to be so that its not regarded as a variant of brute force?

If someone had an algorithm that worked in time n^{100}, it would not be practical, but *the algorithm would not be using any variant of brute force*. In section 0.3 we will define *polynomial time* to make this notion rigorous.

0.3 P: Polynomial Time

Let A be a set of strings. The ***set membership problem*** for A asks for an algorithm that, given an input string x, quickly determines whether $x \in A$. How can we measure speed? Intuitively, we expect that longer strings take more time. Hence, to answer our question, we will study the algorithm's running time as a function of $n = |x|$, the length of the input x.

Definition 0.5.

1. A ***decision problem*** is a set membership problem, or equivalently, a problem for which the answer is either YES or NO. For example, "Given a graph, does it have an Euler circuit?" is a decision problem, equivalent to set membership in the set of all Eulerian graphs.
2. A ***problem*** will usually mean decision problem. It will sometimes mean function evaluation: Given x, compute $f(x)$. In some cases, it will mean a relation: Given x, find *some* y such that $R(x, y)$ is true.
3. We use the term ***instance*** to refer to the input x for a problem.

Definition 0.6.

1. P is the set of decision problems that can be solved in time that is bounded above by a polynomial in n, the length of the input.
2. FP is the set of functions that can be computed in time that is bounded above by a polynomial in n, the length of the input.

For both P and FP we sometimes use another notion of size that is polynomially related to the input length. For example, a dense graph on n vertices takes $\Theta(n^2)$ to represent; however, we will cheat and take n to be the length of the input. For the purpose of telling whether a problem is in P, this cheat does not matter. We note that we cannot always cheat. When discussing more linear and quadratic reductions for problems in NP, which we will do in chapters 2 and 6, we will be very careful about what we do and do not count for length-of-input.

To prove that a problem is in P, one can write an algorithm and show that its running time is bounded above by some polynomial. How do you prove that a problem is *not* in P? This needs a model of computation (as does making the definition of P rigorous). One common formal model is the Turing machine, which we will not define. A more familiar and equivalent model is the ***word RAM***, where an algorithm consists of a sequence of steps like "Add the two integers in these locations in memory and put the result in this memory location" or "If this location in memory stores a positive integer, go to step 5," and each integer is limited to a number of bits equal to the machine word size w. All you need to know is the following:

1. Turing machines and word RAM work in discrete steps. Hence one can measure the number of executed steps and thus running time.
2. Let A be a decision problem. The following are equivalent:
   a. There exists a program M in Python (or whatever your favorite programming language is) and a polynomial p such that, on input x, $M(x)$ will output $A(x)$ and terminate in $\leq p(|x|)$ steps.

b. There exists a word-RAM algorithm M and a polynomial q such that, on input x, $M(x)$ will output $A(x)$, and terminate in $\leq q(|x|)$ steps.

c. There exists a Turing machine M and a polynomial r such that, on input x, $M(x)$ will output $A(x)$, and terminate in $\leq r(|x|)$ steps.

3. We believe that *anything that can be computed at all* can be computed by a Turing machine or a word RAM. This assumption is often called the ***Church–Turing thesis***. See Soare's article [Soa13] for a historical prospective on this thesis.
4. The ***extended Church–Turing thesis*** says furthermore that anything that can be computed in polynomial time can be computed in polynomial time by a Turing machine (or a word RAM). But this stronger thesis might be inconsistent with some modern models of computation, such as quantum computing, leading to other complexity classes than P like BQP.

The Turing machines and word RAM just described are ***deterministic***, which is the default. We now discuss ***nondeterministic Turing machines***. One can similarly define ***nondeterministic Python*** (or your favorite programming language) or ***nondeterministic word RAM***. Nondeterministic Turing machines are only used for set membership. They are not used to compute functions. We will not describe them formally; however, we will give two contrasts to deterministic Turing machines and then discuss their robustness under polynomial time.

1. When a deterministic Turing machine or word RAM operates, given the state of the machine (what is in memory, what the input is, what memory location it is looking at), the next operation it does is determined. For example, on a word RAM, the instruction might be *if memory location 84 has a 1 in it, then erase that 1 and put in a 0.*
2. When a nondeterministic Turing machine or word RAM operates, given the state of the machine (what is in memory, what the input is, what memory location it is looking at), there is a set of next operations it can do. For example, on a word RAM, the instruction might be *if memory location 84 has a 1 in it, then set that memory location to 0,* or *set memory location 83 to 1.*

1. A deterministic Turing machine accepts an input x if when it is run it halts and outputs 1.
2. A nondeterministic Turing machine accepts an input x if there is some sequence of choices of instructions such that using those choices, it will halt and output 1.

Nondeterministic Turing machines are not actual devices. They are a mathematical construct.

For deterministic Turing machines, polynomial time is a robust notion in the sense that polynomial time on a deterministic Turing machine, a word RAM, a Python program, are equivalent. The same holds for nondeterministic Turing machines, nondeterministic word RAM, and nondeterministic Python programs.

0.4 Reductions

To show that a problem is not in P (or likely to not be in P) we need a notion of the following: *If B can be solved (quickly), then A can be solved (quickly).*

Definition 0.7. Let A and B be decision problems. A ***reduction from A to B*** is a function from instances of A to instances of B such that, for all x,

$$x \in A \text{ if and only if } f(x) \in B.$$

If such a reduction exists, we write $\boldsymbol{A \leq_p B}$.

Exercise 0.8. Prove the following:

1. If $A \leq_p B$ and $B \leq_p C$, then $A \leq_p C$.
2. If $B \in \mathsf{P}$ and $A \leq_p B$, then $A \in \mathsf{P}$.
3. If $A \notin \mathsf{P}$ and $A \leq_p B$, then $B \notin \mathsf{P}$. (This is just the contrapositive of the last item; however, it is very important.)

Exercise 0.8.3 gives a technique to show that a problem is not in P: To show $B \notin \mathsf{P}$, show that $A \leq_p B$ for some $A \notin \mathsf{P}$. Great! But we need some $A \notin \mathsf{P}$ to start with. Alas, proving $A \notin \mathsf{P}$ is hard. We consider a particular problem that will motivate our approach.

The reduction $A \leq_p B$ is defined so that, to answer whether $x \in A$, you get to ask *one* question to B and the answer must be the same. These are called ***polynomial-time many-one reductions***[1] or ***Karp reductions***. Polynomial-time reductions where you are allowed to ask many questions are called ***polynomial-time Turing reductions*** or ***Cook reductions***. In this book we will almost always deal with Karp reductions.

0.5 NP: Nondeterministic Polynomial Time

Recall the problem SAT:

> SAT (short for SATISFIABILITY)
> *Instance:* A Boolean formula $\varphi(x_1, \ldots, x_n)$.
> *Question:* Is there a satisfying assignment for φ? That is, does there exist $(b_1, \ldots, b_n) \in \{\text{TRUE}, \text{FALSE}\}^n$ such that $\varphi(b_1, \ldots, b_n) = \text{TRUE}$?
> *Note:* We will almost always take the length of a Boolean formula to be the number of variables in it.

1. "Many-one" or "many-to-one" means that the reduction can map multiple inputs to the same output. There are also 1-reductions where f has to be one-to-one.

As discussed earlier, the naïve algorithm runs in time roughly 2^n time. But note the following: If you *already had* $b_1, \ldots, b_n$, *it would be easy to tell whether* $\varphi(b_1, \ldots, b_n) = $ *TRUE*. This does not make SAT any easier; however, this motivates our definition of NP below.

First we define some very useful notation and give an example of it.

Notation 0.9. Let $\exists^p y$ denote "there exists y of length $q(|x|)$," where q is a polynomial and x is understood. Similarly, let $\forall^p y$ denote "for all y of length $q(|x|)$," where q is a polynomial and x is understood.

Example 0.10. Let

$$A = \{(G, y) \mid y \text{ is a 3-coloring of } G\}.$$

The set of graphs that are 3-colorable can be expressed as

$$\text{3-COLORING} = \{G \mid \exists y : (G, y) \in A\}.$$

The coloring y can be represented as a string over $\{0, 1\}$ of length $2n$, where n is the number of vertices in G.

Hence

$$\text{3-COLORING} = \{G \mid \exists y : |y| = 2n \wedge (G, y) \in A\}.$$

The length of y does not really matter to us so long as it is a polynomial. Hence

$$\text{3-COLORING} = \{G \mid \exists^p y : (G, y) \in A\}.$$

Exercise 0.11.

1. Show how to represent a (not necessarily proper) 3-coloring of a graph on n vertices with an element of $\{0, 1\}^{2n}$.
2. Let $c \geq 3$. Find r such that any (not necessarily proper) c-coloring of a graph on n vertices can be represented by an element of $\{0, 1\}^{rn}$.

We give two equivalent definitions of NP. We will mostly use the first one. The second one is the original definition. We will use it on occasion.

Definition 0.12.

1. $A \in$ NP if there exists a set $B \in$ P such that

$$A = \{x \mid \exists^p y : (x, y) \in B\}.$$

2. $A \in$ NP if it is accepted by a nondeterministic Turing machine that runs in polynomial time.

The intuition is that

- If $x \in A$, then there is a *short* witness (y) that can be used to *verify* $x \in A$ quickly.
- If $x \notin A$, then there is no such witness.

We list many NP problems. For the first few, we supply the witness. For the rest, we leave finding the witness and, hence, proving the problem is in NP as an exercise.

Example 0.13.

1. 3-COLORING: Given a graph, is it 3-colorable? The 3-coloring is the short quickly verified witness.
2. CLIQUE: Given a graph G and a number k, does G have a clique of size k? (A ***clique*** is a set $U \subseteq V$ such that all distinct $u, v \in U$ satisfy $\{u, v\} \in E$.) The clique is the short quickly verified witness.
3. INDEPENDENT SET: Given a graph G and a number k, does G have an independent set of size k? (An ***independent set*** is a set $U \subseteq V$ such that all distinct $u, v \in U$ satisfy $\{u, v\} \notin E$.)
4. VERTEX COVER: Given a graph $G = (V, E)$ and a number k, does G have a vertex cover of size k? (A ***vertex cover*** is a set $U \subseteq V$ such that every edge $e \in E$ has an endpoint in U.)
5. EULERIAN CYCLE: Given a graph, does it have an ***Eulerian cycle*** (a cycle that visits every edge exactly once)? This problem happens to also be in P because a graph is Eulerian if and only if every vertex has even degree. So the statement EULERIAN CYCLE $\in$ NP is correct but not as interesting as EULERIAN CYCLE $\in$ P.
6. FACTORING: Given a pair of numbers (n, a), is there a nontrivial factor of n that is $\leq a$? The factor is the short quickly verified witness. Note that the length of the input (n, a) is roughly $\lg n + \lg a$ because we represent numbers in binary.
7. GRAPH ISOMORPHISM: Given a pair of graphs (G, H), are they isomorphic? In other words, is there a bijection between the vertices of G and the vertices of H that preserves the existence of edges?
8. HAMILTONIAN CYCLE: Given a graph, does it have a ***Hamiltonian cycle*** (a cycle that visits every vertex exactly once)? (***Hamiltonian cycle*** is named for William Rowan Hamilton, an Irish mathematician who studied Hamiltonian cycles on the dodecahedron. Hamilton commercialized his study as the Icosian Game (so named because the dodecahedron is the dual of the icosahedron]).
9. TSP: Given a weighted graph and a number k, is there a Hamiltonian cycle of weight $\leq k$? Note that HAMILTONIAN CYCLE is a subcase of TSP.
10. SET COVER: Given sets $S_1, \ldots, S_m \subseteq \{1, \ldots, n\}$ and an integer k, are there k of the S_i's whose union is all of $\{1, \ldots, n\}$?
11. SHORTEST PATH: Given a graph G, two vertices s, t, and an integer c, is there a path from s to t that uses $\leq c$ edges? This problem is in P by using Breadth-First Search or Dijkstra's algorithm. So the statement SHORTEST PATH $\in$ NP is correct but not as interesting as SHORTEST PATH $\in$ P.

12. SUBSET SUM: Given integers $(a_1, \ldots, a_n, B)$, does some subset of $a_1, \ldots, a_n$ sum to B? The integers $a_1, \ldots, a_n, B$ are in binary, and hence the length of the input is approximately $\lg a_1 + \cdots + \lg a_n + \lg B$.
13. TETRIS: Given a position of a *Tetris* game, and knowledge of future falling pieces, is there a sequence of moves that does not lose the game? This puzzle or "perfect-information" version of TETRIS is not how the game is usually played, where a player does not know the future falling pieces. Intuitively, though, if this problem is hard, then the version where the player does not know the future pieces is also hard.
14. 0-1 PROGRAMMING: Given a matrix M of integers, vectors $\vec{b}$, $\vec{c}$ of integers, and d an integer, does there exist a vector $\vec{x}$ where each component is 0 or 1 such that $M\vec{x} \leq \vec{b}$ and $\vec{x} \cdot \vec{c} \geq d$?
15. INTEGER PROGRAMMING: Given a matrix M of integers, vectors $\vec{b}$, $\vec{c}$ of integers, and d an integer, does there exist a vector $\vec{x}$ of integers such that $M\vec{x} \leq \vec{b}$, $\vec{x} \geq \vec{0}$, and $\vec{x} \cdot \vec{c} \geq d$? For this problem, membership in NP is more challenging than for any other problem on this list. The reason is that you need to show that, if there is *some* $\vec{x}$, then there is an $\vec{x}$ whose components are not too big.

Some of the problems in NP have been worked on for many years (predating the formal definition of P); however, no polynomial-time algorithm for them is known. We need a way to say "These are the problems in NP that we think are not in P."

Definition 0.14. Let B be a problem.

1. B is ***NP-hard*** if, for all $A \in$ NP, $A \leq_p B$.
2. A problem B is ***NP-complete*** if $B \in$ NP and B is NP-hard.

It seems like it would be hard to find a natural problem that is NP-complete. To show B is NP-complete, one needs to show that $A \leq_p B$ for *every* $A \in$ NP. Every $A \in$ NP! Really! However, there is a natural NP-complete problem. Actually there are thousands! In the early 1970s, Cook [Coo71] and Levin [Lev73] (see [Tra84] for a translation of Levin's article into English and some historical context) independently proved the following:

Theorem 0.15 (Cook–Levin theorem). *SAT is NP-complete.*

Shortly after Cook showed there is one natural NP-complete problem, Karp [Kar72] showed that 21 natural problems are NP-complete. We mention a few to give a sense of the variety of the problems: HAMILTONIAN CYCLE, SUBSET SUM, and 0-1 PROGRAMMING.

The proof of theorem 0.15 codes Turing machine computations into formulas. Once SAT was shown to be NP-complete, Turing machines were no longer needed: To show A is NP-complete, just show SAT $\leq_p A$. Or if B is already known to be NP-complete, then show $B \leq_p A$.

The problems in example 0.13 are NP-complete except (1) EULERIAN CYCLE $\in$ P, (2) SHORTEST PATH $\in$ P, (3) FACTORING and GRAPH ISOMORPHISM are in NP but not known to be either in P or NP-complete. We discuss these two problems later in this chapter.

For all the problems in example 0.13 (except INTEGER PROGRAMMING), the proof that they are in NP is very easy. This is typical. When proving that a problem B is NP-complete, we will usually omit the (easy) proof that $B \in$ NP. In the rare case that proving $B \in$ NP is hard or unknown, we will point it out.

The following notation is probably familiar to you; however, we include them for completeness (not NP-completeness) because they are needed for the next exercise.

Notation 0.16. Let Σ be an alphabet. Let $A, B \subseteq \Sigma^*$.

1. $A \cdot B = \{xy \mid x \in A \text{ and } y \in B\}$. This is called the ***concatenation of A and B***.
2. $A^0 = \{e\}$ where e is the empty string.
3. $A^i = A \cdot A \cdot \cdots \cdot A$ where the A appears i times.
4. $A^* = A^0 \cup A^1 \cup A^2 \cup \cdots$.
5. $\overline{A} = \Sigma^* - A$.

Exercise 0.17. For each of the following statements, either prove it, disprove it, or state that it is unknown to science.

1. If $A, B \in$ P, then $A \cup B \in$ P.
2. If $A, B \in$ P, then $A \cap B \in$ P.
3. If $A, B \in$ P, then $A \cdot B \in$ P.
4. If $A \in$ P, then $\overline{A} \in$ P.
5. If $A \in$ P, then $A^* \in$ P.
6. If $A, B \in$ NP, then $A \cup B \in$ NP.
7. If $A, B \in$ NP, then $A \cap B \in$ NP.
8. If $A, B \in$ NP, then $A \cdot B \in$ NP.
9. If $A \in$ NP, then $\overline{A} \in$ NP.
10. If $A \in$ NP, then $A^* \in$ NP.

0.6 coNP

The complement of SAT is the set of all formulas that have *no* satisfying assignment. We call this set CONTRA (for "contradiction"). Formally:

$$\text{CONTRA} = \{\varphi \mid \forall \vec{b} : \varphi(\vec{b}) = \text{FALSE}\}.$$

In terms of polynomial time, clearly SAT $\in$ P if and only if CONTRA $\in$ P because $\varphi \in$ SAT if and only if $\varphi \notin$ CONTRA. Is CONTRA in NP? Probably not. Contrast the following:

1. For Alice to convince Bob that $\varphi \in$ SAT, Alice can give Bob a satisfying truth assignment. The assignment is short and can be verified by Bob in polynomial time.
2. For Alice to convince Bob that $\varphi \in$ CONTRA, Alice can give Bob the truth table for φ. This is exponential in the size of φ and would take Bob exponential time to verify.

The question "Is CONTRA in NP?" is asking whether there is a short verifiable witness that $\varphi \in$ CONTRA. This does not appear to be the case.

So how to classify CONTRA? We define a class ***coNP*** in two ways. This class is where CONTRA naturally lies.

Definition 0.18. $A \in$ coNP if there exists a set $B \in$ P such that

$$A = \{x \mid \forall^p y : (x, y) \in B\}.$$

(coNP stands for "co-nondeterministic polynomial," which comes from an equivalent definition of NP that we do not use.)

The intuition is that

- If $x \notin A$, then there is a *short* witness (y) that can be used to *verify* $x \notin A$ quickly.
- If $x \in A$, then there is no such witness.

Another definition:

Definition 0.19. $A \in$ coNP if $\overline{A} \in$ NP.

It is widely believed that NP $\neq$ coNP.

We can define ***coNP-hardness*** and ***coNP-completeness*** similar to definition 0.14. CONTRA is coNP-complete. Assuming NP $\neq$ coNP, CONTRA is not NP-complete.

0.7 The Winner Is P $\neq$ NP

Gasarch [Gas19] has done three polls of what theorists think of P vs. NP, and other questions. In the last one, in 2019, 88% thought P $\neq$ NP. Why do theorists largely think P $\neq$ NP? Why do we think so?

1. There are thousands of problems that are NP-complete. For many of them, people have worked on getting fast algorithms for a long time, predating the definition of P and NP. In a nutshell: *If (say) SAT is in P, then someone would have discovered the algorithm for it by now.*

2. Intuitively, *verifying* seems easier than *finding*. As an example: Finding a proof of a theorem is often hard, but verifying a proof is easy (or should be).
3. The assumption $\mathsf{P} \neq \mathsf{NP}$ has great explanatory power. We give one example. Consider the function version of SET COVER, which is, given $S_1, \ldots, S_m \subseteq \{1, \ldots, n\}$, find the minimal k such that k of the S_i's cover $\{1, \ldots, n\}$.
   a. In 1979 Chvatal [Chv79] showed that a simple greedy algorithm yields a polynomial-time $\ln n$-approximation. That is, if the algorithm outputs x, then the answer is $\leq x \ln n$. Much work went into trying to obtain a $(1 - \varepsilon) \ln n$ approximation.
   b. In 2013 Dinur and Steurer [DS13] showed that, if there exists $\varepsilon > 0$ such that there is a $(1 - \varepsilon) \ln n$ approximation, then $\mathsf{P} = \mathsf{NP}$.

   The algorithm and the lower bound (based on $\mathsf{P} \neq \mathsf{NP}$) are independent of each other. This phenomenon—where a long-standing approximation is shown to be optimal assuming $\mathsf{P} \neq \mathsf{NP}$—is common. Hence the assumption $\mathsf{P} \neq \mathsf{NP}$ seems to settle many open problems in the direction we think they should go.

0.8 Strong NP-Completeness—Ch. 5

Recall SUBSET SUM. The input consists of $n+1$ numbers *in binary*. As such the problem is NP-complete. But what if the input numbers were specified *in unary*? Then the problem turns out to be in P:

Exercise 0.20. Show that, if the inputs are in unary, then SUBSET SUM $\in$ P.
Hint: Use dynamic programming.

Hence the NP-completeness of SUBSET SUM is related to the fact that the inputs are in binary. Some problems are NP-hard even when the inputs are in unary:

Definition 0.21.

1. A problem is ***weakly NP-hard*** if it is NP-hard when the inputs are in binary.
2. A problem is ***strongly NP-hard*** if it is NP-hard when the inputs are in unary.

In chapter 5, we will show that many problems are strongly NP-hard.

0.9 Intermediate Problems

In example 0.13, we noted (without proof) that

- EULERIAN CYCLE $\in$ P.
- CLIQUE, SAT, SET COVER, SUBSET SUM, TETRIS, and 0-1 PROGRAMMING are all NP-complete.

The above dichotomy is typical; most problems in NP turn out to be either in P or NP-complete. But there are exceptions (assuming $\mathsf{P} \neq \mathsf{NP}$).

Definition 0.22. A problem is ***NP-intermediate*** if it is in NP but neither in P nor NP-complete.

Ladner [Lad75] showed the following:

Theorem 0.23 (Ladner's theorem). *If $\mathsf{P} \neq \mathsf{NP}$, then there exists an NP-intermediate problem.*

The problem from Ladner's theorem is a contrived set constructed for the sole purpose of being in NP, not NP-complete, and not in P. But there are also a few natural and well-studied problems that are *conjectured* to be NP-intermediate:

1. ***Factoring.*** We approach integer factoring via the set

$$\text{Factoring} = \{(N, a) \mid \text{there is a nontrivial factor of } N \text{ that is } \leq a\}.$$

It is easy to see that, if $\text{Factoring} \in \mathsf{P}$, then the problem of, given a number, find a non-trivial factor of it (or output that there isn't one) would be in FP. This problem is very important because many modern crypto system can be cracked if factoring is easy. Hence it has gotten much attention.

Currently there is no polynomial-time algorithm for Factoring, nor is there a proof that it is NP-complete. Algorithms for factoring are hard to analyze and depend on (widely believed) conjectures in number theory. The fastest known algorithm, the general number field sieve, is believed to have running time roughly $2^{1.93\, L^{1/3}(\lg L)^{2/3}}$, where $L = \lg N$ is the length of the input number N. This bound is small enough that the algorithm is practical for moderately large inputs. The naïve algorithm for factoring takes time $2^{L/2}$. Hence reducing the time to roughly $2^{L^{1/3}}$ is a real improvement. In general, a ***subexponential*** bound of $2^{L^{\alpha}}$ where $0 < \alpha < 1$ is strictly in between any polynomial bound of n^c and any exponential bound of 2^{L^c} where c is a positive integer.

There is no clear consensus on whether Factoring is in P; however, if it is, then proving this will require new techniques. Here is why:

a. The last improvement in factoring algorithms was the number field sieve in 1988.

b. There are reasons to think that the current methods yield algorithms with running times of the form $2^{L^t (\ln L)^{1-t}}$, where $0 < t < 1$. The general number field sieve achieves $t = 1/3$. It is plausible that current techniques will solve Factoring in time (say) $2^{L^{1/10}(\ln L)^{9/10}}$ but not in P.

There is another possible route to fast factoring: Peter Shor [Sho94, Sho99] showed that factoring can be done in polynomial time on a quantum computer. This result has led to an explosion of interest in quantum computing.

So is FACTORING NP-complete? Unlikely: If FACTORING is NP-complete, then NP = coNP. We will guide you through a proof in exercise 0.24. For more about factoring, see Wagstaff's book [Wag13].

2. ***DISCRETE LOG.*** We describe discrete logarithm as a function and leave it to the reader to give the decision version.

   DISCRETE LOG: Given prime p, generator g, and $a \in \mathbb{Z}_p$, find x such that $g^x \equiv a \pmod{p}$.

   Much like factoring, (1) DISCRETE LOG is important because, if it was easy to solve, the Diffie–Hellman key exchange protocol in cryptography would be cracked; (2) DISCRETE LOG is not known to be in P; (3) DISCRETE LOG is not known to be NP-complete; (4) there are algorithms for DISCRETE LOG that are better than the naïve one, though still exponential (see [Wikc]); (5) there has not been a better algorithm for DISCRETE LOG since 1998; (6) Shor [Sho94, Sho99] has shown there is a quantum polynomial-time algorithm for DISCRETE LOG; and (7) If DISCRETE LOG is NP-complete, then NP = coNP; hence DISCRETE LOG is unlikely to be NP-complete.

3. ***GRAPH ISOMORPHISM.*** So far there has been no polynomial-time algorithm for GRAPH ISOMORPHISM; however, there has also been no proof that it is NP-complete. The fastest algorithm for GRAPH ISOMORPHISM, due to Babai [Bab16], has running time $2^{(\lg n)^c}$ for some constant c ($c = 3$ seems likely). It is believed that a ***quasipolynomial*** running time of $2^{(\lg n)^c}$ with $c > 1$ is the best one can do with current methods. So is GRAPH ISOMORPHISM NP-complete? The consensus is "no" for the following reasons:

   a. If GRAPH ISOMORPHISM is NP-complete, then all problems in NP are solvable in quasipolynomial time $2^{(\lg n)^{O(1)}}$, which seems unlikely. In particular, this would violate the EXPONENTIAL TIME HYPOTHESIS described in section 0.10 below.

   b. Boppana et al. [BHZ87] proved that, if GRAPH ISOMORPHISM is NP-complete, then $\Sigma_2 = \Pi_2$ (as defined in section 0.15 below). The proof is slightly beyond the scope of this book.

4. ***Minimum circuit size problem (MCSP)*** is the following problem: Given the truth table of a Boolean function f and an integer s, does f have a circuit with at most s logic gates?

   Kabanets and Cai [KC00] showed that if MCSP $\in$ P, then there is no secure crypto system. We take that as evidence that MCSP $\notin$ P. Can we show that with a reduction? Probably not: Murry and R. Williams [MW17] showed that if MCSP is NP-complete under Karp reductions, then some long-standing open problems would be solved. In addition, Saks and Santhanam [SS20] showed that if MCSP is NP-hard under other reductions (in between Karp and Turing reductions) then some other long-standing open problems would be solved. These results do not indicate that there is no such reduction; however, they indicate that finding one will be hard.

For a survey of a large body of work on this topic, see the papers of Allender [All21] and Hirahara-2022 [Hir22].

Exercise 0.24. Let PRIMES be the problem of, given a number, to determine whether it is prime. Let UFT be the theorem that every number can be factored uniquely into primes.

1. Look up or prove for yourself that PRIMES $\in$ NP. (It turns out that PRIMES $\in$ P [AKS04]; however, all we need for this exercise is that PRIMES $\in$ NP.)
2. Use UFT and PRIMES $\in$ NP to prove that $\overline{\text{FACTORING}} \in$ NP.
3. Use FACTORING $\in$ NP $\cap$ coNP to show that, if FACTORING is NP-complete, then NP $=$ coNP.

0.10 ETH—Ch. 6

So how long does SAT actually take to solve? The hypothesis SAT $\notin$ P still allows for the possibility that, say, SAT $\in n^{O(\log n)}$. It is widely believed that SAT requires exponential time. More precisely, it is believed that there exists a sequence $s_3, s_4, \ldots$ such that kSAT requires $2^{s_k n}$ time. This belief is encapsulated by the Exponential Time Hypothesis (ETH). We will present this hypothesis and its implications in chapter 6. In order to use ETH, we need the reductions to be linear, not just polynomial.

We give one contrast between assuming P $\neq$ NP and assuming ETH.

1. Assuming P $\neq$ NP, VERTEX COVER is not in polynomial time.
2. Assuming ETH, VERTEX COVER requires time $2^{\Omega(n)}$.

The ETH still allows for the possibility that, say, kSAT $\in 2^{n/k}$. It is widely believed that as k gets larger, kSAT gets harder. In fact, it is believed that $\lim_{k \to \infty} s_k = 1$. This belief is encapsulated by the Strong Exponential Time Hypothesis (SETH). We will present this hypothesis and its implications in chapter 6.

ETH and SETH were proposed to get better lower bounds on NP-complete problems. However, perhaps surprisingly, they have also been used on the lower level of showing that some problems (essentially) require quadratic or cubic time. We examine this in chapters 17 and 18.

0.11 FPT: Fixed-Parameter Tractability—Ch. 7

Consider the following two problems:

$$\text{CLIQUE} = \{(G, k) \mid G \text{ has a clique of size } k\},$$

$$\text{VERTEX COVER} = \{(G, k) \mid G \text{ has a vertex cover of size } k\}.$$

As noted after theorem 0.15, both of these problems are NP-complete. Now let's fix k. Define

$$\text{CLIQUE}_k = \{G \mid G \text{ has a clique of size } k\},$$

$$\text{VERTEX COVER}_k = \{G \mid G \text{ has a vertex cover of size } k\}.$$

Both CLIQUE_k and VERTEX COVER_k are in time $O(n^k)$ and, hence, in P. This does not imply that VERTEX COVER is in P because the exponent k is a function of the input size. In order to be in P, VERTEX COVER would need to be in time $17n^2$ or $84n^3$ or some time bound that has a constant factor and a constant exponent, both of which are independent of the size of the input.

So both CLIQUE_k and VERTEX COVER_k seem to have complexity $O(n^k)$. Can this be improved? Does the exponent of the time bound have to depend on k? More to the point, are these problems similar or different? It turns out that they are different:

1. VERTEX COVER_k can be solved in time $O(2^k n)$. (This does not imply that VERTEX COVER $\in$ P because the bound has the 2^k term, which is a function of the input size.)
2. There are reasons to think that CLIQUE_k requires $n^{\Omega(k)}$ time.

Problems like VERTEX COVER_k are called ***fixed-parameter tractable*** or FPT. We will study techniques to show that problems are likely not FPT in chapter 7.

0.12 Complexity of Functions and Approximation—Ch. 8, 9, and 10

So far we have discussed *decision problems*. For example, CLIQUE is the problem of deciding whether a graph has a clique of a given size k. But the real problem people want answered is, given a graph G, to return the size of the largest clique. Or better, return a clique of the maximum size (and because there may be more than one, this would be that rare case where we study the complexity of a relation). If our only question is "Can we solve CLIQUE in polynomial time?" then this turns out to be a minor issue: All of these problems are essentially equivalent.

Exercise 0.25. Recall that FP is the set of all functions that are computable in polynomial time.

1. Let $\text{CLIQUESIZE}(G)$ return the size of the largest clique in G. Show that

   $$\text{CLIQUE} \in \mathsf{P} \text{ if and only if } \text{CLIQUESIZE} \in \mathsf{FP}.$$

2. Let $\text{FINDCLIQUE}(G)$ return a clique of size $\text{CLIQUESIZE}(G)$. Show that

   $$\text{CLIQUE} \in \mathsf{P} \text{ if and only if } \text{FINDCLIQUE} \in \mathsf{FP}.$$

3. For all the decision problems A in example 0.13, define a function version f_A. Show that $A \in \mathsf{P}$ if and only if $f_A \in \mathsf{FP}$.

With regard to polynomial time, computing CLIQUESIZE(G) is as hard as computing CLIQUE(G). But what about *approximating* CLIQUESIZE(G)? We will study the complexity of approximation in chapters 8, 9, and 10.

0.13 Complexity Classes That Use Randomization

In this section we informally describe two randomized classes for, pardon the pun, completeness. Actually, there is an irony here in that these classes do not have complete problems.

Definition 0.26. A decision problem A is in ***randomized polynomial time (RP)*** if there is a polynomial-time algorithm M that can flip coins satisfying the following properties:

- If $x \in A$, then the probability that $M(x)$ says $x \in A$ is $\geq \frac{1}{2}$.
- If $x \notin A$, then the probability that $M(x)$ says $x \notin A$ is 1.

Some facts about RP:

1. G. Miller [Mil76] obtained a polynomial-time algorithm for PRIMES that depends on the ***extended Riemann hypothesis*** being true. Rabin [Rab80] modified the algorithm to be in RP without any hypothesis. The final algorithm is called ***The Miller-Rabin test***. It is actually used by cryptographers. For many years it remained open whether PRIMES $\in$ P until Agrawal et al. [AKS04] showed that it is. The algorithm they obtained is too slow to be useful; however, it is still interesting that PRIMES $\in$ P.
2. There are very few problems that are in RP that are not known to be in P. We state one: Given a polynomial $f(x_1, \ldots, x_n)$ and a prime p, is the polynomial identically zero over $\mathbb{Z}_p$?
3. There are reasons to think that P $=$ RP. See the items about BPP after definition 0.28.

Exercise 0.27.

1. Let $0 < \alpha < \frac{1}{2}$. In the definition of RP, replace $\frac{1}{2}$ with α and call the resulting class RP$_\alpha$. Show that RP$_{1/2}$ = RP$_\alpha$.
2. Let $\alpha(n)$ be a decreasing function from $\mathbb{N}$ to $(0, \frac{1}{2})$. In the definition of RP, replace $\frac{1}{2}$ with $\alpha(|x|)$ and call the resulting class RP$_{\alpha(n)}$. Is RP$_{1/n^2}$ = RP? Is RP$_{1/2^n}$ = RP?

Note that RP was defined to have one-sided error. We now define a randomized class that has two-sided error.

Definition 0.28. A set A is in ***bounded probabilistic polynomial time (BPP)*** if there is a polynomial-time algorithm M that can flip coins satisfying the following properties:

- If $x \in A$, then the probability that $M(x)$ says $x \in A$ is $\geq \frac{3}{4}$.
- If $x \notin A$, then the probability that $M(x)$ says $x \notin A$ is $\geq \frac{3}{4}$.

Some facts about BPP:

1. There are no natural problems that are known to be in BPP but not known to be in RP. There aren't even any known unnatural problems.
2. There are reasons to think that P = BPP. A sequence of papers by Nisan and Wigderson [NW94], Babai et al. [BFNW93], and Impagliazzo and Wigderson [IW97] showed that, assuming a reasonable hardness assumptions for circuits, one can prove that a pseudorandom generator exists and hence that P = BPP. There have been many papers following up on this theme. See the surveys of Kabanets [Kab02], Impagliazzo's [Imp03], and Trevisan [Tre12] for more information.
3. It is not known whether BPP $\subseteq$ NP.
4. Lautemann [Lau83] and Sipser [Sip83] proved that BPP $\subseteq \Sigma_2 \cap \Pi_2$ (as defined in section 0.15 below). Lautemann's proof is simpler; however, Sipser's paper started the field of time-bounded Kolmogorov complexity.

Exercise 0.29.

1. Let $\frac{1}{2} < \alpha < 1$. In the definition of BPP, replace $\frac{3}{4}$ with α and call the resulting class BPP_α. Show that $\mathsf{BPP}_{3/4} = \mathsf{BPP}_\alpha$.
2. Let $\alpha(n)$ be a decreasing function from $\mathbb{N}$ to $(0, \frac{1}{2})$ In the definition of BPP, replace $\frac{3}{4}$ with $\alpha(|x|)$ and call the resulting class $\mathsf{BPP}_{\alpha(n)}$. Is $\mathsf{BPP}_{1-1/n^2} = \mathsf{BPP}$? Is $\mathsf{BPP}_{1-1/2^n} = \mathsf{BPP}$?

0.14 Counting Problems—Ch. 11

Recall that SAT is the problem of, given a Boolean formula, *does there exist* a satisfying assignment? Consider the counting version:

> #SAT
> *Instance:* A Boolean formula $\varphi(x_1, \ldots, x_n)$.
> *Output:* The number of $\vec{b} \in \{\text{TRUE}, \text{FALSE}\}^n$ such that $\varphi(\vec{b}) = \text{TRUE}$. That is, how many satisfying assignments does φ have?

Clearly SAT is easier than (or just as easy) as #SAT. Note that the counting version is a function, not a set. It turns out that #SAT seems to be much harder than SAT. What about counting versions of other problems? For example, the problem #CLIQUE is the following: Given a graph G and a number k, how many cliques of size k are in G?

We study the hardness of counting problems, and related notions, in chapter 11.

0.15 Polynomial Hierarchy

We look at a natural problem that does not seem to be in NP.

> MIN FORMULA
> *Instance:* A formula $\varphi(x_1,\ldots,x_n)$ and a number L.
> *Question:* Is there a formula $\psi(x_1,\ldots,x_n)$ of length $\leq L$ that is equivalent to φ (i.e., always produces the same output)?

Example 0.30.

1. $\varphi(w,x,y,z) = (w \vee x) \wedge (w \vee y) \wedge (w \vee z)$. There is a shorter equivalent formula, namely $w \vee (x \wedge y \wedge z)$.
2. $\varphi(w,x,y,z) = w \vee x \vee y \vee z$. There is no shorter equivalent formula.

Exercise 0.31.

1. Show that, if $\mathsf{P} = \mathsf{NP}$, then MIN FORMULA $\in \mathsf{P}$.
2. Vary the notion of size in various ways (e.g., number of $\wedge$, number of $\neg$) and determine whether $\mathsf{P} = \mathsf{NP}$ implies that this variant of MIN FORMULA is in P.

Consider an alternate formulation of MIN FORMULA that asks whether a formula φ is the smallest formula with the same truth table:

$$\text{MIN FORMULA} = \{\text{Boolean formula } \varphi : \forall \psi, |\psi| < |\varphi| : \exists \vec{b} : \varphi(\vec{b}) \neq \psi(\vec{b})\}.$$

This formulation of MIN FORMULA does not put it into NP. We seem to need a $\forall$ and then a $\exists$.

We now define classes that use more quantifiers, so we will have a place to put MIN FORMULA. Stockmeyer [Sto76] defined the levels of the PH.

Definition 0.32. Let $k \in \mathbb{N}$.

1. $A \in \Sigma_k$ if there exists $B \in \mathsf{P}$ such that

$$A = \{x \mid \exists^p y_1 : \forall^p y_2 : \exists^p y_3 : \forall^p y_4 : \cdots : Q^p y_k : (x, y_1, \ldots, y_k) \in B\}$$

   (where Q is the appropriate quantifier).
   Note that $\Sigma_1 = \mathsf{NP}$.
2. $A \in \Pi_k$ if there exists $B \in \mathsf{P}$ such that

$$A = \{x \mid \forall^p y_1 : \exists^p y_2 : \forall^p y_3 : \exists^p y_4 : \cdots : Q^p y_k : (x, y_1, \ldots, y_k) \in B\}$$

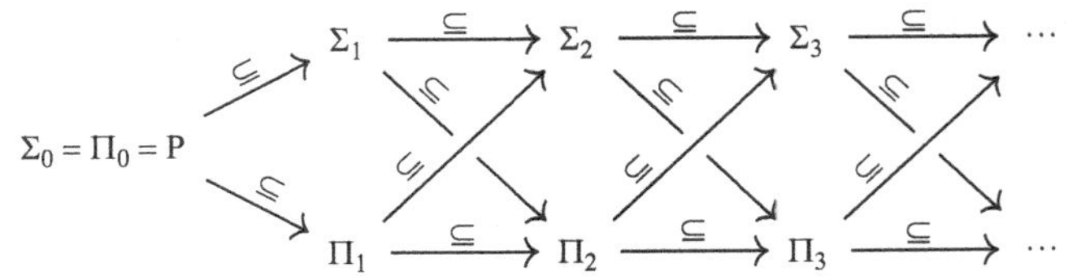

Figure 0.2
The polynomial hierarchy.

(where Q is the appropriate quantifier).

Note that $\Pi_1 = \mathsf{coNP}$.

Definition 0.33. The interwoven hierarchies in figure 0.2 are known collectively as the ***polynomial hierarchy (PH)***:

Note: The notations Σ_i^p and Π_i^p are sometimes used because the notations Σ_i and Π_i are also used in computability theory, where the quantifiers are unbounded. See section 0.19.

Exercise 0.34. Let $i \geq 1$.

1. Show that, if $\Sigma_i = \Pi_i$, then $\Sigma_i = \Pi_j = \Sigma_j$ for all $j \geq i$ ("PH collapse").
2. Define Σ_i-hard, Σ_i-complete, Π_i-hard, and Π_i-complete.

It is believed that the PH is proper—that is, all containment relations in definition 0.33 are strict, though not with the same confidence as the belief that $\mathsf{P} \neq \mathsf{NP}$.

Any NP problem can be modified to form a (possibly contrived) Σ_2 problem. If the original NP problem is NP-complete, then the modified problem is Σ_2-complete. We give one example.

> Σ_2-SAT
> *Instance:* A Boolean formula with two sorts of variables: $\varphi(\vec{x}, \vec{y})$.
> *Question:* Does there exist $\vec{b}$ such that, for all $\vec{c}$, $\varphi(\vec{b}, \vec{c}) = \text{TRUE}$?

Exercise 0.35.

1. For all NP problems presented in this chapter, come up with a modified version that is in Σ_2 and does not seem to be in Σ_1.
2. Define Σ_i-SAT.
3. For all NP problems presented in this chapter, come up with a modified version that is in Σ_i and does not seem to be in Σ_{i-1}.

The following are known:

1. Σ_2-SAT is Σ_2-complete.
2. MIN FORMULA is in Π_2 but is not known to be Π_2-complete.

We can view Σ_2-SAT as a game. The board is the given formula $\varphi(\vec{x},\vec{y})$. First Alice picks an assignment for $\vec{x}$. Then Bob picks an assignment for $\vec{y}$. If the resulting assignment makes φ TRUE, Alice wins; otherwise, Bob wins. Note that $\varphi(\vec{x},\vec{y}) \in \Sigma_2$ if and only if Alice can win.

One can extend this intuition to Σ_k and Π_k. Intuitively, Σ_k corresponds to a k-move game where you move first, and Π_k corresponds to a k-move game where your opponent moves first. An example of a real-world video game where puzzles become Σ_2-complete is "The Witness" with clues [ABC+20]. We will also encounter Σ_2 in section 11.7.

Exercise 0.36. Skim the paper by M. Schaefer and Umans [SU08], which presents over 80 problems complete on some level of the PH.

0.16 EXPTIME, PSPACE, and EXPSPACE—Ch. 13, 14, and 15

Definition 0.37.

1. EXPTIME is the set of problems that can be solved in time that is exponential in the size n of the problem instance—that is, in $2^{n^{O(1)}}$ time.
2. PSPACE is the set of problems solvable in polynomial *space* (memory), without any bound on running time.
3. EXPSPACE is the set of problems that can be solved in space that is exponential in the size n of the problem instance—that is, in $2^{n^{O(1)}}$ space.

Note: By a simple diagonalization argument, EXPTIME $-$ P $\neq \emptyset$ and EXPSPACE $-$ PSPACE $\neq \emptyset$. Hence, if a problem A is EXPTIME-complete, then $A \notin$ P, and similarly A being EXPSPACE-complete implies $A \notin$ PSPACE. This is in contrast to A being NP-complete, which we think implies $A \notin$ P, but we do not know. These separations are corollaries to the more general ***time hierarchy theorem*** and the ***space hierarchy theorem***.

Example 0.38.

1. CHESS is the following problem: Given a position in chess (on an $n \times n$ board), and assuming both players play perfectly, will white win? For most natural versions of chess, this problem is EXPTIME-complete.
2. In chapters 13, 14, and 16, we will discuss more games that are complete or hard in these classes.

3. Let QSAT (Quantified SAT) consist of all expressions of the form

$$\exists\vec{x}_1 : \forall\vec{x}_2 : \exists\vec{x}_3 : \forall\vec{x}_4 : \cdots : \exists\vec{x}_{k-1} : \forall\vec{x}_k : \varphi(\vec{x}_1, \ldots, \vec{x}_k)$$

that are true. QSAT is PSPACE-complete.

PSPACE thus contains every level of PH as PSPACE allows for an arbitrary (polynomial) number of alternations in quantifiers.

0.17 NSPACE: Nondeterministic Space

Our definition of NP involves quantifiers and a witness y. Our definition of NSPACE will need to use nondeterministic Turing machines.

Definition 0.39. $A \in$ NPSPACE if there exists a nondeterministic Turing machine or word-RAM algorithm M, and a polynomials p such that, on input x, any choice of instructions forms a computation that takes $\leq p(|x|)$ space.

More generally, for any function $S : \mathbb{N} \to \mathbb{N}$, we can define SPACE$(S(n))$ and NSPACE$(S(n))$ to allow $S(n)$ space on an input of size n. Savitch [Sav70] showed the following beautiful relation between SPACE and NSPACE:

Theorem 0.40 (Savitch's theorem). *Let $f : \mathbb{N} \to \mathbb{N}$ be a function, with $f(n) \geq n$. Then NSPACE$(f(n)) \subseteq$ SPACE$(f(n)^2)$. In particular, NPSPACE = PSPACE.*

Exercise 0.41. Either prove Savitch's theorem or look up its proof, read it, and understand it.

If $S(n)$ is sublinear in n, then we can still define SPACE$(S(n))$ and NSPACE$(S(n))$ in a meaningful way. Specifically, we restrict the input to be read only, and allow the algorithm only $S(n)$ read/write memory.

Notation 0.42. We define two sublinear space classes. For both of them, the input is on a separate read-only tape.

1. L is SPACE$(\log n)$.
2. NL is NSPACE$(\log n)$.

NL has a notion of NL-completeness. For example, the following problem is NL-complete: Given a directed graph G and two vertices s, t, is there a directed path from s to t? The proof is easy because one can view a nondeterministic logarithmic-space computation as a directed graph. There is a belief that L $\neq$ NL, but it is not as strong as the belief that P $\neq$ NP.

0.18 R: Decidable Sets—Ch. 15 and 16

Definition 0.43. R is the set of decision problems that can be solved by a Turing machine or word-RAM algorithm, with no limit on how long it runs other than "finite." Problems in R are also called ***decidable***.

The R stands for ***recursive*** because at one time, "decidable" sets were called "recursive." Soare's article [Soa96] has an excellent historical perspective on the reason to change the terminology from "recursive" to "decidable." (Sometimes in the literature R stands for randomized polynomial time. When this comes up, we will use RP.)

Essentially all problems in this book are decidable—that is, in R. In chapters 15 and 16, we will discuss problems that are undecidable.

0.19 Arithmetic Hierarchy

Given two problems that are undecidable, is there a way of saying that one is more undecidable? What measure of complexity can you use? The ***arithmetic hierarchy (AH)*** is a way to classify problems using the number of quantifiers. Given the order we presented the material in, you might think that the AH is modeled after the PH from section 0.15; however, it is the other way around. The AH was defined by Kleene [Kle43] in 1943.

More generally, much of the early work in complexity theory was modeled after earlier work in computability theory.

Definition 0.44. We give four equivalent definitions of when a decision problem A is ***computably enumerable***:

1. If there exists a Turing machine or word-RAM algorithm M such that

$$A = \{x \mid \exists y : M(y) \text{ halts and outputs } x\}.$$

   (So A is the *range* of a computable function. Note that M might not halt on some inputs.)
2. If either $A = \emptyset$, or there exists a Turing machine or word-RAM algorithm M that halts on all inputs and satisfies

$$A = \{x \mid \exists y : M(y) \text{ (halts and) outputs } x\}.$$

   (So A is empty or the *range* of a total computable function.)
3. If there exists a Turing machine or word-RAM algorithm M such that

$$A = \{x \mid M(x) \text{ halts}\}.$$

   (So A is the *domain* of a computable function.)

4. If there exists a $B \in \mathsf{R}$ such that

$$A = \{x \mid \exists y : (x, y) \in B\}.$$

Note: What we call ***computably enumerable (c.e.)*** is also called ***recursively enumerable (r.e.)***. Soare [Soa96] has tried to get the community to change from "r.e." to "c.e." and gives good arguments for the change.

Exercise 0.45. Show that all four definitions of c.e. in definition 0.44 are equivalent.

The fourth definition of c.e. in definition 0.44 looks like NP, which is identical to Σ_1. The next exercise asks you to define Π_1, Σ_2, Π_2, etc. in the context of decidability.

Exercise 0.46.

1. In section 0.15 on the polynomial hierarchy, we defined Σ_i and Π_i by adding alternating polynomial-bounded quantifiers to P. Define similar classes by adding alternating unbounded quantifiers to R. Call these classes Σ_i and Π_i also (they will only be used within this problem so there should be no confusion). This is the ***AH***.
2. Let $M_0, M_1, \ldots$ be a list of *all* Turing machines or word-RAM algorithms in some order. For each of the following decision problems, determine which Σ_k or Π_k it is in. Try to make k as low as possible.

$$\text{FIN} = \{i \mid M_i \text{ halts on an infinite number of inputs}\}$$

$$\text{TOT} = \{i \mid M_i \text{ halts on all inputs}\}$$

In section 15.5 we will briefly discuss one problem that does not involve Turing machines yet is not in AH.

0.20 A Few Separations of Complexity Classes

Separating P from NP seems quite hard. Now that we have introduced more complexity classes, though, there are a few pairs of classes that we know are different. In all cases the proof is by a simple diagonalization argument that we omit.

Theorem 0.47.

1. $\mathsf{P} \subsetneq \mathsf{EXPTIME} \subsetneq \mathsf{R}$.
2. $\mathsf{P} \subseteq \mathsf{NP} \subseteq \mathsf{EXPTIME} \subsetneq \mathsf{R}$.
3. $\mathsf{P} \subseteq \mathsf{NP} \subseteq \mathsf{PSPACE} \subsetneq \mathsf{EXPSPACE} \subsetneq \mathsf{R}$.
4. $\mathsf{P} \subseteq \mathsf{NP}$ *and* $\mathsf{NP} \subseteq \mathsf{EXPTIME}$ *so, by part 1, one of these two is proper.*
5. $\mathsf{P} \subseteq \mathsf{PSPACE}$ *and* $\mathsf{PSPACE} \subseteq \mathsf{EXPTIME}$ *so, by part 1, one of these two is proper.*

Refer back to figure 0.1 for a visual overview.

0.21 Polynomial Lower Bounds—Ch. 17 and 18

Once we know that a problem is in P, the question arises: What is its running time? Can we have a notion similar to NP-hardness to show that problems within P are unlikely to have subquadratic algorithms? Subcubic algorithms?

To show that problems are unlikely to be in subquadratic time (that is, $O(n^{2-\delta})$ for some δ), we need a base problem (similar in spirit to SAT) that is unlikely to be in subquadratic time. There is such a problem.

> 3SUM
> *Instance:* n integers.
> *Question:* Do three of the integers sum to 0?
> *Note:* We will consider any arithmetic operation to have unit cost.

Despite enormous effort, nobody has obtained a subquadratic algorithm for 3SUM in a reasonable model of computation. In chapter 17 we show that many problems are unlikely to be in subquadratic time by (1) assuming that 3SUM is not in subquadratic time and (2) using an appropriate notion of reduction.

To show that problems are unlikely to be in subcubic time (that is, $O(n^{3-\delta})$ for some δ), we need a base problem (similar in spirit to SAT) that is unlikely to be in subcubic time. There is such a problem.

> All Pairs Shortest Paths (APSP)
> *Instance:* A weighted directed graph $G = (V, E, w)$ with weights in $\mathbb{N}$.
> *Output:* For all pairs of vertices x, y, $\text{dist}_G(x, y)$.
> *Note:* We will consider any arithmetic operation to have unit cost.

Despite enormous effort, nobody has obtained a subcubic algorithm for APSP. In chapter 18 we show that many problems are unlikely to be in subcubic time by (1) assuming that APSP is not in subcubic time and (2) using an appropriate notion of reduction.

0.22 Online Algorithms—Ch. 19

An ***online problem*** is one where (1) the input is given in pieces (e.g., requests for memory access), and (2) the answer is a sequence of answers (e.g., assignments of whether to put a page in cache or memory), where each answer must be given each time you get a piece of the input *before* the rest of the input is given.

For such problems the issue is *not* how fast the algorithm runs. The algorithm needs to, as soon as it gets a new piece of information, give a response quickly. There is another goal in mind. For example, in the case of memory access, the goal is to minimize the number of times that a page that is not in the cache is requested. The issue is then how well the algorithm does on its goal *compared to what the optimal would be if we knew all future information.*

We study lower bounds for online algorithms in chapter 19. The main tool used is adversary arguments, where an adversary will, given the algorithm, find a way to input the data that causes the algorithm to do badly. These lower bounds are unconditional. There is no need to assume some problem is hard.

0.23 Streaming Algorithms—Ch. 20

Streaming algorithms are similar to online algorithms in that the data comes in pieces. A ***streaming problem*** is one where (1) the input is given in pieces (e.g., edges of a graph), (2) we may allow several passes through the data, and (3) the answer is a string (e.g., an answer to "Does the graph have a connected component of size 100?").

For such problems, the issue is *not* how fast the algorithm runs. We note that the algorithms involved are usually quite fast. The issue is twofold: How much memory does the algorithm use (it cannot store all of the data as we think of the data stream as being enormous), and how many passes over the data does the algorithm use? There is often a trade-off between these two parameters.

We study lower bounds for streaming algorithms in chapter 19. The main tool used is communication complexity. These lower bounds are unconditional: There is no need to assume some problem is hard.

0.24 Parallel Algorithms—Ch. 21

A ***parallel algorithm*** is one where many machines work on a problem at the same time. In chapter 21, we study a recent model of parallelism called ***massively parallel computation (MPC)***. The key parameters in this model are (1) the number of rounds a computation takes, (2) the memory each machine has, and (3) the number of processors. The lower bounds are often trade-offs between these three parameters.

There are two kinds of lower bounds: ***unconditional***, which need no assumptions, and ***conditional***, which need to assume some problem is hard. The problem often used is 1vs2-CYCLE (defined below), which seems to be hard to solve in parallel.

1vs2-Cycle
Instance: An undirected graph $G = (V, E)$. We are promised that it is either one cycle or the union of two cycles.
Question: Is the graph one cycle or the union of two cycles?

0.25 Game Theory—Ch. 22

Imagine that Alice and Bob are playing a game. It is a simple game: Each player simultaneously chooses 1 out of n options, and that pair (Alice's Choice, Bob's Choice) determines how many points each person gets. They will play this game many times.

Nash showed that, if they are both allowed to have a probabilistic strategy (often called a ***mixed strategy***), then there exists a pair (Alice's strategy, Bob's strategy) such that both players know that, if they deviate from the strategy but their opponent does not, then they will do worse than if they stuck with their strategy. Such an ordered pair is called a ***Nash equilibrium***.

Nash's proof that a Nash equilibrium exists does not give a method for finding it. This puts us in a curious position:

1. The problem "Is there a Nash equilibrium?" is trivial: The answer is YES.
2. The problem "Find the Nash Equilibrium" seems hard.

There are reasons to think that finding the Nash equilibrium is not NP-hard. Hence we need another way to prove that it is hard.

There are other problems where a solution always exists, but finding it seems to be hard. To show that problems of this type are unlikely to be in polynomial time, we need a base problem (similar in spirit to SAT) that is unlikely to be in polynomial time. There is such a problem.

Unbalanced
Instance: A directed graph G and an unbalanced node v.
Question: Is there another unbalanced vertex?

This problem is in P for the odd reason that, by theorem 22.2, there is *always* another unbalanced vertex. What if we actually want to *find* that other unbalanced vertex? We certainly can find it in polynomial time by looking at every vertex. But note that the algorithm does not use the proof of theorem 22.2.

What if we were given a short representation (size poly in n) of a large directed graph (size 2^n)? Then you cannot just look at every vertex. We will also restrict the graph to have both in-degree and out-degree ≤ 1. We now define this formally.

END OF LINE (EOL)

Instance: Two circuits P (for previous) and N (for next) on $\{0, 1\}^n$ such that for all $x \in \{0, 1\}^n$ the following holds.

- $P(x)$ returns either NO or an element of $\{0, 1\}^n$.
- $N(x)$ returns either NO or an element of $\{0, 1\}^n$.

We interpret the circuits as describing a directed graph by the following:

- If $P(x) = y$, then there is an edge from x to y.
- If $N(x) = y$, then there is an edge from y to x.

Output: If exactly one of $P(0^n)$, $N(0^n)$ is not NO, then find another unbalanced node. (Such an unbalanced node exists. We will state this in theorem 22.2.)

Note: This is quite different from *finding* the unbalanced node that has a path to 0^n. Papadimitriou [Pap94, Theorem 2] showed that problem is PSPACE-hard. Subtle changes in a problem definition may alter things tremendously.

Despite some effort, nobody has obtained a polynomial time for EOL. In chapter 22 we show that some problems, including finding a Nash equilibrium, are unlikely to be in polynomial time by (1) assuming the EOL $\notin$ P and (2) using an appropriate notion of reduction.

I

NP and Its Variants

1

NP-Hardness via SAT

1.1 Introduction

In this chapter we look at many variants of SAT and show how to use the NP-hard versions to prove various problems NP-hard.

Chapter Summary

1. *Some variants of SAT are in P and some are NP-complete. A small change in the definition can cause a large change in the complexity.*
2. *Different versions of SAT are useful for proving different problems NP-complete. We give examples.*
3. *Curiously, all of the versions of SAT we analyze are either in P or NP-complete. None are intermediate. We will state a theorem that explains this phenomenon. Theorems of this type are called* ***dichotomy theorems****.*

1.2 Variants of SAT

We list many variants of SAT and specify which are in P and which are NP-complete. We mostly do not provide proofs; however, you should consider each statement to be an exercise.

1.2.1 CNF SAT, DNF SAT, and Circuit SAT

We start with three core versions (indeed, special cases) of SAT.

Definition 1.1. Let φ be a Boolean formula over variables $x_1, \ldots, x_n$.

1. The terms x_i and $\neg x_i$ are called ***literals***.
2. φ is in ***conjunctive normal form (CNF)*** if it is an AND of m terms:

$$\varphi = C_1 \wedge \cdots \wedge C_m,$$

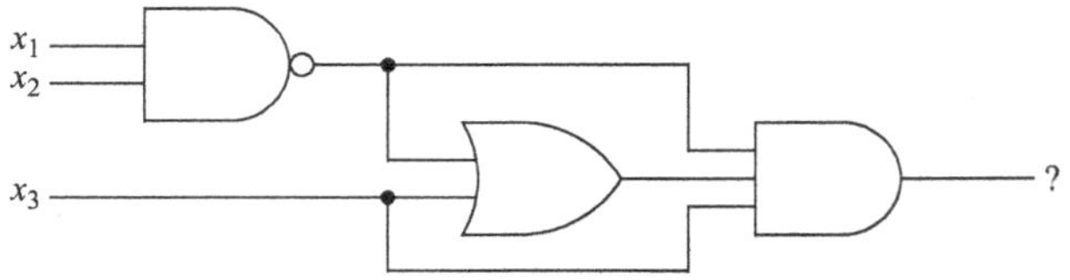

Figure 1.1
A Boolean circuit, with information flowing from left to right. The gates with a straight-line base are AND; the gates with a curved-line base are OR; and the circle is a negation (NOT).

where each C_i is an OR of literals, such as $x_3 \vee \neg x_7 \vee x_8 \vee \neg x_{11}$. The C_i's are called ***clauses***.

3. φ is in ***disjunctive normal form (DNF)*** if it is an OR of m terms:

$$\varphi = D_1 \vee \cdots \vee D_m,$$

where each D_i is an AND of literals.

4. A ***Boolean circuit*** is a circuit with (1) inputs $x_1, \ldots, x_n$; (2) one output; (3) some gates, each of which is AND, OR, or NOT; and (4) acyclic connections, with each connection from a gate output (or an input) to a gate input (or the output). See figure 1.1 for an example.

5. Every Boolean circuit computes a Boolean formula. For example, the circuit of figure 1.1 computes

$$\neg(x_1 \wedge x_2) \wedge (\neg(x_1 \wedge x_2) \vee x_3) \wedge x_3. \tag{1.1}$$

Crucially, though, a circuit can split one output into inputs to multiple gates, whereas the straightforward conversion to a formula would have to repeat those intermediate computations. For example, the straightforward conversion of figure 1.1 into formula (1.1) repeats $\neg(x_1 \wedge x_2)$ twice, once for its use in the OR gate and once for its use in the final AND gate.

6. For any class of formulas X (e.g., CNF), the decision problem ***X SAT*** (e.g., ***CNF SAT***) asks: Given a formula in the class X, is it satisfiable? In other words, can the variables be set to make the formula evaluate to TRUE?

7. For Boolean circuits, we define the analogous decision problem ***Circuit SAT*** (and ***Circuit X SAT***): Given a circuit (in the class X), can the inputs be set to make the circuit output TRUE?

Theorem 1.2.

1. *CNF SAT is NP-complete. Cook [Coo71] and Levin [Lev73][1] proved this, though they did not state it in this form. In particular, they gave polynomial-time algorithms that convert an arbitrary Boolean formula into a CNF formula (with additional variables) having the same satisfiability.*
2. *DNF SAT is in P. This is an easy exercise: Check whether any term can be satisfied—that is, has no contradicting literals.*
3. *Circuit SAT is NP-complete. This is an easy consequence of SAT being NP-complete.*

Exercise 1.3. Since CNF SAT is NP-complete and DNF SAT is in P, one approach to proving P = NP is to find a function $f \in$ FP that will take a CNF formula φ and return an equivalent DNF formula ψ. Alas this cannot work:

Prove that any DNF formula that is equivalent to

$$(x_1 \vee y_1) \wedge (x_2 \vee y_2) \wedge \cdots \wedge (x_n \vee y_n)$$

is of the form $D_1 \vee \cdots \vee D_L$, where (1) each D_i is the $\wedge$ of literals and (2) $L \geq 2^n$.

1.2.2 2SAT Versus 3SAT

How many literals need to be in the clauses of a CNF formula to make CNF SAT NP-complete? The answer turns out to be 3.

Definition 1.4. ***2SAT*** is CNF SAT restricted to formulas that have ≤2 literals per clause. ***3SAT*** is CNF SAT restricted to formulas that have ≤3 literals per clause. We will soon generalize this concept.

Example 1.5. Consider the 2SAT formula

$$(x \vee y) \wedge (\neg x \vee z) \wedge (\neg y).$$

Is it satisfiable, that is, in 2SAT? In any satisfying assignment the following happens: $\neg y$ has to be TRUE, so y is set to FALSE. Since $x \vee y$ has to be TRUE and y is FALSE, x must be set to TRUE. Since $\neg x \vee z$ has to be TRUE, and $\neg x$ is FALSE, z has to be true. Hence $(x, y, z) =$ (TRUE, FALSE, TRUE) satisfies the formula. Of more interest is that many variable settings were forced. This is the intuition as to why 2SAT is in P.

We will sketch the proof that 2SAT $\in$ P. This was first proven by Krom [Kro67]—a general result that implies it was proven by Aspvall et al. [APT79]. We present the proof by Apsvall et al. [APT79]. See also the exposition on Wikipedia [Wikb].

1. See [Tra84] for a translation of Levin's article [Lev73] into English and some historical context.

Theorem 1.6. *2SAT is in P.*

Proof. Here is the algorithm for 2SAT. We assume that the input has exactly two literals in each clause. We leave it as an exercise to extend the algorithm to handle the case where some clauses have one literal.

1. Input a formula φ of the form $C_1 \wedge \cdots \wedge C_m$, where each C_i has exactly two literals.
2. Form a directed graph G as follows. The vertices are all of the variables that occur in the formula and their negations. For each clause $(A \vee B)$ where A, B are literals, add the directed edges $(\neg A, B)$ and $(\neg B, A)$. (If A is $\neg x$, then replace $\neg\neg x$ with x.) These edges are intended to represent the two implications $\neg A \rightarrow B$ (if A is false, then B must be true) and $\neg B \rightarrow A$ (if B is false, then A must be true), which together are equivalent to $A \vee B$.
3. Using breath-first or depth-first search, determine whether there is some variable x so that (1) there is a directed path in G from x to $\neg x$, and (2) there is a directed path in G from $\neg x$ to x. If so, then 2SAT is not satisfiable and the algorithm returns NO. If not, then 2SAT is satisfiable and the algorithm returns YES.

The algorithm clearly works in polynomial time since breath-first and depth-first search are fast. We show that the algorithm is correct.

Assume that the algorithm returns YES. Let (u, v) be any edge in the graph. Set the literal represented by vertex u to TRUE, which sets the corresponding variable to either TRUE or FALSE (depending on whether u is negative). Now reduce the formula to remove these assigned variables: If a set-to-TRUE literal occurs in a clause, then remove the clause (as it is already satisfied); and if a set-to-FALSE literal (i.e., the negation of a set-to-TRUE literal) occurs in a clause, then remove the literal from the clause, leaving a single literal, which forces that literal to also be set to true (which may, in turn, cause more clause reduction and forcing). In particular, this will set all literals corresponding to vertices reachable from u (including v) to be set to TRUE. Because there are no directed paths from a literal to its negation, the assignment to these variables can be carried out without contradiction. If, when all of the forcing is done, there are still some variables that are not assigned, then repeat the process until all variables are assigned or the graph is empty. In the latter case, the remaining variables can be set arbitrarily.

Assume that the algorithm returns NO. Let x be such that there is a directed path from x to $\neg x$ and from $\neg x$ to x. Assume, by way of contradiction, that there is a satisfying assignment. Assume that x is assigned TRUE (the case where it is assigned FALSE is symmetric). Look at the path $L_1, \ldots, L_k$ from x to $\neg x$:

$L_1 = x$	L_2	$\cdots$	L_{k-1}	$L_k = \neg x$
TRUE				FALSE

Let i be the least index such that L_i is assigned FALSE. There is an edge from (L_{i-1}, L_i). This is a contradiction since L_{i-1} is TRUE and L_i is FALSE. □

Next we sketch the proof that 3SAT is NP-complete, by a reduction from CNF SAT. Cook [Coo71] proved the following theorem.

Theorem 1.7. *3SAT is NP-complete.*

Proof. By theorem 1.2, CNF SAT is NP-complete. We show that CNF SAT $\leq_p$ 3SAT.

Before giving the proof, we give an instructive example. Consider the one-clause formula $\varphi = (L_1 \vee L_2 \vee L_3 \vee L_4)$, where each L_i is a literal. Let z be a new variable (not in any L_i). Look at

$$\psi = (L_1 \vee L_2 \vee z) \wedge (\neg z \vee L_3 \vee L_4).$$

We claim that $\varphi \in$ SAT if and only if $\psi \in$ SAT. In one direction, if φ has a satisfying assignment, then we know it sets some L_i to TRUE, which satisfies one of the two clauses in ψ; we can fully satisfy ψ by setting z to satisfy the other clause (TRUE to satisfy the first clause, or FALSE to satisfy the second clause). In the other direction, if ψ has a satisfying assignment, then z's setting satisfies at most one clause (the first if TRUE and the second if FALSE), so the other clause must be satisfied from some L_i being set TRUE; thus φ is also satisfied.

Based on this example, we describe the general reduction. We are given a CNF formula φ. If φ is not already 3CNF, then it has some clause $C = (L_1 \vee \cdots \vee L_k)$, where $k \geq 4$. We can reduce this clause's size by replacing it with two clauses:

$$C' = (L_1 \vee L_2 \vee z) \vee (\neg z \vee L_3 \vee \cdots \vee L_k).$$

By a similar argument to the example, C' is satisfiable if and only if C is; essentially, z satisfies one clause while the other must be satisfied by one of the original literals $L_1, \ldots, L_k$. The reduction repeats this process, finding a clause C of size $k \geq 4$ and replacing it with two clauses C' sizes 3 and $k-1$, until all clauses have size 3. This process terminates in linear time because each replacement reduces by 1 the number of literals beyond 3 in each clause, summed over clauses. □

1.2.3 Maximization Makes 2SAT Hard

What if the goal is not to satisfy *all* of the clauses, but instead to satisfy as many as possible? We prepend the word MAX to indicate that goal. So MAX 2SAT is the function that will, given a 2CNF formula φ, returns the maximum number of clauses that can be simultaneously satisfied. To talk about NP-completeness, we need to define the associated decision problem: Given a 2CNF formula φ and a positive integer k, is there an assignment that satisfies at least k clauses?

Exercise 1.8. Prove that the decision version of MAX 2SAT is NP-complete.

Exercise 1.9. Show that the following problem is NP-complete: Given m clauses, each of which is the AND of three literals, is there an assignment to the variables that satisfies a majority of the clauses—that is, at least $m/2$ of the clauses have all three literals set to TRUE?

We will return to MAX 2SAT in the context of inapproximability in chapter 9.

1.2.4 Restricting Clause Size and Variable Counts

There are many variations of SAT. As we saw with 2SAT and 3SAT, one can restrict the number of literals in a clause. There are some subtleties in the definition, though. Must a 3SAT clause have *exactly* three literals, or at most three? Related, can the same variable appear multiple times in a clause? Different papers unfortunately use the same notation (3SAT) to mean different things (with different answers to these questions). Fortunately, Filho [Fil19], in his master's thesis, devised a unifying system of notation that we use. He also unified many old results and proved some new ones; his system encompassed far more variants of SAT than we will discuss.

Complementary to restricting the size of clauses, we can look to restrict the number of times a variable can appear in the formula. We define many variants on SAT according to these two aspects of restriction.

Definition 1.10. Let $a, b \in \mathbb{N}$. In all of the problems below, the goal is the same as CNF SAT: Given a CNF formula, does it have a satisfying assignment? What varies is the form of the input formula.

1. ***aSAT***: Every clause has $\leq a$ literals per clause. For example, if $a \geq 3$, then $(x \vee x \vee y)$ and $(x \vee \neg x \vee y)$ are allowed, as is $(x \vee y)$.
2. ***EaSAT***: Every clause has *exactly a* literals per clause. For example, if $a \geq 3$, then again $(x \vee x \vee y)$ and $(x \vee \neg x \vee y)$ are allowed, but $(x \vee y)$ is not. This problem is equivalent to aSAT but becomes different when we add other modifiers below.
3. ***EUaSAT***: Every clause has *exactly a* literals per clause, and every variable within a clause occurs *uniquely*. For example, if $a \geq 3$, then $(x \vee x \vee y)$, $(x \vee \neg x \vee y)$, and $(x \vee y)$ are *not* allowed.
4. ***SAT-b***: Every variable occurs $\leq b$ times. (If x occurs once and $\neg x$ occurs once, then that is two occurrences.)
5. ***SAT-Eb***: Every variable occurs *exactly b* times. (If x occurs once and $\neg x$ occurs once, then that is two occurrences.)
6. ***SAT-EUb***: Every variable occurs *exactly b* times, and every occurrence is in a different clause. (This variant has not actually appeared before but is natural given EUaSAT.)
7. We leave it to the reader to define all of the pairwise combinations:

SAT	SAT-b	SAT-Eb	SAT-EUb
aSAT	aSAT-b	aSAT-Eb	aSAT-EUb
EaSAT	EaSAT-b	EaSAT-Eb	EaSAT-EUb
EUaSAT	EUaSAT-b	EUaSAT-Eb	EUaSAT-EUb

For example, in aSAT-Eb, every clause has $\leq a$ literals per clause and every variable occurs *exactly* b times. Each of the three outlined regions indicates problems that are trivially equivalent to each other—for example, by padding clauses.

We state and prove a theorem, due to Tovey [Tov84], that shows that these seemingly small differences in the form of the formula create big differences in complexity.

Theorem 1.11.

1. *3SAT-3 is NP-complete, and thus aSAT-3 is NP-complete for any* $a \geq 3$.
2. *aSAT-2* $\in$ *P for any a.*
3. *Let* φ *be a CNF formula such that (a) there are exactly 3 literals per clause, and (b) every variable occurs* ≤ 3 *times. Then* φ *is satisfiable.*
4. *E3SAT-3* $\in$ *P (and thus EU3SAT-3* $\in$ *P) since every formula of this form is satisfiable by part 3.*[2]

Proof. 1) We show 3SAT $\leq_p$ 3SAT-3; 3SAT-3 is a special case of aSAT-3 for any $a \geq 3$.

Given a formula φ in 3CNF, we produce a formula φ' such that (1) every variable occurs ≤ 3 times and (2) $\varphi \in$ SAT if and only if $\varphi' \in$ SAT. For each variable x occurring $k > 3$ times in φ (as either x or $\neg x$), we do the following:

- Introduce new variables $x_1, \ldots, x_k$.
- Replace the ith occurrence of x with x_i.
- Form the chain of implications $x_1 \to x_2 \to \cdots \to x_m \to x_1$, so that the x_i's all have the same value. More precisely, for $1 \leq i \leq k-1$, add the clause $\neg x_i \vee x_{i+1}$ and add the clause $\neg x_m \vee x_1$.

We leave it as an exercise to prove that (1) φ' is of the correct form and (2) $\varphi \in$ 3SAT if and only if $\varphi' \in$ 3SAT.

2) Let φ be a CNF formula where every variable occurs at most twice.

First, if any clause has only a single literal, then that variable's assignment is forced (TRUE for x and FALSE for $\neg x$). We can then eliminate that variable from all other clauses: If the variable assignment satisfies the clause, then we can remove the clause altogether; otherwise, we just remove the literal from the clause. If any clause is now empty, we

2. Tovey [Tov84] claimed only that EU3SAT-3 $\in$ P, but as we show, his proof applies to E3SAT-3 as well.

know that the formula is unsatisfiable. Otherwise, we claim that it is satisfiable, as follows.

If any variable x occurs only positively (as x) in clauses, then set x to TRUE; and if any variable x occurs only negatively (as $\neg x$) in clauses, then set x to FALSE. In either case, we can remove all clauses containing x or $\neg x$, respectively, as they have all been satisfied. Because each variable occurs at most twice, we have arrived at a formula where every (remaining) variable occurs exactly once positively (as x) and exactly once negatively (as $\neg x$). And by the first simplification, all (remaining) clauses have at least two literals.

If every clause has a positive literal, we can satisfy the formula by setting all variables to TRUE. Otherwise, let N be the set of clauses whose literals are all negative. Consider any clause $C \in N$. If C has a literal $\neg x$ such that the occurrence of x is in a clause with at least one other positive literal, then we can ***flip*** x (replace x with $\neg x$ and vice versa, effectively negating the eventual assignment of x) and obtain an equivalent formula where $|N|$ decreases by 1. Otherwise, flip any literal $\neg x$ in C, and then consider the clause C' that contains the literal x. Repeating this process, we follow a sequence of clauses with exactly one positive literal, and entering from that positive literal while flipping it, so we can never repeat a clause. Therefore we eventually decrease $|N|$, and once it reaches zero, we can assign all variables to TRUE.

3) Let φ be a formula that satisfies the premise. Consider the bipartite multigraph with (1) a vertex on the left for each clause, (2) a vertex on the right for each variable, and (3) an edge between clause C and variable x for each occurrence of x or $\neg x$ in C. This multigraph has duplicate edges when a variable occurs multiple times in a single clause. We analyze this multigraph using Hall's theorem [Hal35]:

Theorem 1.12. *Let $G = (A, B, E)$ be a bipartite (multi)graph. Define a* ***matching from A to B*** *is a disjoint set of edges where every element of A is an endpoint of an edge. There is a matching from A to B if and only if, for all $A' \subseteq A$, $|N(A')| \geq |A'|$, where $N(A')$ represents the* ***neighbor set****:*

$$N(A') = \{b \in B \mid \exists a \in A' : (a, b) \in E\}.$$

Consider a subset A' of clauses; we will argue that $N(A') \geq |A'|$. Every clause has degree exactly 3, so the total number of *edges* incident to A' is $3|A'|$. For each such edge (a, b) where $a \in A'$, and thus $b \in N(A')$, add a mark to the corresponding variable y. Conversely, every variable in $N(A')$ has degree ≤ 3, so it received at most three marks. Hence, $N(A')$ must contain at least $|A'|$ distinct variables in order to have room for the exactly $3|A'|$ marks.

By theorem 1.12, there is a matching from clauses to variables—that is, a set of disjoint edges where every clause is the endpoint of one edge. For each matched edge from clause C to variable x, (1) set x TRUE if $x \in C$, (2) set x FALSE if $\neg x \in C$. (If both x and $\neg x$ are in C, or if x is not the endpoint of any matched edge, then we can set x to TRUE or FALSE.) This is a satisfying assignment because every clause has a literal that satisfies it. □

Exercise 1.13. Show that, in the proof of theorem 1.11(1), φ and φ' have the same number of satisfying assignments.

Theorem 1.11 gives us a dichotomy for aSAT-b:

1. aSAT-b is NP-complete if $a \geq 3$ and $b \geq 3$ (theorem 1.11(1)).
2. aSAT-$b \in$ P otherwise: If $a \leq 2$, then the problem is 2SAT so in P (theorem 1.6), and if $b \leq 2$, then the problem is in P by theorem 1.11(2).

For EaSAT-b and EUaSAT-b, though, we do not yet have such a dichotomy:

Open Problem 1.14. *For which $a, b \in \mathbb{N}$ is EaSAT-b (or EUaSAT-b) in P or NP-complete?*

Kratovíl, Savický, and Tuza [KST93] showed that EUaSAT-b is either always satisfiable (and thus in P), or NP-complete. Thus there is a critical value $b(a)$ for which EUaSAT-$b(a)$ is always satisfiable and EUaSAT-$(b(a)+1)$ is NP-complete. By the same argument as theorem 1.11(2), Tovey [Tov84] proved that $b(a) \geq a$. Tovey [Tov84] conjectured more strongly that $b(a) \geq 2^{a-1} - 1$, but this conjecture is false [Dub90]. In fact, Gebauer, Szabó, and Tardos [GST16] showed that $b(a) \sim \frac{2}{e} \cdot 2^a / a$, where $e \approx 2.71828$ is Euler's constant.

1.2.5 Monotone and Positive Formulas

Another restriction we can make is on which literals can be negated.

Definition 1.15.

1. A formula is ***monotone*** if, for every clause, either all of its literals are positive or all of its literals are negative.
2. To restrict to monotone formulas in a SAT problem, we add the MONOTONE prefix. For example:

> MONOTONE aSAT
> *Instance:* A formula φ such that (1) every clause has $\leq a$ literals (and a clause can have the same variable twice) and (2) every clause has either all the literals positive or all of the literals negative.
> *Question:* Is φ satisfiable?

3. A formula is ***positive*** if, for every clause, all of its literals are positive (not negated).

Theorem 1.16.

1. *(Gold [Gol78]) MONOTONE 3SAT is NP-complete.*
2. *(Darmann et al. [DDD18]) MONOTONE 3SAT-3 is NP-complete.*
3. *(Darmann and Döcker [DD21]) MONOTONE EU3SAT-E4 is NP-complete.*

On the other hand, POSITIVE is such a strong constraint that CNF SAT becomes easy: Setting all variables to TRUE satisfies the formula. In fact, several weaker versions can also be solved in P:

Theorem 1.17. *The following versions of CNF SAT are in P.*

1. *POSITIVE CNF SAT. All literals are positive.*
2. *(Horn [Hor51]) HORN SAT. Each clause has ≤ 1 positive literal.*
3. *(T. Schaefer [Sch78]) DUAL HORN SAT. Each clause has ≤ 1 negative literal. DUAL HORN SAT $\in$ P follows easily from HORN SAT $\in$ P.*
4. *(Lewis [Lew78]) RENAMEABLE HORN SAT. There is a set of clauses $C_1, \ldots, C_k$ such that the following holds: If you take every literal L that appears in $C_1 \cup \cdots \cup C_k$ and, in the entire formula, replace L with $\neg L$ and vice versa (and simplify double negations), then the formula is Horn.*

1.2.6 Other NP-Hard Clause Types

What if the formula is still an AND of clauses, but each clause is a Boolean constraint other than an OR of literals? Here we define some common types of clauses for which SAT is also NP-complete.

Definition 1.18.

1. A ***NAE*** (not all equal) clause is satisfied when not all of the literals in the clause get the same truth value. Equivalently, the clause must have at least one TRUE literal and at least one FALSE literal. NAE assignments are symmetric between TRUE and FALSE: Negating all variables also produces a NAE assignment.
2. A ***1-IN-*** clause is satisfied when exactly 1 literal in the clause is true. One could replace 1 with 2 or any number.
3. To indicate the variation of SAT where every clause is of a particular type (instead of the usual OR of literals), place the condition at the beginning of the problem name. For example:

1-IN-EUaSAT-b
Instance: A formula φ such that (1) every clause has exactly a literals, (2) every variable occurs $\leq b$ times, and (3) every variable in a clause occurs uniquely.
Question: Is there a satisfying assignment of φ where every clause has exactly 1 literal set to true?

These clause types turn out to be NP-complete, even with the strong POSITIVE constraint on literals:

Theorem 1.19. *In all of the following problems, the given formula is an* AND *of clauses, and each clause has at most three literals.*

1. *(T. Schaefer [Sch78]) 1-IN-3SAT. We seek a satisfying assignment where* exactly *one literal per clause is* TRUE. *This problem is NP-complete.*
2. *(T. Schaefer [Sch78]) POSITIVE 1-IN-3SAT.*[3] *There are no negation signs, and we seek a satisfying assignment where* exactly *one literal per clause is* TRUE. *This problem is NP-complete.*
3. *(P. Laroche [Lar92]) POSITIVE 1-IN-EU3SAT. There are no negation signs, there are exactly three distinct variables in each clause, and we seek a satisfying assignment where* exactly *one literal per clause is* TRUE. *This problem is NP-complete.*
4. *(T. Schaefer [Sch78]) POSITIVE NOT-EXACTLY-1-IN-3SAT. There are no negation signs, and we seek an assignment where either 0, 2, or 3 of the variables in each clause are* TRUE. *This problem is trivially in P.*
5. *(T. Schaefer [Sch78]) NAE 3SAT. We seek an assignment where every clause has at least one* TRUE *and at least one* FALSE. *This problem is NP-complete.*
6. *(T. Schaefer [Sch78]) POSITIVE NAE 3SAT.*[4] *The formula has no negations, and we seek an assignment where every clause has at least one* TRUE *and at least one* FALSE. *This problem is NP-complete.*
7. *POSITIVE NAE EU3SAT. The formula has no negations, every clause has exactly three distinct variables, and we seek an assignment where every clause has at least one* TRUE *and at least one* FALSE. *This problem is NP-complete.*

3. Schaefer actually called this problem "One-in-Three Satisfiability." However, some papers use this term to mean either 1-IN-3SAT or POSITIVE 1-IN-3SAT, so we use explicit POSITIVE prefixes to distinguish the two.
4. Similarly, Schaefer actually called this problem "Not-All-Equal Satisfiability." However, some papers use this term to mean either NAE 3SAT or POSITIVE NAE 3SAT, so we use explicit POSITIVE prefixes to distinguish the two.

We prove the one version that does not seem to appear in the literature:

Proof. 7) We give a simple reduction from POSITIVE NAE 3SAT to POSITIVE NAE EU3SAT. A clause NAE(x), NAE(x,x), or NAE(x,x,x) with one unique variable can never be satisfied, so if such a clause exists, we can replace the entire formula with any NO instance (such as all $\binom{5}{3} = 10$ NAE EU3SAT constraints applied to 5 variables). Each clause NAE(x,y), NAE(x,x,y), or NAE(x,y,y) with two unique variables x, y can be converted into

$$\text{NAE}(a,b,c) \wedge \text{NAE}(x,y,a) \wedge \text{NAE}(x,y,b) \wedge \text{NAE}(x,y,c)$$

for three newly added variables a, b, c. The added variables a, b, c will be forced to be equal (and violate the clause NAE(a,b,c)) exactly when x, y are equal, thus correctly simulating NAE(a,b). □

The problems 3SAT, POSITIVE 1-IN-3SAT, and POSITIVE NAE 3SAT are especially important for proving problems NP-complete. We will see several examples in section 1.3.

1.2.7 Schaefer's Dichotomy Theorem

Every variant of SAT we have looked at so far has either been in P or NP-complete. This is not a coincidence. Schaefer's dichotomy theorem says that, with the right setup, all versions of SAT are either NP-complete or in P.

We motivate the approach by looking at 1-IN-3SAT. We want to express the question: *Given $\varphi = C_1 \wedge \cdots \wedge C_k$, is there an assignment that satisfies exactly one literal in each clause C_i?* Define the following Boolean formulas:

1. $R_1(x_1,x_2,x_3) = \text{1-IN-3}(x_1,x_2,x_3) = (x_1 \wedge \neg x_2 \wedge \neg x_3) \vee (\neg x_1 \wedge x_2 \wedge \neg x_3) \vee (\neg x_1 \wedge \neg x_2 \wedge x_3)$.
2. $R_2(x_1,x_2,x_3) = R_1(x_1,x_2,\neg x_3) = (x_1 \wedge \neg x_2 \wedge x_3) \vee (\neg x_1 \wedge x_2 \wedge x_3) \vee (\neg x_1 \wedge \neg x_2 \wedge \neg x_3)$.
3. Similarly, $R_3, \ldots, R_8$ denote the remaining cases for which of x_1, x_2, x_3 are negated.

We can view an instance of 1-IN-3SAT as being given a conjunction of R_i's (repeats allowed) applied to sets of three variables. For example,

$$\text{1-IN-3}(x_1,x_2,x_3) \wedge \text{1-IN-3}(x_1,x_3,\neg x_4) \in \text{1-IN-3SAT}$$

if and only if

$$R_1(x_1,x_2,x_3) \wedge R_2(x_1,x_3,x_4) \in \text{3SAT}.$$

Note that this is just a way to express the problem; it is not a reduction.

We call the R_i's ***relations***. More generally, a relation on m variables is just a truth table or formula on those literals. Each R_i represents a possible clause in the SAT problem. We will view variants of the SAT problem as being defined by a set of relations, where the formula is a conjunction (AND) of relations.

Example 1.20. We give an absurd example. In CRAZY SAT, the formula is a conjunction of R_1's, R_2's, and R_3's applied to various variables, where

1. $R_1(x_1,x_2,x_3,x_4)$ is TRUE if exactly 2 of the x_i's are TRUE.
2. $R_2(x_1,x_2,x_3,x_4)$ is TRUE if either 1 or 4 of the x_i's are TRUE.
3. $R_3(x_1,\ldots,x_{12})$ is TRUE if either 0 or 7 of the x_i's are TRUE.

Is CRAZY SAT in P? NP-complete? After we state and understand Schaefer's dichotomy theorem, this will be possible to determine. (We invented CRAZY SAT as an example. We suspect it has never appeared in the literature before and will never appear in the literature again.)

Definition 1.21. A ***SAT-type problem*** is defined by a set of relations $R_1,\ldots,R_m$, where each $R_i \subseteq D^{a(i)}$ is a relation of ***arity*** $a(i)$ over a finite ***domain*** D, typically {TRUE, FALSE}. A ***formula*** in this context is a conjunction of these relations applied to various subsets of n variables $x_1,\ldots,x_n$. The decision problem to solve is as follows: Given a formula, does there exist an assignment to the variables that satisfies all of the relations in it?

We now have a way of talking about many variants of SAT. Schaefer's dichotomy theorem [Sch78] will tell us *exactly* which of these variants are in P and which are NP-complete, for a binary domain {TRUE, FALSE}. It is remarkable that every variant defined in this way is one or the other.

In fact, we have already seen most of the polynomial-time variants: 2SAT (theorem 1.6), Horn SAT, and dual-Horn SAT (theorem 1.17). There is one other polynomial-time variant: A system of linear equations modulo 2 can be solved by Gaussian elimination.

Definition 1.22.

1. The ***binary XOR*** operator $\oplus$ is defined by the following truth table.

x	y	$x \oplus y$
TRUE	TRUE	TRUE
TRUE	FALSE	FALSE
FALSE	TRUE	FALSE
FALSE	FALSE	TRUE

   It is easy to see that $\oplus$ is associative, so one can write $x \oplus y \oplus z$. Note that if we think of TRUE as 1 and FALSE as 0, then $\oplus$ is addition modulo 2.

2. An ***affine formula*** is a formula of the form $x_1 \oplus \cdots \oplus x_n = c$, where $c \in$ {TRUE, FALSE}.

Theorem 1.23 (Schaefer's dichotomy theorem). *Let $R_1,\ldots,R_k$ be relations over {TRUE, FALSE}, each with at least one satisfying assignment. If any of the following occur, then the SAT-type problem $\{R_1,\ldots,R_k\}$ is in P. If not, then it is NP-complete.*

1. *For all* $1 \le i \le k$, $R_i(\textsc{true}, \ldots, \textsc{true}) = \textsc{true}$ *(setting all variables* TRUE *satisfies the formula).*
2. *For all* $1 \le i \le k$, $R_i(\textsc{false}, \ldots, \textsc{false}) = \textsc{true}$ *(setting all variables* FALSE *satisfies the formula).*
3. *For all* $1 \le i \le k$, $R_i(x_1, \ldots, x_{m_i})$ *is equivalent to a conjunction of clauses with at most 2 literals (2SAT).*
4. *For all* $1 \le i \le k$, $R_i(x_1, \ldots, x_{m_i})$ *is equivalent to a conjunction of Horn clauses.*
5. *For all* $1 \le i \le k$, $R_i(x_1, \ldots, x_{m_i})$ *is equivalent to a conjunction of dual-Horn clause.*
6. *For all* $1 \le i \le k$, $R_i(x_1, \ldots, x_{m_i})$ *is equivalent to a conjunction of affine formulas.*

1.2.8 Polymorphism View of Schaefer's Dichotomy Theorem

How do we determine whether any of the six cases in theorem 1.23 hold? This question is answered by a more modern view of Schaefer's dichotomy theorem, using the idea of "polymorphisms":

Definition 1.24. An a-ary relation $R \subseteq D^a$ is ***closed*** (or ***preserved***) under a b-ary function $f : D^b \to D$ if, for any b assignments $(x_1, \ldots, x_a), (y_1, \ldots, y_a), \ldots$ satisfying R (i.e., $R(x_1, \ldots, x_a) = R(y_1, \ldots, y_a) = \cdots = \textsc{true}$), applying f elementwise preserves satisfaction of R:

$$R(f(x_1, y_1, \ldots), \ldots, f(x_a, y_a, \ldots)) = \textsc{true}.$$

We also call f a ***polymorphism*** on R.

This definition is a bit complicated to understand in general, but it is easier in the context of examples where b is small.

Example 1.25. Consider the binary Boolean function $\textsc{and}(x, y) = x \wedge y$. We can simplify the definition to the binary $(b = 2)$ case as follows: A relation R is closed under AND if, for any two assignments $(x_1, \ldots, x_a), (y_1, \ldots, y_a)$ satisfying R (i.e., $R(x_1, \ldots, x_a) = R(y_1, \ldots, y_a) = \textsc{true}$), the elementwise AND also satisfies R:

$$R(x_1 \wedge y_1, \ldots, x_a \wedge y_a) = \textsc{true}.$$

Now consider the Horn clause $R(x_1, x_2, x_3) = x_1 \vee \neg x_2 \vee \neg x_3$. We claim that R is closed under AND. Consider two assignments $(x_1, x_2, x_3), (y_1, y_2, y_3)$ satisfying R. If $x_2 = \textsc{false}$ or $y_2 = \textsc{false}$, then $x_2 \wedge y_2 = \textsc{false}$, so $R(x_1 \wedge y_1, x_2 \wedge y_2, x_3 \wedge y_3) = \textsc{true}$. Similarly, if $x_3 = \textsc{false}$ or $y_3 = \textsc{false}$, then $x_3 \wedge y_3 = \textsc{false}$, so $R(x_1 \wedge y_1, x_2 \wedge y_2, x_3 \wedge y_3) = \textsc{true}$. If none of these cases hold, then the only remaining possibility satisfying R is $x_1 = y_1 = \textsc{true}$, in which case $x_1 \wedge y_1 = \textsc{true}$, so $R(x_1 \wedge y_1, x_2 \wedge y_2, x_3 \wedge y_3) = \textsc{true}$.

Exercise 1.26. Prove that any Horn clause is closed under AND and any dual-Horn clause is closed under OR.

Horn and dual-Horn sounds a lot like theorem 1.23! Indeed, this connection deepens:

Example 1.27. Consider the ternary Boolean function MAJORITY$(x_1,x_2,x_3)=(x_1 \wedge x_2) \vee (x_1 \wedge x_3) \vee (x_2 \wedge x_3)$, which is TRUE when at least two of the inputs are TRUE and FALSE when at least two of the inputs are FALSE. The definition in this ternary case: A relation R is closed under MAJORITY if, for any three assignments $(x_1,\ldots,x_a),(y_1,\ldots,y_a),(z_1,\ldots,z_a)$ satisfying R (i.e., $R(x_1,\ldots,x_a)=R(y_1,\ldots,y_a)=R(z_1,\ldots,z_a)=$ TRUE), we have the elementwise MAJORITY is also in R:

$$R(\text{MAJORITY}(x_1,y_1,z_1),\ldots,\text{MAJORITY}(x_a,y_a,z_a)) = \text{TRUE}.$$

Now consider the positive 2SAT clause $R(x_1,x_2)=x_1 \vee x_2$. We claim that R is closed under MAJORITY. Because $(x_1,x_2),(y_1,y_2),(z_1,z_2)$ all satisfy R, each of x,y,z sets either the first or second variable to TRUE. If we pick one TRUE variable for each of x,y,z, then some variable $i \in \{1,2\}$ gets picked at least twice, which means that MAJORITY$(x_i,y_i,z_i)=$ TRUE. Thus $R(\text{MAJORITY}(x_1,y_1,z_1),\text{MAJORITY}(x_2,y_2,z_2))=$ TRUE.

Exercise 1.28. Prove that any 2SAT clause is closed under MAJORITY and any affine clause is closed under MINORITY, where MINORITY$(x_1,x_2,x_3)=x_1 \oplus x_2 \oplus x_3$ is TRUE when at most one of the inputs is TRUE and FALSE when at most one of the inputs is FALSE.

Now we can describe the polymorphism view of Schaefer's dichotomy theorem, first proved by Jeavons [Jea98] and described more recently by Chen [Che06, Che09]:

Theorem 1.29. *Let $R_1,\ldots,R_k$ be relations over {TRUE, FALSE}, each with at least one satisfying assignment. If any of the following occur, then the SAT-type problem $\{R_1,\ldots,R_k\}$ is in P. If not, then it is NP-complete.*

1. *For all $1 \le i \le k$, R_i is closed under BOT() = FALSE (setting all variables FALSE satisfies the formula).*
2. *For all $1 \le i \le k$, R_i is closed under TOP() = TRUE (setting all variables TRUE satisfies the formula).*
3. *For all $1 \le i \le k$, R_i is closed under AND (corresponding to Horn).*
4. *For all $1 \le i \le k$, R_i is closed under OR (corresponding to dual-Horn).*
5. *For all $1 \le i \le k$, R_i is closed under MAJORITY (corresponding to 2SAT).*
6. *For all $1 \le i \le k$, R_i is closed under MINORITY (corresponding to affine).*

In particular, given the truth tables for each R_i, we can check the closure properties in polynomial time (in the size of the truth tables) by trying all triples of assignments that satisfy R_i.

Exercise 1.30.

1. Determine whether CRAZY SAT is in P or NP-complete.
2. Write a program that will, given a set of relations, determine whether the SAT-type problem they define is in P or NP-complete.

Note: Theorems 1.23 and 1.29 say that, up to polynomial-time reductions and assuming $\mathsf{P} \neq \mathsf{NP}$, there are two complexity classes that a SAT-type problem could be: in P or NP-complete. Allender et al. [ABI$^+$09] showed that, if we use a more refined notion of reduction, then the problems in P can be further differentiated. There end up being six classes total: five within P, and of course NP-complete.

1.2.9 Further Results

1. If we allow non-Boolean variables (taking more than two possible values FALSE and TRUE), and consider relations over a larger domain D, Zhuk [Zhu20] proved a dichotomy theorem for SAT (usually called *constraint satisfaction problem* in this setting). Again the problem is in P or NP-complete, with polynomial time occurring whenever the relations are all preserved by a common "weak near-unanimity polymorphism" f, meaning that $f(y, x, \ldots, x) = f(x, y, \ldots, x) = \cdots = f(x, \ldots, x, y)$ for all $x, y \in D$.
2. Lewis [Lew79] proved a dichotomy theorem for SAT over formulas with a specified set of allowed Boolean connectives (like AND or NOR). Again the problem is in P or NP-complete, depending on whether the function $\text{NIMPLIES}(x, y) = x \wedge \neg y$ can be represented.
3. Bodirsky and Pinsker [BP15] proved a dichotomy theorem for CNF-like formulas where *literals* are instead of the form $x = y$, called "graph formulas."
4. There has been a lot of work on SAT questions for sentences with quantifiers. Such a formula is satisfiable if there is a domain and an interpretation of the predicates that makes it true. These types of questions have a rich history going back to Turing [Tur37]. The book by Börger et al. [BGG97] is a good source for information.

1.3 Example Reductions from SAT and Its Variants

In this section, we show that several problems are NP-complete by reducing from variants of SAT.

Most of the reductions follow a common pattern of constructing two ***gadgets***, which are pieces of an instance that can be duplicated and combined to form the full instance. For example, when we are reducing to a graph problem, a gadget is a small graph that will be a piece of the overall graph; in general, it has no formal definition. For reductions from SAT variants, we need two types of gadgets. A ***variable gadget*** is a piece of an instance that represents a variable of the given SAT instance; it typically can be locally solved in two different ways, representing the choice of whether the variable is TRUE or FALSE. A ***clause gadget*** is a piece of an instance that represents the constraint of a clause in the given SAT instance; for example, for a 3SAT clause $x \vee \neg y \vee z$, it should force at least one of the three variables x, y, z to have the setting that satisfies the clause ($x =$ TRUE, $y =$ FALSE, or $z =$ TRUE). By combining n instances of the variable gadget and m instances of the clause gadget, we can represent a SAT instance with n variables and m clauses.

1.3.1 CLIQUE, INDEPENDENT SET, and VERTEX COVER

We will prove that CLIQUE is NP-complete by a reduction from 3SAT. We will then prove that (1) INDEPENDENT SET is NP-complete by a reduction from CLIQUE, and (2) VERTEX COVER is NP-complete by a reduction from INDEPENDENT SET. The last two reductions are very easy.

CLIQUE and INDEPENDENT SET
Instance: Graph $G = (V, E)$ and a $k \in \mathbb{N}$.
Question: For CLIQUE (INDEPENDENT SET): Are there k points such that every pair (no pair) has an edge between them?

VERTEX COVER
Instance: Graph $G = (V, E)$ and a $k \in \mathbb{N}$.
Question: Is there a $V' \subseteq V$ of size k such that every $e \in E$ has some $v \in V'$ as an endpoint?

Karp [Kar72] proved that CLIQUE, VERTEX COVER, and INDEPENDENT SET are NP-complete. For CLIQUE, we give a different (folklore) proof that has parsimony properties we will use in theorem 11.15. For INDEPENDENT SET and VERTEX COVER, we give Karp's proofs.

Notation 1.31. If $G = (V, E)$ is a graph, then its ***complement*** $\overline{G}$ is $(V, \overline{E})$, where $\overline{E} = \{(v_1, v_2) \mid v_1, v_2 \in V, v_1 \neq v_2, (v_1, v_2) \notin E\}$. In other words, (v_1, v_2) is an edge in $\overline{G}$ if and only if (v_1, v_2) is not an edge in G.

Theorem 1.32.

1. *E3SAT* $\leq_p$ *CLIQUE, so CLIQUE is NP-complete.*
2. *CLIQUE* $\leq_p$ *INDEPENDENT SET, so INDEPENDENT SET is NP-complete.*
3. *INDEPENDENT SET* $\leq_p$ *VERTEX COVER, so VERTEX COVER is NP-complete.*

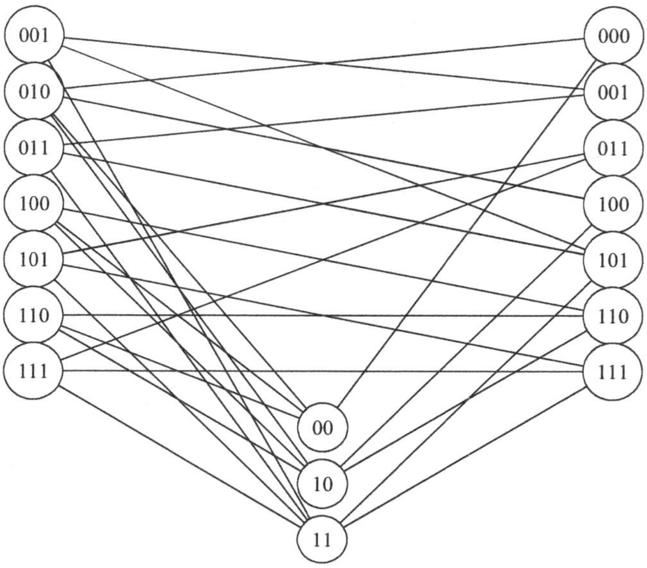

Figure 1.2
Reduction from 3SAT to CLIQUE.

Proof. 1) We give an example of our reduction using a formula that has two size-3 clauses and one size-2 clause in figure 1.2. While this is an 3CNF formula, not an E3CNF formula, it is a good example for readability.

Here is the reduction from E3SAT. (More generally, the same idea works for any version of SAT where the formula is a conjunction of clauses.)

1. Input $\varphi = C_1 \wedge \cdots \wedge C_k$, where each C_i is a size-3 clause.
2. We create a graph G with $7k$ vertices as follows: For each clause, we make seven vertices. Label them with the seven ways to set the three variables to make the clause satisfied. For example, for the clause $x \vee y \vee \neg z$, we have seven vertices:
   - $(x = \text{TRUE}, y = \text{TRUE}, z = \text{TRUE})$
   - $(x = \text{TRUE}, y = \text{TRUE}, z = \text{FALSE})$
   - $(x = \text{TRUE}, y = \text{FALSE}, z = \text{TRUE})$
   - $(x = \text{TRUE}, y = \text{FALSE}, z = \text{FALSE})$
   - $(x = \text{FALSE}, y = \text{TRUE}, z = \text{TRUE})$
   - $(x = \text{FALSE}, y = \text{TRUE}, z = \text{FALSE})$
   - $(x = \text{FALSE}, y = \text{FALSE}, z = \text{FALSE})$

3. There are no edges between vertices associated with the same clause. We put an edge between vertices associated with different clauses if the assignments do not conflict. For example, vertex ($x=$ TRUE, $y=$ TRUE, $z=$ TRUE) will have an edge to the vertex ($w=$ FALSE, $x=$ TRUE, $z=$ TRUE) but not to the vertex ($w=$ FALSE, $x=$ FALSE, $z=$ TRUE).

We leave it to the reader to show that the $\varphi \in$ E3SAT if and only if $(G, k) \in$ CLIQUE.

2) G has a clique of size k if and only if $\overline{G}$ has an independent set of size k.

3) G has an independent set of size k if and only if G has a vertex cover of size $n-k$.

We will return to VERTEX COVER in section 2.5.1, where we will show that it remains NP-complete even when restricted to planar graphs.

Exercise 1.33. Let MASTERMIND be the following problem: Given a position in the game *Mastermind*, is there a solution? Read Stuckman and Zhang's paper [SZ06] on the complexity of *Mastermind*. Rewrite their proof that VERTEX COVER $\leq_p$ MASTERMIND in your own words. (For more on the complexity of MASTERMIND, see the papers of Goodrich [Goo09], Viglietta [Vig12], Ben-Ari [BA18], and Doerr et al. [DDST16]. This is not a complete list of papers.)

> SET PACKING
> *Instance:* $n, k \in \mathbb{N}$ and $X_1, \ldots, X_m \subseteq \{1, \ldots, n\}$.
> *Question:* Are there k X_i's that are all disjoint?

Exercise 1.34. Show that INDEPENDENT SET $\leq_p$ SET PACKING, hence SET PACKING is NP-complete.

1.3.2 2-COLORABLE PERFECT MATCHING Is NP-Complete

We show that 2-COLORABLE PERFECT MATCHING is NP-complete by a reduction from POSITIVE NAE 3SAT.

> 2-COLORABLE PERFECT MATCHING
> *Instance:* A graph G.
> *Question:* Is there a 2-coloring of the vertices such that every vertex has exactly 1 neighbor of the same color? Figure 1.3 gives an example of a graph with a 2-colorable perfect matching.

Theorem 1.35. *(T. Schaefer [Sch78]) 2-COLORABLE PERFECT MATCHING is NP-complete.*

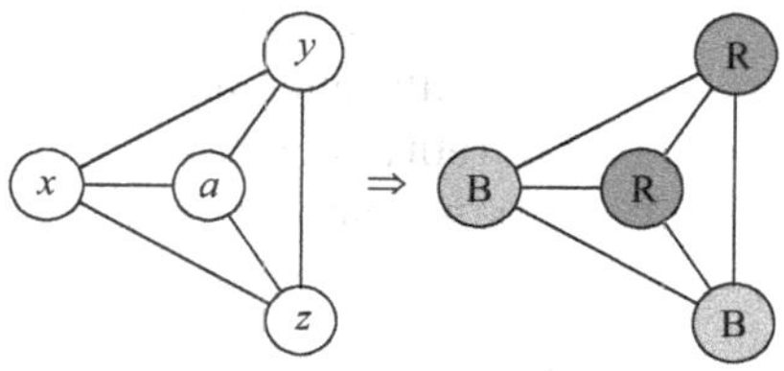

Figure 1.3
A graph that has a 2-colorable perfect matching.

Proof. We show POSITIVE NAE EU3SAT $\leq_p$ 2-COLORABLE PERFECT MATCHING. Given φ, we find a graph G such that $\varphi \in$ POSITIVE NAE EU3SAT if and only if $G \in$ 2-COLORABLE PERFECT MATCHING. We will color the vertices of the graph TRUE and FALSE. Since the goal is to have no clause have all TRUE's or all FALSE's, TRUE and FALSE are symmetric here.

1. For each variable x, we make a corresponding vertex x. This single vertex acts as a ***variable gadget***, as it can be colored in two ways.
2. For each clause NAE(x, y, z), we make the ***clause gadget*** in figure 1.3 with a new vertex a. The vertices x, y, z are shared by other clause gadgets, while a appears only in this clause gadget. This gadget makes sure that, in any 2-coloring satisfying the condition, X, Y, Z cannot all be the same color. Hence, any 2-coloring satisfying the condition will be a not-all-equal truth assignment.

(Schaefer's original reduction [Sch78] was from POSITIVE NAE 3SAT instead of POSITIVE NAE EU3SAT, which required an additional gadget to make the graph simple.) □

Schaefer [Sch78] claimed without proof that 2-COLORABLE PERFECT MATCHING remains NP-complete when restricted to planar 3-regular graphs. Demaine, Karntikoon, and Pitimanaaree [DKP25] filled in a proof and further showed that the graph can be assumed to be 2-connected.

1.3.3 CRYPTARITHMS Is NP-Complete

Cryptarithms are classic puzzles involving arithmetic on words.

Example 1.36. Dudeney [Dud24] posed this classic cryptarithm with a unique solution:

$$\begin{array}{r} S\ E\ N\ D \\ +\ M\ O\ R\ E \\ \hline M\ O\ N\ E\ Y \end{array}$$

The goal is to replace each letter with a digit; no digit is assigned to two different letters, such that the resulting sum works out.

Here is how one might begin solving the cryptarithm in example 1.36:

1. A carry can be at most 1. Because we would not write $MONEY$ with a leading 0, we have $M=1$.
2. Because the S, M, O column contributes a carry and $M=1$, the digit O must be either 0 or 1. Because $O \neq M$, we have $O=0$.
3. Look at the E, O, N column. Because $O=0$ and $E \neq N$, the N, R, E column must contribute a carry. So we have $E+1 \equiv N \pmod{10}$.
4. The E, O, N column cannot contribute a carry. Suppose that it did. Because $O=0$, we must have $E=9$. But for the $N, R, E=9$ column to sum correctly, we must have N or R equal to $9=E$, a contradiction.
5. Because $M=1$, $O=0$, and the E, O, N column does not contribute a carry, we have $S=9$.
6. We have $E, N \notin \{O, M, S\} = \{0, 1, 9\}$. Because $N \neq 9$, we have $E \neq 8$. Because $E \neq 1$, we have $N \neq 2$. In summary, $E \notin \{0, 1, 8, 9\}$ and $N \notin \{0, 1, 2, 9\}$. Using this restriction on E, N and that $E+1 \equiv N \pmod{10}$, we get:

$$(E, N) \in \{(2,3), (3,4), (4,5), (5,6), (6,7), (7,8)\}.$$

We stop here; however, notice that we may end up encountering many possibilities. If you work them through, then you will find that the answer is unique:

$$S=9, E=5, N=6, D=7, M=1, O=0, R=8, Y=2.$$

$$\begin{array}{ccccc} & 9 & 5 & 6 & 7 \\ + & 1 & 0 & 8 & 5 \\ \hline 1 & 0 & 6 & 5 & 2 \end{array}$$

Is there a trick to these puzzles so that they can always be solved fast and avoid having too many cases? Likely no: David Eppstein [Epp87] showed

$$\text{3SAT} \leq_p \text{CRYPTARITHMS},$$

hence CRYPTARITHMS is NP-complete. We will show

$$\text{POSITIVE 1-IN-3SAT} \leq_p \text{CRYPTARITHMS}.$$

By theorem 1.19(2), we have that CRYPTARITHMS is NP-complete. Our reduction is easier than that of Eppstein; however, we need that POSITIVE 1-IN-3SAT is NP-complete.

First we define the problem rigorously.

CRYPTARITHMS
Instance:

1. $B, m \in \mathbb{N}$. Let Σ be an alphabet of B letters.
2. $x_0, \ldots, x_{m-1} \in \Sigma$.
3. $y_0, \ldots, y_{m-1} \in \Sigma$.
4. $z_0, \ldots, z_m \in \Sigma$. The symbol z_m is optional.

Question: Is there a bijection from Σ to $\{0, \ldots, B-1\}$ so that the following arithmetic statement is true in base B?

$$\begin{array}{ccccc} & x_{m-1} & \cdots & x_0 \\ + & y_{m-1} & \cdots & y_0 \\ \hline z_m & z_{m-1} & \cdots & z_0 \end{array}$$

Theorem 1.37. *CRYPTARITHMS is strongly NP-complete. (We discussed* strongly *NP-complete briefly in section 0.8 and will discuss it at length in chapter 5.)*

Proof. We show POSITIVE 1-IN-3SAT $\le_p$ CRYPTARITHMS. Given a formula $\varphi = C_1 \wedge \cdots \wedge C_k$ with n variables and k clauses, where every literal is positive, we want to create an instance R of CRYPTARITHMS such that the following are equivalent:

- There exists an assignment that satisfies exactly one literal per clause.
- There exists a solution to R.

1) *Constants:* We need two letters that we suggestively call "0" and "1" that we force to map to the numbers 0 and 1. To achieve this, we use the following ***constants gadget***.

$$\begin{array}{ccc} 0 & p & 0 \\ 0 & p & 0 \\ \hline 1 & q & 0 \end{array}$$

This gadget uses four letters $0, 1, p, q$, and three columns. It is easy to see that these columns force (1) the letter "0" to have the value 0, and (2) the letter "1" to have the value 1. (You will need to use that carries are either 0 or 1.)

2) *Variables:* Let v be a variable in φ. We represent it by a single letter, which we also call v. We consider v to be *true* if $v \equiv 1 \pmod 4$ and *false* if $v \equiv 0 \pmod 4$.

Hence we need to ensure that $v \equiv 0 \pmod 4$ or $v \equiv 1 \pmod 4$. This constraint is accomplished by the following ***variable gadget***, which multiplies an unknown value a by 4 and allows for adding one carry (which is either 0 or 1):

0	b	c	0	a	0
0	b	c	0	a	0
0	v	d	0	b	0

$$b = 2a$$
$$2c = d + C \qquad C = \text{carry}(c+c) \in \{0, 1\}$$
$$v = 2b + C$$
$$= 4a + C \equiv C \pmod 4$$

We do not have to consider the negation $\neg v$ because our formula has all positive literals.

This gadget uses five letters a, b, c, d, v, all unique to this variable, and six columns. (C is not a letter in the puzzle.) Over all n variables, the variable gadgets use $5n$ letters and $6n$ columns.

3) *Clauses:* Consider a clause $C = 1\text{-in-}3(x, y, z)$. (Here x, y, z need not be distinct.) As above, x, y, z also denotes the letters representing the variables (called v in the variable gadget). The goal of a clause gadget is to enforce the 1-in-3 constraint. Given our 0-or-1-mod-4 representation of FALSE and TRUE respectively, this constraint is equivalent to $x + y + z \equiv 1 \pmod 4$.

First we construct a letter d that can be any number $\equiv 1 \pmod 4$, by multiplying an unknown number a by 4 and adding 1. Then we set $x + y + z = d$, using an intermediate variable i for $x + y$. The following ***clause gadget*** accomplishes all of this:

0	i	0	x	0	1	0	b	0	a	0
0	z	0	y	0	c	0	b	0	a	0
0	d	0	i	0	d	0	c	0	b	0

$$b = 2a$$
$$c = 2b$$
$$= 4a$$
$$d = c + 1$$
$$= 4a + 1$$

$$x + y = i$$
$$i + z = d$$

This gadget uses five letters a, b, c, d, i unique to this clause, and 11 columns. (Letters x, y, and z are from variable gadgets.) Over all k clauses, the clause gadgets use $5k$ letters and $11k$ columns.

The overall instance R of CRYPTARITHMS consists of one constants gadget, n variable gadgets, and k clause gadgets. In total, it uses $5n + 5k + 4$ letters and $6n + 11k + 3$ columns. However, we cannot just take $B = 5n + 5k + 4$. We need enough numbers so that, for example, the clause gadget does not have $b + b$ equal to $x + y$. We revisit choosing the base B soon.

Clearly, a solution to the cryptarithm R gives a solution to the POSITIVE 1-IN-3SAT problem φ. The other direction is less clear. Assume we have a solution to the POSITIVE 1-IN-3SAT problem φ. This assigns TRUE or FALSE to each of the variables $v_1, \ldots, v_n$. We translate this to an assignment of the letters $v_1, \ldots, v_n$ to numbers. We do so inductively. Assume that the letters in the variable gadgets for $v_1, \ldots, v_{i-1}$ have been assigned, along with some letters in some of the clause gadgets.

1. If variable v_i is TRUE, then we will assign letter v_i to a number $\equiv 1 \pmod 4$.
2. If variable v_i is FALSE, then we will assign letter v_i to a number $\equiv 0 \pmod 4$.
3. Assign letter v_i to be the least number that has the right congruence mod 4, has not been assigned to any other letter, and does not cause any letter to be assigned to an already-used number. This arises both with the variable gadgets, because once you assign v, you need to assign a, b, c, d; and with any clause gadget that contains v, where the other variables in the gadget have been assigned.

We leave it to the reader to determine how large B must be to accommodate all these numbers.

- B will be large enough so that many numbers will not be used. Hence, there will be letters that do not map to any number. Note that in the SEND + MORE = MONEY puzzle, many letters (e.g., Z) do not map to any number.
- B will be bounded by a polynomial in k, n. Hence CRYPTARITHMS is strongly NP-complete. □

Exercise 1.38. Do a direct reduction from 3SAT to show that 3SAT $\leq_p$ CRYPTARITHMS.

Exercise 1.39.

1. Write an algorithm for CRYPTARITHMS and analyze its run time.
2. How fast is your algorithm when B is a constant?

1.3.4 Pushing 1 × 1 Blocks

In this section we look at the PUSH family of puzzle games. The inspiration for PUSH comes from the video game *Sokoban* (literally *warehouseman* in Japanese). In *Sokoban*, you are a 1×1 player, and you can move and push around 1×1 blocks. Your job is to push k blocks into k target locations. Some blocks you cannot push; we call them ***fixed walls***. You can only push one block at a time, by one space, and only if the block would move into an empty space. For example, a block in a corner cannot be pushed.

Lots of video games contain pushing-block puzzles in many variants. For example, the Legend of Zelda series has several. In *The Legend of Zelda: Minish Cap*, the blocks are on ice, so blocks will fly off to infinity until they encounter an obstacle (a wall or another block); the goal is to cover one target space with any block.

We abstract these many types of puzzles into the PUSH family of puzzles [DDO00, DHH02]. Figure 1.4 illustrates the following variants:

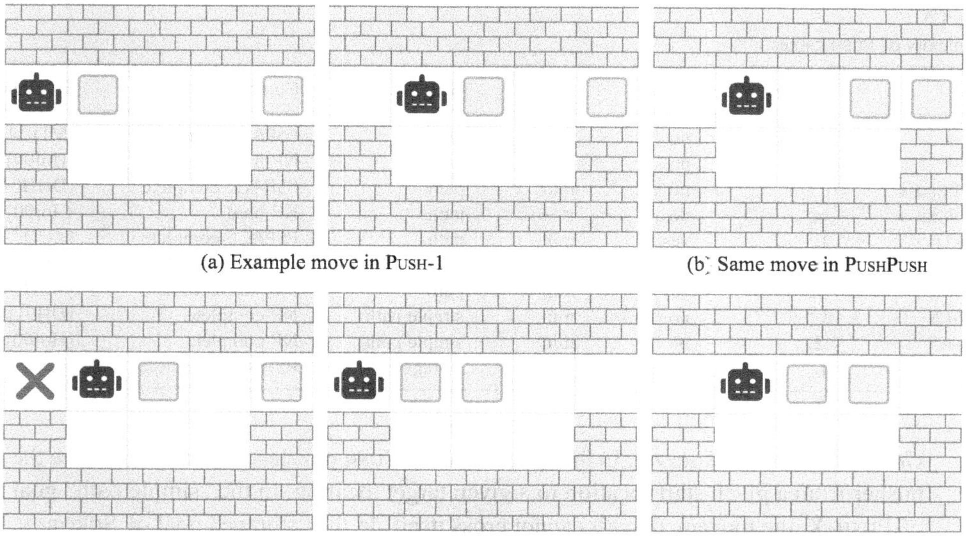

(a) Example move in PUSH-1 (b) Same move in PUSHPUSH

(c) Same move in PUSH-1X (d) Example move in PUSH-2 (invalid in PUSH-1)

Figure 1.4
PUSH and its variants.

1. In PUSH (figure 1.4a), when you push a block, the block moves one space (as in *Sokoban*).
2. In PUSHPUSH (figure 1.4b), when you push a block, it slides until it hits another block or wall (as in *The Legend of Zelda: Minish Cap*).
3. In PUSH-F, some squares are fixed walls (drawn as bricks) that cannot be pushed. Without this suffix, all occupied squares within the rectangular puzzle are pushable blocks.
4. In PUSH-k (figure 1.4a vs. 1.4d), k represents the strength of the pusher: You can push at most k blocks at a time. PUSH-$*$ represents unlimited strength.
5. In PUSH-X (figure 1.4c), you are not allowed to revisit a square. In many video games, this is visualized by spaces falling away as you move.

Table 1.1 summarizes the known complexity results for many of these variants. It also summarizes the rules of each variant according to the following parameters:

1. *Strength*: how many blocks can be pushed (designated by $k \in \mathbb{N} \cup \{\infty\}$).
2. *Fixed*: whether fixed blocks are available. (In any case, the puzzle is bounded within a rectangle.)
3. *Slide*: how far blocks slide when pushed.

Table 1.1
Complexity of PUSH games.

Name	Strength	Fixed	Slide	Goal	Complexity	Reference
PUSH-k	$k \geq 1$	no	min	path	PSPACE-complete	[MIT24c]
PUSH-$*$	∞	no	min	path	NP-hard	[Hof00]
PUSHPUSH-k	$k \geq 1$	no	min	path	PSPACE-complete	[DHH04]
PUSHPUSH-$*$	∞	no	max	path	NP-hard	[Hof00]
PUSH-1F	1	yes	min	path	PSPACE-complete	[ACD+22]
PUSH-kF	$k \geq 2$	yes	min	path	PSPACE-complete	[DHH02]
PUSH-$*$F	∞	yes	min	path	PSPACE-complete	[ACD+24]
PUSH-kX	$k \geq 1$	no	min	simple path	NP-complete	[DH01]
PUSH-$*$X	∞	no	min	simple path	NP-complete	[Hof00]
SOKOBAN	1	yes	min	storage	PSPACE-complete	[Cul98]

4. *Goal*: the goal of the puzzle. In the "path" goal of PUSH puzzles, the goal is to move the pusher from a given start location to a given target location. In the "simple path" goal of PUSH-X, the movement path cannot cross itself. In the "storage" goal of SOKOBAN, the goal is to cover k given storage locations with the k movable blocks.

PUSH(PUSH)-k(F)(X)
Instance:

- A board with empty squares, movable blocks, and (if -F) fixed walls.
- A start square and a target square.

Question: Is there a path for the pusher from start to target that pushes at most k blocks in each step, pushes minimally or maximally according to PUSH(PUSH), and (if -X) does not cross itself?

All of these variants are NP-hard. Some are in NP (and hence NP-complete), but most are PSPACE-complete. None are harder than that. In the rest of this section, we describe some of the earlier NP-hardness proofs.

PUSH-$*$ is NP-Hard

Theorem 1.40. *(Hoffmann [Hof00]) PUSH-$*$ in a fixed box is NP-hard.*

Proof sketch. Recall that, in PUSH-$*$, the pusher can push an arbitrary number of blocks, and there are no fixed walls. We assume that the pusher and the blocks are confined to a rectangle.

We reduce from 3SAT. Refer to figures 1.5 and 1.6.

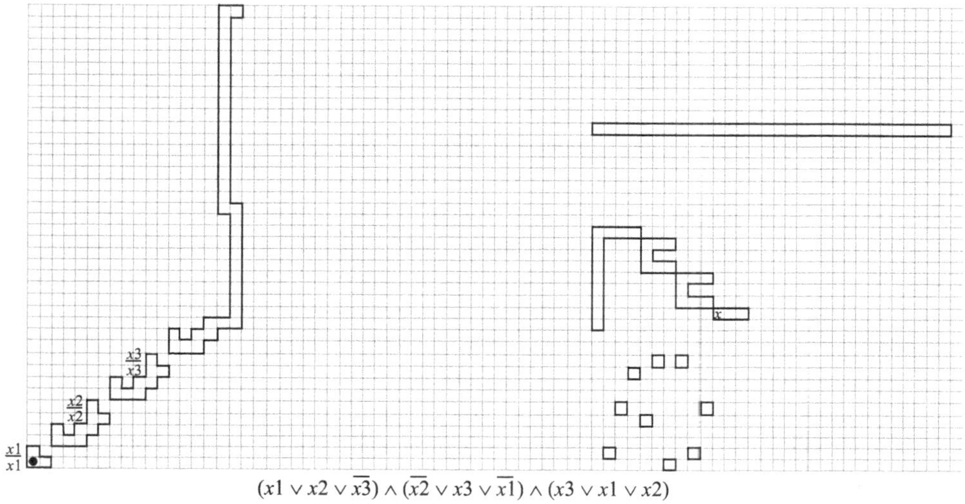

Figure 1.5
PUSH-∗ NP-hardness reduction from 3SAT.

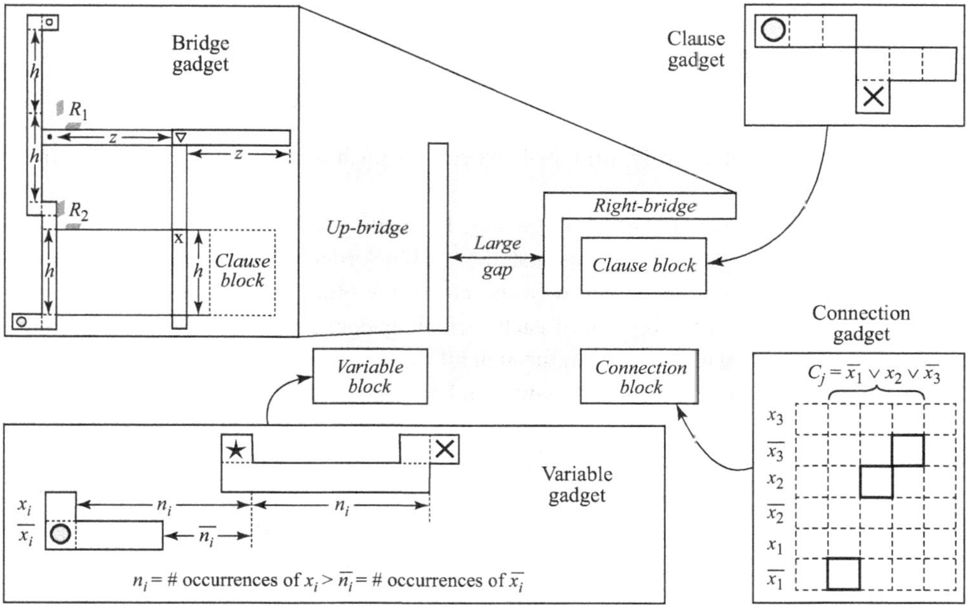

Figure 1.6
PUSH-∗ gadgets and overview of how they fit together.

Most of the space (everything not outlined in figure 1.5) is occupied by (movable) blocks. The pusher is therefore very constrained in its path.

First, the pusher walks through the Variable Block (bottom left). For each Variable Gadget, the pusher has an option to push one of two rows to the right (either a positive or negative instance) into free space in the Connection Block. Pushing both positive and negative instances of a variable only makes things worse further down the line (in the Clause Block).

The Connection Block (bottom right) encodes the bipartite graph of variables and clauses into a matrix with open spaces when variables (rows) connect to clauses (columns), and filled with blocks otherwise. Pushing a satisfying sequence of variables to the right will leave enough space in the Connection Block to maneuver the Clause Block later.

The Bridge Gadget forces the transition from variables to clauses to be one-way, and to fill in the space to the left and top of the Clause Block.

Each Clause Gadget in the Clause Block allows the pusher to make progress only when there is space below any of the variables in that clause. (This is why pushing both instances of a variable only hurts you, because it may block off needed empty squares.) Therefore successful traversals correspond to satisfying solutions to the 3SAT formula. □

Exercise 1.41. Fill in the details of the proof of theorem 1.40.

Open Problem 1.42. *Is PUSH-$*$ in NP or PSPACE-complete?*

PUSHPUSH-1 is NP-Hard

Theorem 1.43. *(Demaine et al. [DDO00]) PUSHPUSH-1 is NP-hard.*

Proof sketch. Recall that, in PUSHPUSH-1, when you push a block, it slides until it hits another block or wall.

Figure 1.7 gives an example of the reduction. The paths are width-1 tunnels, denoted by lines. At each variable gadget, you can push the top block (denoted by a circle) in the TRUE or FALSE direction, which opens one way and closes the other way, never to be traversed again. The two blocks at the bottom of each variable gadget keep you from backtracking. After the last variable gadget, you run through all the clause gadgets. If any of the literals that satisfy a clause were visited, then you could have pushed the corresponding block of the clause gadget to the right, blocking off the left vertical tunnel in the clause gadget and allowing the top block of the clause gadget to be pushed down while keeping the bottom tunnel of the clause gadget open for traversal. This reduction is the basis for many of the proofs of this type; it is a very straightforward reduction from 3SAT.

At this point we have proved NP-hardness of PUSHPUSH-1 in 3D. In 3D, we can route all the needed connections between variables and clauses without collisions. To extend the result to 2D, figure 1.8 shows a ***crossover gadget*** where the pusher can traverse from left to right and/or from top to bottom. The main point here is that this gadget is very complicated; chapter 2 will explore ways to avoid such crossover gadgets. □

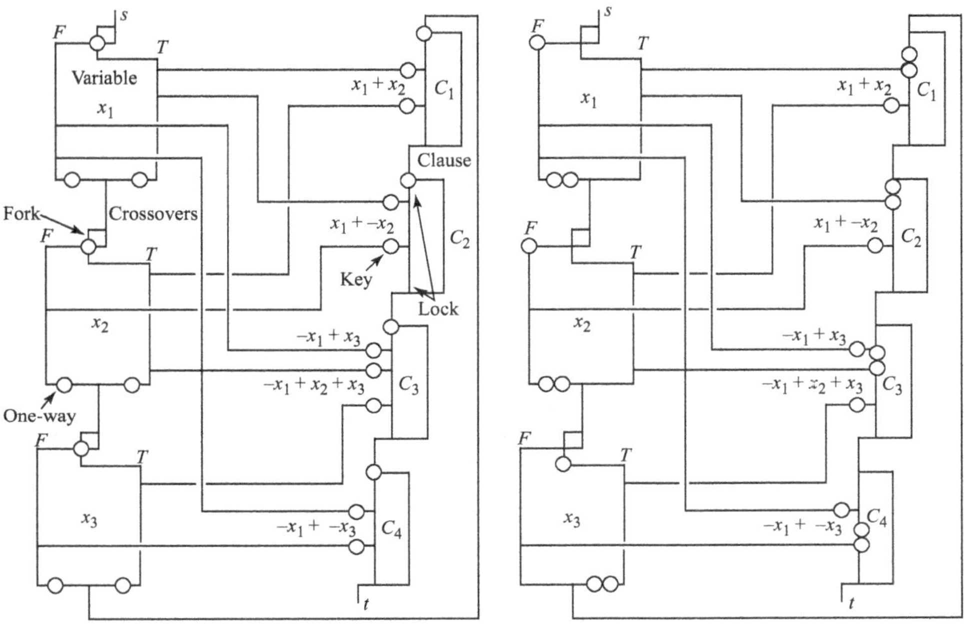

Figure 1.7
PUSHPUSH-1 in 3D is NP-hard.

1.3.5 *SUPER MARIO BROS.*

In this section, we will look at the video game *Super Mario Bros*. Aloupis et al. [ADGV15] showed that this game is NP-hard. In that same paper, they showed that several other Nintendo games—*Donkey Kong Country*, *The Legend of Zelda*, *Metroid*, and *Pokémon*—are NP-hard. Later, Demaine et al. [DVW16] showed that *Super Mario Bros*. is PSPACE-complete. We will briefly discuss that result in section 13.6.3.

Super Mario Bros. is a platform video game. The player controls a character who can run, jump, and climb between suspended platforms while avoiding various obstacles. As usual, SUPER MARIO BROS. means the problem of, given a position in the game, deciding whether the player win the level by reaching a specified target location.

Theorem 1.44. *(Aloupis et al. [ADGV15]) SUPER MARIO BROS. is NP-hard.*

Proof sketch. We show 3SAT $\leq_p$ SUPER MARIO BROS. Given an instance of 3SAT, which has some variables, literals, and clauses, we will make gadgets to represent them in SUPER MARIO BROS. Refer to figure 1.9.

A ***variable gadget*** is a place where the player can fall in one direction or the other, which we think of as corresponding to setting the variable TRUE or FALSE. Because the player cannot jump too high, if we put a long fall after the decision, the decision cannot be undone. Each

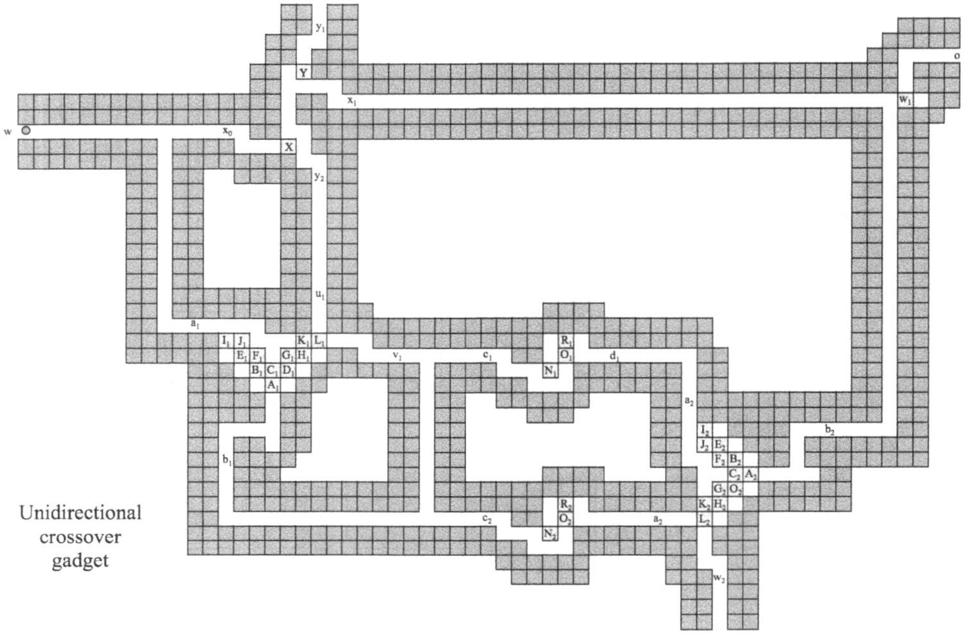

Figure 1.8
Unidirectional crossover gadget.

choice (TRUE or FALSE) corresponds to a literal (x or $\neg x$), and it connects via a collection of paths to the clauses that contain that literal.

A ***clause gadget*** (figure 1.9b) is a sequence of fire bars for the player to run through, just after three question blocks in the floor containing invincibility stars, any one of which lasts long enough to let the player run through the fire bars. Each question block can release its invincibility star only if the player enters the gadget from the corresponding literal below. Thus the player can release at least one invincibility star in the clause gadget if and only if at least one of that clause's variable gadgets the player made the choice that satisfies the clause.

The gadgets are arranged so that you first set each variable (and can release all the corresponding stars), then run through the clause gadgets in order. Traversing the clause gadgets in this way is possible if and only if one literal from each clause was satisfied—that is, if and only if there is a satisfying assignment to the original 3SAT problem.

The graph of paths between the variable gadgets and clause gadgets is not planar, so the reduction also has a ***crossover gadget***. The original crossover gadget by Aloupis et al. [ADGV15] is complicated. Figure 1.9c shows a simpler crossover gadget from the later SUPER MARIO BROS. PSPACE-hardness proof [DVW16]. The player can run at full speed

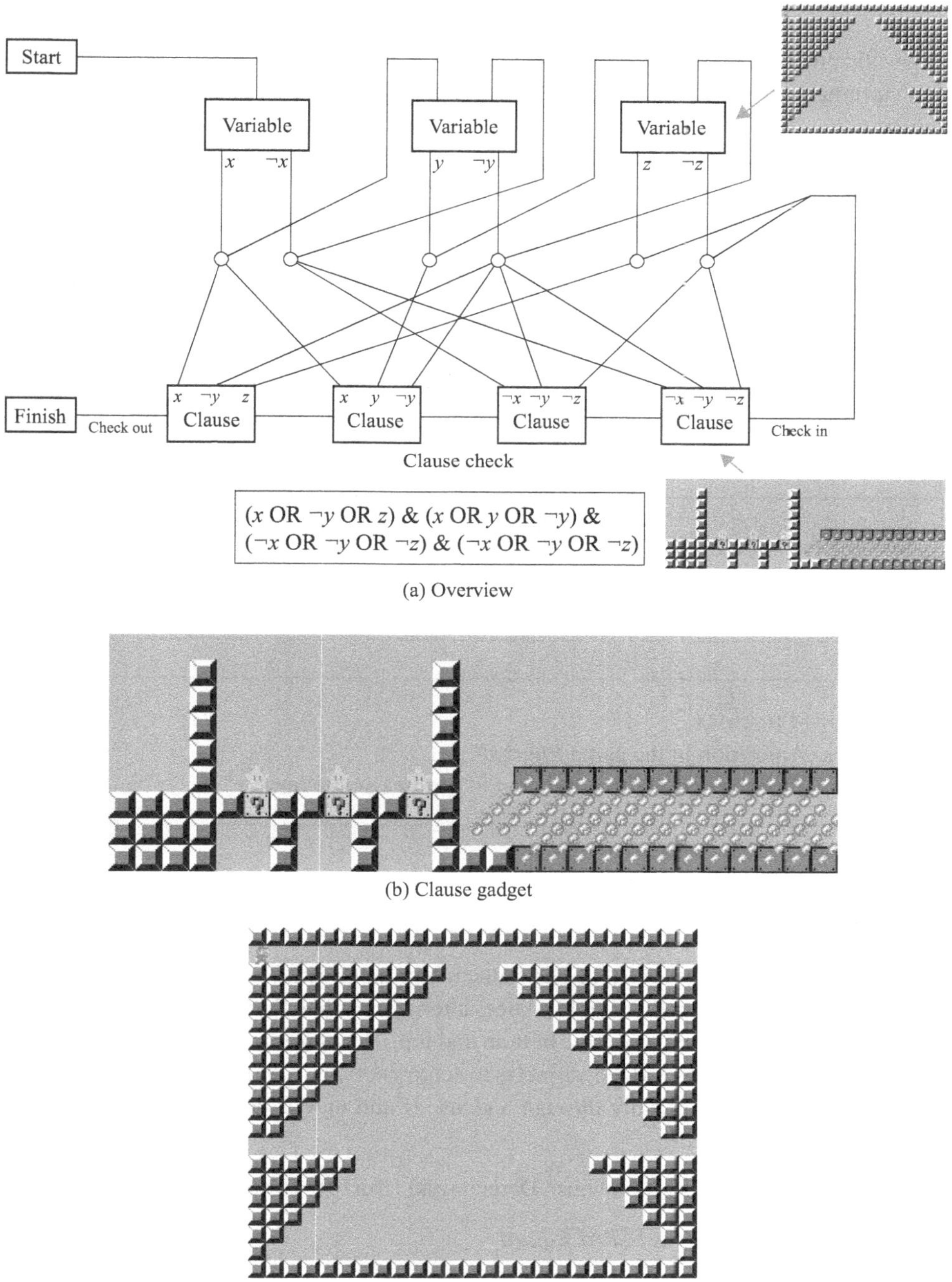

(a) Overview

(b) Clause gadget

(c) Crossover gadget from [DVW16]

Figure 1.9
NP-hardness reduction for Super Mario Bros., based on [ADGV15, DVW16].

from either top entrance to the opposite bottom entrance, but momentum prevents the player from turning around mid-air, without falling into the pit at the bottom from which there is no escape. □

1.3.6 *Phutball* (Philosopher's Football)

Conway [BCG82] devised the game *Phutball* which is short for *Philosopher's Football*. *Phutball* is a two-player game played with many white stones and one black stone on a square lattice (or Go board). Players move by either adding another white stone to the board or moving the black stone by a sequence of jumps over contiguous groups of white stones, which are immediately removed (before the next jump in the sequence). The goal is to move the black stone to one of the player's goals at one end of the board.

While the most natural question is "Who wins?" we will also consider the simpler "mate-in-1" question.

> PHUTBALL
> *Instance:* A position in the game *Phutball*.
> *Question:* Can the player whose move it is win?

> MATE-IN-1 PHUTBALL
> *Instance:* A position in the game *Phutball*.
> *Question:* Can the player whose move it is win in a single turn?

Demaine et al. [DDE02] showed the following:

Theorem 1.45. *MATE-IN-1 PHUTBALL is NP-complete.*

Proof sketch. Refer to figure 1.10. The reduction lays out variables and clauses on a large two-dimensional board, with variable choices alternating along the left and right edges, and clauses verified alternating along the bottom and top. The player makes a sequence of long horizontal jumps to choose which variables to set to TRUE (top side) or FALSE (bottom side), and then can traverse vertically through a clause if and only if none of the crucial black stones have been cleared. □

What about determining who wins? Dereniowski [Der10] showed the following:

Theorem 1.46. *PHUTBALL is PSPACE-hard.*

Open Problem 1.47. *Pin down the complexity of PHUTBALL.*

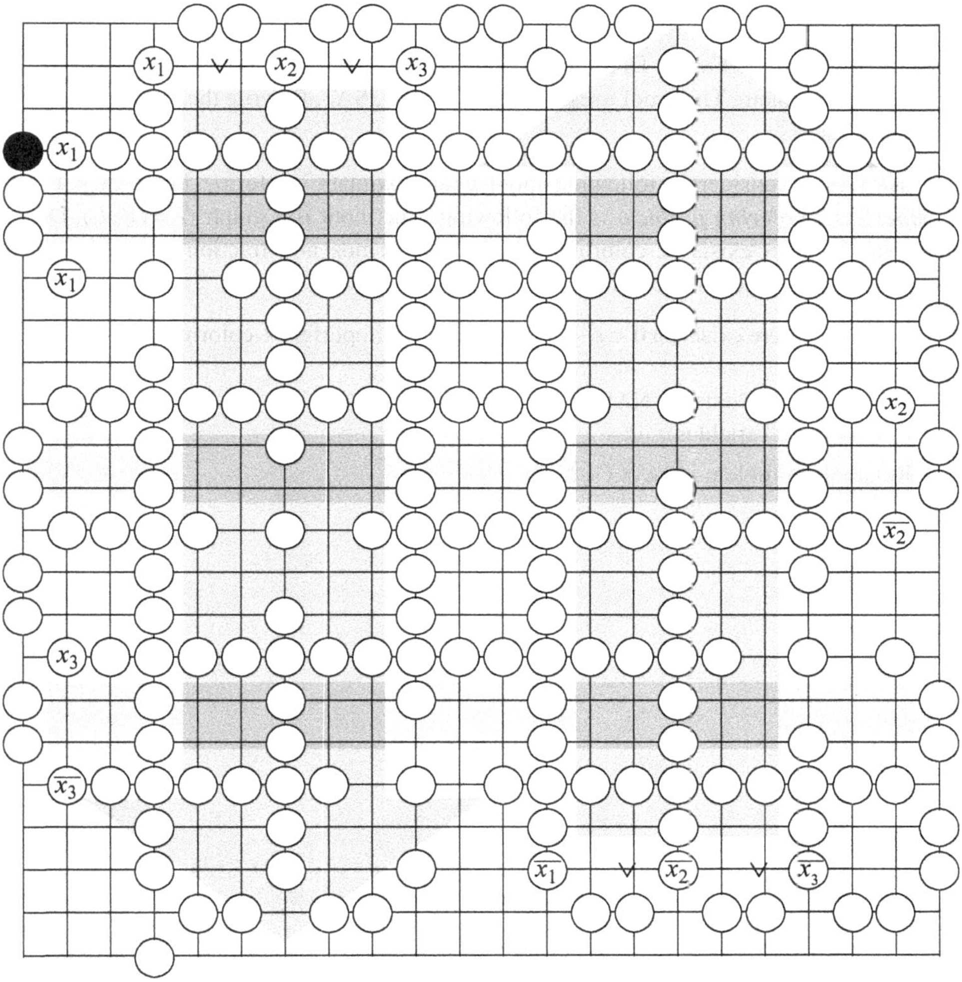

Figure 1.10
NP-hardness reduction from 3SAT to PHUTBALL. Variable values are set by going left to right and vice versa. Clauses are verified by going bottom to top and vice versa.

1.3.7 Exercises

Exercise 1.48. Read Franck Dernoncourt's paper [Der14] on the NP-completeness of the *TrackMania* problem. The proof uses a reduction from 3SAT. Rewrite the reduction in your own words.

Exercise 1.49. Consider the following notion of approximate 2-coloring. Let $0 < \varepsilon < 1$. The ***ε-imperfect 2-coloring problem*** is the following: The input is a graph $G = (V, E)$. Determine whether there exists a 2-coloring of V such that at most an ε fraction of the edges have endpoints that are the same color.

1. Show that there exists an $0 < \varepsilon < 1$ such that the ε-imperfect 2-coloring problem is NP-complete.
   Hint: Reduce POSITIVE NAE 3SAT to this problem (you can assume every clause has exactly three variables).
2. Relate this problem to MAX CUT.

2

NP-Hardness via Planar SAT

2.1 Introduction

Many problems that we would like to prove NP-hard are *planar*: They take place on a planar graph or in the two-dimensional Euclidean plane. In this case, it is helpful to reduce from a similarly planar problem. In this chapter, we introduce one important such problem, *planar 3SAT*, and use it to prove that many graph problems remain NP-complete when restricted to planar graphs, which may also be useful for further reduction.

Chapter Summary

1. *We show 3-Coloring is NP-complete. By developing a crossover gadget, we reduce 3-Coloring to Planar 3-Coloring. Hence Planar 3-Coloring is also NP-complete.*
2. *We could prove many problems NP-complete with crossover gadgets. We take a different approach. We define Planar 3SAT, which is an NP-complete variant of SAT, and then use Planar 3SAT to show that many planar problems are NP-complete.*
3. *We use the planar graph problems we proved NP-complete to prove that other planar problems, including 2D geometric problems, are NP-complete.*

Why do we use Planar 3SAT rather than devise different crossover gadgets?

1. Rather than devise many crossover gadgets for many problems, we essentially devise only one, the one used to show Planar 3SAT is NP-complete.
2. Crossover gadgets are often very complicated. We saw one example in theorem 1.43.
3. Gurhar et al. [GKM$^+$12], and independently Burke [Bur], showed that, for the planar Hamiltonian cycle problem, it is literally *impossible* to create crossover gadgets. This may be true for other problems as well.

In this chapter, we also often want to restrict the degrees of vertices in a graph. This can be useful in a planar context when working on a square grid, for example, where we can reasonably represent only four edges incident to a vertex.

Definition 2.1. A graph has ***maximum degree d*** if every vertex has degree at most d. A graph is ***d-regular*** if all vertices have the same degree d.

Some of our reductions map an input of length n to an output of length $O(n)$ or of length $O(n^2)$. This will be useful for chapter 6.

2.2 Planar Graph Coloring

Graph 2-Coloring can be solved in polynomial time by a simple greedy algorithm. Karp [Kar72] showed that, if the number of colors is allowed to vary, then Graph Coloring is NP-complete (it was one of his original 21 problems). What if the number of colors is fixed at three?

> 3-Coloring, Max-Degree-4 3-Coloring, Planar 3-Coloring, and Planar Max-Degree-4 3-Coloring
>
> *Instance:* A graph $G=(V,E)$. In 3-Coloring, the input is the graph without any conditions. In Max-Degree-4 3-Coloring, the graph has maximum degree 4. In Planar 3-Coloring, the graph is planar. In Planar Max-Degree-4 3-Coloring, the graph is planar and has maximum degree 4.
>
> *Question:* Does there exist a 3-coloring of G, which is an assignment of V to $\{1,2,3\}$ (the colors) such that all adjacent vertices have different colors?

Garey et al. [GJS76] showed that these problems are all NP-complete. Their proof did not use Planar 3SAT, but rather a custom crossover gadget for coloring. We present their proof.

Theorem 2.2.

1. *3-Coloring is NP-complete.*
2. *Planar 3SAT is NP-complete.*
3. *Max-Degree-4 3-Coloring is NP-complete.*
4. *Planar Max-Degree-4 3-Coloring is NP-complete.*

Proof. 1) We show 3SAT $\leq_p$ 3-Coloring.

1. Input $\varphi = C_1 \wedge \cdots \wedge C_m$.
2. Create a graph G as follows.
   a. The ***colors gadget*** consists of a single triangle, whose vertices must be colored distinctly. We pretend that we know which vertex is colored which, and refer to the

BLUE/TRUE, RED/FALSE, and GREEN vertices. Any actual coloring is the same up to renaming of colors.

b. For every variable x, we have a ***variable gadget*** consisting of a vertex x, a vertex $\overline{x}$, an edge between them, and an edge from the GREEN colors vertex to both of them. This gadget forces one of x and $\overline{x}$ to be colored BLUE/TRUE and the other to be colored RED/FALSE.

c. For every clause $x_i \vee x_j \vee x_k$, we have a ***clause gadget*** as drawn in figure 2.1. The rightmost vertex must be colored BLUE/TRUE because it is connected to the RED/FALSE and GREEN colors vertices. Thus one of the other vertices in that triangle is colored RED/FALSE and the other is colored GREEN. The RED/FALSE vertex's neighbor is either x_k, in which case it forces x_k to be colored BLUE/TRUE; or it is a vertex of the leftmost triangle, in which case one of the remaining vertices must be not RED/FALSE, and thus the non-triangle neighbor (either x_i or x_j) must be colored

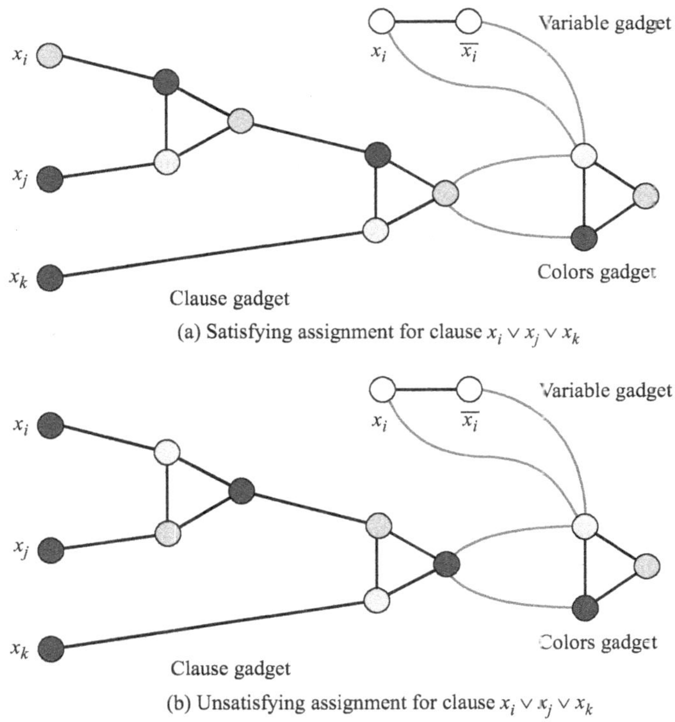

(a) Satisfying assignment for clause $x_i \vee x_j \vee x_k$

(b) Unsatisfying assignment for clause $x_i \vee x_j \vee x_k$

Figure 2.1
Gadgets for 3-COLORING.

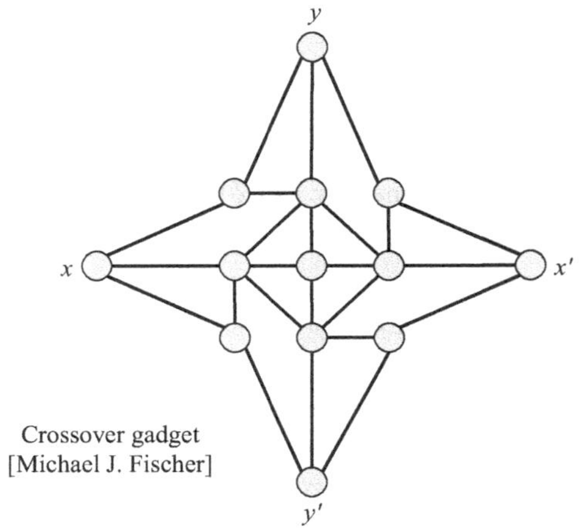

Figure 2.2
3-COLORING crossover gadget, attributed to Michael J. Fischer [GJS76].

BLUE/TRUE. Thus at least one of x_i, x_j, x_k is colored BLUE/TRUE. Indeed any such coloring of x_i, x_j, x_k can be completed to a coloring of the clause gadget.

2) We show that 3-COLORING $\leq_p$ PLANAR 3-COLORING. The high-level idea of our reduction is to replace each crossing with the ***crossover gadget*** in figure 2.2.

Figure 2.3 shows the two distinct ways (up to permutation of colors) that the crossover gadget can be colored. Both cases result in $x = x'$ and $y = y'$, with independent choices of the colors for x and y.

Use of this crossover gadget has some subtleties. For starters, suppose we have a graph with one crossing between edges (a, b) and (c, d). Then we do *not* want to apply the crossover gadget with $x = a$, $x' = b$, $y = c$, and $y' = d$: This would force the colors of a and b (likewise, c and d) to be the same. But we want the colors of a and b (likewise, c and d) to be *different* in 3-COLORING. To fix this problem, we add a vertex a' near and connected to b, and a vertex c' near and connected to d, and apply the crossover gadget with $x = a$, $x' = a'$, $y = c$, and $y' = c'$. This forces a and a' (likewise, c and c') to have the same color, and then the edge between a' and b (likewise, c' and d) forces a/a' and b (likewise, c/c' and d) to have different colors.

A second issue is that each crossover gadget adds edges and so, without care, could *increase* the number of crossings, and we would never finish the reduction. To fix this problem, we add vertices to localize each crossover gadget as follows. Given an instance of 3-COLORING, we draw the graph in the plane, allowing edges to cross. For each edge crossed by at least one other edge, we add subdivision vertices between consecutive crossing points, and between the last crossing point and the corresponding endpoint of the edge (like the

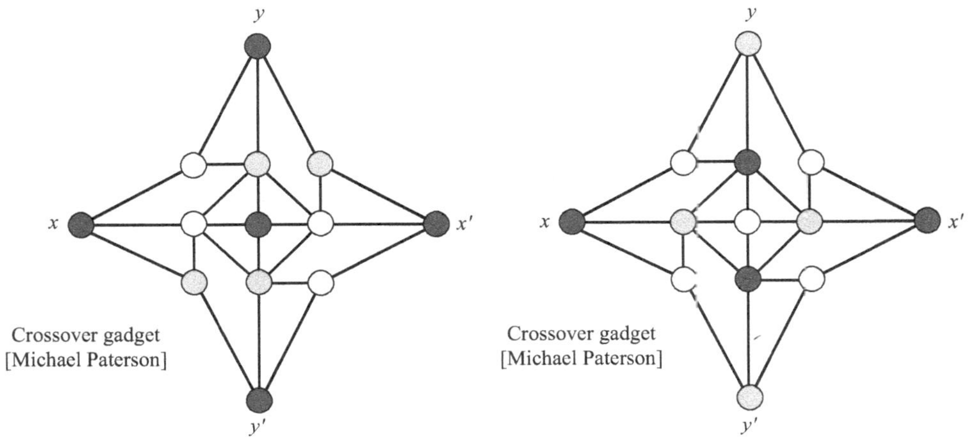

Figure 2.3
Two ways to 3-color the crossover gadget.

a' and b' vertices above). This subdivision expands each edge into a path of subedges. Then we replace every crossing, between subedges (x, x') and (y, y'), by the crossover gadget in figure 2.2.

We call the new graph G'. We leave it to the reader to show that $G \in$ 3-COLORING if and only if $G' \in$ PLANAR 3-COLORING.

3 and 4) We show that MAX-DEGREE-4 3-COLORING (PLANAR MAX-DEGREE-4 3-COLORING) is NP-complete by reducing 3-COLORING (PLANAR 3-COLORING) to it. For each vertex x of degree $k \geq 5$, we replace the vertex with the gadget in Figure 2.4, resulting in k copies of x that must all have the same color, so act like a single vertex. For each of the k neighbors of x, we attach it to a unique copy of x, which has degree 2 in figure 2.4 and so ends up with degree 3. The resulting maximum degree is 4. If we preserve the cyclic order of vertices around x, this reduction preserves planarity. □

Theorem 2.2 raises the question of which variants of graph coloring are in P. The following are known.

1. 3-COLORING restricted to graphs of maximum degree 3 is in P since, by Brook's theorem [Bro41], such graphs are always 3-colorable.
2. 4-COLORING of planar graphs is in P since, by the celebrated four-color theorem of Appel et al. [AH77a, AHK77, AH77b] (and later a simpler proof by Robertson et al. [RSST97]), all planar graphs are 4-colorable.

To state some open problems, we need the following definitions.

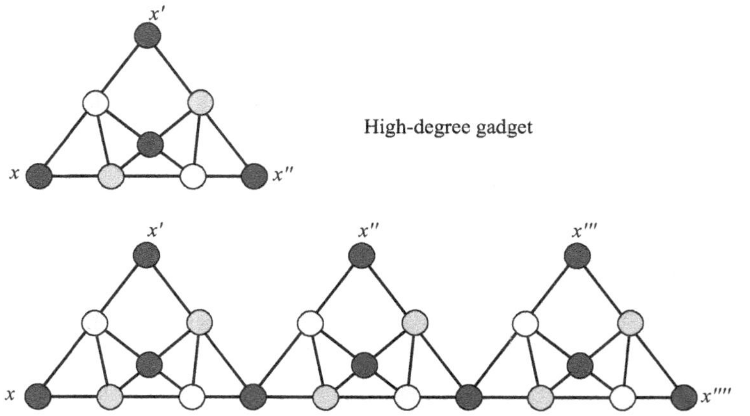

Figure 2.4
Gadget for reducing maximum degree to 4.

Definition 2.3. Let $g, k \geq 0$.

1. A graph G has ***genus g*** if it can be drawn on a sphere with g handles attached without any edges crossing. Note that genus 0 is the same as planar.
2. A graph has ***crossing number k*** if it can be drawn in the plane with $\leq k$ crossings.

Open Problem 2.4. *For the following problems, either show they are in P or that they are NP-complete or (very unlikely) that they are neither.*

1. *Given a graph of genus 1, is it 4-colorable?*
2. *Given a graph of crossing number 1, is it 4-colorable?*
3. *Gasarch et al. [GHOP22] summarize what is known and not known about the complexity of graph coloring when the graph is restricted to graphs of bounded genus or bounded crossing number. Read this paper and solve some of its open problems.*

Exercise 2.5. Show that the CLIQUE problem restricted to planar graphs is in P.

2.2.1 Homomorphism Generalization of 3-COLORING

Theorem 2.2.1 showed that 3-COLORING is NP-complete. We discuss a generalization of this result.

Definition 2.6. Let $G = (V_G, E_G)$ $H = (V_H, E_H)$ be graphs. A ***homomorphism of H into G*** is a function $f : V_H \to V_G$ such that, for all $x, y \in V_H$, if $(x, y) \in E_H$ then $(f(x), f(y)) \in E_G$. Note that this definition works for undirected graphs as well.

> $\mathrm{HOM}_U(H)$, $\mathrm{HOM}_D(H)$
> *Instance:* ($\mathrm{HOM}_U(H)$) An undirected graph G.
> *Instance:* ($\mathrm{HOM}_D(H)$) A directed graph G.
> *Question:* Is there a homomorphism from H into G?
> *Note:* If $H = K_c$, then $G \in \mathrm{HOM}_D(H)$ if and only if G is c-colorable.

What happens if H is not a complete graph? Hell and Nesetril [HN90] showed the following dichotomy theorem for $\mathrm{HOM}_U(H)$.

Theorem 2.7.

1. *If H is bipartite, then $HOM_U(H)$ is in P.*
2. *If H is not bipartite, then $HOM_U(H)$ is NP-complete.*

The case of $\mathrm{HOM}_D(H)$ is much harder. Zhuk [Zhu20] and Bulatov [Bul17] independently showed the following.

Theorem 2.8. *If H is a directed graph, then $HOM_D(H)$ is either in P or NP-complete. (They did give a criteria on H, but it is very complicated.)*

2.3 Planar 3SAT

A CNF formula can be viewed as a bipartite graph called the ***variable–clause graph***, with (1) variables on one side and clauses on the other and (2) a ***positive edge*** (drawn solid) between variable x and clause C if x is in C, a ***negative edge*** (drawn dashed) between x and C if $\neg x$ is in C, and no edge between x and C if x does not appear in C. Figure 2.5 gives an example.

Definition 2.9. A formula is ***planar*** if its associated variable–clause graph is planar.

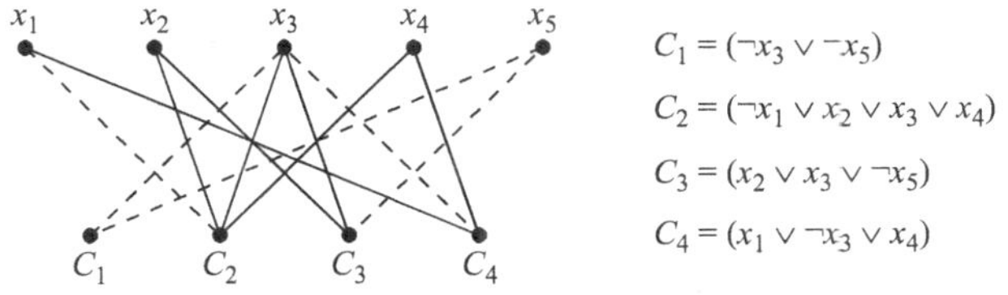

Figure 2.5
Bipartite graph associated with a CNF formula.

PLANAR 3SAT

Instance: A planar formula φ given as a variable–clause graph (with planar embedding).

Question: Is φ satisfiable?

Note: Hopcroft and Tarjan [HT74] showed that determining whether a graph is planar, and if so finding a planar embedding, can be solved in polynomial time (actually linear time). Hence it does not matter whether a planar embedding is given; if not, we can find one, or if the graph turns out not to be planar, output NO.

David Lichtenstein [Lic82] showed that PLANAR 3SAT is NP-complete.

Theorem 2.10. *3SAT $\leq_p$ PLANAR 3SAT, so PLANAR 3SAT is NP-complete.*

Proof. We show 3SAT $\leq_p$ PLANAR 3SAT.

1. Input $\varphi = C_1 \wedge \cdots \wedge C_k$, where each C_i is a clause with at most three literals.
2. Construct the variable–clause graph that represents φ. If the graph is planar, then we are done and we can just output that graph. More likely it is not.
3. Draw the graph in the plane, allowing edges to cross.

   For example, as shown in figure 2.7 (left), we could put the clauses on the left edge of the drawing, and the variables on the bottom edge, and connect each appropriate clause–variable pair via a one-bend edge consisting of a horizontal segment from the clause and a vertical segment from the variable.
4. Replace every crossing in the drawing with the ***crossover gadget*** shown in figure 2.6. In this figure, the small vertices represent clauses and big vertices represent variables. The blue connections are positive, while red are negative.

   We leave it as an exercise that the truth value of a is the same as that of a_1 and the truth value of b is the same as that of b_1. Thus we successfully copy variables a and b to the other side of the crossover gadget. The drawn clauses at the very top and the very left of figure 2.6 are present only in the first gadget along a path connecting a clause to a variable; all subsequent gadgets connect variables to variables.
5. The central clause in the crossover gadget of figure 2.6 has size 4 instead of 3. To fix this, we use the reduction from 4SAT to 3SAT described in theorem 1.7, which replaced $(L_1 \vee L_2 \vee L_3 \vee L_4)$ with

$$(L_1 \vee L_2 \vee z) \wedge (\neg z \vee L_3 \vee L_4).$$

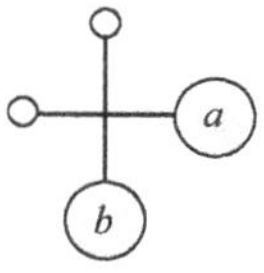

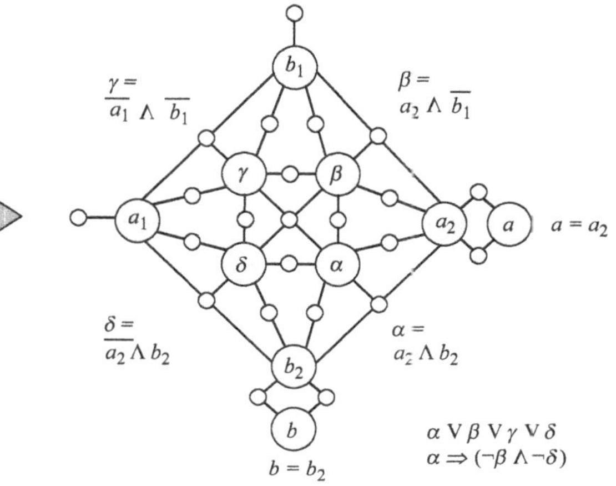

a_2	b_2	α	β	δ	γ	a_1	b_1
0	0	0	0	0	1	0	0
0	1	0	0		1?	0?	0?
1	0	0		0	1?	0?	0?
1	1	1	0	0		1	1

Figure 2.6
PLANAR 3SAT crossover gadget.

Plugging this into the degree-4 vertex in figure 2.6 results in two degree-3 clause vertices, connected via an intermediate variable vertex for z, which preserves planarity of the figure. □

Exercise 2.11. Show that the gadget in figure 2.6 works.

2.4 LINKED PLANAR 3SAT

David Lichtenstein [Lic82], in fact, showed NP-completeness of two stronger forms of PLANAR 3SAT. The first variant requires more planarity:

Definition 2.12. A formula is ***variable-linked planar***[1] if its associated variable–clause graph, plus a cycle through its clause vertices $C_1, \ldots, C_m$ in order, is planar.

> VARIABLE-LINKED PLANAR 3SAT
> *Instance:* A variable-linked planar formula φ, given as an embedded variable–clause graph with variable cycle.
> *Question:* Is φ satisfiable?

1. This terminology, or more specifically "var-linked," was introduced by Fellows et al. [FKMP95].

The second variant provides a separation between positive and negative edges:

> SIDED VARIABLE-LINKED PLANAR 3SAT
> *Instance:* A variable-linked planar formula φ, given as an embedded variable–clause graph with variable cycle, where all positive (solid) connections are on one side of the cycle and all negative (dashed) connections are on the other side of the cycle.
> *Question:* Is φ satisfiable?
> *Note:* SIDED VARIABLE-LINKED PLANAR 3SAT is a special case of MONOTONE 3SAT because each clause is on one side of the variable cycle, so it can have only positive or only negative edges.

Theorem 2.13. *(Lichtenstein [Lic82])*

1. *VARIABLE-LINKED PLANAR 3SAT is NP-complete.*
2. *SIDED VARIABLE-LINKED PLANAR 3SAT is NP-complete.*

Exercise 2.14.

1. Prove theorem 2.13.1 by taking the variable–clause graph that results from the reduction in the proof of theorem 2.10, and showing that you can put a cycle through all of the variable vertices while preserving planarity. Figure 2.7 is a hint.
2. Prove theorem 2.13.2 by showing that the construction can be further modified to be sided. Figure 2.8 is a hint.

As you may have guessed from the name "VARIABLE-LINKED PLANAR 3SAT," we can similarly define CLAUSE-LINKED PLANAR 3SAT and SIDED CLAUSE-LINKED PLANAR 3SAT. CLAUSE-LINKED PLANAR 3SAT was proved NP-complete by Kratochvíl et al. [KLN91]; see also [FKMP95]. SIDED CLAUSE-LINKED PLANAR 3SAT was proved NP-complete only recently [Pil18], along with many other variants such as MONOTONE CLAUSE-LINKED PLANAR 3SAT.

Exercise 2.15. This exercise serves as a warning about further variants of PLANAR 3SAT.

1. Consider the variation where we insist that the all the variable vertices are connected in a cycle (as in VARIABLE-LINKED PLANAR 3SAT) *and* that every clause vertex is inside that cycle. Show that this variation of PLANAR 3SAT is, in fact, solvable in polynomial time.
   Hint: Use dynamic programming over the "tree" of clauses.
2. Consider the variation where we insist that all the variable vertices are connected in a cycle (as in VARIABLE-LINKED PLANAR 3SAT) *and* that the clause vertices are connected

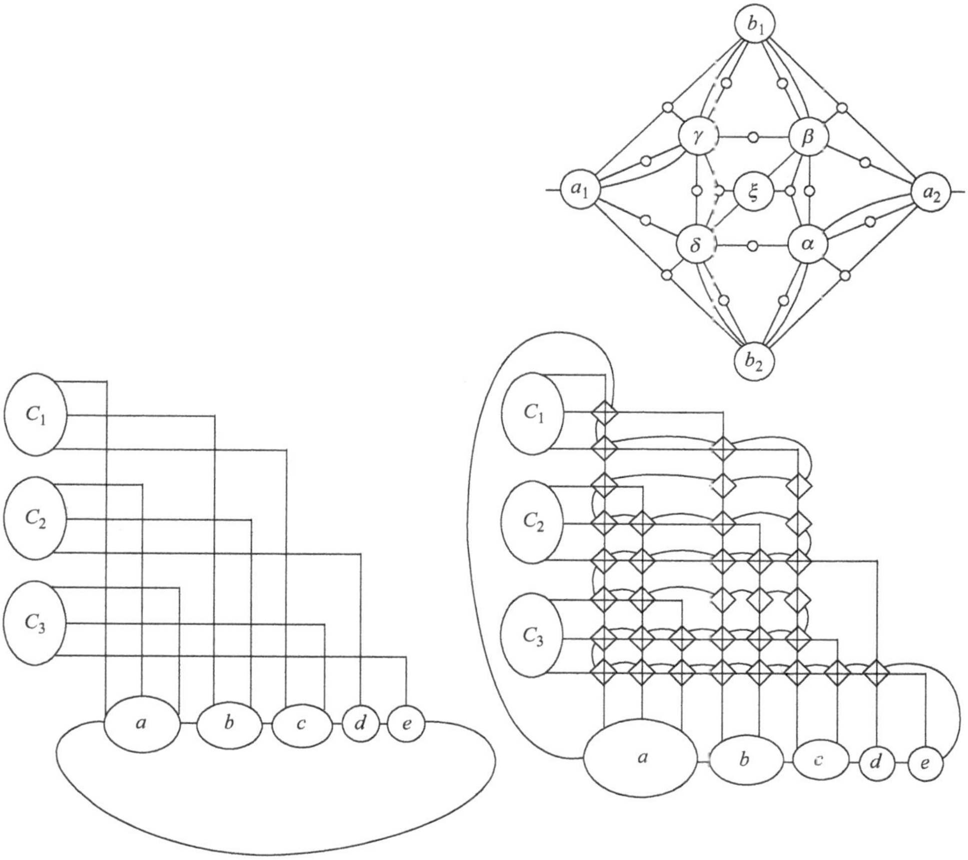

Figure 2.7
PLANAR 3SAT with a cycle through variable nodes. Top right: the path through the crossover gadget. Left: the initial arrangement of the bipartite graph, with the variables in a cycle. Bottom right: the arranged graph, with crossover gadgets inserted to preserve planarity.

in a path. Show that this variation of PLANAR 3SAT is, in fact, solvable in polynomial time.

Hint: Reduce to the previous part.

Although we will not use it in this book, there is a stronger form of PLANAR 3SAT that is NP-complete, despite being quite similar to exercise 2.15.2, which is polynomial:

Definition 2.16. A formula is ***linked planar*** if its associated variable–clause graph, plus a cycle in order through its variable vertices $x_1, \ldots, x_n$ and then its clause vertices $C_1, \ldots, C_m$, is planar.

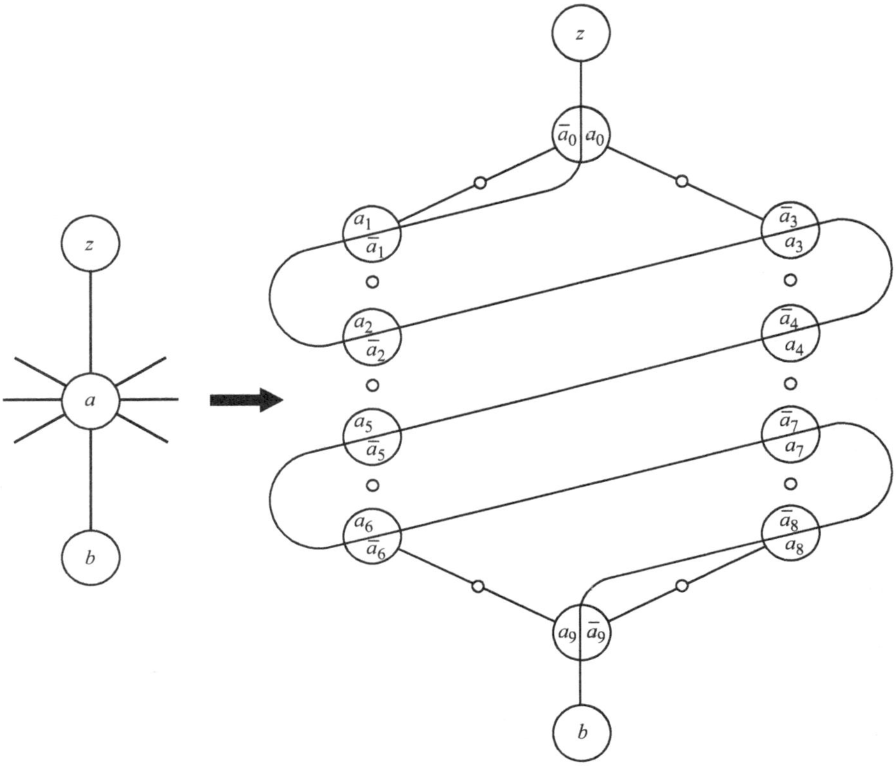

Figure 2.8
PLANAR 3SAT with variables divided to separate positive and negative connections.

LINKED PLANAR 3SAT
Instance: A linked planar formula φ, given as an embedded variable–clause graph with Hamiltonian cycle.
Question: Is φ satisfiable?

SIDED LINKED PLANAR 3SAT
Instance: A linked planar formula φ, given as an embedded variable–clause graph with Hamiltonian cycle, where all positive (solid) connections are on one side of the cycle and all negative (dashed) connections are on the other side of the cycle.
Question: Is φ satisfiable?

The added cycle includes a path connecting all the clause vertices (in some order) and a path connecting all the variable vertices. As we saw in exercise 2.15.2, requiring one more edge that connects the clause vertices into a cycle makes the problem easy. Nonetheless, Pilz [Pil18] proved the following (among many other variations):

Theorem 2.17.

1. *PLANAR 3SAT* $\leq_p$ *LINKED PLANAR 3SAT, so LINKED PLANAR 3SAT is NP-complete.*
2. *PLANAR 3SAT* $\leq_p$ *SIDED LINKED PLANAR 3SAT, so SIDED LINKED PLANAR 3SAT is NP-complete.*

2.5 More Planar Graph Problems

Next we show that a few more graph problems remain NP-complete when restricted to planar graphs. The reductions are from PLANAR 3SAT, so they serve as examples of such reductions.

2.5.1 PLANAR MAX-DEGREE-3 VERTEX COVER and INDEPENDENT SET

VERTEX COVER, MAX-DEGREE-3 VERTEX COVER, PLANAR VERTEX COVER, PLANAR MAX-DEGREE-3 VERTEX COVER

Instance: A graph $G=(V,E)$ and a number k. In MAX-DEGREE-3 VERTEX COVER, the graph has maximum degree 3. In PLANAR VERTEX COVER, the graph is planar. In PLANAR MAX-DEGREE-3 VERTEX COVER, the graph is planar and has maximum degree 3.

Question: Is there a $V' \subseteq V$ of size k such that for every $e \in E$ has some $v \in V'$ as an endpoint?

Lichtenstein [Lic82] proved PLANAR MAX-DEGREE-3 VERTEX COVER is NP-complete by giving a reduction from 3SAT to MAX-DEGREE-3 VERTEX COVER that, when restricted to planar formulas, outputs planar graphs. This result also translates to the analogous restrictions on INDEPENDENT SET.

Theorem 2.18.

1. *MAX-DEGREE-3 VERTEX COVER is NP-complete.*
2. *PLANAR MAX-DEGREE-3 VERTEX COVER is NP-complete.*
3. *PLANAR MAX-DEGREE-3 INDEPENDENT SET is NP-complete.*

Proof. 1) We give a reduction from 3SAT to MAX-DEGREE-3 VERTEX COVER. Figure 2.9 gives an example of the reduction.

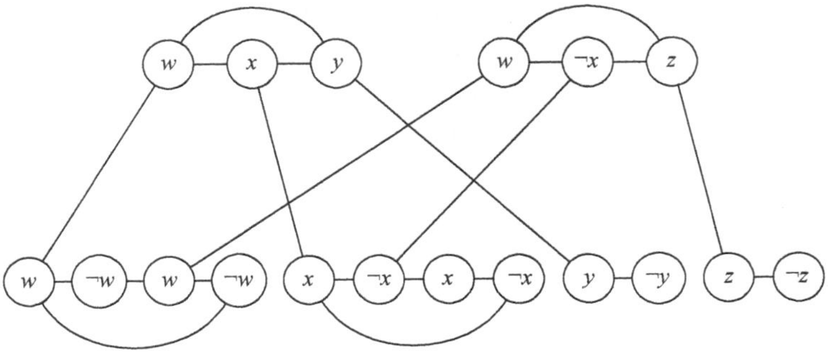

Figure 2.9
Graph from $(w \vee x \vee y) \wedge (w \vee \neg x \vee z)$.

1. Input $\varphi = C_1 \wedge \cdots \wedge C_m$.
2. Form a graph $G = (V, E)$ as follows:
   a. For every clause C_i, the ***clause gadget*** consists of a triangle whose three vertices are labeled with the three literals in C_i. Any vertex cover of the graph will need at least two vertices of a triangle, so every vertex cover has at least $2m$ vertices within the clause gadgets.
   b. For every variable x, let m_x be the number of times x or $\bar{x}$ appears in φ. The ***variable gadget*** for x consists of a cycle of length $2m_x$ where the vertices are alternatively labeled $x, \bar{x}, \ldots$. (If $m_x = 1$, then we just have a path graph, not a cycle, with two vertices labeled x and $\neg x$.) Any vertex cover of this gadget will need at least m_x vertices to cover the cycle, so every vertex cover has at least $\sum_x m_x = 3m$ vertices within the variable gadgets.
   c. For every literal L (either variable x or its negation $\neg x$) in every clause C, draw an edge between a vertex labeled L in the variable gadget for x and a vertex labeled L in the clause gadget for C. Because we have m_x vertices labeled L in the variable gadget for x, we can make these connections without reusing a variable-gadget vertex (or a clause vertex), so that the resulting graph has maximum degree 3.
3. Output $(G, 5m)$.

We leave the proof that $(G, 5m) \in$ VERTEX COVER if and only if $\varphi \in$ 3SAT to the reader.

2) To show that PLANAR 3SAT $\leq_p$ PLANAR MAX-DEGREE-3 VERTEX COVER, we show that the reduction from part 1 preserves planarity—that is, if φ is planar, then G is planar. The only edges that might cross are the edges from the variable gadgets to the clause gadgets. Let G' be the graph obtained by shrinking each variable gadget to one point and shrinking each clause gadget to one point. We need G' to be planar. It is! G' is exactly the variable–clause graph associated with φ. Because φ is planar, G' is planar, so G is planar.

3) We show that Vertex Cover $\leq_p$ Independent Set while preserving all graph properties, so in particular Planar Max-Degree-3 Vertex Cover $\leq_p$ Planar Max-Degree-3 Independent Set. A graph G has a vertex cover C of size k if and only if the same graph G has an independent set of size $|V(G)| - k$, namely $V(G) \setminus C$; see, for example, [GJ79, Lemma 3.1]. Thus the reduction converts (G, k) into $(G, |V(G)| - k)$. □

2.5.2 Planar Dominating Set

Dominating Set and Planar Dominating Set
Instance: A (planar) graph $G = (V, E)$ and a number k. k is the parameter.
Question: Is there a dominating set D of size k? (Every $v \in V$ is either in D or has an edge to some $u \in D$.)

Garey and Johnson [GJ79, GT2] mention without proof that Dominating Set is NP-complete, even when restricted to planar graphs of maximum degree 3, or to 4-regular planar graphs. We give a simple reduction without any guarantees on vertex degrees.

Theorem 2.19.

1. *Vertex Cover $\leq_p$ Dominating Set, so Dominating Set is NP-complete.*
2. *Planar Vertex Cover $\leq_p$ Planar Dominating Set, so Planar Dominating Set is NP-complete.*

Proof sketch. We give a reduction for Vertex Cover $\leq_p$ Dominating Set that preserves planarity, so it is also a reduction for Planar Vertex Cover $\leq_p$ Planar Dominating Set. Figure 2.10 gives an example of the reduction.

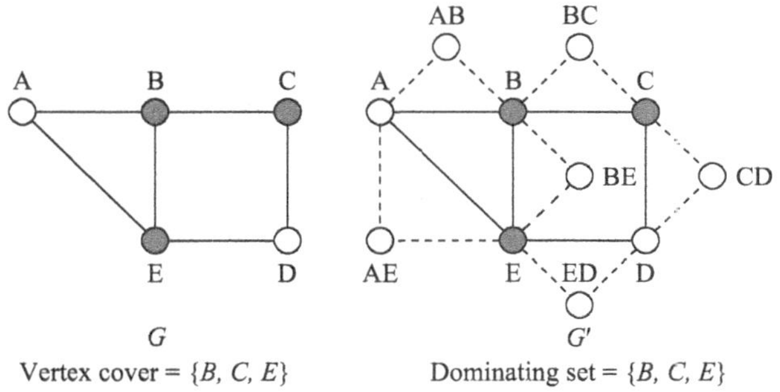

Figure 2.10
Reduction from Vertex Cover to Dominating Set.

1. Input graph $G = (V, E)$ and number k.
2. Create a graph $G' = (V', E')$ by starting with G, for every edge (u, v) in G create a new vertex uv, and then putting a new edge from uv to both u and v. □

Exercise 2.20. Prove that PLANAR MAX-DEGREE-3 DOMINATING SET and PLANAR 4-REGULAR DOMINATING SET are NP-complete.

2.5.3 PLANAR DIRECTED HAMILTONIAN CYCLE

DIRECTED HAMILTONIAN CYCLE, PLANAR DIRECTED HAMILTONIAN CYCLE
Instance: A (planar) directed graph.
Question: Is there a Hamiltonian cycle? A Hamiltonian cycle is a cycle that visits every vertex exactly once, and respects edge directions: If (u, v) is an edge but (v, u) is not, then the cycle cannot use (v, u).

David Lichtenstein [Lic82] showed that PLANAR DIRECTED HAMILTONIAN CYCLE is NP-complete.

Theorem 2.21.

1. *3SAT $\le_p$ DIRECTED HAMILTONIAN CYCLE, so DIRECTED HAMILTONIAN CYCLE is NP-complete.*
2. *PLANAR 3SAT $\le_p$ PLANAR DIRECTED HAMILTONIAN CYCLE, so PLANAR DIRECTED HAMILTONIAN CYCLE is NP-complete.*

Proof sketch. We give the reduction 3SAT $\le_p$ DIRECTED HAMILTONIAN CYCLE. We leave to the reader the other parts, including that the reduction can preserve planarity.

1. Input a 3CNF formula φ.
2. For each variable a in φ, form the ***variable gadget*** shown in figure 2.11 (top). If m is the number of times a occurs in φ, then the gadget has $4m$ rungs. Any directed Hamiltonian cycle enters the gadget from the left and then goes either up or down, which corresponds to setting a to TRUE or FALSE, respectively, and then it alternates down and up in a zigzag. Thus we can label each edge with the literal a or $\neg a$, and either all a edges or all $\neg a$ (and never both types) will be visited by the cycle.
3. For each clause C in φ, form the ***clause gadget*** shown in figure 2.11 (bottom left), for the example $C = a \vee \neg b \vee c$. This gadget consists of a single vertex, attached to three variable gadgets via length-2 paths as alternatives to appropriate literal edges. This vertex can be visited by a directed Hamiltonian cycle if and only if at least one of the corresponding literal edges can be visited—that is, at least one of the corresponding literals has a TRUE assignment (i.e., the clause C is satisfied). □

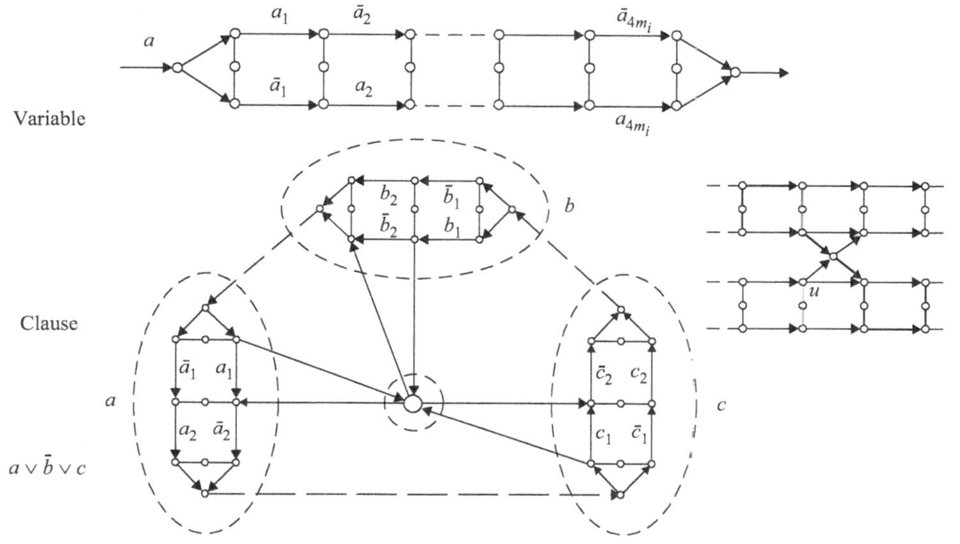

Figure 2.11
Hamiltonian Cycle gadgets. Undirected edges represent edges in both directions.

Exercise 2.22.

1. Show that the reduction in the proof of theorem 2.21 works. In particular, show that it never helps to go from one variable gadget to a clause gadget to another variable gadget, as in figure 2.11 (right).
2. Show that, if the reduction given in the proof of theorem 2.21 started with a planar formula, then with some care the output is a planar graph.

We will see many more versions and applications of Hamiltonian Cycle in chapter 4.

2.6 Reducing from Planar 3SAT to Geometric Problems

To reduce to a problem in the 2D plane, we usually need to define explicit coordinates for variable and clause gadgets. A useful starting point is the following theorem, due to Schnyder [Sch90].

Theorem 2.23. *Every n-vertex planar graph can be drawn with its vertices placed at grid points of an $n \times n$ grid, where edges are drawn as straight lines that do not cross. Such a drawing can be computed in $O(n)$ time.*

In many situations, however, it is difficult to design the wires implementing the edges of the graph that connect two arbitrary grid points. Often it is simpler to design ***wire gadgets*** that only work in certain directions (e.g., horizontal and vertical), together with ***turn gadgets*** that allow the wire to change direction (e.g., by 90°). In this case, we need edges to be drawn as orthogonal polygonal lines instead of single line segments. A useful starting point for this type of reduction is the following:

Theorem 2.24. *Every n-vertex planar graph of maximum degree 4 can be drawn on an $n \times n$ grid, with its vertices placed at grid points and its edges routed along grid edges [BK98a]. Furthermore, each edge has at most two bends, and such a drawing can be computed in $O(n)$ time.*

When embedding on a grid, wires often have a parity constraint where they must repeat a module of size k, so their size is always of a fixed length modulo k. This can be a problem when not all gadgets are aligned modulo k. In this case, we often need a ***shift gadget*** to adjust a wire's position modulo k—for example, by ± 1. Applying a shift gadget enough times, we can adjust a wire's end positions modulo k to match whatever gadgets they need to attach to.

In the rest of this section, we give some examples of such reductions.

2.6.1 Shakashaka

Shakashaka is a Nikoli puzzle game [Nik08, Nik]. (Nikoli is a Japanese publisher of pencil-and-paper puzzles that are popular around the world.) The game begins with a board like the left part of figure 2.12. The goal is to half fill in some of the white squares (with constraints we will discuss soon) so that the white squares form a disjoint set of rectangles, as in the right part of figure 2.12. Some of the black squares have numbers. A black square with number x must have exactly x half-filled white squares adjacent to it. If a black square has

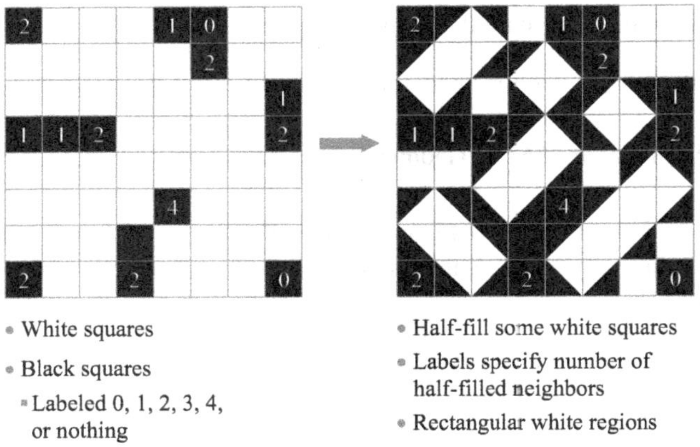

Figure 2.12
Example of a *Shakashaka* puzzle.

no number, then there is no constraint on the number of half-filled white squares adjacent to it.

> SHAKASHAKA
> *Instance:* A *Shakashaka* board.
> *Question:* Can the player win—that is, successfully half fill in squares to satisfy the given constraints?

Demaine, Okamoto, Uehara, and Uno [DOUU14] showed that SHAKASHAKA is NP-complete.

Theorem 2.25. *SHAKASHAKA is NP-complete.*

Proof sketch. The NP-hardness reduction is from PLANAR 3SAT. There are several gadgets needed in the reduction from Planar 3SAT, illustrated by example in figure 2.13. These gadgets include a wire gadget (representing the value of a variable over a long distance), a clause, and a parity shift gadget. □

2.6.2 PLANAR RECTILINEAR MONOTONE 3SAT and MAP LABELING

While reading these definitions, refer to figures 2.14 and 2.15.

Definition 2.26.

1. A ***rectilinear formula*** is a CNF formula with the following representation. Each variable is a horizontal segment on the x axis, while each clause is a horizontal segment above or below the x axis with vertical connections to the variables it includes on the x axis. Each connection can be marked either positive or negative.
2. A ***planar rectilinear formula*** is a rectilinear formula where none of the segments "cross," or more precisely, segments intersect only at intended connections between a vertical segment and the corresponding variable and clause horizontal segments.

> PLANAR RECTILINEAR 3SAT
> *Instance:* A planar rectilinear 3CNF formula φ given as an embedded graph.
> *Question:* Is φ satisfiable?

The following problem combines (is a special case of) PLANAR RECTILINEAR 3SAT and MONOTONE 3SAT from section 1.2.5.

Figure 2.13
A *Shakashaka* puzzle encoding a Planar 3SAT instance using variable, clause, and wire gadgets.

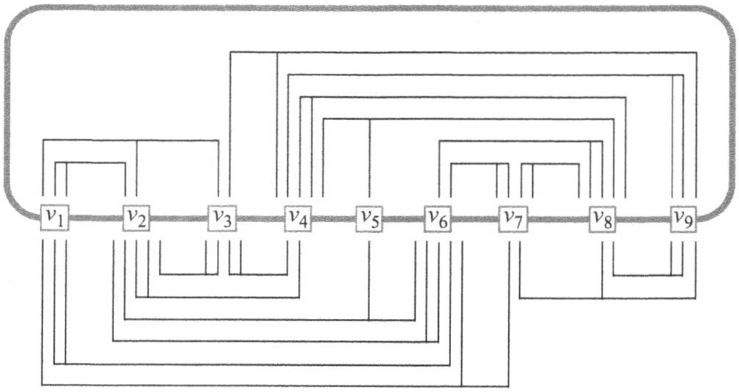

Figure 2.14
Planar Rectilinear 3SAT.

PLANAR RECTILINEAR MONOTONE 3SAT
Instance: A planar rectilinear 3CNF formula φ given as an embedded graph such that (1) every clause has either all positive or all negative literals and (2) the positive clauses are above the x axis and the negative clauses are below the x axis.
Question: Is φ satisfiable?

Theorem 2.27.

1. *(Knuth and Raghunathan [KR92]) PLANAR RECTILINEAR 3SAT is NP-complete.*
2. *(De Berg and Khosravi [dBK12]) PLANAR RECTILINEAR MONOTONE 3SAT is NP-complete.*

Proof sketch. Reducing VARIABLE-LINKED PLANAR 3SAT to PLANAR RECTILINEAR 3SAT is a relatively easy problem in graph drawing. The gadget in figure 2.15 (bottom) gives a reduction from PLANAR RECTILINEAR 3SAT to PLANAR RECTILINEAR MONOTONE 3SAT. □

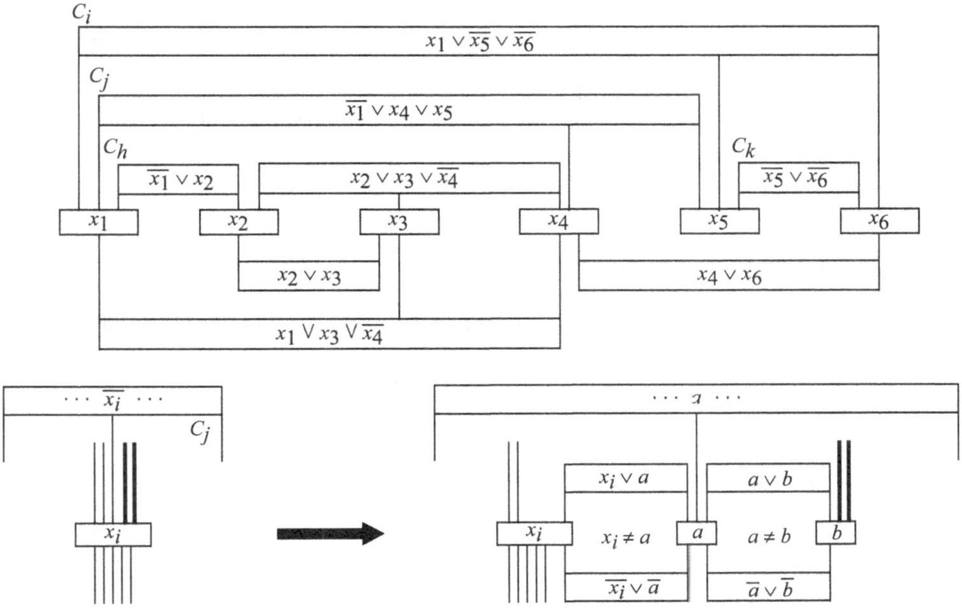

Figure 2.15
PLANAR RECTILINEAR MONOTONE 3SAT.

Knuth and Raghunathan [KR92] invented PLANAR RECTILINEAR 3SAT and showed that it is NP-complete. They used its complexity as a lemma to show that the following map labeling problem is NP-complete.

Suppose we are given a map of locations (e.g., cities), where each location is a point p_i in the plane, and we would like to add text labels to those locations (e.g., city names). We would like each label to be near its corresponding city—say, at a unit distance away. We would also like each label to be strictly farther from other cities so that it is clear which city it belongs to. And we would like the labels not to overlap each other.

> SIMPLE MAP LABELING
> *Instance:* A set of lattice points in the plane $(p_1, \ldots, p_n)$.
> *Question:* Is there a set of lattice points in the plane $(x_1, \ldots, x_n)$ such that the following hold?
>
> 1. For all $1 \le i \le n$, $|x_i - p_i| = 1$.
> 2. For all $1 \le i < j \le n$, $|x_i - p_j| > 1$.
> 3. For all $1 \le i < j \le n$, $|x_i - x_j| \ge 2$.

De Berg and Khosravi [dBK12] invented PLANAR RECTILINEAR MONOTONE 3SAT and showed it is NP-complete. They used its complexity as a lemma to show that a problem involving binary space partitions of the plane is NP-complete. We omit the formal definition of their problem.

2.6.3 PLANAR 1-IN-3SAT, PLANAR X3C, PLANAR 3DM

We can extend the definitions of variable–clause graphs and planar formula to work for a large class of formulas. In particular, to formulas formed by taking the AND of the types of clauses described in chapter 1. This gives us a planar version of each type of SAT. For example:

> PLANAR 1-IN-E3SAT
> *Instance:* A planar 1-IN-E3SAT formula φ.
> *Question:* Is there an assignment satisfying φ?

Dyer and Frieze [DF86] defined PLANAR 1-IN-E3SAT, proved it NP-complete, and used this lemma to analyze the complexity of the following problems.

X3C (EXACT COVER BY 3-SETS) and PLANAR X3C

Instance: (X3C) A set X where $|X| \equiv 0 \pmod 3$ and a collection of size-3 sets $E_1, \ldots, E_m \subseteq X$. Equivalently, the input is a hypergraph where the vertex set has size a multiple of 3 and every edge has size 3.

Instance: (PLANAR X3C) The input is restricted to the case where the following bipartite graph is planar: The left vertices are the elements of X, the right vertices are the sets E_i, and there is an edge between v and E_i if $v \in E_i$.

Question: Does there exist a set of $|X|/3$ E_i's such that every element of X occurs in exactly one of the E_i's?

Note: We can equivalently think of X3C as 1-IN-E3SAT-E3, where each element represents a clause that must be covered/satisfied exactly once, and each set represents a variable that can be chosen (set to TRUE) or not (set to FALSE).

A special case of X3C has the elements colored one of three colors so that each set has all three colors:

3D MATCHING and PLANAR 3DM

Instance: (3D MATCHING) A hypergraph H with vertices partitioned into three disjoint sets A, B, C, where $|A| = |B| = |C| = n$, and hyperedges $E \subseteq A \times B \times C$ (with exactly one vertex from each set). (See figure 2.16.)

Instance: (PLANAR 3DM) The input is restricted to the case where the following bipartite graph is planar: The left vertices are the vertices $A \sqcup B \cup C$ of H, the right vertices are the hyperedges E of H, and there is an edge between v and e if $v \in e$.

Question: Are there n disjoint edges that cover all of the vertices?

Dyer and Frieze [DF86] showed the following.

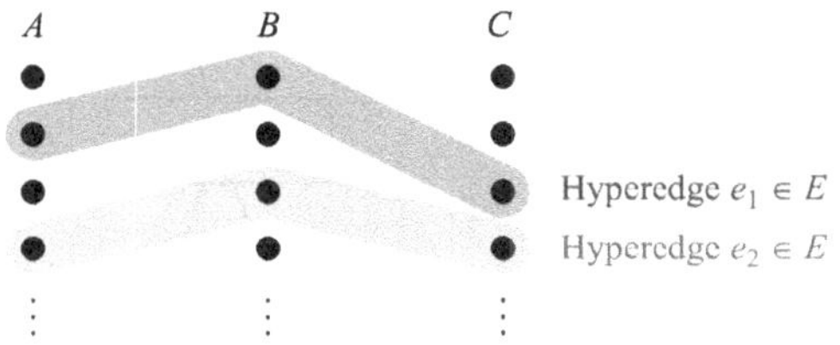

Figure 2.16
3D MATCHING.

Theorem 2.28.

1. *PLANAR 3SAT* $\leq_p$ *PLANAR 1-IN-E3SAT, so PLANAR 1-IN-E3SAT is NP-complete.*
2. *PLANAR 1-IN-E3SAT* $\leq_p$ *PLANAR X3C, so PLANAR X3C is NP-complete.*
3. *PLANAR 1-IN-E3SAT* $\leq_p$ *PLANAR 3DM, so PLANAR 3DM is NP-complete.*

Proof sketch. 1) Here is the reduction from PLANAR 3SAT to PLANAR 1-IN-E3SAT:

1. Input a planar 3CNF formula φ.
2. We can assume each clause has exactly three literals by duplicating literals in clauses as necessary.
3. For each clause $C = L_1 \vee L_2 \vee L_3$, we replace the clause C with the graph of three clauses and four new variables shown in figure 2.17 (right).
4. The resulting formula is φ'.

We leave it to the reader to show that φ' is planar and that $\varphi \in$ PLANAR 3SAT if and only if $\varphi' \in$ PLANAR 1-IN-3SAT.

2) To give intuition, figure 2.18 shows a reduction from PLANAR 1-IN-3SAT to PLANAR X≤3C, a weaker version of PLANAR X3C where each set has size *at most* 3 instead of *exactly* 3. The ***variable gadget*** consists of a cycle alternating between elements and sets, of length divisible by 4 so that it has an even number of elements and an even number of sets. Every other set in this cycle (either the even or odd indexed sets) must be chosen,

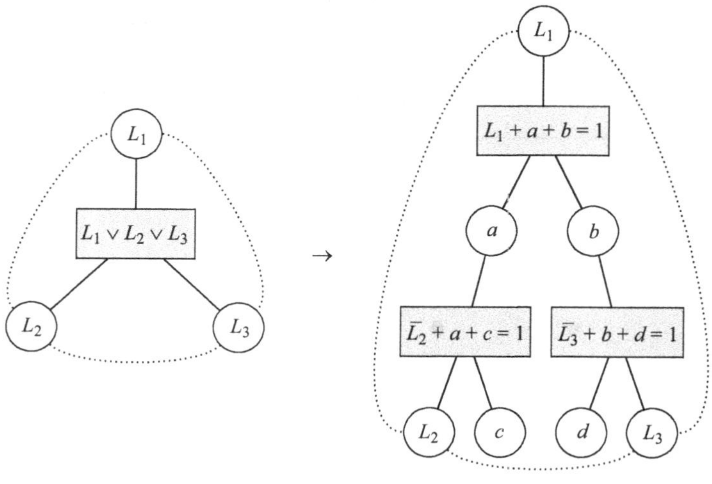

Figure 2.17
Reduction from PLANAR 3SAT to PLANAR 1-IN-3SAT.

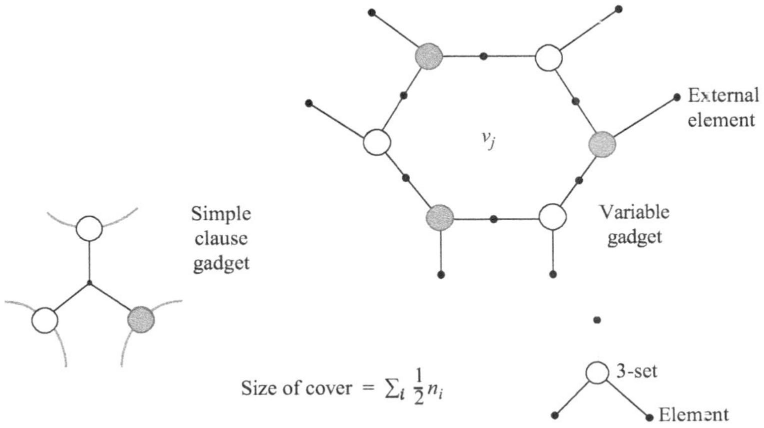

Figure 2.18
Reduction from PLANAR 1-IN-3SAT to PLANAR X≤3C.

corresponding to setting the variable TRUE or FALSE. The ***clause gadget*** consists of a single element connected to three sets, each representing the corresponding literal from the corresponding variable gadget. The constraint that this element must be covered by exactly one of these sets gives us the 1-IN-3SAT clause constraint. The variable gadget gives us enough copies of each literal to use a different one in each clause, but unused literals will end up creating sets of size 2. A reduction to PLANAR X3C follows from part 3.

3) Figure 2.19 sketches the reduction from PLANAR 1-IN-3SAT to PLANAR 3DM. □

Exercise 2.29. Show that the reductions given in the sketch of the proof of theorem 2.28 work.

Exercise 2.30. Call a graph $G=(V,E)$ ***beautiful*** if we can color the vertices of G with either blue or red such that every vertex has *exactly one* blue neighbor. The BEAUTIFUL problem is, given a graph G, to determine whether G is beautiful.

1. Show that BEAUTIFUL is NP-complete.
   Hint: Use X3C.
2. What happens if we restrict BEAUTIFUL to planar graphs?

Exercise 2.31. The CONNECTED BISECTION PROBLEM is as follows. The input is a graph $G=(V,E)$ with n vertices. Determine whether V can be partitioned into two sets, each of size $n/2$, such that each part induces a connected subgraph. Show that 3D MATCHING $\leq_p$ CONNECTED BISECTION PROBLEM, and hence CONNECTED BISECTION PROBLEM is NP-complete.

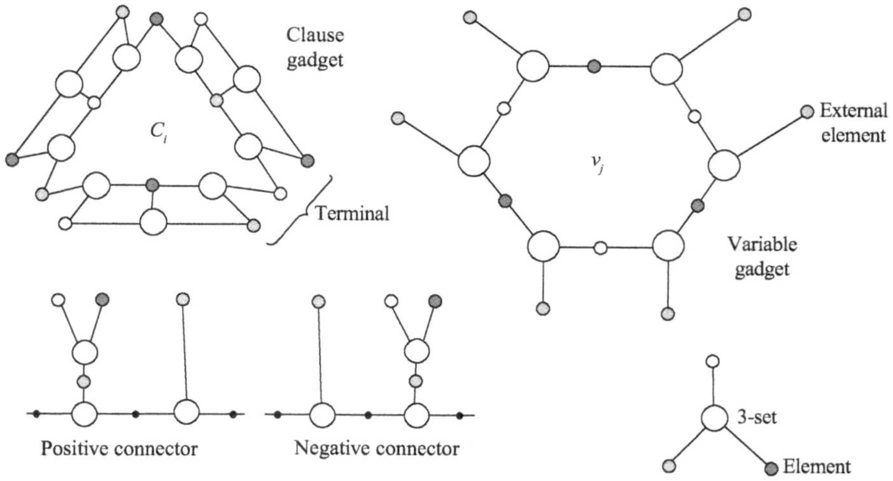

Figure 2.19
Reduction from PLANAR 1-IN-3SAT to PLANAR 3DM.

2.6.4 POSITIVE PLANAR 1-IN-3SAT and TRIANGULATION

The reduction in theorem 2.28 showing that PLANAR 1-IN-E3SAT is NP-complete required us to negate literals. But we know from theorem 1.19 that POSITIVE 1-IN-E3SAT is NP-complete. What about PLANAR POSITIVE 1-IN-E3SAT? It turns out we can avoid negation, avoid duplicate variables, as well as get the rectilinear structure from definition 2.26:

> PLANAR RECTILINEAR POSITIVE 1-IN-EU3SAT
> *Instance:* A planar rectilinear POSITIVE 1-IN-EU3SAT formula φ.
> *Question:* Is there an assignment satisfying φ?

Mulzer and Roe [MR08] defined this problem and showed that it is NP-complete. P. Laroche [Lar92] proved a version of this result earlier.

Theorem 2.32. *PLANAR RECTILINEAR 3SAT reduces to PLANAR RECTILINEAR POSITIVE 1-IN-EU3SAT. Hence PLANAR RECTILINEAR POSITIVE 1-IN-EU3SAT is NP-complete.*

Proof sketch. The proof is similar to theorem 2.28.1, except now we cannot use negations, so we use different gadgets. We reduce from PLANAR RECTILINEAR 3SAT. Figure 2.20 shows how to force two variables x, y on the x axis to have different assignments ($x \neq y$) or the same assignment ($x = y$). figure 2.21 (top) uses these gadgets to remove all negations from the problem. Then figure 2.21 (bottom) shows how to simulate a 3SAT clause. All of the 1-IN-3SAT clauses consist of exactly three distinct variables, unnegated. □

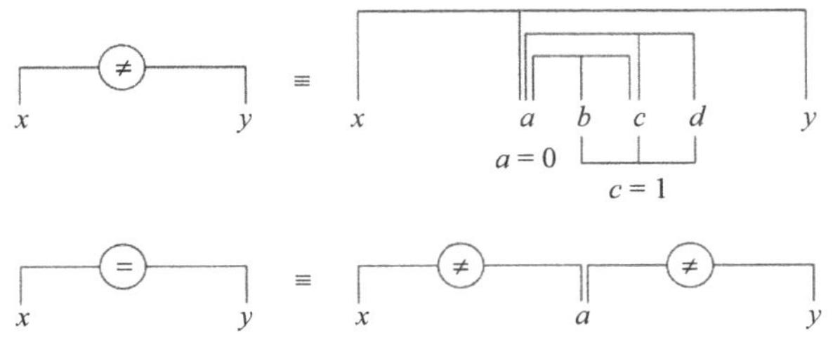

Figure 2.20
PLANAR RECTILINEAR POSITIVE 1-IN-EU3SAT gadgets.

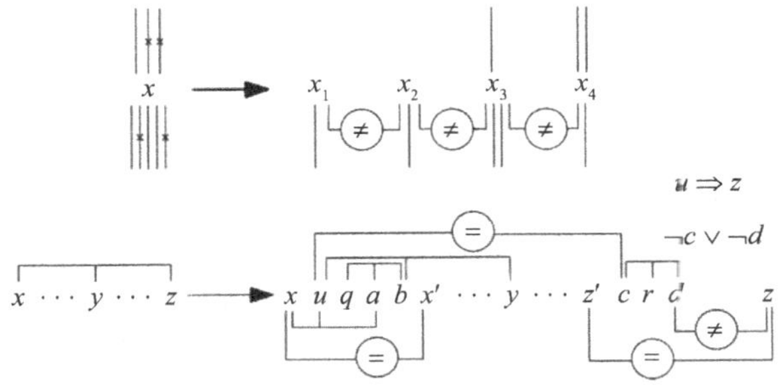

Figure 2.21
Reduction from PLANAR RECTILINEAR 3SAT to PLANAR RECTILINEAR POSITIVE 1-IN-EU3SAT.

Exercise 2.33. Complete the proof of theorem 2.32.

Mulzer and Roe [MR08] used PLANAR RECTILINEAR POSITIVE 1-IN-EU3SAT to solve a famous problem in computational geometry and complexity.[2]

Definition 2.34. Let S be a set of points in the plane. A ***triangulation*** of S is an embedded planar graph that has S as the set of vertices such that (1) all of the edges are straight lines, (2) the graph is planar, and (3) if any more edges were added, then the graph would no longer be planar.

> MINIMUM-WEIGHT TRIANGLUATION
> *Instance:* A set S of n points in the plane.
> *Output:* A triangulation of S that minimizes the sum of the lengths of the edges.

2. At one point, there were plans to write a second edition of Garey and Johnson's book [GJ79], which would be centered around a few of the major open problems. MINIMUM-WEIGHT TRIANGULATION was on that short list.

Theorem 2.35. *MINIMUM-WEIGHT TRIANGLUATION is NP-hard.*

We omit the proof, which involves complicated computer-checked gadgets.

2.6.5 FLATTENING FIXED-ANGLE CHAINS

Molecular geometry (also called ***stereochemistry***) studies 3D geometry of the atoms and the bonds between them that form a molecule. An atom is a vertex and a bond between atoms is an edge. A molecule is a graph but drawn in 3D instead of 2D.

Some molecules can be drawn in 2D. Which ones? We formalize this problem and (as usual) state that it is NP-complete.

Definition 2.36.

1. A ***linkage*** is an edge-weighted graph $G=(V,E,\ell)$ where the nonnegative edge weight $\ell(u,v)\geq 0$ represents the geometric length of edge (u,v).
2. A ***configuration*** of a linkage in d dimensions is a mapping $C: V\to\mathbb{R}^d$ assigning coordinates to vertices satisfying the specified edge lengths: For every edge $(u,v)\in E$, the distance between points $C(u)$ and $C(v)$ is exactly $\ell(u,v)$.
3. A configuration is ***non-crossing*** if the line segments representing two edges intersect only when the two edges share a common vertex, and in that case, only at the corresponding endpoint of the segments. If $d=2$, then non-crossing is the same as planar.
4. A ***fixed-angle linkage*** is a linkage with an additional constraint: a function Θ that takes every (v_1,v_2,v_3), where (v_1,v_2) and (v_2,v_3) are edges, and returns the angle between them at v_2.
5. An ***open chain of length n*** is a linkage whose graph is a path graph, consisting of vertices $V=\{v_1,\ldots,v_n\}$ and edges $(v_1,v_2),(v_2,v_3),\ldots,(v_{n-1},v_n)$.
6. A ***flat state*** of a fixed-angle linkage is a non-crossing 2D configuration of the linkage.

> FLATTENING FIXED-ANGLE CHAINS
> *Instance:* A fixed-angle linkage that is an open chain.
> *Question:* Does it have a flat state?

Demaine and Eisenstat [DE11] prove the following:

Theorem 2.37. *FLATTENING FIXED-ANGLE CHAINS is strongly NP-hard. (We discuss* strongly NP-complete *briefly in section 0.8 and at length in chapter 5.)*

Proof sketch. Figure 2.22 shows an example of the reduction, which is from PLANAR RECTILINEAR MONOTONE 3SAT. The variable gadgets are chained together in the middle of the

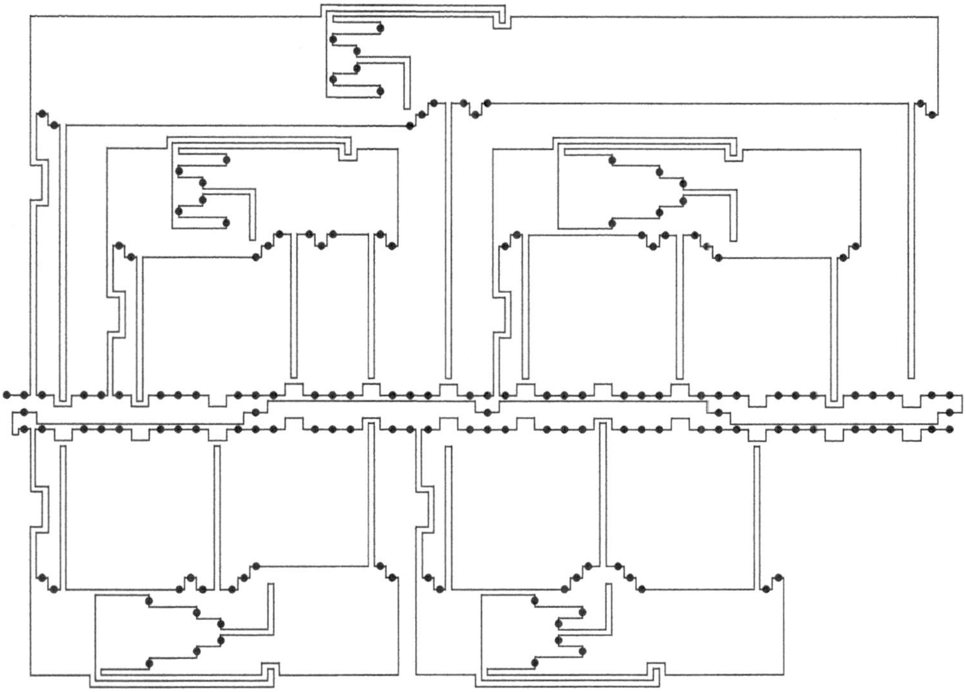

Figure 2.22
Reduction from PLANAR RECTILINEAR MONOTONE 3SAT to FLATTENING FIXED-ANGLE CHAINS.

diagram. Each long horizontal bar can flip up or down, corresponding to setting the variable TRUE or FALSE. A clause has three possible positions, one each with a protrusion at a point corresponding to a variable, thus forcing the variable chain to take one position or the other. □

2.7 PLANAR NAE 3SAT

From the results in this chapter, one might think that all planar SAT problems are NP-complete. Not so: NAE 3SAT becomes easy in the planar case. To build intuition, we start by analyzing the POSITIVE case:

Theorem 2.38. *Every instance of PLANAR POSITIVE NAE EU3SAT is satisfiable. Thus PLANAR POSITIVE NAE EU3SAT is in P.*

Proof. This proof is based on [Pil18, theorem 11]. Start with the (planar) variable–clause graph G. For each clause vertex C incident to variable vertices x, y, z, replace C with a 3-cycle connecting x, y, z. By routing this 3-cycle near C, we see that the resulting graph G'

is also planar. By the four-color theorem [AH77a, AHK77, RSST97], we can color the vertices of G' (which are all variables) with four colors. (We can even do so in quadratic time [RSST97].) Assign two colors of variables to TRUE, and assign the other two colors of variables to FALSE. For each clause NAE(x,y,z), G' has a 3-cycle/clique on x,y,z, so x,y,z receive three different colors. Thus at least one of the variables is TRUE and at least one is FALSE. □

To solve non-POSITIVE cases of PLANAR NAE 3SAT, we introduce another problem:

> MAX CUT and PLANAR MAX CUT
> *Instance:* A (planar) graph $G=(V,E)$ and a $k \in \mathbb{N}$.
> *Question:* Is there a partition $V = V_1 \cup V_2$ such that the number of edges from V_1 to V_2 is at least k?

MAX CUT is NP-complete [GJS76], as we will show in theorem 2.39 below. In chapter 9, we will show that even approximating MAX CUT is NP-hard. But PLANAR MAX CUT turns out to be easy, and we can use this to solve PLANAR NAE 3SAT.

Theorem 2.39.

1. *(Hadlock [Had75]) PLANAR MAX CUT* $\in$ P.
2. *(Moret [Mor88]) PLANAR NAE 3SAT* $\leq_p$ *PLANAR MAX CUT, hence, combined with part 1, PLANAR NAE 3SAT* $\in$ P.

Proof sketch. We sketch the reduction of part 2, which looks surprisingly like an NP-hardness reduction: It has both a variable gadget and a clause gadget. In fact, it is an NP-hardness reduction NAE 3SAT $\leq_p$ MAX CUT, which also preserves planarity.

1. Input a planar NAE 3SAT formula φ.
2. If any clauses have size less than 3, we can simplify as follows. Clauses of size 1 are never satisfied. Clauses of size 2 force two variables to have the opposite assignment, so we can combine the two variables, negating all instances of one of them.
3. For each variable x_i in φ, we have the ***variable gadget*** shown in figure 2.23 (left), which is an even cycle. If x_i or its negation occurs m_i times in clauses, then the cycle has length $2m_i$. We add $2m_i$ to the desired size of the cut, which forces the even and odd vertices in the cycle to be assigned to opposite sides of the cut. Thus we can think of the even vertices as representing x_i and the odd vertices as representing $\neg x_i$, and the two sides of the cut as TRUE and FALSE.
4. For each clause NAE(L_i, L_j, L_k), we have the ***clause gadget*** shown in figure 2.23 (right), which is a triangle where the three vertices are connected to corresponding

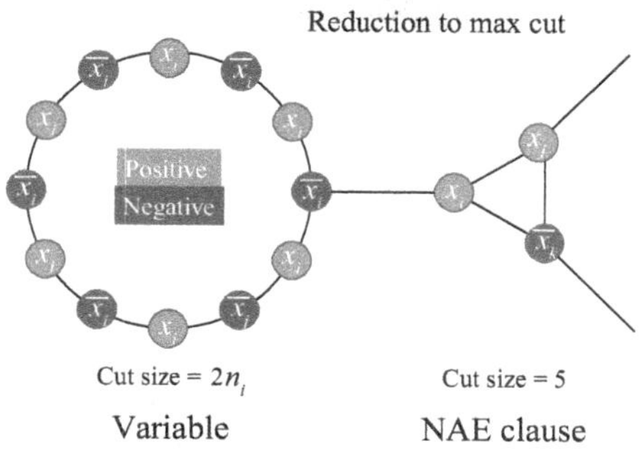

Figure 2.23
Reduction from PLANAR NAE 3SAT to PLANAR MAX CUT.

instances of the literals L_i, L_j, L_k in the variable gadget for x_i, x_j, x_k, respectively. Because the variable gadgets are large enough, we can use a vertex of each variable gadget that is unique to this clause. We add 5 to the desired size of the cut, which forces the vertices in the triangle to receive opposite assignments to the literals they represent, and forces the vertices in the triangle to have not-all-equal assignments. Thus this gadget enforces the constraint NAE($\neg L_i, \neg L_j, \neg L_k$), which is equivalent to NAE(L_i, L_j, L_k).

5. The resulting instance of PLANAR MAX CUT is the constructed graph together with the desired cut size $k = \sum_i 2m_i + 5m$, where m is the number of clauses. □

Exercise 2.40.

1. Prove that the reduction in the proof of theorem 2.39 preserves planarity.
2. Prove PLANAR MAX CUT $\in$ P by yourself or look it up.
3. (Warning: This question might be ill-defined.) Prove PLANAR NAE 3SAT $\in$ P directly, not through a reduction.

2.8 Schaefer-Like Dichotomy Theorem

Is PLANAR NAE 3SAT the only planar SAT variant that is easier than the nonplanar case? Among 3-literal clauses, yes. This fact follows from a general dichotomy theorem that characterizes the complexity of PLANAR SAT-type problems as either polynomial or NP-complete, for every clause type over binary variables.

Definition 2.41. A PLANAR SAT-type problem is a SAT-type problem (definition 1.21) restricted to instances where the variable–clause graph is planar.

Dvořák and Kupec [DK15] showed that Schaefer's dichotomy theorem (from sections 1.2.7 and 1.2.8) applies in all cases except one new type of clause, which we will define next. Kazda, Kolmogorov, and Rolínek [KKR18] proved that this case can be solved in polynomial time, completing the dichotomy theorem.

Definition 2.42.

1. An a-ary relation R is ***self-complementary*** if it is closed under negation of all a variables—that is, $R(x_1, \ldots, x_a)$ implies $R(\neg x_1, \ldots, \neg x_a)$. In other words, R treats TRUE and FALSE symmetrically. For example, NAE satisfies this condition.
2. A self-complementary relation R can be thought of as depending on just the vector of consecutive XOR of its variables:

$$d(x_1, \ldots, x_a) = (x_1 \text{ XOR } x_2, x_2 \text{ XOR } x_3, \ldots, x_a \text{ XOR } x_1).$$

   Specifically, whenever $R(x_1, \ldots, x_a) = \text{TRUE}$, we define $dR(d(x_1, \ldots, x_a)) = \text{TRUE}$ (and dR is FALSE otherwise).
3. The relation dR is an ***even* $\boldsymbol{\Delta}$*-matroid*** if, for any two assignments $(x_1, \ldots, x_a)$, $(y_1, \ldots, y_a)$ satisfying dR (i.e., $dR(x_1, \ldots, x_a) = dR(y_1, \ldots, y_a) = \text{TRUE}$), and for any index i for which they differ ($x_i \neq y_i$), there is another index j for which they differ ($x_j \neq y_j$), such that flipping variables x_i and x_j in x satisfies R (i.e., $dR(x_1, \ldots, \neg x_i, \ldots, \neg x_j, \ldots, x_a) = \text{TRUE}$).

Example 2.43. POSITIVE NAE 3SAT is defined by the relation R with the following TRUE assignments:

$$\{(\text{TRUE}, \text{FALSE}, \text{FALSE}), (\text{FALSE}, \text{TRUE}, \text{FALSE}), (\text{FALSE}, \text{FALSE}, \text{TRUE}),$$
$$(\text{FALSE}, \text{TRUE}, \text{TRUE}), (\text{TRUE}, \text{FALSE}, \text{TRUE}), (\text{TRUE}, \text{TRUE}, \text{FALSE})\},$$

which is self-complementary: Each TRUE assignment in the first row has a corresponding complemented TRUE assignment in the second row. Taking the consecutive XOR of each of these assignments, the relation dR has the following TRUE assignments:

$$\{(\text{TRUE}, \text{FALSE}, \text{TRUE}), (\text{TRUE}, \text{TRUE}, \text{FALSE}), (\text{FALSE}, \text{TRUE}, \text{TRUE}),$$
$$(\text{TRUE}, \text{FALSE}, \text{TRUE}), (\text{TRUE}, \text{TRUE}, \text{FALSE}), (\text{FALSE}, \text{TRUE}, \text{TRUE})\}$$
$$= \{(\text{TRUE}, \text{FALSE}, \text{TRUE}), (\text{TRUE}, \text{TRUE}, \text{FALSE}), (\text{FALSE}, \text{TRUE}, \text{TRUE})\}.$$

This relation dR requires exactly two of the three variables to be TRUE.

We claim that dR is an even Δ-matroid. Suppose dR is satisfied by two different assignments (x_1, x_2, x_3) and (y_1, y_2, y_3). By inspection of the three satisfying assignments for dR, these assignments x, y must differ in exactly two indices a, b where $x_a = y_b = \text{TRUE}$ and $x_b = y_a = \text{FALSE}$ (preserving the number of TRUE variables at 2). Given one of these indices

$i \in \{a, b\}$, we can set j to be the other index from $\{a, b\}$, and we know that flipping assignment x at indices i and j results in assignment y, and thus satisfies dR.

The following dichotomy theorem has all the easy cases from theorem 1.29, plus one more.

Theorem 2.44. *(Dvořák and Kupec [DK15]; Kazda, Kolmogorov, and Rolínek [KKR18]) Let $R_1, \ldots, R_k$ be relations over {TRUE, FALSE}, each with at least one satisfying assignment. If any of the following occur, then the PLANAR SAT-type problem $\{R_1, \ldots, R_k\}$ is in P. If not, then it is NP-complete.*

1. *For some $f \in$ {BOT, TOP, AND, OR, MAJORITY, MINORITY}, for all $1 \le i \le k$, R_i is closed under f.*
2. *For all $1 \le i \le k$, R_i is self-complementary and dR_i is an even Δ-matroid.*

There are myriad other variants of PLANAR 3SAT. Tippenhauer's thesis [Tip16] looks at versions where the number of variables are bounded and the formulas are monotone. Filho's thesis [Fil19] gives an extensive survey.

2.9 Further Results

Here is a partial list of NP-hard (usually NP-complete) problems that were proven so using a known NP-complete planar problem.

- ***PLANAR k-MEANS***: Given a finite set $S = \{p_1, p_2, \ldots, p_n\}$ of points with rational coordinates in the plane, an integer $k \ge 1$, and a bound $R \in \mathbb{Q}$ determine whether there exists k centers $\{c_1, \ldots, c_k\}$ in the plane such that

$$\sum_{i=1}^{n} \left(\min_{1 \le j \le k} [d(p_i, c_j)]^2 \right) \le R.$$

  ($d(p, c)$ is the Euclidean distance from p to c.) Mahajan et al [MNV12] showed

$$\text{PLANAR 3SAT} \le_p \text{PLANAR } k\text{-MEANS},$$

  so the problem is NP-hard, though it is not known to be in NP.
- ***Multi-robot path planning problems on planar graphs***: Given a planar graph, robots (start vertices), and destinations (end vertices), and a number k, is there a set of paths along the graph such that no two paths lead to a collision with arrival time $\le k$? Yu [Yu16] showed this problem is NP-hard using a reduction from Monotone PLANAR 3SAT. For another problem from robotics that is proven hard by a reduction from PLANAR 3SAT, see P. Agarwal et al. [AAGH21].

3

NP-Hardness via CIRCUIT SAT

3.1 Introduction

In this chapter, we describe some NP-hardness reductions from CIRCUIT SAT. Recall the problem as defined in section 1.2.1:

> CIRCUIT SAT
> *Instance:* A Boolean circuit $C(x_1, x_2, \ldots, x_n)$ with n inputs.
> *Question:* Does C ever output TRUE? That is, does there exist $(b_1, b_2, \ldots, b_n) \in \{\text{TRUE}, \text{FALSE}\}^n$ such that $C(b_1, b_2, \ldots, b_n) = \text{TRUE}$?

Chapter Summary

1. *Prove that PLANAR CIRCUIT SAT is NP-complete.*
2. *Use that PLANAR CIRCUIT SAT is NP-complete to show many puzzles are NP-complete.*

3.2 Gate Sets for CIRCUIT SAT

Instead of using the general form of CIRCUIT SAT, which allows any Boolean gate, it suffices to restrict the gates to be ***functionally complete***—that is, capable of representing all Boolean functions. For example, the following gate sets are functionally complete:

- {NAND}
- {NOR}
- {AND, NOT}
- {OR, NOT}
- {IMPLIES, BOT} (where $\text{IMPLIES}(x, y) = \neg x \vee y$ and the BOT gate always outputs FALSE)

- Any set given by Post's functional completeness theorem (see Post [Pos41] or Pelletier and Martin [PM90] for a modern approach).

Even stronger, Harry Lewis [Lew79] proved that CIRCUIT SAT is NP-complete for any gate set that can build the function $\text{NIMPLIES}(x,y) = \neg\text{IMPLIES}(x,y) = x \wedge \neg y$.

3.3 PLANAR CIRCUIT SAT

We define a planar version of CIRCUIT SAT. Surprisingly, the terminology PLANAR CIRCUIT SAT is not used often.

> PLANAR CIRCUIT SAT
> *Instance:* A Boolean circuit $C(x_1, x_2, \ldots, x_n)$ represented by an embedded planar directed acyclic graph. Each node in the graph is either a source (of in-degree 0, representing an input), a NAND gate (of in-degree 2), or a sink (out-degree 0, representing the output). There must be exactly one sink.
> *Question:* Does C ever output TRUE? That is, does there exist $(b_1, b_2, \ldots, b_n) \in \{\text{TRUE}, \text{FALSE}\}^n$ such that $C(b_1, b_2, \ldots, b_n) = \text{TRUE}$?

Equivalently, the question is whether each edge (wire) of the graph can be labeled with a value of either TRUE or FALSE such that the following hold:

- The value of the edge into the sink is TRUE.
- The value of any edge out of a NAND gate is the NAND of the truth values of the two edges going into the gate.
- (The value of an edge out of a source can be anything.)

Theorem 3.1. *PLANAR CIRCUIT SAT is NP-complete. The problem is also NP-complete if we replace the word "NAND" with the word "NOR everywhere."*

Proof. We prove NP-hardness by a reduction from CIRCUIT SAT, using the crossover gadgets in figure 3.1. The base gadget (a) is built out of XOR gates. Each XOR gate can be replaced by a subcircuit of NAND gates (b) to prove PLANAR CIRCUIT SAT hard, and each NAND gate can be replaced by a subcircuit of NOR gates (c) to prove the NOR analog hard. Crucially, all of these subcircuits have no crossings. Hence we can remove all crossings by replacing each crossing with a crossover gadget of the appropriate type. □

More generally, McColl [McC81] showed that any functionally complete set of gates can build a planar crossover gadget.

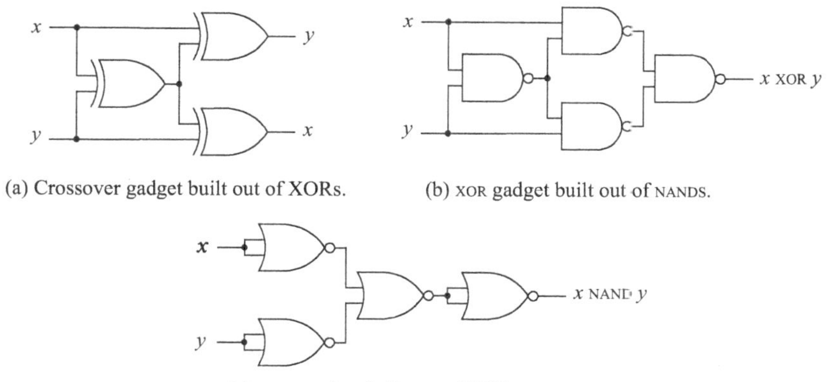

(a) Crossover gadget built out of XORs.

(b) XOR gadget built out of NANDS.

(c) NAND gadget built out of NORs.

Figure 3.1
Crossover gadgets for Circuit SAT.

Because we have a crossover gadget, Planar Circuit SAT could require additional planarity properties of the circuit. For example, we can require that all sources and the sink be on the same face of the planar embedding. Equivalently, we can require that the graph remains planar if we add a cycle connecting the sources and sink in some order.

3.4 Reductions to Puzzles

Next we will demonstrate the hardness of several puzzles. These Circuit SAT reductions develop the following gadgets:

1. Most obviously, we need a ***gate gadget*** to represent each type of gate.
2. Next we need a ***wire gadget*** to connect the output of one gate to the input of another gate. Typically, each gate gadget only has one copy of its output, so we need a ***split gadget*** for duplicating a gate output (given by one wire) so that we can route it to every gate input where it is needed (via additional wires).
3. To represent the inputs, we need a ***terminator gadget*** that ends a wire without constraining the wire. Thus the wire is free to choose whether it carries a value of TRUE or FALSE, just like an x_i input in the Circuit SAT problem.
4. To represent the desire of the final output to be TRUE, we need a ***TRUE terminator gadget*** that ends a wire and forces its value to be TRUE. All the gadgets can be satisfied if and only if the circuit is satisfiable.

Exercise 3.2. Read Erich Friedman's paper [Fri02] on the NP-completeness of the spiral galaxies problem. The proof uses a reduction from CIRCUIT SAT. Rewrite the reduction in your own words.

3.4.1 LIGHT UP

An instance of the puzzle *Light Up* or *Akari* consists of a grid of black and white squares where some black squares have a number between 0 and 4 written on them. The goal of the puzzle is to place lights on the white squares of the board so that the following rules are satisfied:

- Each black square that has a number written on it must have that number of lights in a horizontally or vertically adjacent square.
- Each white square should be lit up by exactly one light. A light lights up a square if both the light and the square are in the same row or column and if there are no black squares between them.

There have been entire puzzle books on this game, for example, published by Nikoli [Nik08, Nik].

> LIGHT UP
> *Instance:* A board position for the game *Light Up*.
> *Question:* Can the lights be placed to attain the goal?
> *Note:* This game is also called *Akari*.

Brandon McPhail [McP05] has shown that LIGHT UP is NP-complete.

Theorem 3.3. *LIGHT UP is NP-complete.*

Proof sketch. The hardness proof is a reduction from PLANAR CIRCUIT SAT (NOR variation). Like most such proofs, the reduction starts with the wire gadget. A basic wire gadget consists of a repeating sequence of a black square with a 1 followed by two white squares. This forces exactly one of the two white squares to have a light in a way such that the choice of using the white square on the right or on the left propagates down the wire. Alternatively, by lining a corridor with 0's, one can have a different wire because the light must be at one of the ends of the corridor. Making turns is also easy. All of these gadgets are depicted in figures 3.2 and 3.3.

The negation/split gadget in figure 3.3 (top) outputs a single negated copy and two regular copies of the incoming wire. By using a terminator gadget, you can use this gadget as either the split gadget or the not gadget.

Together, the OR and NOT gates create a NOR gate.

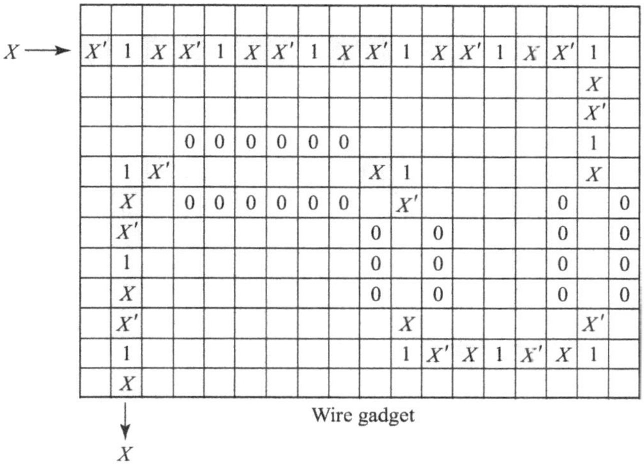

Figure 3.2
Wire gadget.

The above gadgets, together with a terminator gadget (which requires simply placing a 0 or 1 at the end of a wire to set its value), are sufficient to prove that LIGHT UP is NP-complete. □

Exercise 3.4. Formalize the proof of theorem 3.3 and construct some (very small) examples of the reduction.

3.4.2 MINESWEEPER

Minesweeper is the following puzzle computer game, particularly famous for being included in most versions of Microsoft Windows:

1. The parameters are n, m (the board will be $n \times m$) and b (the number of bombs).
2. The player initially sees an $n \times m$ board of unlabeled spaces.
3. b of the spaces have hidden bombs. Some of the other spaces have numbers that indicate how many bombs are adjacent to that space. Here adjacent means up, down, left, right, or diagonal. Hence most spaces have eight spaces adjacent to them.
4. The player will either left-click or right-click on a space.
   a. If a player left-clicks and the space has a bomb, the game is over and the player loses.
   b. If a player left-clicks and the space has a number, that number is revealed and the player may use that to try to figure out where bombs are.

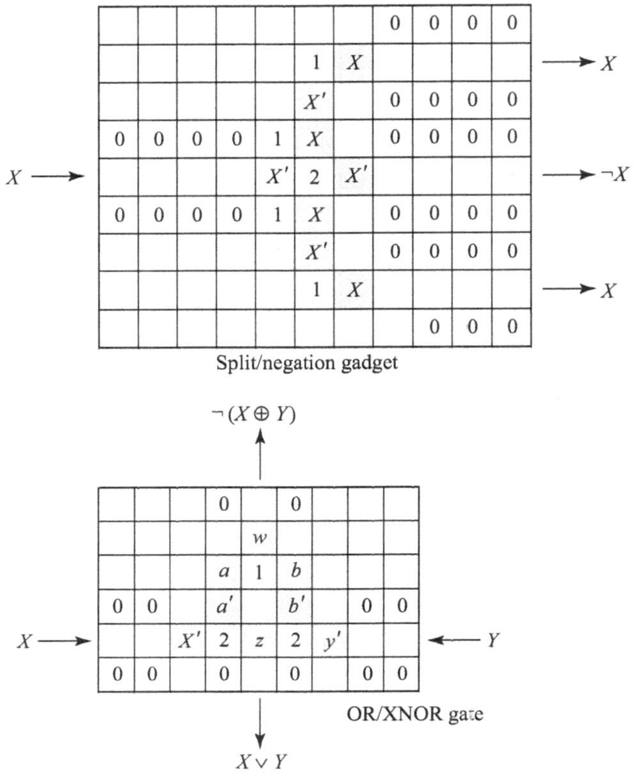

Figure 3.3
Above split/negation gadget; below OR/XOR gadget.

c. If a player left-clicks and the space has nothing, then that space does not have a bomb, nor do any the adjacent spaces. Those adjacent spaces are also uncovered.

d. If a player right-clicks, then a flag is put on that space. The player does not know whether there was a bomb or not. A player is only allowed to do this b times.

5. If the player reveals all non-bomb spaces, then they win. If a player gets blown up, then they lose.

6. Caveat: In some versions, the first space clicked on cannot have a bomb. There are various ways to implement this rule.

Given a grid where some spaces have numbers, is it ways possible to put bombs on the grid so that the numbers are consistent? No. Consider the grid in example 3.5.

Example 3.5. An impossible configuration.

		1		
1		5		1
		1		

Exercise 3.6. Prove that there is no way to plant bombs such that example 3.5 is correct.

> MINESWEEPER CONSISTENCY
> *Instance:* A *Minesweeper* board.
> *Question:* Is that board possible?

The motivation for the MINESWEEPER CONSISTENCY problem is that if we could solve the problem efficiently, then placing a bomb in a location and then checking whether the board is consistent allows a player to check whether that location is safe to click. One of the problems with this approach to playing *Minesweeper* is that there might not be any safe moves. As a result, this is not the best problem to pose, but it is the first one to be proved hard:

Theorem 3.7. *(Richard Kaye [Kay00]) MINESWEEPER CONSISTENCY is NP-complete.*

Proof. Reduction from PLANAR CIRCUIT SAT (NAND variation). The wire gadget for *Minesweeper* in figure 3.4 (top) looks a lot like the *Light-Up* wire. The bombs must all be placed in the right square or the left of the pairs of unspecified squares.

If we just stop a wire gadget like we did in *Light-Up*, we end up forcing a bomb, which gives us a TRUE terminator gadget. For an unconstrained terminator gadget, we need the more complicated figure 3.4 (top).

The gadget in figure 3.5 is a split, NOT, and turn gadget all in one. Each individual gadget can be extracted from this design by adding terminators to the unnecessary wires.

The simpler NOT gate in figure 3.6 can be used twice in order to fix the "mod-3-parity" issue (that wires are always a multiple of 3 length).

The AND gadget in figure 3.7 looks very complicated, but all one has to do to check it is check all four cases.

Together, the AND and NOT gates create a NAND. The above gadgets are enough to prove that the problem is NP-hard. □

While theorem 3.7 is interesting, it is not quite the right question. A player wants to know what their next move should be. Also note that the number of bombs used in each gadget

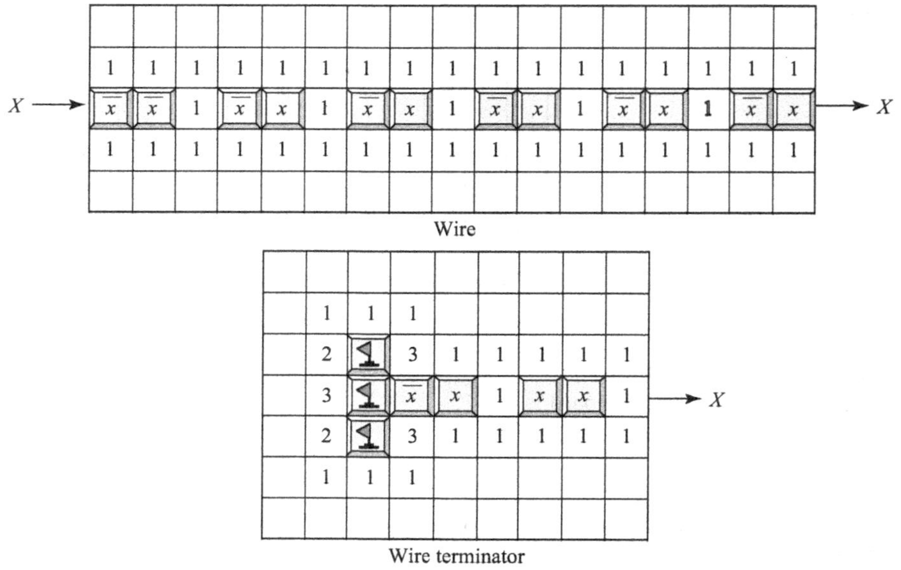

Figure 3.4
Minesweeper wire gadget (top) and terminator gadget (bottom).

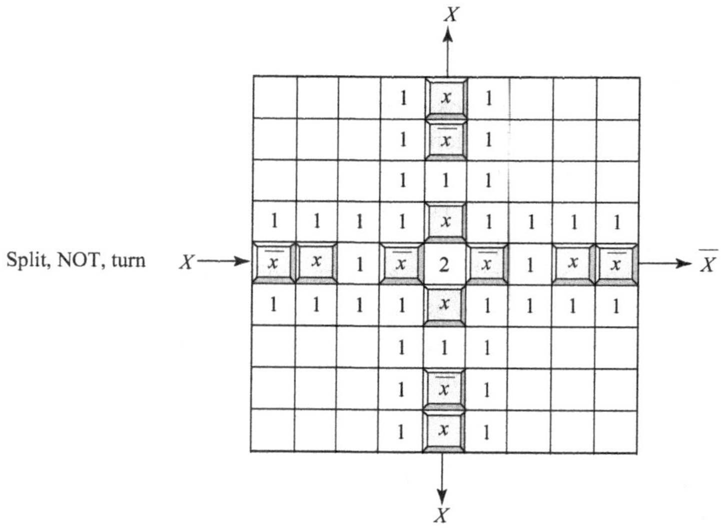

Figure 3.5
Minesweeper split/NOT/turn gadget.

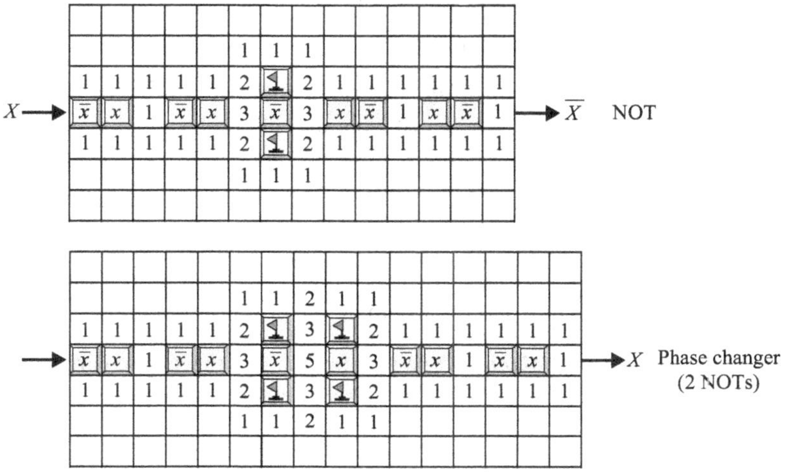

Figure 3.6
Minesweeper NOT gate (top) and parity shift gadget (bottom).

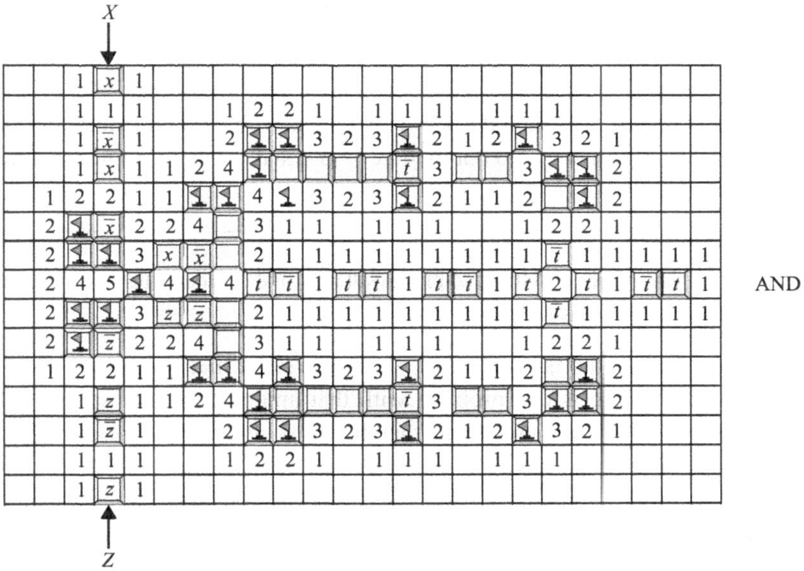

Figure 3.7
Minesweeper AND gate.

varies according to the truth values of the variables. In actual *Minesweeper*, the number of bombs is specified, so that is another problem.

Definition 3.8. A ***Minesweeper position*** is a grid, a number b of bombs, some of the grid squares are uncovered and have a number (the number of bombs around it), and some have a flag (indicating that there is a bomb there). Note that the player is promised that there is a consistent way to place bombs.

For most of the problems in this book there are no preconditions on the input. You can be given *any* formula or graph. For *Minesweeper*, we will need to promise the player that the game is consistent. We take this opportunity to define promise problems.

Definition 3.9. A ***promise problem*** is two sets (A, B). A is the set we care about and B is the promise. A promise problem is in P if there is a polynomial-time algorithm M that does the following:

1. If $x \in B$ and $x \in A$, then $M(x) = 1$.
2. If $x \in B$ and $x \notin A$, then $M(x) = 0$.
3. If $x \notin B$, then $M(x) \in \{0, 1\}$ (arbitrary behavior).

> MINESWEEPER INFERENCE
> *Instance:* A position in *Minesweeper* that you are promised is consistent.
> *Question:* Is there a square such that one can deduce with certainty either (a) there is a bomb there or (b) there is no bomb there? Note that you may deduce that there is no such square.

Exercise 3.10. Show that, if MINESWEEPER INFERENCE $\in$ P, then there is a strategy for *Minesweeper* that safely wins if possible.

To prove that *Minesweeper* is hard to play, we want to prove that MINESWEEPER INFERENCE is NP-complete. But there is one problem with that approach: MINESWEEPER INFERENCE does not seem to be in NP. To demonstrate that there is a safely deducible square, we would need to show that *all* solutions have the same behavior in that square. But this is a statement over exponentially many solutions, so it is not clear how to make it into a polynomial certificate. Rather, the problem is more naturally in coNP: We can demonstrate that there is *no* safely deducible square by, for every square, giving two solutions that differ on that square. Because there are only a linear number of squares, and thus solutions are of linear size, this gives us a polynomial certificate for coNP. It turns out that coNP is the right complexity class for MINESWEEPER INFERENCE:

Theorem 3.11. *MINESWEEPER INFERENCE is coNP-complete.*

Scott, Stege, and van Rooij [SSvR11] first introduced MINESWEEPER INFERENCE, and claimed that MINESWEEPER INFERENCE is coNP-complete. Unfortunately, their hardness proof has a bug [MIT24b]. The first correct proof is by the MIT Hardness Group, namely Della Hendrickson and Andy Tockman [MIT24b]. They also showed coNP-completeness of an even more relevant problem to playing *Minesweeper*:

MINESWEEPER SOLVABILITY
Instance: A position in *Minesweeper* and a corresponding full solution.
Question: Can the player make a sequence of guaranteed-safe moves that bring the position to the solution?

4

NP-Hardness via Hamiltonian Cycle

4.1 Introduction

Recall that in the preface we contrasted the following two problems.

1. Given a graph G is there an ***Eulerian cycle***, which is a cycle that visits every *edge* exactly once? We call this problem EULERIAN CYCLE.
2. Given a graph G, is there a ***Hamiltonian cycle***, which is a cycle that visits every *vertex* exactly once? We call this problem HAMILTONIAN CYCLE.

We noted there, informally, that EULERIAN CYCLE is easy and HAMILTONIAN CYCLE is thought to be hard. We can now state this formally: EULERIAN CYCLE $\in$ P and HAMILTONIAN CYCLEis NP-complete.

Chapter Summary

1. *We define many variants of* HAMILTONIAN CYCLE *and show or state that they are NP-complete.*
2. *Use these variants of* HAMILTONIAN CYCLE *to show that other problems are NP-complete. These other problems will mostly be about games.*
3. *Present metatheorems that allow for more games to be shown NP-complete using* HAMILTONIAN CYCLE *and variants.*

4.2 Variants of HAMILTONIAN CYCLE

In this section, we discuss many variants of HAMILTONIAN CYCLE. We start with the original problem and some basic variants:

HAMILTONIAN CYCLE, HAMILTONIAN PATH, DIRECTED HAMILTONIAN CYCLE, DIRECTED HAMILTONIAN PATH

Instance: (HAMILTONIAN CYCLE) A graph.

Question: Is there a Hamiltonian cycle? A Hamiltonian cycle is a cycle that visits every vertex exactly once.

Instance: (HAMILTONIAN PATH) A graph and two vertices a, b.

Question: Is there a Hamiltonian path from a to b? A Hamiltonian path from a to b is a path that begins at a, ends at b, and visits every vertex exactly once.

Instance: (DIRECTED HAMILTONIAN CYCLE) A directed graph.

Question: Is there a Hamiltonian cycle? The cycle must respect edge directions: If (u, v) is an edge but (v, u) is not, then the cycle cannot use (v, u). (The same holds for DIRECTED HAMILTONIAN PATH.)

Instance: (DIRECTED HAMILTONIAN PATH) A directed graph and two vertices a, b.

Question: Is there a Hamiltonian path from a to b?

Note: One can define planar versions of these problems by restricting to planar graphs or planar directed graphs.

We state the following NP-complete results without proof.

- HAMILTONIAN CYCLE is NP-complete. This was one of Karp's original 21 problems shown to be NP-complete [Kar72].
- DIRECTED HAMILTONIAN CYCLE, HAMILTONIAN PATH, DIRECTED HAMILTONIAN PATH are all NP-complete.
- We proved that PLANAR DIRECTED HAMILTONIAN CYCLE is NP-complete in theorem 2.21.
- HAMILTONIAN CYCLE restricted to planar 3-regular 3-connected graphs with minimum face degree 5 is NP-complete. This was proven by Garey et al. [GJT76].
- HAMILTONIAN CYCLE restricted to bipartite graphs is NP-complete. This was proven by Krishnamoorthy [Kri75].
- If G is a graph, then the ***square*** of G is the graph with edges added between all vertices connected by a path of length 2. HAMILTONIAN CYCLE on squares of graphs is NP-complete. This was proven by Underground [Und78].

Here are some trivial instances of HAMILTONIAN CYCLE, though *proving* they are trivial is difficult.

- Tutte [Tut56] proved that all planar 4-connected graphs are Hamiltonian. Hence the problem restricted to planar 4-connected graphs is trivial: Just say yes.

- Karaganis [Kar68] showed that the cube of a graph (the graph with edges added between all vertices connected by a path of length 3 or less) is always Hamiltonian. Hence the problem restricted to graphs that are cubes of other graphs is trivial: Just say yes.

Note that Tutte proved his result in 1956, and Karaganis proved his result in 1968, before the notion of NP-complete was known.

4.2.1 Maximum-Degree-3 Planar Graphs: Directed and Undirected

Plesník [Ple79] showed the following:

Theorem 4.1. *HAMILTONIAN CYCLE restricted to planar maximum-degree-3 directed graphs is NP-complete. Here the degree of a vertex is the sum of its in-degree and out-degree.*

Proof sketch. The proof is by reduction from 3SAT. Figure 4.1 illustrates the construction. The XOR gadget enforces that exactly one of two edges is in the cycle. The variable gadget then consists of pairs of doubled edges linked by an XOR gadget, forcing the pairs to have opposite membership, representing TRUE and FALSE assignments. The clause gadget contains a large cycle that can be in the Hamiltonian cycle if and only if one of the incoming literals (connected to the clause by XOR gadgets) is TRUE. This reduction also requires a crossover gadget to allow the XOR gadgets connecting variables to clauses to cross; the crossover gadget itself uses the XOR gadgets, essentially exploding them. □

Itai et al. [IPS82] showed the following:

Corollary 4.2. *HAMILTONIAN CYCLE restricted to planar undirected bipartite maximum-degree-3 graphs is NP-complete.*

Proof. The proof is by reduction from the directed graph problem of theorem 4.1. Figure 4.2 illustrates the proof. Given a degree-3 directed graph G, first check whether any vertex has in-degree 3 or out-degree 3. If so, then G cannot have a Hamiltonian cycle and we are done (formally, output a bipartite undirected maximum-degree-3 graph that does not have a Hamiltonian cycle). Henceforth, we can assume that every vertex has either in-degree 1 or out-degree 1.

Look at a vertex v in the directed graph. If it has in-degree 1 (out-degree 1), then the edge coming into v (going out of v) must be in any Hamiltonian cycle of G. We subdivide that edge by adding a degree-2 vertex in the middle of it. Any Hamiltonian cycle in the resulting graph will alternate between vertices from the original graph and vertices introduced to model forced edges, so the resulting graph is bipartite. If we remove edge directions, this alternation remains forced, because once we are adjacent to an added degree-2 vertex, we must visit it, or else we would never be able to visit it. □

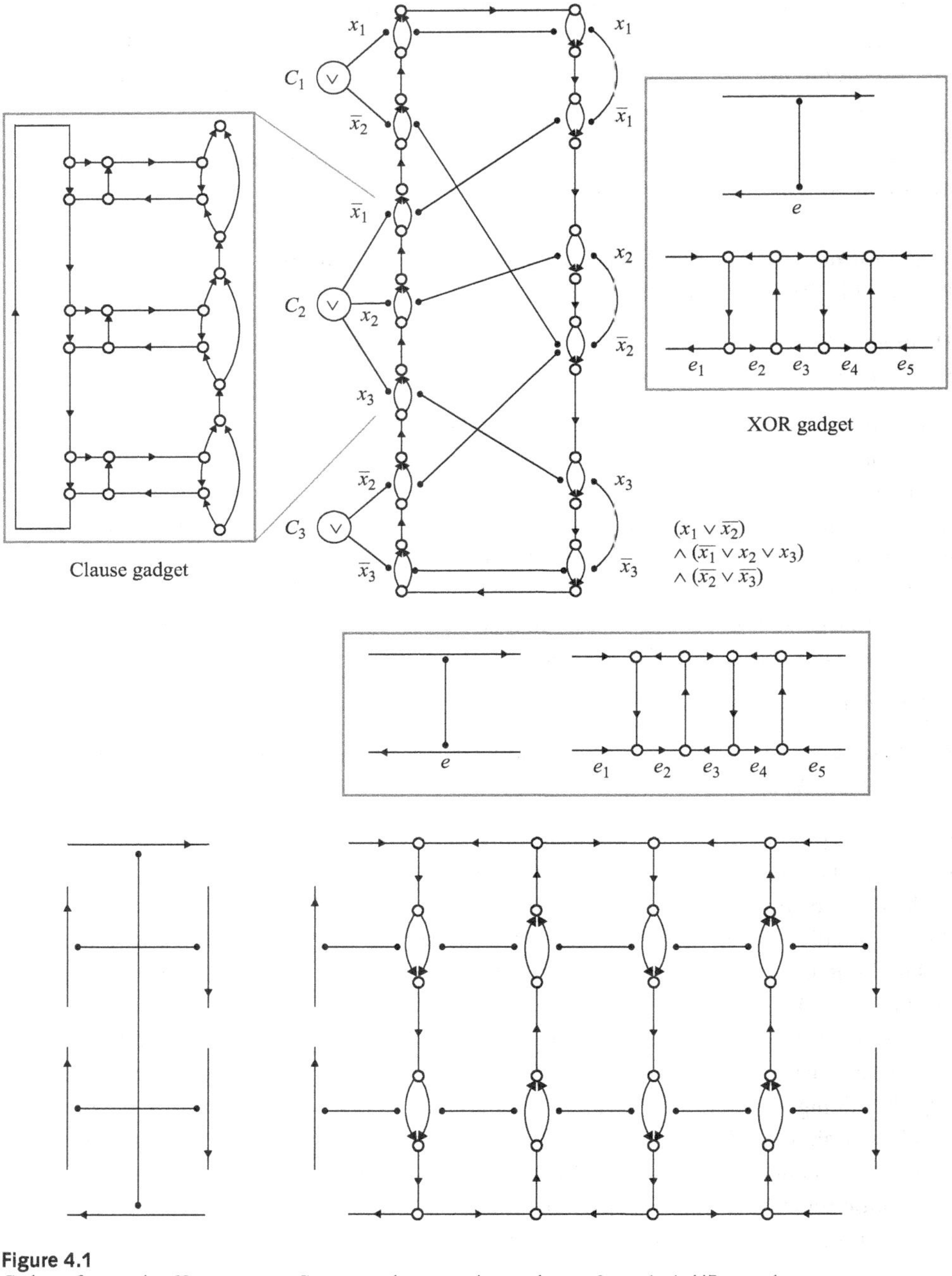

Figure 4.1
Gadgets for proving Hamiltonian Cycle on planar maximum-degree-3 graphs is NP-complete.

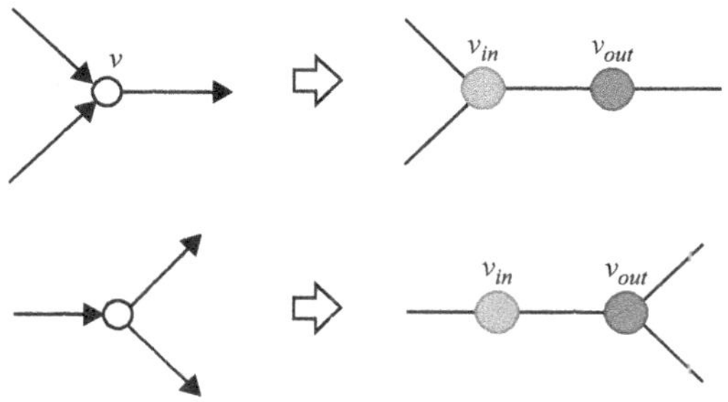

Figure 4.2
Reduction from HAMILTONIAN CYCLE on planar directed maximum-degree-3 graphs to HAMILTONIAN CYCLE on planar *undirected* maximum-degree-3 *biparitite* graphs.

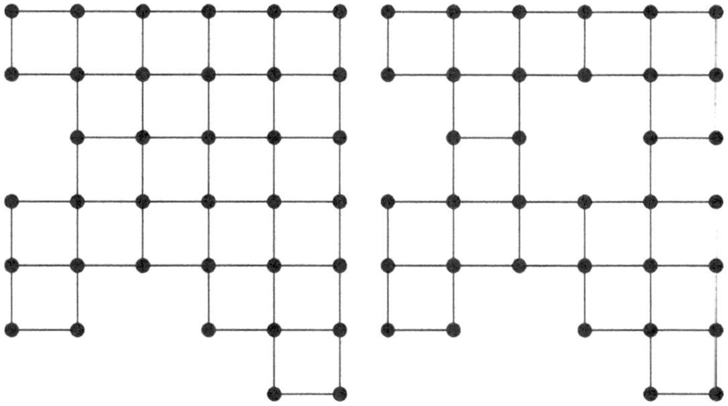

Figure 4.3
Left: A solid grid graph. Right: a non-solid grid graph.

4.2.2 HAMILTONICITY IN GRID GRAPHS and Applications

Definition 4.3. Refer to figure 4.3.

1. ***Grid graphs*** are graphs with vertices on a (subset of a) lattice and edges between all pairs of vertices at unit distance. Unqualified, grid graphs are usually on the square lattice, but the triangular and hexagonal lattices can also be considered.
2. Faces of grid graphs of unit area are called ***pixels***, while faces with greater area (i.e., containing at least one lattice point that is not a vertex) are called ***holes***.
3. A ***solid grid graph*** has no holes—that is, every face of bounded area is a pixel.

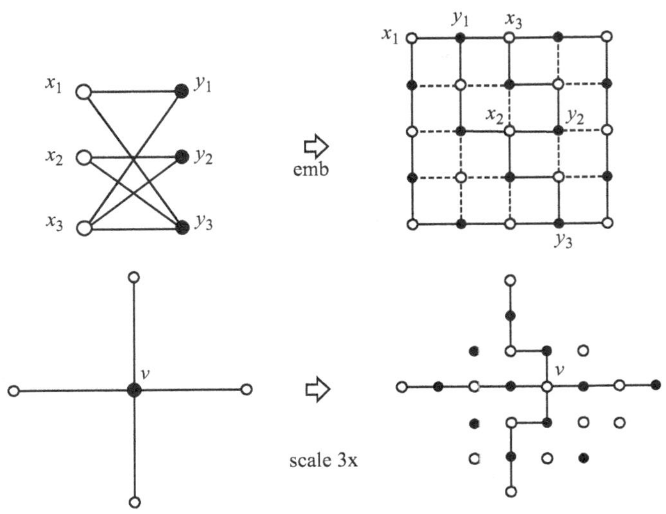

Figure 4.4
Planar bipartite graph drawing.

Umans and Lenhart [UL97] showed that HAMILTONIAN CYCLE can be solved in polynomial time on solid grid graphs. By contrast, Itai et al. [IPS82] proved the following:

Theorem 4.4. *HAMILTONIAN CYCLE restricted to grid graphs is NP-complete.*

Proof sketch. The proof is by reduction from Hamiltonian cycle on planar undirected degree-3 bipartite graphs, which we proved NP-hard in corollary 4.2. First we draw the input bipartite graph on the grid, routing the edges as orthogonal polygonal lines, via theorem 2.24. Then we modify the drawing to respect the grid graph's inherent 2-coloring, as shown in figure 4.4 (bottom): We scale the drawing by a factor of 3, and if a vertex's color on the grid does not match the color in the given bipartite graph, then we shift it 1 to the right and add local wiggles to accommodate.

We start with an ***edge gadget***, which consists of a path of pixels, as shown in figure 4.5. The vertices of this gadget can be visited in two ways by a Hamiltonian cycle: in a zigzag from one end to the other or by a border traversal that returns to the same end. These visits correspond to a Hamiltonian cycle in the original graph either using the edge (and thus ending at the other endpoint of the edge) or not using the edge (and thus not moving from the start vertex). Recall that a Hamiltonian cycle in the original graph needs to visit every vertex, but not every edge; but because our edge gadget adds vertices, we need to show how to visit the added vertices. The edge gadget accomplishes this goal.

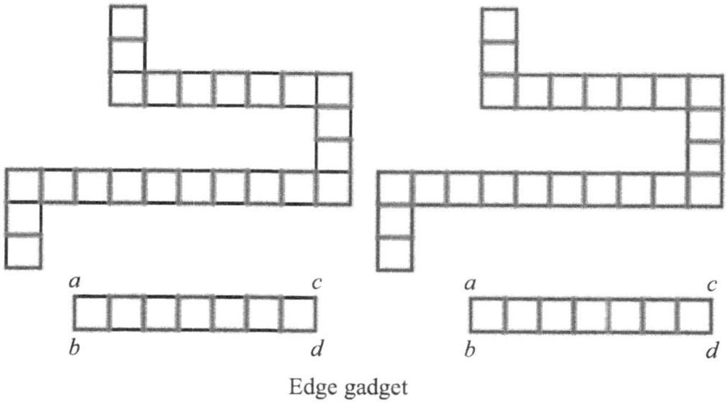

Figure 4.5
Edge gadget.

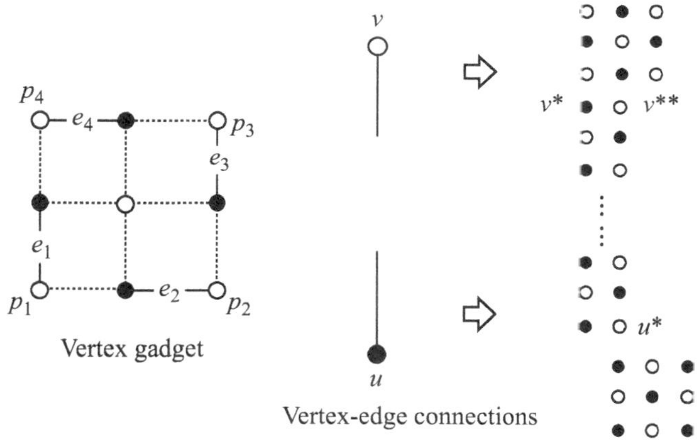

Figure 4.6
Vertex gadget.

The ***vertex gadget*** is a 3×3 square of vertices or 2×2 square of pixels, as shown in figure 4.6. The corners of the gadget are chosen to have the same color in the grid as the corresponding vertex in the input graph. This gadget has a Hamiltonian path from every corner to every other corner, and they all traverse the four marked edges e_1, e_2, e_3, e_4.

Figure 4.7 shows how to connect the edge and vertex gadgets. An edge gadget connects to white vertices in the original graph via a full pixel in the expanded graph, but connect to black vertices at a single point. This arrangement allows the edge to be used or not while preserving the parity of the cycle. □

Next we describe two simple applications of this result.

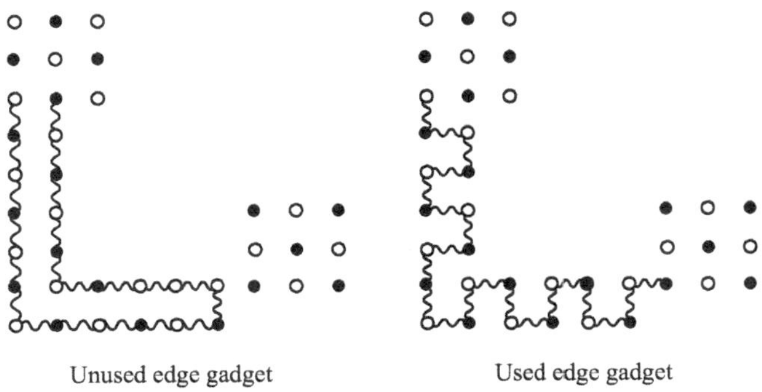

Figure 4.7
Hamiltonian grid graph.

Definition 4.5. Given a set of points in the plane, the ***Euclidean Traveling Salesman Problem (Euclidean TSP)*** asks for a tour through the points with Euclidean length less than k (or in the optimization problem, with minimum length).

Corollary 4.6. *The Euclidean TSP problem is NP-hard. (Note that we did not say NP-complete.)*

Proof. We show that Hamiltonian Cycle restricted to grid graphs reduces to Euclidean TSP.

1. Input G, a grid graph. We can assume it is in the plane and the vertices are lattice points. Let n be the number of vertices in G.
2. The instance I of Euclidean TSP is just the points in the grid graph with the goal of having a Hamiltonian cycle of length n.

Any tour that visits all the points in I must be a cycle in G, because if the tour visits two nonadjacent points it will not have cost $\leq n$. □

Note: We did not state that Euclidean TSP is NP-complete. It is not known whether Euclidean TSP is in NP. Why is this? Let the input be a set of points in the plane forming a weighted graph, and a number k. If you are given a Hamiltonian cycle that is a proposed witness, to see whether the Hamiltonian cycle has weight $\leq k$ involves asking whether a sum of square roots is $\leq k$. This problem is not known to be in NP because of issues of precision. We discuss this more in section 12.7.2.

4.2.3 Maximum-Degree-3 Grid Graphs

Papadimitriou and U. Vazirani [PV84] proved the following stronger form of theorem 4.4:

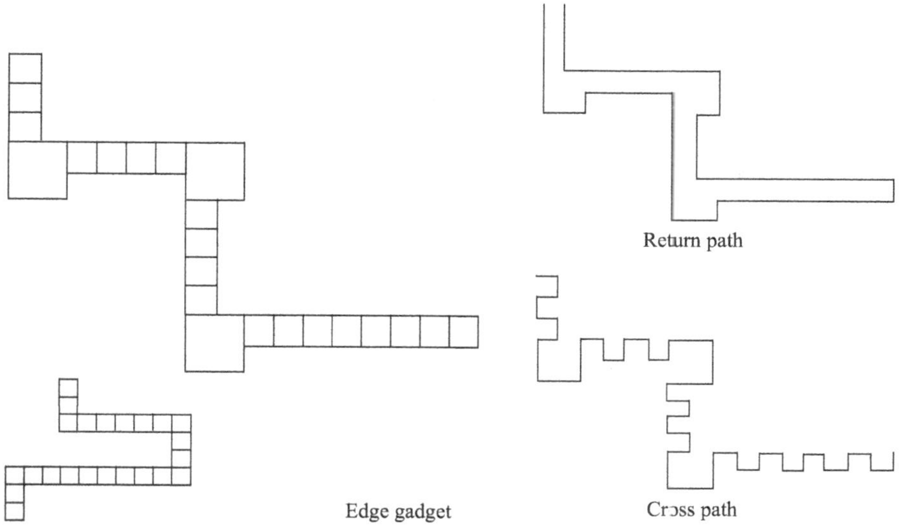

Figure 4.8
Edge gadget for maximum-degree-3 grid graphs.

Theorem 4.7. *Hamiltonian cycle restricted to maximum-degree-3 grid graphs is NP-complete.*

Proof sketch. The reduction is similar to that of theorem 4.4, but with the gadgets modified to avoid degree-4 vertices. To avoid the degree-4 vertex at each reflex corner of the edge gadget, we expand each turn pixel into a 3×3 square with a hole in the middle, as shown in figure 4.8, The vertex gadget is now a "dumbbell," as shown in figure 4.9. Degree-2 vertices have their edge gadgets connected to opposite ends of the dumbbell. For degree-3 vertices, we connect the forced degree-2 neighbor on one end of the dumbbell and the other two neighbors at the other end. □

The minimum spanning tree problem (MST) can be solved in polynomial time by either Kruskal's or Prim's algorithm. However, there are variants that are NP-complete. For example, if we require that the tree have maximum degree 2, then the problem is NP-complete: This is just the TSP problem. What about degree 3? Papadimitriou and U. Vazirani [PV84] proved the following:

Corollary 4.8. *MST restricted to Euclidean grid graphs, where we require the tree to have maximum degree 3, is NP-complete.*

Proof. The proof is a reduction from Hamiltonian path on maximum-degree-3 grid graphs. The reduction adds new vertices very close to each existing vertex in the grid graph,

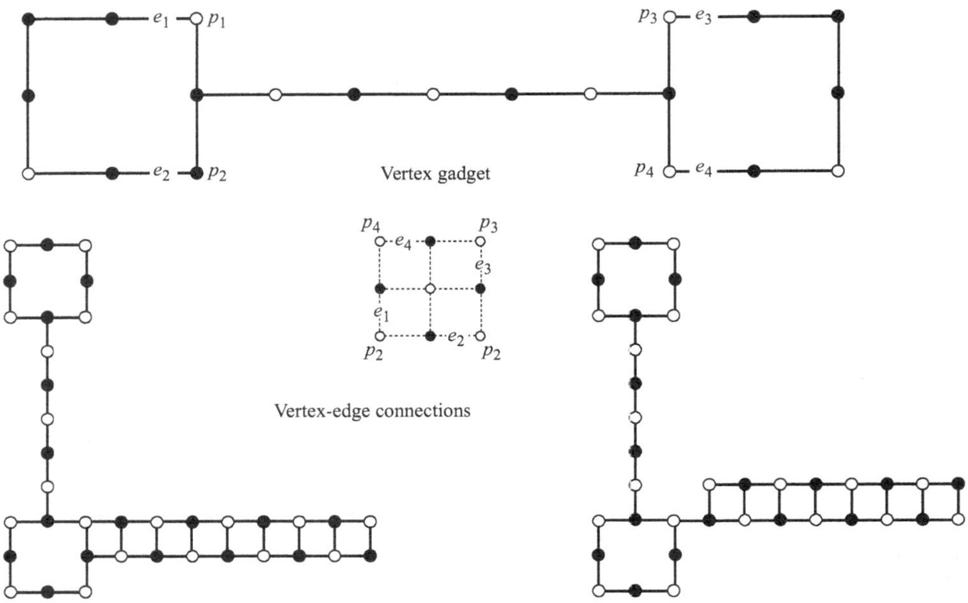

Figure 4.9
Vertex gadget for maximum-degree-3 grid graphs.

forcing the edge between that vertex and the new vertex to be in the minimum spanning tree. The remainder of the tree is then finding a Hamiltonian path in the grid graph. □

4.2.4 Grid Graph Hamiltonicity Taxonomy

Many other restrictions on grid graphs have been studied:

Definition 4.9.

1. ***Superthin*** grid graphs have no pixels; all faces are holes or the outside face.
2. ***Thin*** grid graphs have all vertices on the boundary.
3. ***Polygonal*** grid graphs have no shared edges between holes or the outside face; polygonal graphs could be considered anti-superthin.

With these definitions, we taxonomize Hamiltonicity over various kinds of grid graphs in table 4.1. Most of the results are from Arkin et al. [AFI$^+$09] or Demaine and Rudoy [DR17].

Figure 4.10
Polygonal tiling.

Table 4.1
Taxonomy of the complexity of Hamiltonicity

	triangular	square	hexagonal
general	NP-complete	NP-complete	NP-complete
max degree 3	NP-complete	NP-complete	NP-complete
thin	NP-complete	NP-complete	NP-complete
superthin	NP-complete	P	P
polygonal	P	NP-complete	NP-complete
thin polygonal	P	P	P
solid	P	P	open

Recently, the MIT Hardness Group [MIT24a] gave simpler and stronger proofs of NP-hardness of Hamiltonian Cycle in (square) grid graphs, including when restricted to maximum degree 3. They also proved NP-hardness of the following new cases:

1. Maximum-degree-3 *subgraphs* of grid graphs where the vertex set is the *entire rectangle* $\{1, \ldots, m\} \times \{1, \ldots, n\}$. This result is stronger because it requires every vertex in the rectangle, but weaker because it allows subgraphs, so we no longer have that there is an edge between every pair of vertices at unit distance.
2. *Directed* grid graphs of maximum degree 3.
3. *Directed* grid graphs where the vertex set is the *entire rectangle* $\{1, \ldots, m\} \times \{1, \ldots, n\}$.

4.3 Reductions from Hamiltonian Cycle

4.3.1 Settlers of Catan Longest Road Card

Hamiltonian Cycle restricted to hexagonal grid graphs is NP-complete [AFI$^+$09, DR17]. We use that to prove the following.[1]

Definition 4.10. Let G be a game and $i \in \mathbb{N}$. The ***Mate-in-i problem for G*** is the problem of, given a game position, can the player who is about to move get a win within i moves. Note that ***Mate-in-0*** means the player about to move has already won; however, checking this might still be hard.

Theorem 4.11. *Mate-in-1 and mate-in-0 are NP-complete for Settlers of Catan.*

Proof sketch. In Settlers of Catan, the player with the ***Longest Road card*** gets two victory points. This card is awarded to the player whose built roads can form the longest simple path in the hexagonal grid of the game. If the opponent has a road of length $n-2$, then determining whether the player has the Longest Road requires checking for a Hamiltonian path of length $n-1$. If the player is otherwise only two victory points away from winning, then to determine whether they have already won from the Longest Road requires solving Hamiltonicity on their roads. □

4.3.2 Slitherlink

Definition 4.12. *Slitherlink* is a Nikoli game [Nik08, Nik] (Nikoli is a publisher of puzzles that are culture-independent.) played on a grid graph in which some pixels are labeled with the numbers 0 through 4. The objective of the game is to find a cycle (not necessarily Hamiltonian) along the grid lines such that the numbered pixels are bordered by that number of edges in the cycle. (See figure 4.11 for an example.)

1. This result is from unpublished work by Kyle Burke, Erik Demaine, Gabe van Eycke, and Neil McKay (2011).

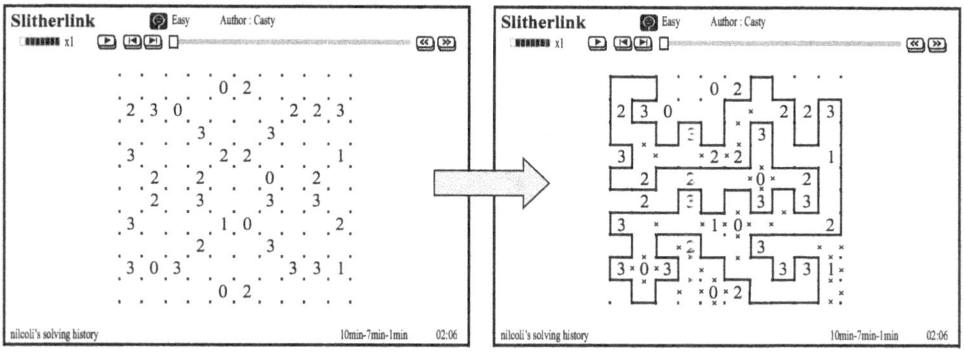

Figure 4.11
Example of Slitherlink.

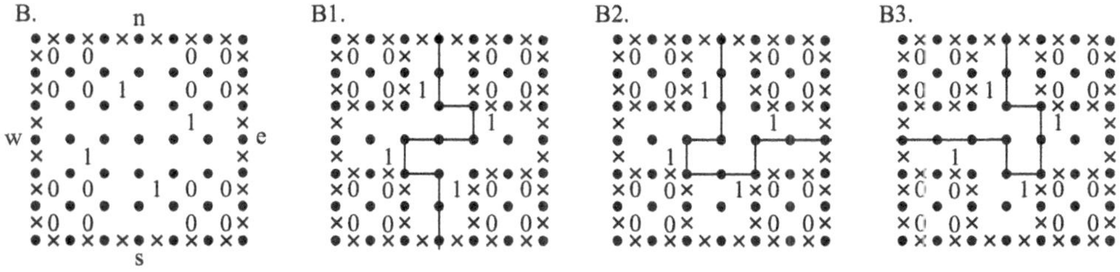

Figure 4.12
SLITHERLINK is NP-complete.

> SLITHERLINK
> *Instance:* An instance of the *Slitherlink* Game.
> *Question:* Can the player win?

Yato [Yat00] proved the following:

Theorem 4.13. *SLITHERLINK is NP-complete.*

Proof sketch. The proof is a reduction from HAMILTONIAN CYCLE on grid graphs. Figure 4.12 shows the vertex gadget. A non-vertex can be represented by a square of 0 clues. (The original proof [Yat00] was from HAMILTONIAN CYCLE on planar maximum-degree-3 graphs, so needed more gadgets.) □

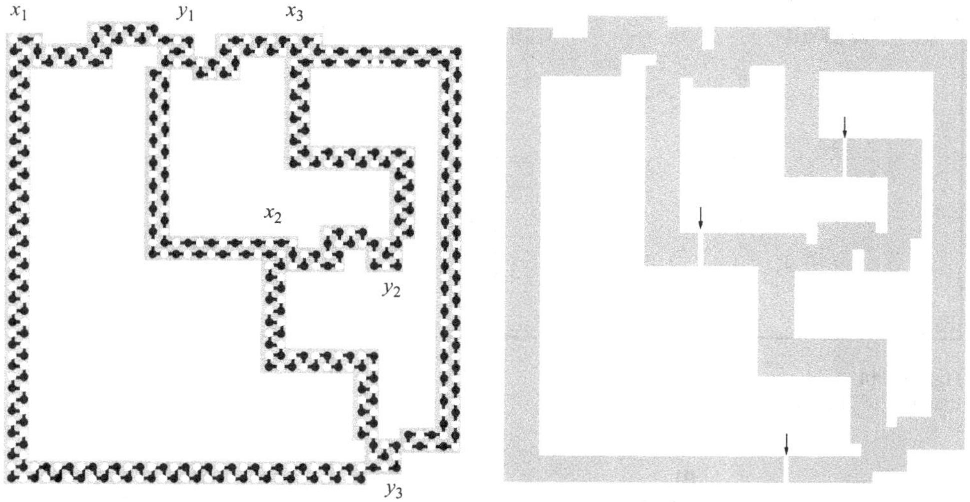

Figure 4.13
Milling and lawn mowing are NP-complete.

4.3.3 Lawn Mowing and Milling

The lawn mowing and milling problems both involve cutting a specified region with a tool (we are not going to define them rigorously). In lawn mowing problems, the tool path can go outside the region, while in milling it must stay inside. The goal is generally to find the shortest path. Milling problems arise in actual physical milling, while lawn mowing problems arise in laser and waterjet cutting and in each layer of 3D printing by deposition. Arkin et al. [AFM00] proved the following:

Theorem 4.14. *Milling and lawn mowing are NP-hard for grid polygons and a unit square tool.*

Proof sketch. The proof is a reduction from Hamiltonicity in grid graphs; see figure 4.13. For milling, we just thicken the grid graph into a unit square for each vertex. For lawn mowing, we can add thin slits to make the target region a simple polygon without holes.

□

4.4 NP-Hardness Metatheorems from Hamiltonicity

Forišek [For10] and Viglietta [Vig14a] proved "metatheorems" that serve as tools for proving specific games hard. These metatheorems can be applied to a lot of games. The term "metatheorem" here denotes an informal theorem, because it is difficult to state all the specific assumptions for all games. Rather, a metatheorem gives a general setup for the proof.

Here we will cover their metatheorems for NP-hardness that are based on Hamiltonicity. In chapter 13, we will describe Viglietta's metatheorems for PSPACE-hardness.

In all cases, we assume that the player controls a character in the game by moving it around an environment to complete objectives. The environment and initial location of the player are given as input. One basic assumption we make is that the player can, at most times, choose their direction of movement.

4.4.1 Forišek's Metatheorems 1 and 2

These metatheorems consider the case where the goal of the player is ***location traversal***, meaning that specified locations need to be traversed in order to win the level (e.g., to collect all items).[2]

Forišek's first metatheorem states that, without a time limit, such games are easy. We can perform a depth-first search over the graph G of possible moves to decompose into strongly connected components (consisting of pairwise reachable vertices) and construct a directed acyclic graph G' on those connected components. The level is solvable if and only if there is a path in G' starting at the starting node that visits all the nodes with location-traversal constraints, which is easy to test via dynamic programming. If the level has a designated exit location that must be visited last, we can incorporate that constraint as well: The path must end at the exit node.

Forišek's second metatheorem states that, with a specified time limit, such games are NP-hard. The idea of the proof is to reduce from Hamiltonian Path in grid graphs, as most games take place on a grid. The reduction simply requires location traversal at all the grid graph vertices, and it sets the timer so that you barely have enough time to follow a Hamiltonian path of $n - 1$ edges. In particular, moving between any nonadjacent pair of vertices would slow the player down and end up violating the time limit. If there is a specified exit location, we can instead reduce from Hamiltonian Cycle in grid graphs.

In particular, this metatheorem implies that speedrunning in most video games is NP-hard. In games with a time limit (e.g., *Super Mario Bros.*), it also immediately gives NP-hardness of collecting all items (e.g., coins).

4.4.2 Viglietta's Metatheorem 1

This metatheorem states that a game is NP-hard if it involves a player traversing a 2D environment with a start location and the following features:

2. Forišek called this property "item collection," while Viglietta called it "location traversal." We use the latter term for consistency.

1. Location traversal, meaning specific locations need to be traversed in order to win the level.
2. Single-use paths, which can be traversed only once.

The idea for the proof is to reduce from HAMILTONIAN PATH on planar maximum-degree-3 graphs. The reduction transforms each vertex in the graph into a location that must be traversed and each edge into a single-use path. Because each vertex has degree 3, and each edge is single-use, once we traverse two edges to visit the vertex and leave, those edges will disappear, leaving just one edge. Now the vertex is unreachable because, if we use the third edge, then the player will be trapped in that location, as there is no fourth edge out of the vertex. We conclude that there will be a way to clear the game if and only if there is a Hamiltonian path.

Using this metatheorem, we can prove NP-hardness for many games.

Boulder Dash Biasi [Bia11] first proved *Boulder Dash* is NP-hard; however, Viglietta's metatheorems give a simpler proof.

In this game, the player has a starting position and can walk around without being affected by gravity and dig in adjacent cells if they are made of earth. There are boulders that are influenced by gravity, and if they fall on the avatar, the player dies. The goal is to collect all the diamonds and get to the end location. You only need two gadgets: location traversal (shown in figure 4.14 left) and single-use path (shown in figure 4.14 center).

Figure 4.14 (right) demonstrates the single-use path gadget after it has been traversed from left to right: We push the first boulder into the pit in order to clear the obstacle; we then push the lower of the two stacked boulders over to the other pit, then push the last boulder into the final pit as part of our traversal. During this time, the higher of the two stack boulders falls down and blocks our path, since there is no pit to push it into.

(One small note is that the boulder can "rest" on the player without killing him, so pushing the lower boulder to the right and causing the player to stand under the higher boulder is permissible.)

The conclusion is that *Boulder Dash* is NP-hard.

LODE RUNNER *Lode Runner* was proven NP-hard by Viglietta's metatheorems [Vig14a].

In this game, the player has to collect all the coins and avoid enemies. They can dig a hole on the ground and the monsters can fall in this hole. Eventually, the hole will refill (after some specific time), releasing the monster. The avatar cannot jump, and this property is exploited in the Single-use path gadget. Figure 4.15 shows the gadgets required for metatheorem 1. The conclusion is that *Lode Runner* is NP-hard.

ZELDA II Another application of Viglietta's metatheorems is to *Zelda II* [ADGV15]. Part of this game is a platform game. The location traversal is done by placing keys at certain

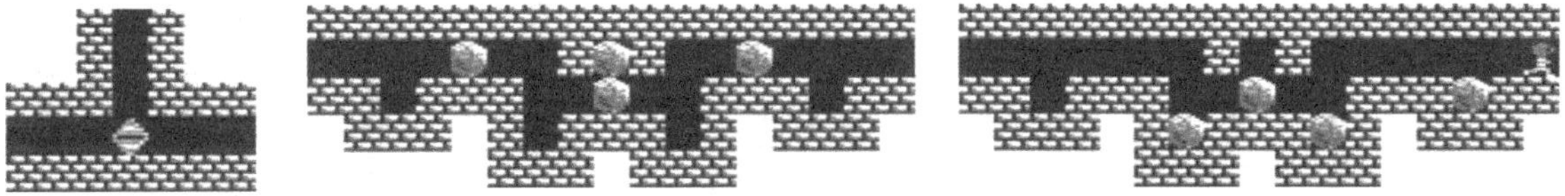

Figure 4.14
(Left) Location traversal. (Center) Single-use path gadgets for *Boulder Dash*. (Right) Single-use gadget after traversing from left to right.

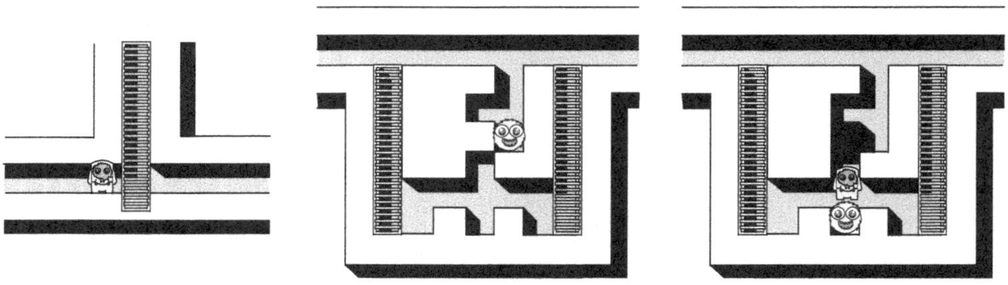

Figure 4.15
(Left) Location traversal gadget. (Center) Single-use path gadget. (Right) Single-use gadget being used.

locations. The keys are used to open doors. At the end of the level, there will be exactly the same number of doors as the number of keys in the level, so in order to clear the level, we must collect all keys. The single-use paths are bridges that disappear when Link walks over them. The conclusion is that *Zelda II* is NP-hard.

4.4.3 Viglietta's Metatheorem 2

This is a slight variation of metatheorem 1. It starts from the same kind of setup (traversing a 2D environment with location traversal) but requires "tokens" and "toll roads" instead of single-use paths, where a ***token*** is something that can be picked up and a ***toll road*** requires a token to be traversed. The idea is to simulate a single-use path with this kind of mechanism:

- place a token at each vertex; and
- transform every (single-use) edge into a toll road that requires one token.

Thus, to get from one vertex to the next, the player must pay the token that was just acquired; if a vertex is visited more than once, there will be no token at that vertex, and the player is unable to cross over any edge. The reduction from Hamiltonicity follows in the same way as in metatheorem 1.

Pac-Man We can use metatheorem 2 to prove *Pac-Man* NP-hard. To win in *Pac-Man*, you have to collect all the dots, which will also function as our tokens. Figure 4.16 shows the degree-3 vertex with the token (dot) and the toll road (path with ghosts). Eating a token causes the ghosts to change state so that Pac-Man can eat the ghosts; after some time, though, the ghosts revert to becoming harmful to Pac-Man. Also, whenever a ghost changes its state, it will change the direction of movement. The ghosts that are "eaten" revive after a while and appear inside a "cage," which they then leave and continue moving around in their original paths.

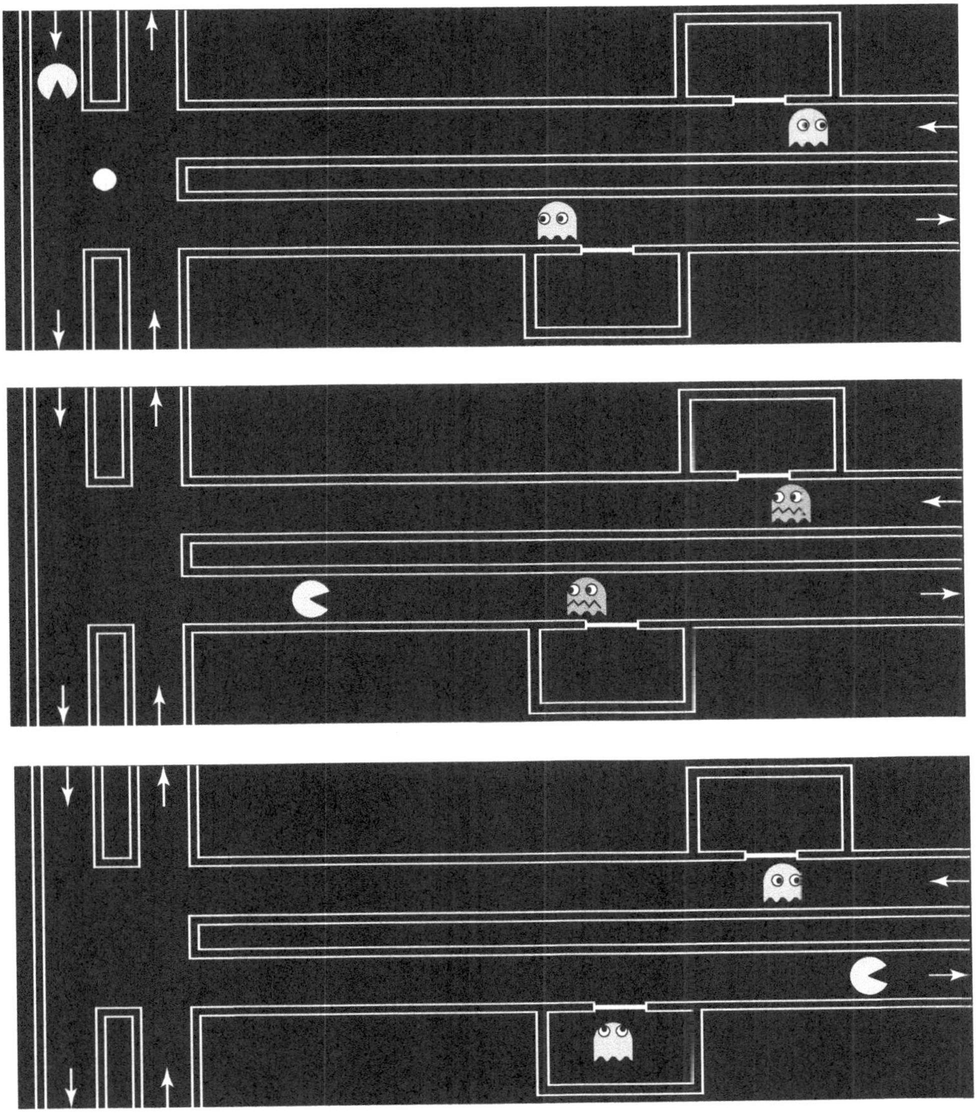

Figure 4.16
Token+Toll road gadget for *Pac-Man*.

Thus, the only way to traverse an edge is to consume the token and the ghost along the chosen exit edge. For the sake of argument, we will assume that the ghosts change state long enough for Pac-Man to exit safely, but short enough so that if Pac-Man even comes back to the same area of the map, the ghosts will have reverted states or respawned. Thus each edge can be traversed if and only if Pac-Man consumes the token; once the token is consumed, Pac-Man cannot return to this vertex again, because he will not be able to pay the toll (eat the dot) anymore.

Therefore, *Pac-Man* is also NP-hard.

5

NP-Hardness via 3-PARTITION

5.1 Introduction

Many problems have numbers in their input. For some of these problems, it matters how the numbers are represented. If the numbers are represented in unary, then *polynomial in the length of the input* is much longer than if the numbers are represented in binary.

Chapter Summary

1. *We define two variations of NP-complete—**weakly NP-complete** and **strongly NP-complete**—and give examples of both. We are most interested in problems that are strongly NP-complete because that lets us represent the numbers in unary, which is useful for reductions, and still the problem is NP-complete.*
2. *We will use the strongly NP-hard problems to show that problems in the following fields are NP-hard: scheduling, graph theory, packing, disk packing, puzzles, computer games, and three-dimensional geometry. All of the problems are also NP-complete except for the packing problems whose status in NP is unknown.*

5.2 Types of NP-Hardness

Consider a problem whose input includes one or more integers (which we refer to simply as ***numbers***). The complexity of the problem may depend on how the integers are represented. More precisely, a problem involving numbers can be interpreted as *two different problems* involving bits: one using a unary encoding for the numbers and one using a binary encoding for the numbers. These two problems have two different notions of the "size of the input," which affects the notion of "polynomial" for both running times and allowed reductions. We rephrase definition 0.21 of weak and strong NP-hardness, and extend it to weak and strong NP-completeness. These notions may have been introduced by Garey and Johnson [GJ79, page 90].

Definition 5.1. Let A be a problem that has numbers in the input. Define A_{unary} to be the version of A where the numbers are given in unary, and define A_{binary} to be the version of A where the numbers are given in binary.

1. A is ***weakly NP-hard*** if A_{binary} is NP-hard.
2. A is ***weakly NP-complete*** if A_{binary} is NP-complete.
3. A is ***strongly NP-hard*** if A_{unary} is NP-hard. (In this case, A is also weakly NP-hard, but we do not usually call it that.)
4. A is ***strongly NP-complete*** if A_{unary} is NP-complete.

We can similarly differentiate the notion of polynomial-time algorithms for the two cases. Figure 5.1 summarizes the relationships between these hardness types and algorithm strengths.

Definition 5.2. Let A be a problem that has numbers $a_1, \ldots, a_n$ in the input. Assume that these numbers form most or all of the input. Let $a_{\max}$ denote the largest number.

1. An algorithm for A is ***pseudopolynomial*** if it is polynomial in n and $a_{\max}$ (so the input can be viewed as being in unary). This often arises in dynamic programming, where the

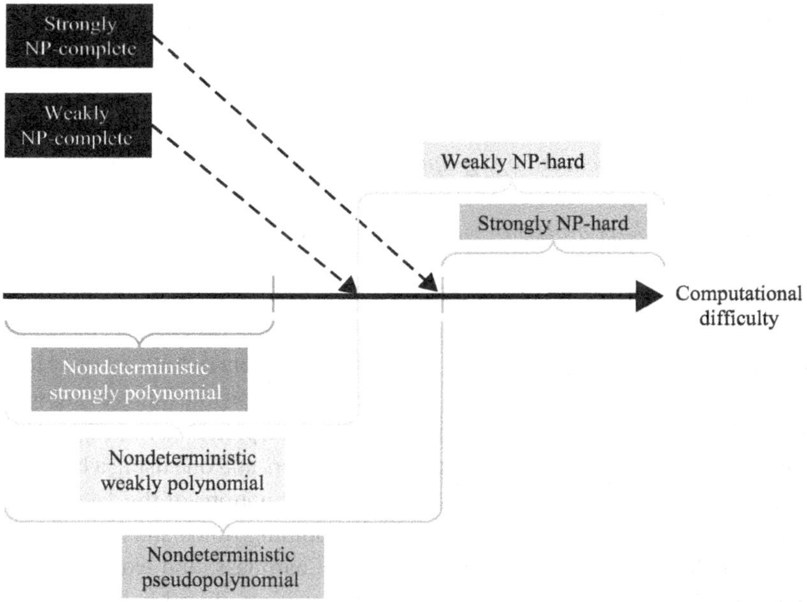

Figure 5.1
Relationships between types of NP-hardness and nondeterministic polynomial time (NP).

table grows proportionally with the integer values. Equivalently, this is a polynomial-time algorithm for A_{unary}.

2. An algorithm for A is ***weakly polynomial*** if it is polynomial in n and $\log(a_{\max})$ (so the input can be viewed as being in binary). This is the typical meaning of "polynomial time." Equivalently, this is a polynomial-time algorithm for A_{binary}.
3. An algorithm for A is ***strongly polynomial*** if it is polynomial in n alone. (This notion makes sense on models such as the word RAM, but not on a Turing machine where a tape cell has just one bit.)

What we care about most is the distinction between pseudopolynomial and weakly polynomial. Assuming $\mathsf{P} \neq \mathsf{NP}$:

- If a problem is weakly NP-hard, then there is no weakly polynomial algorithm. There might be a pseudopolynomial algorithm, in which case the problem is strongly NP-complete.
- If a problem is strongly NP-hard, then there is no pseudopolynomial algorithm (and therefore no weakly polynomial algorithm).

Thus there is practical significance in showing strong vs. weak NP-hardness, as only the former rules out pseudopolynomial algorithms.

5.3 Partition Problems

Most of the proofs in this chapter will be reductions from 3-PARTITION. But before that, we'll introduce the better known problem PARTITION.[1]

Notation 5.3. If A is a multiset of numbers, then $\sum_{a\in A} a$ is the sum of all the numbers in A, counting multiplicities. For example, if $A=\{1,1,2\}$, then $\sum_{a\in A} a=4$.

5.3.1 PARTITION

We start with a weakly NP-complete problem, which was one of Karp's original 21 NP-complete problems [Kar72].

> PARTITION
> *Instance:* Multiset of n integers $A=\{a_1,a_2,\ldots,a_n\}$.
> *Question:* Can A be partitioned into two sets A_1 and A_2 that have the same sum? Formally: $\sum_{a\in A_1} a=\sum_{a\in A_2} a=\frac{1}{2}\sum_{a\in A} a$.

1. PARTITION is sometimes called 2-PARTITION, though we will *never* call it that because 3-PARTITION is very different: the "2" and "3" would be measuring very different (roughly opposite) things.

We can also generalize PARTITION to an arbitrary specified target sum:

> SUBSET SUM
> *Instance:* Multiset of n integers $A = \{a_1, a_2, \ldots, a_n\}$, and a target sum t.
> *Question:* Is there a subset $S \subseteq A$ such that $\sum_{a \in S} = t$?

Karp [Kar72] showed the following; see also Garey and Johnson [GJ79, page 60].

Theorem 5.4. *PARTITION is weakly NP-complete, hence SUBSET SUM is weakly NP-complete.*

Is PARTITION strongly NP-complete? Unlikely, as exercise 5.5 shows that PARTITION with numbers in unary is in P.

Exercise 5.5.

1. Design a pseudopolynomial-time algorithm for PARTITION.
   Hint: Use dynamic programming.
2. Show that PARTITION is weakly NP-complete.

Exercise 5.6. Alfonsín [Alf98] looked at variants of SUBSET SUM. We look at one of them. The basic idea is that instead of being able to use a_i just once, you can use it $\leq r_i$ times where r_i is part of the input.

> SUBSET SUM WITH REPETITION
> *Instance:* Positive integers $a_1, \ldots, a_n$ and $r_1, \ldots, r_n$, and a target t.
> *Question:* Do there exist $0 \leq x_i \leq r_i$ such that $\sum_{i=1}^{n} x_i a_i = t$?

1. Show that SUBSET SUM WITH REPETITION is weakly NP-complete.
2. Show that SUBSET SUM restricted to the case where the elements are superincreasing (every element is greater than or equal to the sum of all of the previous elements) can be solved in weakly polynomial time.
3. Show that SUBSET SUM WITH REPETITION restricted to the case where the elements are superincreasing is weakly NP-complete.
4. Show that SUBSET SUM WITH REPETITION restricted to the case where the elements are superincreasing and every $r_i \in \{1, 2\}$ is weakly NP-complete.

5.3.2 3-PARTITION

3-PARTITION is a very useful strongly NP-hard problem:

> 3-PARTITION
> *Instance:* A multiset of n integers $A=\{a_1,a_2,\ldots,a_n\}$, where n is divisible by 3 and $\frac{t}{4}<a_i<\frac{t}{2}$ for all i.
> *Question:* Can A be partitioned into $n/3$ sets $A_1,\ldots,A_{n/3}$ such that they all have equal sums? Formally, for all $1\le i\le n/3$, $\sum_{x\in A_i}x=t$, where we define $t=(\sum_{a\in A}a)/(n/3)$. Each A_i has size exactly 3 (see exercise 5.8).

Garey and Johnson [GJ79, page 96] showed the following:

Theorem 5.7. *3-PARTITION is strongly NP-complete.*

Exercise 5.8.

1. Show that, in any solution of 3-PARTITION, every A_i has size exactly 3.
2. Prove theorem 5.7.
3. Theorem 5.7 restricts the a_i's by $\frac{t}{4}<a_i<\frac{t}{2}$. Show that 3-PARTITION remains NP-complete if use the restriction $\frac{7t}{24}<a_i<\frac{10t}{24}$. Show that, for all $\delta>0$, 3-PARTITION remains NP-complete if we use restriction $(\frac{1}{3}-\delta)t<a_i<(\frac{1}{3}+\delta)t$.

Exercise 5.9. Give a direct reduction from 3-PARTITION to PARTITION.
Hint: First reduce directly from 3-PARTITION to SUBSET SUM, then modify the proof to work with PARTITION.

Next we define a special case of 3-PARTITION that is also strongly NP-complete:

> NUMERICAL 3D MATCHING
> *Instance:* Three multisets $A=\{a_1,\ldots,a_n\}, B=\{b_1,\ldots,b_n\}, C=\{c_1,\ldots,c_n\}$ of n integers each.
> *Question:* Is there a partition of $A\cup B\cup C$ into n sets $D_1,\ldots,D_n$ such that (1) each set has exactly one element from A, one from B, and one from C; and (2) each set has the same sum? Specifically, the sum must be $t=(\sum_{x\in A\cup B\cup C}x)/n$.

Garey and Johnson [GJ79, page 224] noted that their proof of theorem 5.7 can be easily modified to show the following:

Theorem 5.10. *NUMERICAL 3D MATCHING is strongly NP-complete.*

To explain the terminology NUMERICAL 3D MATCHING, we recall the related problem 3D MATCHING from section 2.6.3, where we showed that 3D MATCHING is NP-complete even in the planar version.

> 3D MATCHING
> *Instance:* (3D MATCHING) A hypergraph H with vertices partitioned into three disjoint sets A, B, C, where $|A| = |B| = |C| = n$, and hyperedges $E \subseteq A \times B \times C$ (with exactly one vertex from each set).
> *Question:* Are there n disjoint edges that cover all of the vertices?

NUMERICAL 3D MATCHING is the special case of 3D MATCHING where A, B, C contain numbers, and there is a hyperedge $(a, b, c) \in E$ if and only if $a + b + c = t$.

Exercise 5.11. Give an easy proof (not going through the Cook–Levin theorem whose proof uses Turing machines) that NUMERICAL 3D MATCHING reduces to 3-PARTITION.
Hint: Let N be a large number. Let a, b, c be numbers you find. Add aN to all the elements of A, bN to all the elements of B, and cN to all the elements of C.

Exercise 5.12. For each of the following problems, either (1) show that the problem is in P by giving a polynomial-time algorithm or (2) show that the problem is NP-hard by reducing one of the following to it: (a) 3-PARTITION, (b) 3D MATCHING, or (c) NUMERICAL 3D MATCHING. In the following problems the numbers are given in binary.

1. Given a set of numbers $A = \{a_1, \ldots, a_{2n}\}$ that sum to $t \cdot n$, find a partition of A into n sets $S_1, \ldots, S_n$ of size 2 such that each set sums to t.
2. Given a set of numbers $A = \{a_1, \ldots, a_{2n}\}$ that sum to $t \cdot n$, find a partition of A into n sets $S_1, \ldots, S_n$ of any size such that each set sums to t.
3. Given a set of numbers $A = \{a_1, \ldots, a_{2n}\}$ and a sequence of target numbers $\langle t_1, \ldots, t_n \rangle$, find a partition of A into n sets $S_1, \ldots, S_n$ of size 2 such that for each $i \in \{1, \ldots, n\}$, the sum of the elements in S_i is t_i.

5.4 Reductions from 3-PARTITION

In the rest of this chapter, we use 3-PARTITION to prove NP-completeness of several problems. Some of these involve numbers, but others do not, instead relying on the ability to represent the numbers in the 3-PARTITION instance in unary and still get a polynomial reduction.

5.4.1 Scheduling

We begin with a trivial reduction to a basic scheduling problem.

MULTIPROCESSOR SCHEDULING
Instance: n jobs with completion times $a_1, \ldots, a_n$, and p identical sequential processors.
Question: Can all jobs be scheduled to execute on the processors, where each job must be executed entirely by one processor without interruption (consuming a_i time), a processor can execute only one job at once (sequential), such that all processors finish in time $\leq t$?
Note: Figure 5.2 shows an example schedule.
Note: The amount of time it takes for all jobs/processors to finish is called the ***makespan***.

Theorem 5.13. *MULTIPROCESSOR SCHEDULING is strongly NP-complete.*

Proof. Clearly MULTIPROCESSOR SCHEDULING $\in$ NP. We show 3-PARTITION $\leq_p$ MULTIPROCESSOR SCHEDULING.

1. Input $(a_1, \ldots, a_n)$, an instance of 3-PARTITION.
2. Output $(a_1, \ldots, a_n), p = n/3, t = \sum_{i=1}^{n} a_i / p$.

The proof that this reduction works is trivial. □

5.4.2 1-PLANAR

Definition 5.14. A ***1-planar graph*** is a graph that can be drawn in the plane with each edge crossing at most one other edge.

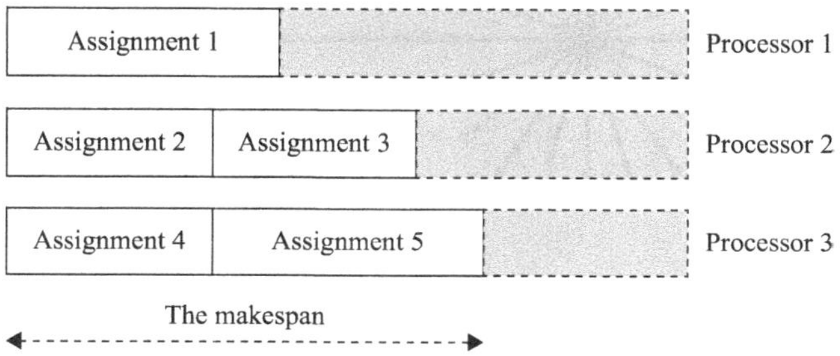

Figure 5.2
An assignment of processes to 3 processors.

> 1-PLANAR
> *Instance:* A graph G.
> *Question:* Is G 1-planar?

Grigoriev and Bodlaender [GB07] showed the following:

Theorem 5.15. *1-PLANAR is NP-complete.*

The 1-PLANAR problem does not have numbers in it and, hence, we do not call it weakly or strongly NP-complete. This will be our first example where it is helpful to represent the numbers in a 3-PARTITION instance in unary.

Proof sketch. Clearly 1-PLANAR $\in$ NP. We reduce from 3-PARTITION to 1-PLANAR. Refer to figure 5.3.

One gadget in this construction is an ***uncrossable edge***, shown in figure 5.3 (top). This gadget is a graph, namely K_6, with the property that any edge passing through it must cross an edge that already has a crossing, which is forbidden. Instances of this gadget are denoted by bold edges in the rest of the figure.

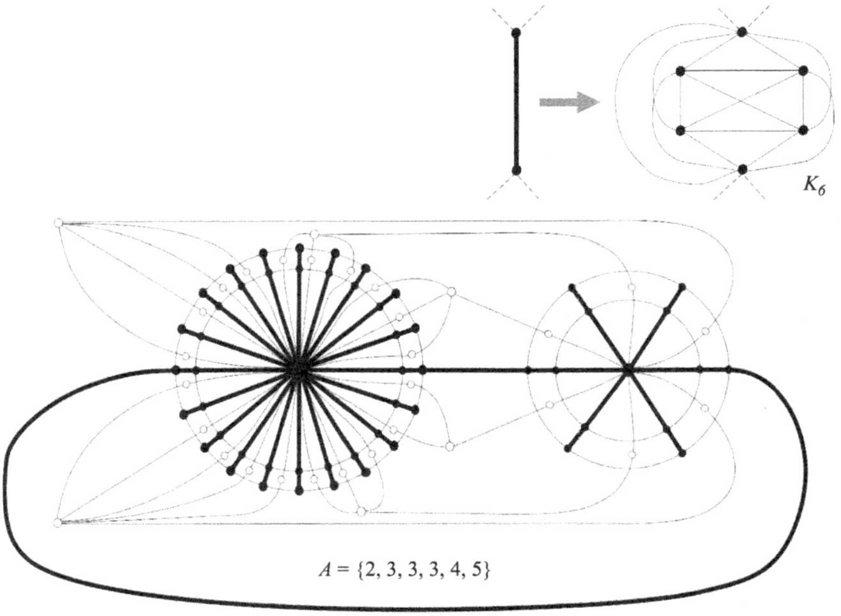

Figure 5.3
Reduction from 3-PARTITION to 1-PLANAR.

Now suppose we are given an instance $A = \{a_1, \ldots, a_n\}$ of 3-PARTITION with target sum $t = (\sum_{a \in A} a)/(n/3)$. We build two "double wheels" where the spokes are uncrossable edges, which has a unique planar drawing. The first wheel has $\sum_{a \in A} a$ spokes, and the second wheel has n parts. Then we represent each a_i by a vertex of degree $a_i + 1$ connected via a_i length-2 paths to the center of the first wheel, and one length-2 path to the center of the second wheel. Finally, we add uncrossable edges between the edges of the two wheels to decompose the plane into $n/3$ regions, where each region has t sectors of the first wheel and three sectors of the second wheel. Each a_i must be placed in one such region, and because there are n such a_i's, we must have exactly three a_i's in each region, whose total value is exactly t. □

5.4.3 Packing Rectangles

In a packing problem, the goal is to pack some given shapes (e.g., rectangles) into a bigger given shape (e.g., rectangles).

> RECTANGLE-RECTANGLE PACKING
> *Instance:* n integer rectangles $R_1, \ldots, R_n$ and a target integer rectangle T.
> *Question:* Can rectangles $R_1, \ldots, R_n$ be packed into T with no overlap? Rotation and translation are allowed. They do not need to cover all of T. (See figure 5.4.)

Packing problems are unusual in that they are *not known to be in NP*. This is because of the continuous degree of freedom in rotation, as illustrated in figure 5.4, so it may be difficult to encode a solution. If we disallow rotation other than multiples of 90° and require integer translation, then RECTANGLE-RECTANGLE PACKING is in NP. Similarly, if the areas match perfectly so that there can be no gaps in the packing, then we cannot possibly rotate by anything other than a multiple of 90°, so again RECTANGLE-RECTANGLE PACKING is in NP. We call the latter problem ***tiling***:

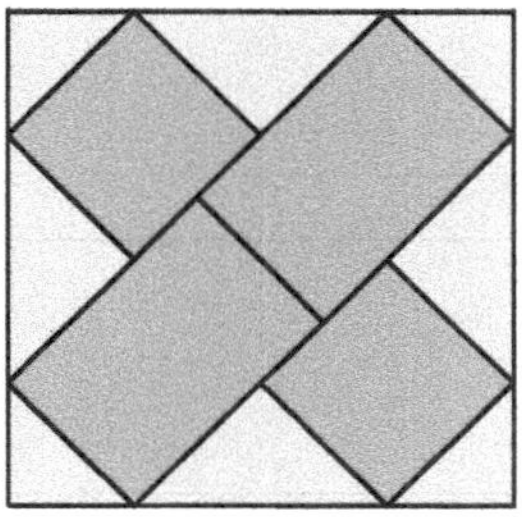

Figure 5.4
An example of rectangle packing.

> RECTANGLE-RECTANGLE TILING
> *Instance:* n integer rectangles $R_1, \ldots, R_n$ and a target integer rectangle T, where $\sum_{i=1}^{n} \text{area}(R_i) = \text{area}(T)$.
> *Question:* Can the n rectangles be packed into T with no overlap and no gaps? Rotation and translation are allowed.

Eternity [Wike] is a complicated tiling puzzle that launched in 1999 with an offer of £1,000,000 for the first solution. It sold around 500,000 copies, at £35 each. It was solved by two Cambridge mathematicians (working together) Alex Selby and Oliver Riordan. Hence, one reason to study this topic is in case more money is offered for harder puzzles. Alas, we will show that such problems are generally hard.

Demaine and Demaine [DD07] showed that RECTANGLE-RECTANGLE TILING is strongly NP-complete:

Theorem 5.16. *RECTANGLE-RECTANGLE TILING is strongly NP-complete.*

Proof. We reduce from 3-PARTITION. For illustrative purposes, we first reduce to a variant of RECTANGLE-RECTANGLE TILING where rotation is forbidden. Refer to figure 5.5.

1. Input $A = \{a_1, \ldots, a_n\}$. Let $t = \sum_{i=1}^{n} a_i/(n/3)$.
2. For $1 \le i \le n$, we make an $1 \times a_i$ rectangle R_i, where 1 is the height.
3. The target rectangle T is $\frac{n}{3} \times t$.

We view the target rectangle T as being divided into $\frac{n}{3}$ subrectangles of dimensions $1 \times t$. Without rotation, each $1 \times a_i$ rectangle R_i must be placed in one of these subrectangles. Because the areas forbid any gaps, every subrectangle must be filled exactly by such rectangles R_i, R_j, R_k such that $a_i + a_j + a_k = t$. Thus a tiling is equivalent to a 3-partition.

What if we allow rotation in the problem? Then we can effectively prevent rotation in the reduction by scaling the 3-PARTITION instance so that every a_i is bigger than $n/3$. Then each

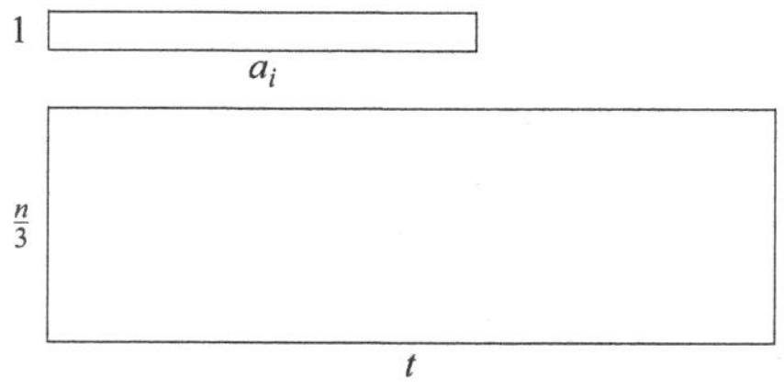

Figure 5.5
Reduction from 3-PARTITION to RECTANGLE-RECTANGLE PACKING without rotation.

$1 \times a_i$ rectangle must remain horizontal, because it does not fit vertically (and the lack of gaps force it to be horizontal or vertical). To achieve this property, we can multiply all the a_i's (and t) by $n/3+1$. Alternatively, because we know every group of a_i's summing to t must have size 3, we can add $n/3$ to every a_i (and add n to t). □

In a similar vein: What about packing squares into a rectangle? Or packing squares into squares?

> SQUARE-RECTANGLE PACKING
> *Instance:* n squares and a target rectangle R.
> *Question:* Can the squares be packed into R without overlap? Rotation is forbidden (though we will allow it later). The squares do not need to cover the entire area of R.

Li and Cheng [LC89] showed the following.

Theorem 5.17. *SQUARE-RECTANGLE PACKING is strongly NP-hard.*

Proof sketch. We reduce from 3-PARTITION, as illustrated in figure 5.6.

1. Input $A = \{a_1, \ldots, a_n\}$. Let $t = (\sum_{i=1}^{n} a_i)/(n/3)$.
2. Let B be a large number to be named later.
3. The n squares have side lengths $a_1 + B, \ldots, a_n + B$.
4. The target rectangle T is $(3B+t) \times (B+t)\frac{n}{3}$. We think of T as consisting of $\frac{n}{3}$ subrectangles that are $(3B+t) \times (B+t)$.

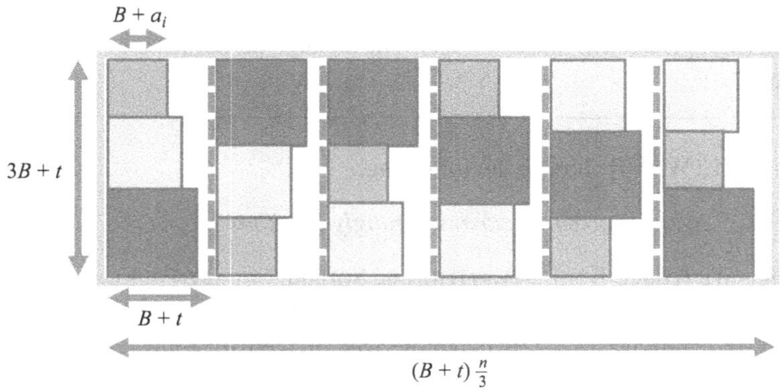

Figure 5.6
Reduction from 3-PARTITION to SQUARE-RECTANGLE PACKING.

We sketch why this reduction works.

Suppose that A is in 3-PARTITION. By renumbering, we can assume that the partition is $\{a_1, a_2, a_3\}$, $\{a_4, a_5, a_6\}$, ..., $\{a_{n-2}, a_{n-1}, a_n\}$. Because $a_1 + a_2 + a_3 = t$, we have $(a_1 + B) + (a_2 + B) + (a_3 + B) = 3B + t$. Thus we can pack the three squares of side lengths $a_1 + B, a_2 + B, a_3 + B$ into the first $(3B + t) \times (B + t)$ subrectangle. Repeating this process for each subrectangle, we obtain a packing.

Now suppose that there is a way to pack the squares into rectangle T. We set B to be much larger than $\sum_{a \in A} a = t \cdot \frac{n}{3}$. Thus each square is roughly $B \times B$, and the target rectangle is roughly $3B \times B \cdot \frac{n}{3}$. Because the number of squares is n, the packing must roughly consist of a $3 \times \frac{n}{3}$ grid of $B \times B$ squares. The columns of this grid correspond to the subrectangles. Thus we can assume that each subrectangle contains three full squares, stacked on top of each other. By renumbering, assume that the first subrectangle contains the $a_1 + B$, $a_2 + B$, and $a_3 + B$ squares. Because the height of the subrectangle is $3B + t$, we have that $(B + a_1) + (B + a_2) + (B + a_3) \leq 3B + t$, which is equivalent to $a_1 + a_2 + a_3 \leq t$. Because this reasoning holds for all of the $\frac{n}{3}$ subrectangles, and the sum of a_i's is $t \cdot \frac{n}{3}$, we must in fact have $a_1 + a_2 + a_3 = t$.

Warning: Squares may not be cleanly split between subrectangles, and instead span two adjacent subrectangles. For example, the first subrectangle might have part of a fourth rectangle in it, using some of the leftover space on the right of the subrectangle. The space savings here will be minor, though, because B is large, and it is possible to rearrange any valid packing to put each square in just one subrectangle. Details are left to the reader. □

> SQUARE-SQUARE PACKING
> *Instance:* n squares $S_1, \ldots, S_n$ and a target square T.
> *Question:* Can squares $S_1, \ldots, S_n$ be packed into T without overlap? Rotation is forbidden (though we will allow it later). The squares do not need to cover the entire area of T.

Leung et al. [LTW+90] showed the following:

Theorem 5.18. *SQUARE-SQUARE PACKING is strongly NP-hard.*

Proof sketch. We reduce from 3-PARTITION to SQUARE-SQUARE PACKING, building on the reduction from 3-PARTITION to SQUARE-RECTANGLE PACKING of theorem 5.17. The main construction, shown in figure 5.7, is parameterized by a positive integer x:

1. One large square of side length $x(x + 1) - 1$.
2. x squares of side length $x + 1$.

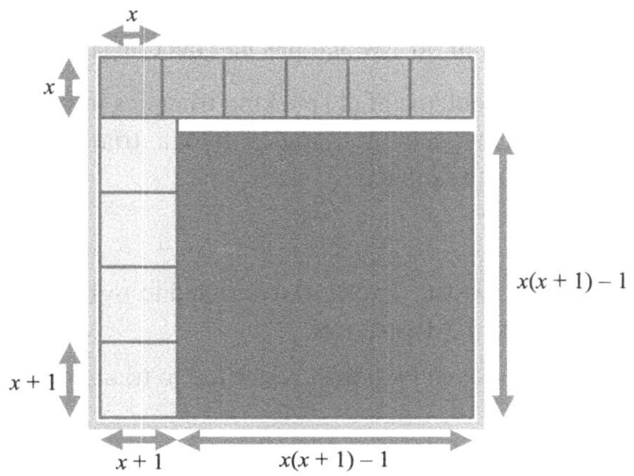

Figure 5.7
Reduction from 3-PARTITION to SQUARE-SQUARE PACKING.

3. $x+2$ squares of side length x.
4. The target square has side length $x(x+2)$.

For the purposes of packing remaining squares into the target square, we claim that the best packing of these squares is to leave a $1 \times x(x+1)-1$ empty rectangle, as shown in figure 5.7. Scaling this construction by $3B+t$, the empty rectangle now has dimensions $(3B+t) \times (3B+t)(x(x+1)-1)$. By setting x to roughly $\sqrt{n}/3$, we get an empty rectangle of rough dimensions $(3B+t) \times (B+t)\frac{n}{3}$. Then the construction and proof of theorem 5.17 applies. □

Exercise 5.19. Fill in the details of the proof of theorem 5.18.

We can allow rotation by changing packing into tiling. This also gives us a problem in NP.

> SQUARE-SQUARE TILING
> *Instance:* n squares $S_1, \ldots, S_n$ and a target square T, where $\sum_{i=1}^{n} \text{area}(S_i) = \text{area}(T)$.
> *Question:* Can squares $S_1, \ldots, S_n$ be packed into T with no overlap and no gaps? Rotation is allowed.

Corollary 5.20. *SQUARE-SQUARE TILING is strongly NP-complete.*

Proof. We reduce from SQUARE-SQUARE PACKING. We compute $d = \sum_{i=1}^{n} \text{area}(S_i) - \text{area}(T)$, and add d squares of dimension 1×1. These squares can fill the gaps in any packing in the SQUARE-SQUARE PACKING instance. The matching areas force a tiling, which prevents rotation other than by multiples of 90°, which has no effect on squares. □

Chou [Cho16] studied the problems of (1) packing triangles into a rectangle (TRIANGLE-RECTANGLE PACKING) and (2) packing triangles into a triangle (TRIANGLE-TRIANGLE PACKING). The triangles cannot be rotated.

Exercise 5.21.

1. Show that TRIANGLE-RECTANGLE PACKING, restricted to right triangles, is NP-hard.
   Hint: Use a reduction from 3-PARTITION.
2. Show that TRIANGLE-TRIANGLE PACKING, restricted to triangles being right or equilateral, is NP-hard.
   Hint: Use a reduction from 3-PARTITION.

5.4.4 DISK PACKING

> DISK PACKING
> *Instance:* A set of disks and a target square.
> *Question:* Can you place the disks so that they do not overlap and all of their centers lie within the square?

This problem is interesting in and of itself. It also comes up as a special case of computational origami design, which is how Demaine, Fekete, and Lang [DFL10] came to study it. They proved the following:

Theorem 5.22. *DISK PACKING is strongly NP-hard.*

Proof sketch. We give a reduction from 3-PARTITION to DISK PACKING. Suppose the 3-PARTITION instance is $A=\{a_1,\ldots,a_n\}$, and let $t=(\sum_{a\in A} a)/(n/3)$. Allowing rational numbers instead of just integers, we scale the instance by $1/t$ so that $t=1$ in the new instance. We will output an instance of DISK PACKING.

As a warm-up, we consider packing an equilateral triangle rather than a square, as illustrated in figure 5.8. We set up a triangular grid of large equal-radius disks and then build 3-PARTITION gadgets in the space between them, each of which will represent one of the subsets in our partition. There are three kinds of disks:

1. Large disks. In the example of figure 5.8, there are ten of them, but in general their number is a triangular number of the form $k(k+1)/2$. These disks are all the same size, chosen so that they are forced to be packed in a triangular grid as in figure 5.8.

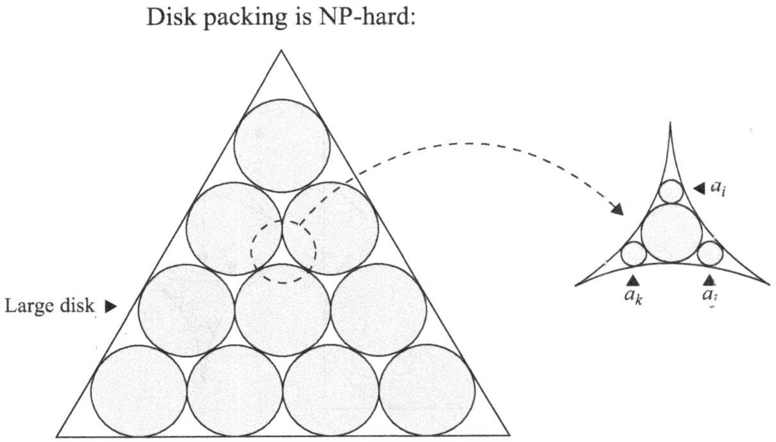

Figure 5.8
Disk packing in a triangle.

We scale the packing so that the large disks have radius 1. We call the circular-arc triangular regions in the cracks between three large disks ***pockets***.

2. Medium disks. For each pocket, we place a medium disk in the center. These disks are all the same size, almost large enough that they are forced to touch all three large disks of the pocket (and thus be in the center of the pocket). In fact, we reduce the medium-disk radius by $1/n$ to leave a little bit of wiggle room.
3. Small disks. For each a_i, we have a disk designed to go in a space between two large disks and one medium disk (as shown in figure 5.8. If the medium disk's radius was not reduced, the small disk would have a natural size that forces it to touch all three disks. We *increase* the radius of the a_i small disk by $\frac{a_i}{n} - \frac{1}{n^2}$. This ends up forcing that the three small disks (say, a_i, a_j, a_k) in the same pocket as a medium disk satisfy that $a_i + a_j + a_k = 1 = t$, essentially because the medium disk is $1/n$ smaller than natural while the small disks are a_i/n larger than natural.

   Working this out requires some subtlety because circles are not straight, so sums do not map so clearly to the dimensions of circles. But everything is linear to the first order, and we can use Taylor expansions to calculate the proper radii. In particular, we shrink each small circle by an amount $-1/n^2 + a_i/n$ because the $1/n^2$ term cancels higher-order terms in the expansion.

Finally, we extend to packing into a square instead of a triangle, as illustrated in figure 5.9. The idea is to fill up the grid with lots of circles of decreasing sizes in $O(\log n)$ steps, which constrains the gadgets to sit in the right place. By considering circles in decreasing

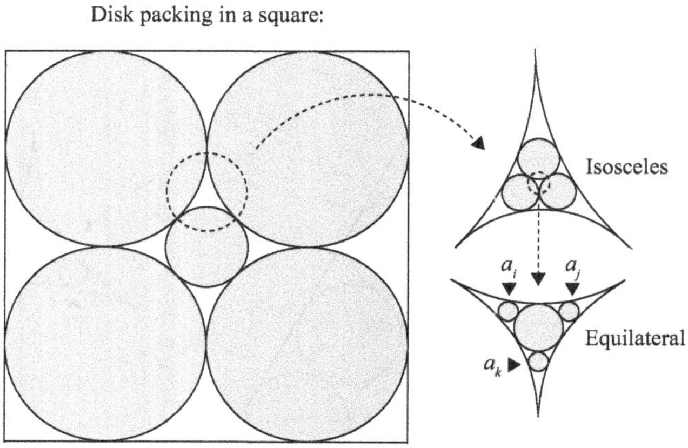

Figure 5.9
How to adapt disk packing to a square instead of a triangle.

size, we can be convinced at every stage that infrastructure circles are constrained. In this way, we again obtain $n/3$ identically sized pockets, as needed for the rest of the proof. □

5.5 Puzzles

In this section, we show that four types of puzzles are equivalent and all strongly NP-complete. These results are by Demaine and Demaine [DD07].

5.5.1 Edge Matching

In an edge-matching puzzle, you need to arrange several given pieces to match up the patterns across edges where pieces meet. The first edge-matching puzzle was patented by E. L. Thurston in 1892 for use as advertising [SB86, Ste07].

Eternity II [Wikd] was a complicated edge-matching puzzle that launched in 2007 with an offer of £2,000,000 for the first solution before December 31, 2010 (it was made by the same people who made *Eternity*). We have not been able to find out how much it sold for or how many sold, but it is available on eBay for around $40. *Eternity II* was never solved (though of course the designer knows how to solve it). However, Louis Verhaard found a partial solution where he placed 467 of the 480 pieces, for which he received $10,000. So a good reason to study these puzzles is that they may be lucrative. On the other hand, we will show that solving them is hard.

> Edge Matching
>
> *Instance:* A multiset of n unit squares with the edges colored, and a target rectangle T of area n. (Such tiles are called ***Wang Tiles***; see also section 15.8.) We also require that there are perimeter(T) edge colors that appear exactly once, so must appear on the boundary of T.
>
> *Question:* Is there a packing of the n squares into T (filling the entire space) such that all tiles sharing an edge have matching colors? There are two versions, depending on whether tiles can be rotated, but this will not affect the complexity.

Because all the squares are of equal unit size, we cannot immediately reduce from 3-Partition. Instead, we will first construct gadgets composed of tiles that are forced to be joined together in a particular way.

Theorem 5.23. *Edge Matching is NP-complete, with or without rotation.*

Proof. Clearly Edge Matching ∈ NP. We show 3-Partition $\leq_p$ Edge Matching.
More specifically, here is the reduction:

1. Input $a_1, \ldots, a_n$. Let $t = (\sum_{i=1}^{n} a_i)/(n/3)$.
2. Construct the ***frame gadget*** shown in figure 5.10 to have a hole of dimensions $\frac{n}{3} \times t$. The boundary edges of all frame tiles use uniquely occurring colors, so all remaining

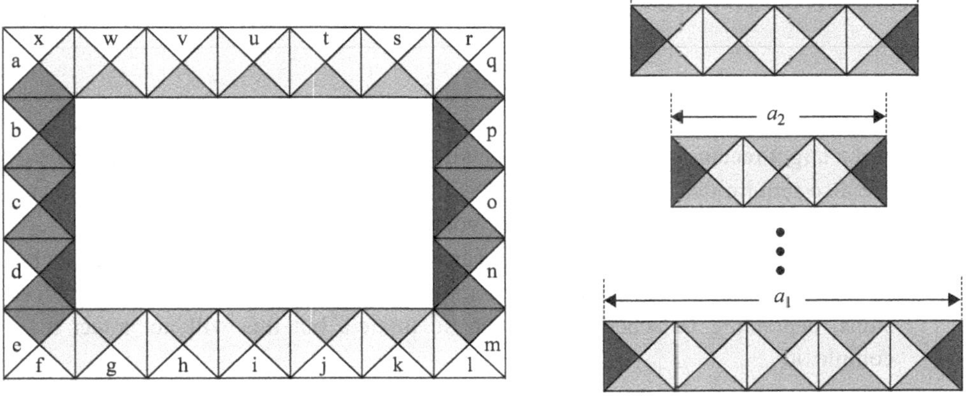

Figure 5.10
The frame gadget and the a_i gadgets for Edge Matching.

edges *must* be matched with another tile. The internal edges of the frame use colors that appear in pairs, so they must be glued to each other, which forces the shape of the frame. The frame's inner horizontal edges have a single color h, and the inner vertical edges have another single color v. This coloring will prevent tiles from rotating even if it is allowed.

3. For each $1 \leq i \leq n$, construct the ***a_i gadget*** consisting of a_i colored tiles as in the right part of figure 5.10. Color the tiles so that they must be joined together into a $1 \times a_i$ rectangle, by setting the inner edge colors to appear exactly twice, so they must be glued to each other. The outer horizontal edges are all colored h, while the outer vertical edges are all colored v, so they match each other and the frame gadget.
4. The dimensions of T are that of the frame, $(\frac{n}{3} + 2) \times (t + 2)$.

Thus we effectively simulate the reduction from 3-PARTITION to RECTANGLE-RECTANGLE TILING without rotation, from theorem 5.16.

The constructed instance of EDGE MATCHING has $\Theta(\sum_i a_i)$ tiles. For the reduction to remain polynomial, we need that each a_i is polynomial in the input size n. The strong NP-hardness of 3-PARTITION guarantees that it is still NP-hard under this restriction. (If we tried to reduce from PARTITION, it would not work because a_i could be exponential.) □

5.5.2 SIGNED EDGE MATCHING

In SIGNED EDGE MATCHING, we again must pack unit tiles with colored edges into a target rectangle. However, now colors come in pairs: a matches A, b matches B, and so on. A color no longer matches with itself, but instead must be matched with a buddy color. This setup is common in real-world edge-matching puzzles, where color a might be the head of a lizard and color A is the tail of the same lizard.

> SIGNED EDGE MATCHING
>
> *Instance:* A set of n unit squares with the edges colored, a perfect pairing of the colors, and a target rectangle T of area n. We will call two paired colors ***complementary***. We also require that there are perimeter(T) edge colors whose complement does not appear, so they must appear on the boundary of T.
>
> *Question:* Is there a packing of the n squares into T (filling the entire space) such that all tiles sharing an edge have complementary colors? There are two versions, depending on whether tiles can be rotated, but this will not affect the complexity.

Theorem 5.24. *EDGE MATCHING $\leq_p$ SIGNED EDGE MATCHING, so SIGNED EDGE MATCHING is NP-complete.*

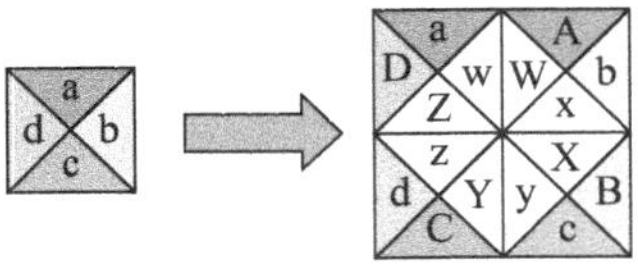

Figure 5.11
Creating a 2 × 2 supertile of signed tiles from an unsigned tile. The internal colors are unique to this supertile.

Proof. Here is the reduction.

1. Input a set of colored unit tiles and a rectangle T. We refer to these unit tiles as ***unsigned tiles***.
2. For each unsigned tile, create a 2 × 2 ***supertile*** of signed tiles as shown in figure 5.11. These will be the tiles for our instance of SIGNED EDGE MATCHING. The internal edges of the supertile use paired colors that appear only once, so the tiles are forced to assemble into the supertiles (because the perimeter is already taken).
3. The output target rectangle is T with each side length doubled. □

5.5.3 JIGSAW

Consider a classic jigsaw puzzle, except with no image on the pieces, and ambiguous mates: There may be more than one piece that fits any particular tab or pocket. Such a jigsaw puzzle is very similar to a signed color-matching puzzle. In the jigsaw puzzle, instead of pairs of colors, we have tabs and pockets. A tab matches a pocket if and only if its shape is the exact inverse of the other. Some of the pieces have flat edges on one or two sides, which must be placed at the rectangular border, just like the edge colors whose complement does not appear.

Exercise 5.25.

1. Define JIGSAW rigorously.
2. Prove theorem 5.26.

Theorem 5.26. *SIGNED EDGE MATCHING $\leq_p$ JIGSAW, so JIGSAW is strongly NP-complete.*

5.5.4 POLYOMINO TILING

Definition 5.27. A ***polyomino*** is a finite set of unit squares glued together edge to edge to form a connected shape. See figure 5.12 for some examples.

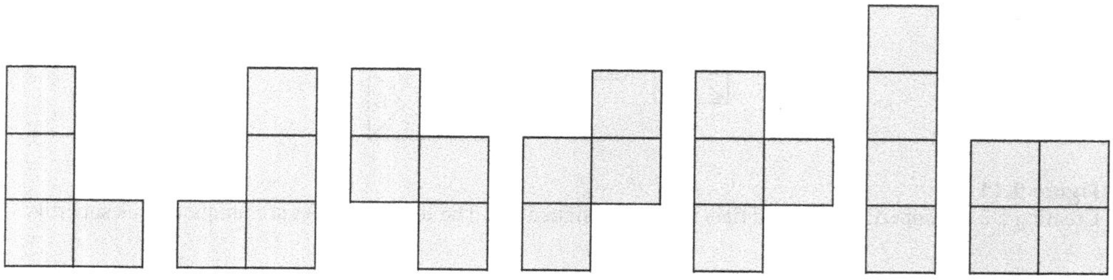

Figure 5.12
The seven polyominoes with four squares that are used in *Tetris*, called ***tetrominoes***.

> POLYOMINO TILING
> *Instance:* A multiset of n polyominoes $P_1, \ldots, P_n$ and a target rectangle T, where $\sum_{i=1}^{n} \text{area}(P_i) = \text{area}(T)$.
> *Question:* Can the polyominoes be packed into T with no overlap and no gaps? Rotations are allowed.

POLYOMINO TILING is a generalization of RECTANGLE-RECTANGLE TILING to where each piece is a polyomino. We already know POLYOMINO TILING is NP-hard because it contains RECTANGLE-RECTANGLE PACKING as a special case. However, the following reduction shows the same hardness even when all polyominoes are *small*.

Theorem 5.28.

1. *JIGSAW* $\leq_p$ *POLYOMINO TILING with pieces of area* $O(\log^2 n)$.
2. *POLYOMINO TILING with pieces of area* $O(\log^2 n)$ *is strongly NP-complete. (This follows from part 1 and that POLYOMINO TILING is clearly in NP.)*

Proof. Here is the reduction.

1. Input a set of jigsaw pieces (or equivalently, signed tiles) and a target rectangle T.
2. Map each tab/pocket shape pair to a unique binary string b of length L. There are $O(n)$ such shape pairs, so $L = O(\log n)$ suffices.
3. For each jigsaw piece, create a polyomino as follows; figure 5.13 gives an example. Start with a solid square of side length $4 + L$. For jigsaw sides with tab type b, append unit squares along the polyomino side corresponding to the position of 1's in b, going in a clockwise direction. For pockets of type b, make unit square cuts into the polyomino, going in a counterclockwise direction. Leave each 2×2 square at the corner alone as padding. In this way, two polyomino edges fit together with no blank space if and only if their corresponding jigsaw edges fit together. Because the areas enforce

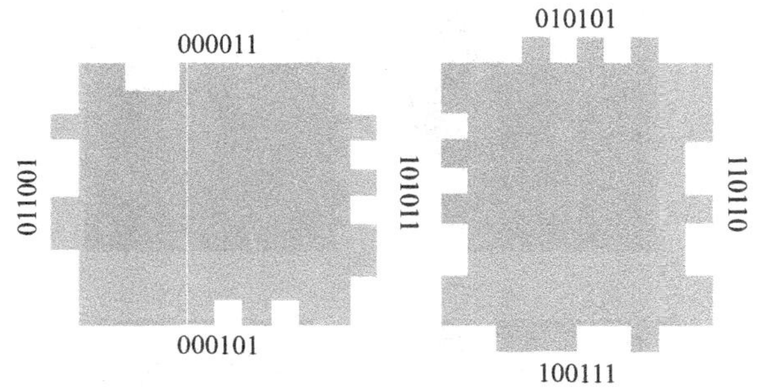

Figure 5.13
Encoding of jigsaw pieces into polyominoes.

a tiling, any blank space left between two polyominoes will be unfillable by our large polyominoes. Flat jigsaw edges map to flat polyomino edges. The polyominoes have dimensions $O(\log n) \times O(\log n)$, and thus area $O(\log^2 n)$.

4. Scaling the target rectangle's dimensions by a factor of $4+L$ completes the reduction. □

A very similar reduction was given earlier by Solomon Golomb [Gol70].

Exercise 5.29. Show that JIGSAW $\leq_p$ POLYOMINO TILING with pieces of area $O(\log n)$.
Hint: See figures 5.14. In figure 5.14b, the gray area represents free pixels which may be used to encode bits of the pocket. The black areas always remain solid and guarantee that the polyominoes are connected. Likewise the white areas determine the shape of the "key" and also are shaped to remain connected. In the scheme depicted, a square with side length $8r+5$ has $8r^2$ gray bits available for each edge.

5.5.5 Closing the Loop

Though it does not give any extra complexity results, for fun, we close the loop of puzzle reductions:

Theorem 5.30.

1. *POLYOMINO TILING $\leq_p$ EDGE MATCHING.*
2. *3-PARTITION $\leq_p$ EDGE MATCHING $\leq_p$ SIGNED EDGE MATCHING $\leq_p$ JIGSAW $\leq_p$ POLYOMINO TILING $\leq_p$ EDGE MATCHING where all reductions have constant-factor blowup except for JIGSAW $\leq_p$ POLYOMINO TILING which adds a logarithmic factor.*

Proof. The following is a slight modification of the reduction from 3-PARTITION to SIGNED EDGE MATCHING. As in theorem 5.23, we create a frame gadget and a gadget for

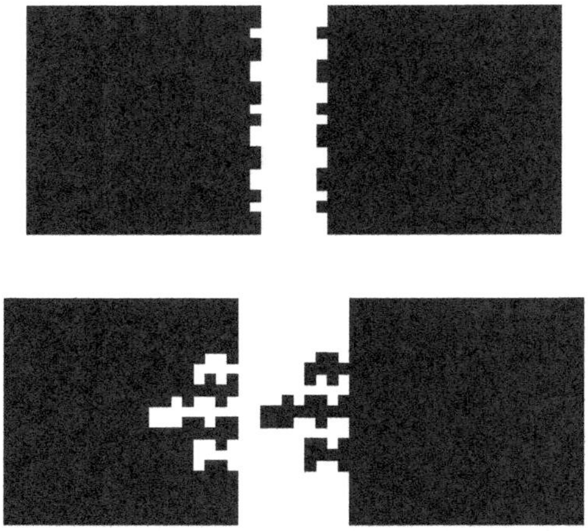

(a) Surface keys versus deeper keys

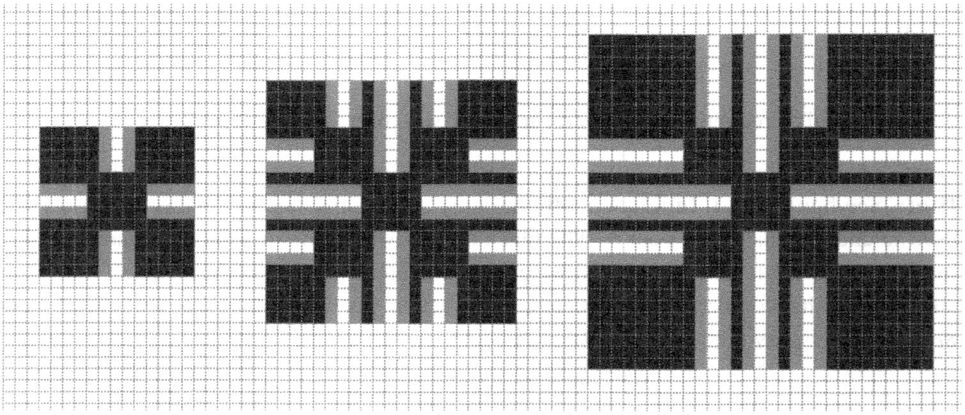

(b) Schematic of pocket holes

Figure 5.14
Reducing polyomino areas to $O(\log n)$ using deeper keys.

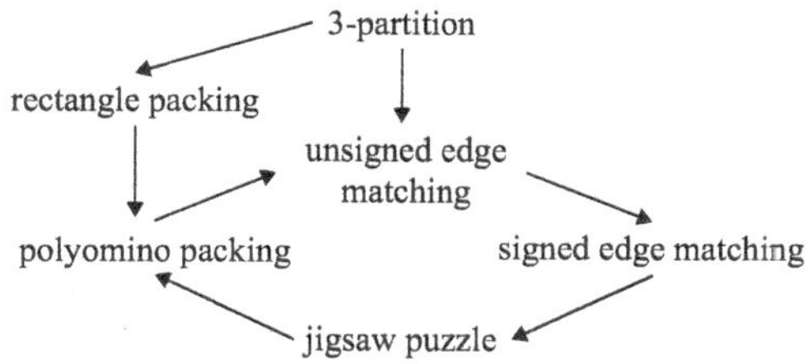

Figure 5.15
Each arrow represents a reduction from one problem to another.

each polyomino piece. Previously, the piece gadgets were $1 \times x$ rectangles. Now, we create arbitrary polyominoes by fusing tiles together in the same shape as the polyomino, with unique color pairs for each internal edge to force them to glue together. Lastly, to allow for polyomino pieces to rotate, we make h and v the same color. □

Thus, all four puzzle types are efficiently reducible to one another and strongly NP-complete. Figure 5.15 summarizes all the reductions in this section.

5.5.6 Further Results

Follow-up work by Bosboom et al. [BDD$^+$17, BCC$^+$20] has extended the edge-matching and jigsaw results in a few directions:

1. If the target rectangle is $1 \times n$ (essentially one-dimensional), then the problem is still NP-complete [BDD$^+$17].
2. It is also NP-complete to find approximate solutions; see section 9.7.1.
3. If edge matching does not require exact matching (=), but instead we have strict inequalities (<) between adjacent edge colors (numbers), then the problem is NP-complete [BCC$^+$20].
4. If edge matching does not require exact matching (=), but instead we have weak inequalities (≤) between adjacent edge colors (numbers), then the problem is in P [BCC$^+$20].
5. Triangular edge-matching puzzles are NP-complete [BCC$^+$20].
6. Two-player versions of edge matching with square tiles are PSPACE-complete [BCC$^+$20].

5.6 Computer Games

5.6.1 CLICKOMANIA

Clickomania [Wikg] is a computer puzzle game. We describe it informally. The board consists of an $m \times n$ rectangle of blocks, where each block has a color. A move consists of clicking on a contiguous group of $n > 1$ blocks of the same color, which is thereby destroyed. Blocks fall column by column to fill empty space, and empty columns are removed by pulling the two sides together. The goal is to remove all blocks. CLICKOMANIA is the problem of determining whether the player can win on a given board.

Biedl et al. [BDD+02] showed that CLICKOMANIA with only one column or only one row is in P, by reduction to a context-free grammar. They also showed:

Theorem 5.31. *3-PARTITION $\leq_p$ CLICKOMANIA, so CLICKOMANIA is NP-complete. This result holds even when CLICKOMANIA is restricted to (a) 2 columns and 5 colors, or (b) 5 columns and 3 colors.*

Proof sketch. We sketch the 2 column/5 color reduction, which is illustrated in figure 5.16. The right column encodes the a_i's, and the left column encodes the desired sum t of each partition. The left column is mostly a symmetric checkerboard pattern, interspersed with $n/3$ red squares; the only way to clear the entire column is to clear all the red squares (and thereby solve 3-PARTITION). Clicking on blocks in the right column corresponds to choosing elements for one subset A_i; when that subset reaches the desired sum, a red block in the left column will line up with a red block in the right column, allowing us to clear both. If we fail to construct sums of t, some red blocks will remain to the end. □

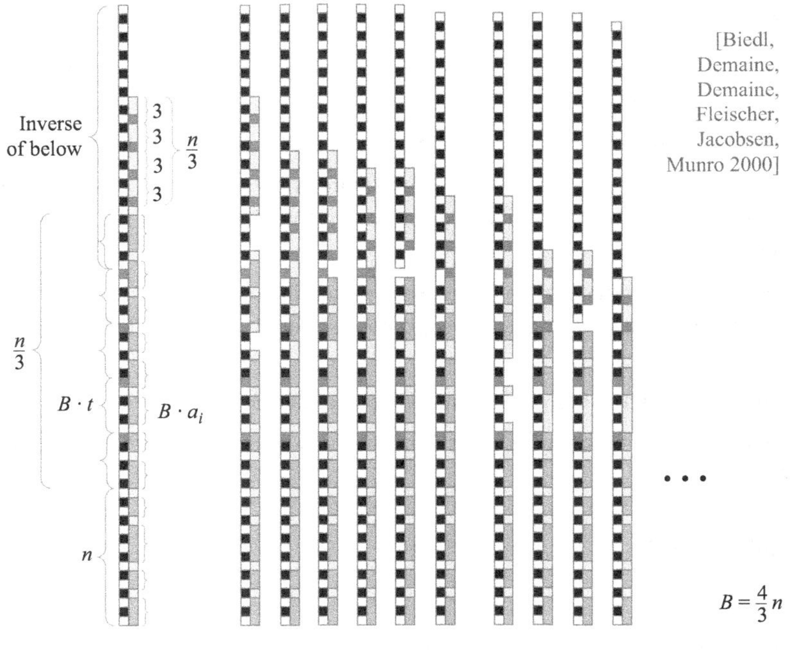

Figure 5.16
Reduction from 3-PARTITION to CLICKOMANIA with two columns and five colors.

More recently, Adler et al. [ADH+17] improved this result to show that CLICKOMANIA is NP-hard even for just two columns and two colors.

5.6.2 TETRIS

Tetris is a computer game played on a rectangular board with falling tetrominoes, which are the polyominoes we saw in figure 5.12. Each block can be rotated as it falls from the sky, and filled rows disappear. You lose if the pile of blocks ever reaches the top edge of the board (the "sky").

> TETRIS
> *Instance:* An initial board and a sequence of future pieces.
> *Question:* Is there a sequence of moves for these pieces that avoids losing?

TETRIS is the perfect-information version of *Tetris*: We assume we know everything about the current and future states of game. The actual video game is an online problem where you must make decisions with incomplete information about the future. Intuitively, TETRIS is easier than the real game. Even so, Breukelaar et al. [BDH+04] showed the following.

Theorem 5.32. *3-PARTITION $\leq_p$ TETRIS, so TETRIS is NP-complete.*

Proof sketch. We reduce from 3-PARTITION.

The initial board configuration is shown in figure 5.17. It contains gadgets called ***buckets***, each $\Theta(t)$ call. Filling these buckets corresponds to choosing a_i to fill your subsets. Once these buckets are filled, a single T piece can unlock an empty column to the right of the playing field, which can be filled by I pieces. Without unlocking this column, it is impossible to survive.

We represent each a_i by the sequence of $2a_i + 3$ pieces shown in figure 5.18. The first piece makes a flat bottom in some bucket, which enables the next $2a_i$ pieces to perfectly fill some of the bucket. The final two pieces reset the bucket to have a bumpy bottom, preventing it from being filled more. By case analysis, we can show that it is not possible to split the pieces in the a_i gadget across multiple buckets without leaving some gaps, which we cannot afford. Thus we need to partition the a_i's among the buckets, which corresponds to a solution to 3-PARTITION. □

Note that this proof could also reduce from PARTITION. But because we end up encoding the a_i in unary (necessarily so because TETRIS does not have any numbers), we need to use the strongly NP-hard 3-PARTITION problem instead of the weakly NP-hard PARTITION problem.

Recently, Asif et al. [ACD+20] adapted this proof to show that TETRIS is NP-complete even when restricted to 4 rows or 8 columns. They also showed that *Tetris* is NP-complete

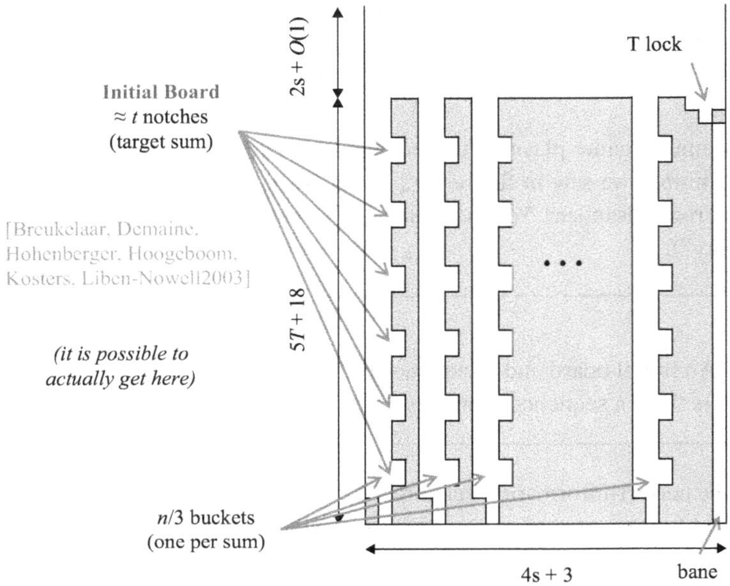

Figure 5.17
Initial board for *Tetris*.

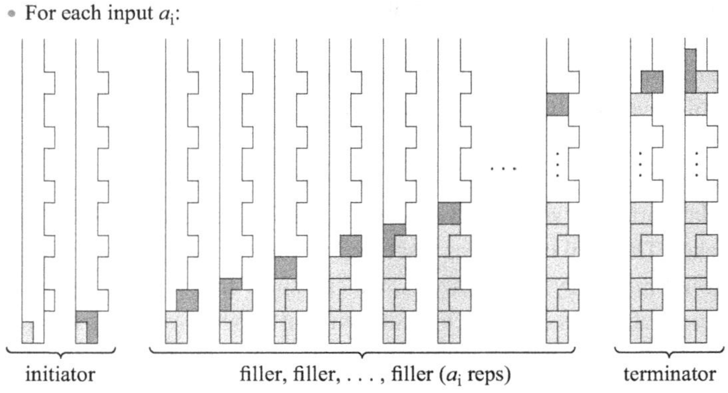

Figure 5.18
Gadget representing a_i in *Tetris*.

starting from an initially empty board, if we generalize to arbitrary polyomino pieces (of size at most 65), even when restricted to 8 columns. Demaine et al. [DDE+17] also analyzed *Tetris* where all pieces are polyominoes of size k, for $k = 1, 2, 3, \ldots$.

More recently, the MIT Hardness Group [MIT24d] analyzed *Tetris* with a more limited set of tetromino piece shapes. In particular, they proved that Tetris remains NP-complete when restricted to any two out of the seven piece shapes from figure 5.12. They also showed how to avoid the last-minute piece slides in the proof above.

5.7 Further Results

We give a partial list problems that were proven strongly NP-complete by a reduction from 3-PARTITION or related problems covered in this chapter. You can either consider these as exercises (so to not look up the references) or recommended reading (so do look up the references). Some we state as function problems; the reader can translate them to the corresponding decision problems for reductions.

1. (Mauro Dell'Amico and Silvano Martello [DM99]) BIN PACKING (Natural Number Version). Given $a_1, \ldots, a_n \in \mathbb{N}$ and bin capacity $C \in \mathbb{N}$, find the least number of bins B so that $a_1, \ldots, a_n$ can be packed into B bins of capacity C.
2. (Bernstein et al. [BRG89]) SCHEDULING PROBLEMS FOR PARALLEL/PIPELINED MACHINES. Imagine that there are two processors P_1, P_2 and two types of jobs, so that P_j can process jobs of type j. The problem: Given n jobs $J = \{J_1, J_2, \ldots, J_n\}$ where each job is represented by $J_i = (K_i, T_i, D_i, R_i)$ where K_i is the job type, T_i is the execution time in unary, D_i is the time a job needs before it begins processing (called the delay time), and R_i resource requirement (this is either 0 or 1 and you cannot have two jobs with resource requirement 1 being processed at the same time). We are also given $G = (J, E)$ be the precedence graph that specifies which job should be executed before the other one. Find a way to schedule the jobs so as to minimize the total completion time of all jobs.
3. (Cieliebak and Eidenbenz [CE04]) APPROXIMATE PARTIAL DIGEST. Given $m \in \mathbb{N}$ and $\delta \in \mathbb{Q}$, $\delta > 0$, and a multiset D of $\binom{m}{2}$ natural numbers, do there exist m points on a line such that the multiset of $\binom{m}{2}$ pairwise distances has a bijection to D such that, if $|p - q|$ maps to d, then $d - \delta \le |p - q| \le d + \delta$? The above problem is additive error. They also showed that the problem for multiplicative error is reducible from 3-PARTITION.
4. (Formann and Wagner [FW90]) VLSI LAYOUT. Given a graph G of maximum degree 4, and $A \in \mathbb{N}$, can G be embedded in a grid with area $\le A$? (There are variants of this problem that are also strongly NP-complete.)
5. (Abel and Demaine [AD11]) EDGE UNFOLDING. Given a polyhedron, can it be cut along its edges to unfold it into a connected flat piece with no overlap? This problem has

numbers in it because the vertices have coordinates. It turns out to be strongly NP-complete by a reduction from SQUARE-SQUARE TILING.

6. SIMPLE FOLDING. Given a crease pattern on a piece of paper, is there a sequence of simple (one-line) folds that ends up folding exactly the creases in the pattern? There are many versions of this problem, depending on the exact definition of "simple fold" and on the allowed shape of the piece of paper. Arkin et al. [ABD+04] showed that many versions are weakly NP-hard by a reduction from PARTITION. Akitaya et al. [ADK17] later improved these results to strong NP-hardness by a reduction from 3-PARTITION. Many additional versions were later shown strongly NP-hard [AAB+20, ABD+23].

6

Exponential Time Hypothesis

6.1 Introduction

If Alice proved that $\mathsf{P} \neq \mathsf{NP}$ by showing that 3SAT required at least $n^{\Omega(\log\log\log(n))}$ (where n is the number of variables), then she would still win the Millennium prize; however, the result would not really tell us what we want to know. All NP-complete problems would be proven to take at least $n^{\Omega(\log\log\log(n))}$ which is a much weaker lower bound than what we believe to be true. In this chapter we will look at conjectures that more precisely tell us how hard we think 3SAT and kSAT are.

Chapter Summary

1. *We will look at the best known algorithms for kSAT and use them to guide us to a conjectures about how fast kSAT can be solved. The conjectures are called the* ***exponential time hypothesis*** *(ETH) and the* ***strong exponential time hypothesis*** *(SETH). We will only use SETH in the Further Results section.*
2. *We will show that, assuming ETH, many problems have run time $2^{\Omega(L)}$, where L is the length of the input. (We do not use n since n is usually the number of variables for a boolean formula.)*
3. *We will show that, assuming ETH, some problems have run time $2^{\Omega(\sqrt{L})}$, where L is the length of the input.*

6.2 Conjectures on the Runtime for kSAT

What is a good conjecture for the runtime of 3SAT? Of kSAT? Let's turn this around: What are the best known algorithms for 3SAT? For kSAT?

1. Makino et at. [MTY13] have obtained a deterministic $O(1.3303^n)$ algorithm for 3SAT.

2. Hertli [Her14] has obtained a randomized $O(1.308^n)$ algorithm for 3SAT.
3. Dantsin et al. [DGH$^+$02] have obtained a deterministic $O((2-\frac{1}{k+1})^n)$ algorithm for kSAT.
4. Paturi et al. [PPZ99] have obtained a randomized $O(2^{n-(n/k)})$ algorithm for kSAT.

(See the book by Schoning and Toran [ST13] for an overview of SAT algorithms.)

It is plausible that the bounds for 3SAT will be improved to $O(\alpha^n)$ for some $1 < \alpha < 1.3$. However, it seems likely that the bounds will always be of the form $O(\alpha^n)$. An algorithm running in time $O(n^{O(\log n)})$ seems unlikely. It is plausible that the bounds for kSAT will be improved to $O(2^{n-(\alpha n/k)})$ for some $\alpha > 1$. However, it seems likely that as k goes to infinity, the exponent of 2 will have limit n.

We are going to hypothesize that the best algorithm for kSAT has time complexity $2^{s_k n}$ for some s_k. But this is not quite right. It is possible (though, we think, very unlikely) that, for example,

- For all i there is an algorithm for 3SAT that runs in time $O(2^{(0.1+(1/10^i)n)})$.
- There is no algorithm for 3SAT that runs in time $O(2^{0.1n})$.

This (remote) possibility leads to the following notation, which we will need to state the hypothesis.

Notation 6.1. For $k \geq 1$,

$$s_k = \inf\{s \mid \text{ there is an } O(2^{sn}) \text{ algorithm for } k\text{SAT}\}.$$

Looking at the algorithms for kSAT mentioned above, note that (1) they all have time of the form 2^{cn}, and (2) as k goes to infinity, the exponent goes to n. Based on these intuitions, Impagliazzo and Paturi [IP01] formulated the following hypotheses:

Hypothesis 6.2.

1. *The **ETH**: For all $k \geq 3$, $s_k > 0$. Since $s_3 \leq s_4 \leq \cdots$ this is equivalent to $s_3 > 0$. It is also equivalent to 3SAT requiring time $2^{\Omega(n)}$.*
2. *The **SETH**: The sequence $s_3, s_4, \ldots$ converges to 1.*

Note that ETH in the form above uses the *number of variables* as the size of a formula. It turns out that it is better to use the *actual length of the formula*. Happily, these two different hypotheses are equivalent. We discuss this issue in section 6.3 and henceforth will use the *length of formula* formulation.

Based on their names, one would think that SETH $\Rightarrow$ ETH. While this is true, it is not obvious. The interested reader should see the paper by Impagliazzo et al. [IPZ01].

Open Problem 6.3. *Prove or disprove that ETH $\Rightarrow$ SETH.*

6.3 Number of Variables Versus Actual Length

In the ETH (SETH) we are using n, the number of variables, as the length of the input. What if we took the real length of the input? Impagliazzo et al. [IPZ01] showed that such a hypothesis would be equivalent to ETH (SETH). We state the key lemma and then use it to prove the equivalence.

Lemma 6.4. (the sparsification lemma) Let $k \in \mathbb{N}$ and $\varepsilon > 0$. There is a constants c (that depends on k, ε) and an algorithm such that the following hold when the input is a kCNF formula φ on n variables:

1. The output is t kCNF formulas $\varphi_1, \ldots, \varphi_t$ where $t \leq 2^{\varepsilon n}$.
2. $\varphi \in$ SAT if and only if there is an i such that $\varphi_i \in$ SAT.
3. The actual length of φ_i is $\leq cn$.
4. The algorithm runs in time $2^{\varepsilon n} n^{O(1)}$.

Theorem 6.5. *Let ETH' (SETH') be ETH (SETH) only instead of n use the actual length of the input.*

1. *ETH' is equivalent to ETH.*
2. *SETH' is equivalent to SETH.*

Proof. Clearly ETH' implies ETH and SETH' implies SETH. We show ETH implies ETH'. The proof that SETH implies SETH' is similar and left for the reader.

We show that ETH implies ETH' by showing its contrapositive. Assume ETH' is false. Let $\varepsilon > 0$. We give an algorithm for 3SAT that runs in time $2^{\varepsilon n}$ where n is the number of variables. We have parameters ε' and ε'', which we will pick later.

1. Input φ, a 3CNF formula with n variables.
2. Apply the algorithm from lemma 6.4 to obtain t 3CNF formulas $\varphi_1, \ldots, \varphi_t$, where $t \leq 2^{\varepsilon' n}$ such that (1) $\varphi \in$ 3SAT if and only if there is an i such that $\varphi_i \in$ 3SAT, (2) each φ_i is of length $\leq cn$. The algorithm runs in time $2^{\varepsilon n} n^{O(1)}$.
3. Since ETH' is false, there is an algorithm for 3SAT that takes time $2^{\varepsilon'' L}$ where L is the actual length of the formula. Apply this algorithm to each φ_i. This takes time $2^{\varepsilon' n} \times 2^{\varepsilon'' cn}$.
4. If there is some i such that in the last step the algorithm returned YES, then output YES. Otherwise output NO.

This algorithm takes time

$$2^{\varepsilon' n} n^{O(1)} + 2^{(\varepsilon' + \varepsilon'' c)n}.$$

We take $\varepsilon' = \varepsilon/3$ and $\varepsilon'' = \varepsilon/3c$ to get running time

$$2^{(\varepsilon n/3)} n^{O(1)} + 2^{(\varepsilon/3 + \varepsilon/3)n} \leq 2^{\varepsilon n}.$$

□

We now state ETH and SETH in an equivalent but more useful form.

Notation 6.6. Let L be the actual length of a formula. For $k \geq 1$,

$$t_k = \inf\{t \mid \text{there is an } O(2^{tL}) \text{ algorithm for } k\text{SAT}\}.$$

Hypothesis 6.7.

1. *The **ETH**: For all $k \geq 3$, $t_k > 0$. Since $t_3 \leq t_4 \leq \cdots$ this is equivalent to $t_3 > 0$. It is also equivalent to 3SAT requiring time $2^{\Omega(L)}$.*
2. *The **SETH**: The sequence $t_3, t_4, \ldots$ converges to 1.*

From now on when we refer to ETH and SETH this is the version we mean.

6.4 Linear Blowup and Quadratic Blowup Reductions

ETH is stronger than P $\neq$ NP; hence we do not think it will be proven anytime soon. However, we believe that it is true. Assuming ETH we can prove strong lower bounds on many problems. We need some more refined notions of reductions.

Definition 6.8. Let $A \leq_p B$ via f.

1. f is a ***linear blowup reduction*** if $|f(x)| \leq O(|x|)$.
2. f is a ***quadratic blowup reduction*** if $|f(x)| \leq O(|x|^2)$.

Note that for both linear and quadratic blowup the reduction can take nonlinear (but polynomial) *time*. It is the length of the output that is constrained.

Theorem 6.9. *In this theorem L is the length of the length of the input.*

1. *If $A \leq_p B$ with a linear blowup reduction, and $B \leq_p C$ with a linear blowup reduction, then $A \leq_p C$ with a linear blowup reduction.*
2. *Assume ETH. Assume that 3SAT $\leq_p B$ by a linear blowup reduction. Then B requires time $2^{\Omega(L)}$.*
3. *Assume ETH. Assume that 3SAT $\leq_p B$ by a quadratic reduction. Then B requires time $2^{\Omega(\sqrt{L})}$. (This we leave to the reader.)*

4. *Assume ETH. Assume that 3SAT* $\le_p B_1 \le_p \cdots \le_p B_m$ *and that all but one the reductions have linear blowup and the one that does not has quadratic blowup. Then* B_m *requires time* $2^{\Omega(\sqrt{L})}$. *(This follows from part 1 and 2.)*

Proof. 1) Assume $A \le_p B$ via linear blowup reduction f and $B \le_p C$ via linear blowup reduction g. Note that both f and g are in time $n^{O(1)}$.

1. Input x.
2. Run $f(x)$ to get y. Note that this takes time $|x|^{O(1)}$ and that $|y| = O(L)$.
3. Run $g(y)$ to get z. Note that this takes time $|y|^{O(1)} = |x|^{O(1)}$ and that $|z| = O(|y|) = O(|x|)$.

Clearly $x \in A$ if and only if $z \in C$. Note that the reduction takes time $|x|^{O(1)}$ and $|z| = O(|x|)$. Hence this is a linear reduction.

2) Assume 3SAT $\le_p B$ with a linear blowup reduction f. The reduction takes time $L^{O(1)}$ but outputs a string y of length $O(L)$.

Assume that B does not have a $2^{\Omega L}$ lower bound on time. Then, for all ε', there exists a $2^{\varepsilon' L}$ algorithm for B (we use ε' since we will soon use ε for something else). We show that, for all ε, there is a $2^{\varepsilon L}$ algorithm for 3SAT, which contradicts ETH.

The algorithm has parameter ε', which we will choose later.

1. Input φ, a formula of length L.
2. Compute $f(\varphi) = y$. This takes time $L^{O(1)}$. y is of length aL for a constant a (which will be important).
3. Use the algorithm for B to determine whether $y \in B$. This takes time $2^{\varepsilon' L}$. Output the answer.

This algorithm takes time

$$L^{O(1)} + 2^{\varepsilon' aL}.$$

Pick $\varepsilon' = \varepsilon / 2a$ so that the above algorithm is

$$L^{O(1)} + 2^{(\varepsilon L)/2} \le 2^{\varepsilon L}.$$

□

Exercise 6.10. Prove part 3 and 4 of theorem 6.9.

There is a difference between how we can use linear blowup reduction and quadratic blowup reductions.

1. By theorem 6.9.1 we can string together a series of linear blowup reductions to get a linear blowup reduction. Hence we can use these strings of reductions to, assuming ETH, obtain a $2^{\Omega(L)}$ lower bound on a problem A that required many reductions to get from 3SAT to that A.

2. Quadratic blowup reductions are not transitive. Hence to get a lower bound of $2^{\Omega(\sqrt{L})}$ on a problem A you cannot string together a series of quadratic blowup reductions. You can string together reductions such that one has quadratic blowup and the rest have linear blowup.

Note the following parallel:

1. By assuming P $\neq$ NP and using polynomial time reductions, we can show that many problems are not in P.
2. By assuming ETH and using linear blowup reductions, we can show that many problems require $2^{\Omega(L)}$ time.

6.5 ETH Yields $2^{\Omega(L)}$ Lower Bounds

By theorem 6.9, if 3SAT $\leq_p A$ by a linear blowup reduction, and we assume ETH, then we get a $2^{\Omega(L)}$ lower bound on A. Many of the reductions we have already done have linear blowup. Hence we have, assuming ETH, many $2^{\Omega(L)}$ lower bounds.

We will need the following in both this section and the next.

1. The length of a formula in 3CNF form is $O(m)$, where m is the number of clauses.
2. There are many ways to represent a graph, and they may affect the length of the representation. We will represent graphs by a list of vertices and a list of edges. Hence $G=(V,E)$ has length $O(|V|+|E|)$.
3. By theorem 6.9.1, the composition of two linear blowup reductions is a linear blowup reduction. We will use this implicitly.
4. The composition of a several linear blowup reductions and one quadratic blowup reduction is a quadratic blowup reduction. They can be composed in any order. The proof of this is left to the reader.

Theorem 6.11. *Assume ETH. Then the following problems require $2^{\Omega(L)}$ time to solve.*

1. *Clique.*
2. *Independent Set.*
3. *Vertex Cover.*
4. *3-Coloring.*
5. *Dominating Set.*
6. *Directed Hamiltonian Cycle.*
7. *Directed Hamiltonian Path, Hamiltonian Cycle, Hamiltonian Path. We leave this to the reader.*

Proof. 1) Theorem 1.32.1 gives a reduction from 3SAT to CLIQUE, which takes a 3CNF formula on n variables and m clauses and produces a graph G on $O(m)$ clauses. This is clearly a linear blowup reduction.

2) Theorem 1.32.2 gives a linear blowup reduction from CLIQUE to INDEPENDENT SET.

3) Theorem 1.32.3 gives a linear blowup reduction from INDEPENDENT SET to VERTEX COVER.

4) Theorem 2.2.1 gives a reduction from 3SAT to 3-COLORING that maps a 3CNF formula on n variables an m clauses to a graph with $O(m)$ vertices and $O(m)$ edges. This is clearly a linear blowup reduction.

5) Theorem 2.19 gives a linear blowup reduction from VERTEX COVER to DOMINATING SET.

6) Theorem 2.21 gives a reduction from 3SAT to DIRECTED HAMILTONIAN CYCLE. The reduction takes a 3CNF formula φ with m clauses and constructs a graph G' that does the following: (a) for every variable there is a gadget on $O(a_v)$ edges and $O(a_v)$ vertices where a_v is the number of times v appears in the formula, (b) for every clause there is an vertex and $O(1)$ edges. G' has $\sum_{v\in V} a_v = 3m$ vertices and edges from the variable-clauses, and $O(m)$ vertices and edges from the clause gadget. Hence G' has $O(m)$ vertices and edges, so the reduction is linear. □

Exercise 6.12. Assume ETH. Prove that the following problems, restricted to planar graphs, require $2^{\Omega(L)}$ time: directed hamiltonian path, Hamiltonian cycle, and Hamiltonian path.

6.6 ETH Yields $2^{\Omega(\sqrt{L})}$ Lower Bounds for Planar Problems

By theorem 6.9, if 3SAT $\leq_p A$ by a quadratic blowup reduction, and we assume ETH, then we get a $2^{\Omega(\sqrt{L})}$ lower bound on A. Some of the reductions we have already done have quadratic blowup. Hence we have, assuming ETH, some $2^{\Omega(L)}$ lower bounds.

By theorem 6.9, if a reduction has quadratic blowup, and we assume ETH, we get a $2^{\Omega(\sqrt{L})}$ lower bound. Some of the reductions in this book have quadratic blowup. Hence, assuming ETH, we have some $2^{\Omega(\sqrt{L})}$ lower bounds.

We first look at planar 3-colorability and variants.

Theorem 6.13. *Assume ETH. Then the following problems require $2^{\Omega(\sqrt{L})}$ time to solve.*

1. *PLANAR 3-COLORING.*
2. *PLANAR MAX-DEGREE-4 3-COLORING.*

Proof. Assume ETH. By theorem 6.11.2, 3-COLORING requires $2^{\Omega(L)}$ time. Let A be a problem. If there is a quadratic blowup reduction from 3-COLORING to A, then A requires $2^{\Omega(\sqrt{L})}$ time. We will use this.

1) The proof of theorem 2.2.2 gives a reduction from 3-COLORING to PLANAR 3-COLORING. The reduction takes a graph $G = (V, E)$ and produces a graph G'. G' is formed by replacing

every crossing in G with a constant-sized gadget. There are at most $|E|^2$ crossings, so G' has at most $|V| + |E|^2$ vertices and $|E| + |E|^2$ edges. Hence the length of G' is $O(|V| + |E|^2) \leq O((|V| + |E|)^2)$. Hence the reduction from 3-COLORING to PLANAR 3-COLORING has quadratic blowup. By theorem 6.11, the reduction from 3-COLORING to PLANAR 3-COLORING has quadratic blowup.

2) The proof of theorem 2.2.4 gives a reduction from PLANAR 3-COLORING to PLANAR MAX-DEGREE-4 3-COLORING. The reduction takes a graph $G = (V, E)$ and replaces every vertex of degree $d \geq 5$ with a gadget. The gadget has $O(d)$ vertices and $O(d)$ edges. By the handshaking lemma, this blows up the size of the graph by a linear factor. Hence the reduction from 3-COLORING to PLANAR MAX-DEGREE-4 3-COLORING has quadratic blowup. □

Recall that in chapter 2 we proved 3SAT $\leq_p$ PLANAR 3SAT and then used PLANAR 3SAT to show many planar problems are NP-complete. We now analyze those reductions more carefully and get, assuming ETH, $2^{\Omega(\sqrt{L})}$ lower bounds.

Theorem 6.14. *Assume ETH. Then the following problems require $2^{\Omega(\sqrt{L})}$ time to solve.*

1. *PLANAR 3SAT.*
2. *PLANAR VERTEX COVER.*
3. *PLANAR DOMINATING SET.*
4. *PLANAR DIRECTED HAMILTONIAN CYCLE.*

Proof. 1) Theorem 2.10 gives a quadratic blowup reduction from 3SAT to PLANAR 3SAT.

2) Theorem 2.18.2 gives a linear blowup reduction from 3SAT to PLANAR VERTEX COVER.

3) Theorem 2.19.2 gives a linear blowup reduction from PLANAR DOMINATING SET to PLANAR VERTEX COVER.

4) Theorem 2.21.2 gives a linear blowup reduction from PLANAR 3SAT to PLANAR HAMILTONIAN CYCLE. □

Exercise 6.15. Assume ETH for this exercise.

1. Investigate the complexity of CLIQUE restricted to planar graphs.
2. Investigate the complexity of variants of DIRECTED HAMILTONIAN CYCLE restricted to planar graphs.

Can we improve the lower bounds in theorem 6.13 or 6.14 from $2^{\Omega(\sqrt{L})}$ to $2^{\Omega(L)}$? This is one of the rare times in this book that we can give an answer without a hypothesis:

All of the problems in theorem 6.13 and 6.14 can be solved in time $2^{O(\sqrt{n})}$.

One way to obtain these algorithms is the theory of bidimensionality. See the survey by Demaine and Hajiaghayi [DH08], the papers of Demaine et al. [DHT04, DFHT05], the references in those papers, or the slides of Marx [Mar13].

6.7 Consequences of SETH

6.7.1 Problems on Vectors

Orthogonal Vectors

> Orthogonal Vectors (OrVecs)
> *Instance:* $A, B \subseteq \{0,1\}^d$, $|A| = |B| = n$.
> *Question:* Does there exist $\vec{a} \in A, \vec{b} \in B$ with $\vec{a} \cdot \vec{b} \equiv 0 \pmod 2$?
> *Note:* It is easy to see that Orthogonal Vectors can be solved in $O(n^2 d)$ time. Hence this *seems* unrelated to ETH or SETH. But things are not always how they seem.

Conjecture 6.16. *The Orthogonal Vectors Conjecture (OVC) is that, for all $\varepsilon > 0$, there is no $O(n^{2-\varepsilon})$ algorithm for Orthogonal Vectors.*

R. Williams [Wil05] showed the following:

Theorem 6.17. *SETH implies OVC.*

We discuss consequences of OVC in chapter 17.

Closest Vector Problem

Definition 6.18. A lattice $\mathcal{L}$ in $\mathbb{R}^n$ is a discrete subgroup of $\mathbb{R}^n$.

> Closest Vector Problem (CVP)
> *Instance:* A lattice $\mathcal{L}$ (specified through a basis) together with a target vector $\vec{v} \in \mathbb{R}^n$.
> *Output:* A vector $\vec{u} \in \mathcal{L}$ that is closest to $\vec{v}$.
> *Note:* What do we mean by closest? Let $1 \leq p \leq \infty$. Then the ***p*-norm** of a vector $(x_1, \ldots, x_n)$ is
>
> - $(|x_1|^p + \cdots + |x_n|^p)^{1/p}$ if $p \neq \infty$.
> - $\max_{1 \leq i \leq n} |x_i|$ if $p = \infty$.
>
> *Note:* A norm p will be specified as well. We will discuss a variety of values for p.

The common case is $p = 2$, which is the standard Euclidean distance. To indicate that the p-norm is being used, the notation CVP_p is the convention. Aggarwal et al. [ABGS21] showed the following:

Theorem 6.19. *Assume SETH. For all $\varepsilon > 0$, for all $p \notin 2\mathbb{Z}$, there is no $2^{(1-\varepsilon)n}$ algorithm for CVP_p.*

It is unfortunate that they do not have the result for $p = 2$, which is the case of most interest. They comment that the gadgets they use do not exist for $p \in 2\mathbb{Z}$.

H. Bennett et al. [BGS20] have a survey of open problems on the complexity of lattice problems. We state one about CVP.

Open Problem 6.20. *Assume SETH. Show there is no $O(2^{0.99n})$ time algorithm for CVP_2.*

SHORTEST VECTOR PROBLEM

SHORTEST VECTOR PROBLEM (SVP)
Instance: A lattice $\mathcal{L}$ (specified through a basis).
Output: A vector $\vec{u} \in \mathcal{L}$ that is of minimal norm.
Note: A norm p will be specified as well. We will discuss a variety of values for p.
Note: SHORTEST VECTOR PROBLEM is NP-complete.
Note: For lower bounds on SHORTEST VECTOR PROBLEM that do not use SETH or other assumptions from this chapter, see theorem 9.31. For upper and lower bounds on approximating SHORTEST VECTOR PROBLEM, see section 9.7.3.

The common case is $p = 2$, which is the standard Euclidean distance. To indicate that the p-norm is being used, the notation SVP_p is the convention. Aggarwal et al. [ABGS21, Theorem A.1], building on the work of Bennett et al. [BGS17] and Aggarwal and Stephens-Davidowitz [AS18], showed the following:

Theorem 6.21. *Assume a randomized version of SETH. There exists $p_0 \sim 2.1357$ and, for every $p \geq p_0$, $p \notin 2\mathbb{Z}$, there is a real C_p, such that the following hold.*

1. *For n large, there is no $2^{n/C_p}$ algorithm for SVP_p.*
2. $\lim_{p\to\infty} C_p = 1$

It is unfortunate that they do not have the result for $p = 2$, which is the case of most interest.

H. Bennett et al. [BGS20] have a survey of open problems on the fine-grained complexity (e.g., uses assumptions like ETH) of lattice problems. In addition, H. Bennett [Ben23] has a survey of open problems about SHORTEST VECTOR PROBLEM that covers both fine-grained complexity (e.g., uses assumptions like ETH) and computational complexity (e.g., uses assumptions like $P \neq NP$). We state one of those.

Open Problem 6.22. *Assume SETH. Show that there is no $O(2^{n/10})$ time algorithm for SVP_2.*

6.7.2 Larger Exponentials

GRAPH HOMOMORPHISM
Instance: Graphs G, H.
Question: Is there a homomorphism from G to H? (A ***homomorphism*** from $G=(V,E)$ to $H=(V',H')$ is a function $f\colon V \rightarrow V'$ such that if $(u,v) \in E$ then $(f(u), f(v)) \in E'$.
Note: If H is K_c, then this question is asking whether G is c-colorable. Hence GRAPH HOMOMORPHISM is NP-complete.

SUBGRAPH ISOMORPHISM
Instance: Graphs G, H.
Question: Is H isomorphic to some subgraph of G?
Note: If $H = K_c$, then this question is asking whether G has a clique of size k. Hence SUBGRAPH ISOMORPHISM is NP-complete.

Cygan et al. [CFG+17] showed that, assuming ETH, the following hold:

1. GRAPH HOMOMORPHISM cannot be solved in $|V(H)|^{O(|V(G)|)}$ time.
2. SUBGRAPH ISOMORPHISM cannot be solved in $|V(H)|^{O(|V(G)|)}$ time.

6.7.3 Fractional Coloring

Definition 6.23. An $(a:b)$ coloring of a graph is a coloring where you assign b colors to each vertex out of a total a colors, so that adjacent vertices have disjoint sets of colors. One may also say informally that G is $\frac{a}{b}$-colorable. This number is called the ***fractional chromatic number***.

FRACTIONAL CHROMATIC NUMBBER(a, b)
Instance: A graph G and a, b.
Question: Is G $(a:b)$-colorable?
Note: If $b=1$, this is asking whether G is a-colorable. Hence this problem is NP-complete.

The study of fractional chromatic number was motivated as follows.

- Appel et al. [AH77a, AHK77] showed that every planar graph is 4-colorable. Their proof made extensive use of a computer program to check a massive amount of cases. Robertson et al. [RSST97] had a simpler proof, though it still needed a computer program. In short, the proof is not *human readable.*
- By contrast, the proof that every planar graph is 5-colorable is easy to follow and is clearly human readable.
- Fractional chromatic number was defined with the goal of finding human-readable proofs that every planar graph is c-colorable for some values of $c < 5$. Cranston and Rabern [CR18] showed that every planar graph is 4.5-colorable. It is open to lower that.
- For more on fractional concepts in graph theory, see the book by Scheinerman and D. Ullman [SU96].

The following lower bounds are known.

Theorem 6.24.

1. *(Grötschel et al. [GLS81]) For all $c > 2$, $c \in \mathbb{Q}$, determining whether G is c-colorable is NP-complete.*
2. *(Bonamy et al. [BKP+19]) Assume ETH. Fix $a, b \in \mathbb{N}$ such that $b \leq a$. For any computable function f,* Fractional Chromatic Numbber(a, b) *cannot be solved in time $O(f(b)2^{o(\log b)n})$.*

6.8 To Be Continued

The assumptions ETH and SETH will be used to show several results in this book:

1. Lower bounds in parameterized complexity: sections 7.1 and 7.14.
2. Lower bounds on the approximate nearest neighbor problem: section 9.7.4.
3. Lower bounds for dSUM: theorem 17.33.
4. Quadratic lower bounds on problems in computational geometry: section 18.9.
5. Lower bounds on the approximate Nash equilibrium problem: theorem 22.40.

7

Parameterized Complexity

7.1 Introduction

In this chapter we will present several results about fixed-parameter tractability and parameterized complexity. For more on these fields there are several good books. We mention four: Cygan et al. [CFK+15], Downey and Fellows [DF99], Flum and Grohe [FG06], and Niedermeier [Nie06].

Recall that the following two problems are NP-complete:

$$\text{VERTEX COVER} = \{(G, k) \mid G \text{ has a vertex cover of size } k\},$$

$$\text{CLIQUE} = \{(G, k) \mid G \text{ has a clique of size } k\}.$$

Fix $k \in \mathbb{N}$. Let

$$\text{VERTEX COVER}_k = \{G \mid G \text{ has a vertex cover of size } k\},$$

$$\text{CLIQUE}_k = \{G \mid G \text{ has a clique of size } k\}.$$

Both of these problems are in time $O(k^2 n^k)$ by a brute-force algorithm ($\binom{n}{k} \sim n^k$ sets of k vertices to check, and each one takes k^2 time to check). Hence they are both in P, for any fixed (constant) k. However, the degree of the polynomial depends on k. Is there a polynomial-time algorithm for either VERTEX COVER$_k$ or CLIQUE$_k$ such that the degree of the polynomial does not depend on k?

1. For VERTEX COVER$_k$ the answer is yes. We give such an algorithm in theorem 7.1 and an even better algorithm in theorem 7.2.
2. For CLIQUE$_k$ we show that, under reasonable assumptions, there is no such algorithm.

The field of *fixed parameter tractability* is the study of what happens to a (usually) NP-complete problem that has a natural parameter k if that parameter is fixed.

Chapter Summary

1. *We give two algorithms for* VERTEX COVER. *One has complexity* $O(2^k n)$ *and the other has complexity* $O(kn + k^2 2^k)$. *These algorithms motivate the definitions of* ***fixed parameter tractable*** *(FPT).*
2. *We define* **FPT** *to capture problems like* VERTEX COVER *that have drastically short running times if a natural parameter k is fixed.*
3. *We define an appropriate notion of reduction so that if we have a problem that is unlikely to be FPT, we can use these reductions to prove that other problems are unlikely to be FPT.*
4. *We present a problem that we will assume is not FPT. We use this one problem, and the notion of reduction in the last item, to define a complexity class W[1]. We also define W[2] though that is used much less.*
5. *We show that a large set of problems are W[1]-hard and, hence, unlikely to be FPT.*

7.2 A Good Algorithm for VERTEX COVER

Mehlhorn [Meh84] showed the following.

Theorem 7.1. *VERTEX COVER can be solved in time* $O(2^k n)$.

Proof. Algorithm for VERTEX COVER:

1. Input $G = (V, E)$ and $k \in \mathbb{N}$.
2. Form a tree as follows. The root has $(G, e_1, \emptyset)$ where $e_1 = (u, v)$ is an edge (chosen arbitrarily) and the set $X = \emptyset$, which we will add to. Note that either u or v is in the vertex cover. We will branch two ways depending on if we choose u or v. The two children of the root are as follows:
   a. The right child is $(G - \{u\}, e_2, \{u\})$ where e_2 is some edge of $G - \{u\}$, together with $X \cup \{u\}$.
   b. The left child is $(G - \{v\}, e_3, \{v\})$ where e_3 is some edge of $G - \{v\}$, together with $X \cup \{v\}$.
3. Keep doing this until you have a tree of height k or every leave has the empty graph associated to it so that no edge can be chosen. One of the following happens.
   a. There is some leaf that is associated to the empty graph. Then there is a vertex cover of size k.
   b. Every leaf is associated to a non-empty graph. Then there is no vertex cover of size k.

There are $O(2^k)$ nodes on the tree. Forming each node takes $O(n)$ time. Hence the algorithm runs in $O(2^k n)$ time. □

Are better results known or likely? The following two theorems give a yes or a no depending on your definitions of "better" and "likely."

1. Chen et al. [CKX10] showed VERTEX COVER is in time $O(kn + 1.2738^k)$.
2. Cai and Juedes [CJ03] showed that if VERTEX COVER is in time $2^{o(k)}n^L$ for some L, then 3SAT is in time $2^{o(n)}$, which violates the ETH. So, for example, it is unlikely that there is an algorithm for VERTEX COVER that runs in time $O(2^{k/\log k}n^{100})$. We will prove this in theorem 7.52.

What if we restrict to planar graphs?

1. Demaine et al. [DFHT05] showed that PLANAR VERTEX COVER can be solved in $2^{O(\sqrt{k})}n^{O(1)}$ time.
2. Cai & Juedes [CJ03] showed that if PLANAR VERTEX COVER is in time $2^{o(\sqrt{k})}n^L$ for some L, then 3SAT is in time $2^{o(n)}$, which violates the ETH. We will prove this in theorem 7.52.

We will later see that there are reasons to think that any polynomial time algorithm for CLIQUE will have degree that depends on k.

7.3 A Better Algorithm for VERTEX COVER

By theorem 7.1 VERTEX COVER is in time $O(2^k n)$. We will show a better algorithm. The key to the algorithm is that, if there is a vertex of degree $k+1$, then it must be in the vertex cover; otherwise, all $k+1$ of its neighbors have to be in the vertex cover.

The following theorem is (correctly) attributed to S. Buss, though it is not published.

Theorem 7.2. *VERTEX COVER can be solved in time $O(kn + k^2 2^k)$.*

Proof. 1. Input $G = (V, E)$ and k.

2. Let $U = \emptyset$. We will be adding vertices of the vertex cover to U. Let $L = k$. We are currently looking for a vertex cover of size $L = k$. At each step we will be looking for a vertex cover of size L. We will always have $|U| + L = k$.
3. If there is a vertex v of degree $\geq L+1$, then (1) $U \leftarrow U \cup \{v\}$, (2) $L \leftarrow L-1$, (3) $G \leftarrow G - \{v\}$. Repeat until the answer is NO. There will be at most k iterations, each taking at most $O(n)$ steps, so $O(kn)$ steps.

4. (This is a comment, not part of the algorithm.) Every vertex of G has degree $\leq L$. Hence if there is a vertex cover of size L, with each vertex covering $\leq L$ edges, there has to be at most L^2 edges. Therefore there are at most $2L^2$ nonisloated vertices (we can ignore the isolates vertices). We will assume the worse case, which is that initially every vertex is of degree $\leq k$.
5. If G has more than k^2 edges, then output NO and stop.
6. (If you got here, then G has $\leq k^2$ edges and hence $\leq 2k^2$ non-isolated vertices.) Do the algorithm from theorem 7.1 to determine whether there is a vertex cover of size k on a graph with $\leq 2k^2$ vertices. This step takes time $O(k^2 2^k)$.

   From the comments in the algorithm it works and it takes time $O(kn + k^2 2^k)$.

□

The algorithm in theorem 7.2 took a VERTEX COVER problem of size n vertices with parameter k and reduced it to a VERTEX COVER problem of size $2k^2$ vertices and parameter k. We will return to this point in section 7.5.

7.4 Fixed-Parameter Tractable

For VERTEX COVER there is a polynomial-time algorithm where the degree of the polynomial does not depend on k. We will define this notion formally. We will look at (usually) NP-complete problems that have a parameter (like k for VERTEX COVER) and ask what happens if k is fixed.

Definition 7.3. A ***parameterized problem*** is a subset of $\Sigma^* \times \mathbb{N}$. An instance of a parameterized problem will be (x, k). We will almost always use k.

Definition 7.4. Let A be a parameterized problem. A is ***FPT*** if there is an algorithm for A and a computable function f such that the algorithm, on input (x, k) runs in time $f(k)|x|^{O(1)}$

Note that (1) $f(k)$ might be quite large, and (2) the degree of the polynomial $f(k) \cdot n^{O(1)}$ does not depend on k.

7.5 Kernelization

Definition 7.5. Let A be a parameterized problem. A has a ***kernelization*** if there exists a polynomial time function $g\colon \Sigma^* \times \mathbb{N} \to \Sigma^* \times \mathbb{N}$ and a computable function $h\colon \mathbb{N} \to \mathbb{N}$ such that the following holds: Let $g(x, k) = (x', k')$.

1. $(x, k) \in A$ if and only if $(x', k') \in A$.
2. $k' \leq k$ and $|x| \leq h(k)$.

(Formally we will need to specify the notion of size that we use. Assume it is the length of the input unless otherwise specified. For graphs we will use $|V|$ and $|E|$ for size and will specify which one when the issue arises.)

Definition 7.6. Let A be a parameterized problem. Assume it has a kernelization via (g, h).

1. Let $g(x,k) = (x',k')$. (x',k') is called ***the kernel of*** (x,k) or just ***the kernel*** if (x,k) is understood.
2. $h(k)$ is called ***the size of the kernel***.

In general $h(k)$ can be exponential, but a good kernel should be polynomial or even linear in k.

By the proof of theorem 7.2, Vertex Cover has a kernel of size $2k^2 = O(k^2)$ vertices. Soleimanfallah and Yeo [SY11] showed that, for all $c \in \mathbb{N}$, there is a kernel of size $2k - c$ vertices. Lampis [Lam11] showed that, for all $c \in \mathbb{N}$, there is a kernel of size $2k - c\log k$ vertices. This is the smallest kernel currently known. How much smaller can the kernel be? We state two theorems about this. The first is an easy exercise for the reader. The second is a difficult result of Dell and van Melkebeek [DvM14].

Theorem 7.7.

1. *If there is a kernelization of Vertex Cover with kernel of size $O(\log k)$ vertices, then $P = NP$.*
2. *If there exists ε and a kernelization of Vertex Cover with kernel of size $O(k^{2-\varepsilon})$ edges, then coNP $\subseteq$ NP/poly (which is thought to be unlikely and implies that $\Sigma_2 = \Pi_2$).*

Niedermeier [Nie02] (see also Flum and Grohe [FG06]) showed the following connection between FPT and kernelization. We omit the proof.

Theorem 7.8. *Let A be a parameterized problem. The following are equivalent.*

1. *$A \in FPT$.*
2. *A is decidable and has a kernelization.*

7.6 Parameterized Reductions

As usual, we'll show problems are hard via reductions. In general, we have a parameterized problem A, a parameterized problem B, and a map from (x,k) to (x',k'). This is going to look similar to Karp-style reductions, but with a few tweaks.

Definition 7.9. Let A and B be parameterized problems. Recall that they are both subsets of $\Sigma^* \times \mathbb{N}$. A is ***parameterized reductible*** to B if there exist computable functions f and g

and a map from A to Σ^* such that the following hold. Assume that (x, k) maps to x' and $g(k) = k'$.

1. $(x, k) \in A$ if and only if $(x', k') \in B$.
2. The map can be computed in time $f(k) \cdot |x|^{O(1)}$.

Theorem 7.10.

1. *If A is parameterized-reducible to B and $B \in \mathsf{FPT}$, then $A \in \mathsf{FPT}$.*
2. *If $A \notin \mathsf{FPT}$ and A is parameterized-reducible to B, then $B \notin \mathsf{FPT}$. (This is just the contrapositive of part 1. However, we will use this to show, under assumptions, that some problems are not FPT.)*

Proof. Assume A is parameterized-reducible to B with computable functions f_1 and g, and polynomial p. Assume that $g(k)$ can be computed in time $h(k)$. Assume that $B \in \mathsf{FPT}$ with computable function $f_2 \geq 1$.

We show $A \in \mathsf{FPT}$.

1. Input (x, k), an instance of A.
2. Compute the mapping on (x, k) to get x'. This takes time $f_1(k) \cdot |x|^{O(1)}$. Note that $|x'| \leq f_1(k) \cdot |x|^{O(1)}$.
3. Compute the function $g(k)$ to get k'. This takes time $h(k)$. Note that $k' \leq h(k)$.
4. Use the FPT algorithm for B on (x', k'). This takes time

$$f_2(k') \cdot |x'|^{O(1)} \leq f_2(h(k)) \cdot (f_1(k) \cdot |x|^{O(1)})^{O(1)} = f_2(h(k)) \cdot f_1(k)^{O(1)} \cdot |x|^{O(1)}.$$

Summing the running times for steps 2–4, we obtain a total running time of

$$f_1(k) \cdot |x|^{O(1)} + h(k) + \underbrace{f_2(h(k))}_{\geq 1} \cdot f_1(k)^{O(1)} \cdot |x|^{O(1)}$$
$$\leq h(k) + f_2(h(k)) \cdot f_1(k)^{O(1)} \cdot |x|^{O(1)}.$$

□

Many reductions that we have seen before *are not* parameterized reductions. We give an example of a reduction that is *not* a parameterized reduction and then an example of one that is.

Example 7.11. We give an example of a reduction that is not a parameterized reduction. The standard reduction from INDEPENDENT SET to VERTEX COVER is to map (G, k) to $(G, n - k)$. For this to be a parameterized reduction, we would need the map from k to $n - k$ be of the form $g(k)$. But it is not since it uses n. Hence this is not a parameterized reduction.

Is there a parameterized reduction from INDEPENDENT SET to VERTEX COVER? Unlikely since (1) VERTEX COVER $\in$ FPT and (2) Chen et al. [CHKX06] showed that if CLIQUE is FPT, then ETH is false.

Example 7.12. The standard reduction from INDEPENDENT SET to CLIQUE is to map (G, k) to $(\overline{G}, k)$ Here $k' = k$, so k' depends on k and is easily bounded by a function of k. Hence this is a parameterized reduction. Note that the reduction from INDEPENDENT SET to CLIQUE is very similar and is also a parameterized reduction.

Notation 7.13. Henceforth, in this chapter, "reduction" means "parameterized reduction."

7.7 W[1]

In order to use reductions to show that problems are not FPT, we need to have some problems that we already think are not FPT. That is, we need an analog of SAT for showing problems not in P. Recall that we think SAT is not in P because (1) by the Cook–Levin theorem, if SAT $\in$ P, then P $=$ NP, (2) people have been trying to get SAT (and many other problems in NP) into P without any success (some of the effort was before P was defined), and (3) it just seems to require brute force.

Is there some problem that we are confident is not in FPT? There is, though frankly, the evidence is nowhere near as strong as for SAT not being in P. The evidence is similar to points (2) and (3) about SAT. Once we have the problem defined, we will define a complexity class based on it.

The problem we discuss involves nondeterministic Turing machines. The reader may take on faith what we say about these devices or look up the concept.

Definition 7.14. Let N be a nondeterministic Turing machine over alphabet Σ and state set Q. Let x be an input to it and let $k \in \mathbb{N}$. Run $N(x)$ for k steps. A ***configuration*** is an element of $\Sigma^* \times \mathbb{N} \times Q$ that represents the content of the tape, the position of the head, and the state of the machine. Note that since the machine has run for $\leq k$ steps, we can assume that all configurations are of length $O(k)$.

> NONDETERMINISTIC TURING MACHINE ACCEPTANCE (NONDET TM ACCEPTANCE)
> *Instance:* A nondeterministic Turing machine M and a number k. We will assume that there are $O(n)$ states, $O(n)$ alphabet size, and $O(n)$ choices at each step.
> *Question:* If M is run on 0^n, is there an accepting path of length k?

NONDET TM ACCEPTANCE can be solved in $O(n^k)$ steps. Fix k. There does not seem to be any way to do this problem in $f(k) \cdot n^{O(1)}$ for some (even quite large) f. In short, it looks like this problem is not in FPT. Indeed, we will assume that it is not.

We now define a complexity class based on NONDET TM ACCEPTANCE. It is called W[1] for reasons we will get into in section 7.11.

Definition 7.15. Let A be a parameterized problem.

1. $A \in$ W[1] if there is a parameter reduction from A to NONDET TM ACCEPTANCE. (This will usually be easy to show.)
2. A is W[1]-hard if there is a parameter reduction from NONDET TM ACCEPTANCE to A.
3. A is W[1]-complete if it is in W[1] and is W[1]-hard.

Some sources, including Niedermeier's book [Nie06], use a different problem, WEIGHTED 2SAT, to define W[1]-complete. We will reexamine WEIGHTED 2SAT problems in section 7.10 as a prelude to defining the W-hierarchy.

WEIGHTED 2SAT
Instance: A 2CNF φ and $k \in \mathbb{N}$. (k is the parameter.)
Question: Is there a satisfying assignment of φ with exactly k variables set to true? This is called ***an assignment of weight k***.
Note: This is a terrible name for this problem since ***weighted SAT*** usually means that the clauses have weights and you want to maximize the sum of the weights of the satisfied clauses.

It turns out that NONDET TM ACCEPTANCE and WEIGHTED 2SAT are reducible to each other by parameterized reductions, and thus the definition of W[1] using NONDET TM ACCEPTANCE and the one using WEIGHTED 2SAT are equivalent. We will use the definition based on NONDET TM ACCEPTANCE.

According to D. Marx [Mar14] there are hundreds of W[1]-complete problems. This seems like an exaggeration; however, there are many W[1]-complete problems. If any of them are FPT, they are all FPT. Many are problems that have been worked on a lot. This is good evidence that (1) none of them are FPT, and (2) hence NONDET TM ACCEPTANCE is not FPT.

7.8 INDEPENDENT SET and CLIQUE are W[1]-Complete

7.8.1 INDEPENDENT SET and CLIQUE

The reader is advised to read about nondeterministic Turing machines in section 0.3.

Theorem 7.16.

1. *Independent Set is W[1]-complete.*
2. *Clique is W[1]-complete. (This is an easy reduction from Independent Set, so we omit it.)*

Proof. We show that Independent Set reduces to Nondet TM Acceptance, which proves Independent Set $\in$ W[1]. We then show that Independent Set reduces to Nondet TM Acceptance, which proves Independent Set is W[1]-hard.

Independent Set reducible to Nondet TM Acceptance: Given (G,k), set up a nondeterministic Turing machine that has G built into it and guesses k vertices (that's k moves) and then checks that each pair is not an edge (that's k^2 moves). The length of the path in the NTM is $O(k^2)$. Thus Independent Set $\in$ W[1].

Nondet TM Acceptance reducible to Independent Set:

1. Input nondeterministic Turing machine N and $k \in \mathbb{N}$. Note that N has $O(n)$ states, $O(n)$ alphabet size, $O(n)$ nondeterministic branching. N has alphabet Σ and state set Q. The alphabet is Σ and the state set is Q.
2. k' will be k^2.
3. Imagine the possible sequence of k configurations of the nondeterministic Turing machine M going for k steps. This sequence will have k^2 cells that are labeled (i,j) for ith row, jth column.
4. For each cell we create a clique (which we describe soon). So far there are k^2 cliques.
5. Each clique is the same: The vertices are $\Sigma \cup \Sigma \times Q$, which are all possible entries in a cell. So there are $O(n^2)$ nodes in the clique.
6. The idea is to put in edges between nodes that cannot both be in the sequence of configurations.
   a. Put an edge between two vertices in the same row that both think the head is there. Formally let $i \in \mathbb{N}$ and $j_1 \neq j_2$. Any vertex of the (i,j_1) clique labeled with an element of $\Sigma \times Q$ (so the head is at j_1) will have an edge to any element of the (i,j_2)-clique labeled with an element of $\Sigma \times Q$ (so the head is at j_2).
   b. Put an edge between two vertices from different rows that clearly cannot occur. Formally let $i,j \in \mathbb{N}$ and look at a vertex of the (i,j)-clique that is labeled with an element of $\Sigma \times Q$, say (a,p). If the Turing machine was in state p and was looking at an a then the transition function constrains what happens next. Put an edge between that vertex and the vertices in the $(i+1,j-1)$-clique, $(i+1,j)$-clique, and $(i+1,j+1)$-clique that are incompatible with the (i,j) cell having (p,a).
7. Call this graph G. We show that G has an Independent Set of size k^2 if and only if there is an accepting path in N of length k.

If G has a INDEPENDENT SET of size k^2 then it has to take one node from each of the (i,j)-cliques. These nodes code an accepting path of length k.

If there is an accepting path of length k, then that tells you how to fill out all the cells, and hence gives an INDEPENDENT SET of size k^2. □

Mathieson and Szeider [MS08] showed that CLIQUE is still W[1]-complete when restricted to regular graphs. We leave this as an exercise.

Exercise 7.17. Prove the following.

1. CLIQUE restricted to regular graphs is W[1]-complete.
2. INDEPENDENT SET restricted to regular graphs is W[1]-complete. (This is a reduction from CLIQUE restricted, so we omit the proof.)

7.8.2 Using INDEPENDENT SET and CLIQUE to Prove Problems W[1]-Hard

> PARTIAL VERTEX COVER
> *Instance:* A graph G and two numbers $k, l \in \mathbb{N}$. We will regard this as a parameterized problem with parameter k.
> *Question:* Is there a set of vertices of size k that covers l edges?

We will show that PARTIAL VERTEX COVER is W[1]-hard. Had we made l the parameter, then this problem would be FPT. Note that if there is more than one choice for a parameter, which one is chosen matters!

Theorem 7.18. *PARTIAL VERTEX COVER is W[1]-hard.*

Proof. By exercise 7.17, INDEPENDENT SET on regular graphs is W[1]-complete. We show that INDEPENDENT SET for regular graphs reduces to PARTIAL VERTEX COVER.

1. Input (G, k), where G is Δ-regular.
2. We will just use G, k, and $l = \Delta k$.

If G has an INDEPENDENT SET of size k, then those k vertices cover Δk edges since no two of the vertices are adjacent to the same edge.

If G has a set of k vertices that cover Δk edges, then they must be independent since otherwise they would cover $< \Delta k$ edges. □

Open Problem 7.19. *Is PARTIAL VERTEX COVER in W[1]?*

> MULTI COLORED CLIQUE (MCLIQUE) and MULTICOLORED INDEPENDENT SET (MIS)
> *Instance:* A graph $G=(V,E)$ and a partition $V=V_1\cup\cdots\cup V_k$. k will be the parameter.
> *Question:* Is there a clique (independent set) with one vertex from each V_i? (It is called "multicolored" since we think of each V_i as a color.)

Pietrzak [Pie03] and independently Fellows et al. [FHRV09] showed the following.

Theorem 7.20.

1. *MULTI COLORED CLIQUE is* W[1]*-complete.*
2. *MULTICOLORED INDEPENDENT SET is* W[1]*-complete (this is an easy reduction from part 1, or a proof similar to part 1, so we omit the proof).*

Proof. MULTI COLORED CLIQUE reduces to CLIQUE (exercise), which is in W[1]; hence MULTI COLORED CLIQUE is in W[1].

We show that CLIQUE reduces to MULTI COLORED CLIQUE.

1. Input $G=(V,E)$ and k.
2. Create a graph G' and a partition of the vertices as follows.
   a. For every $x\in V$ there are k vertices $x_1,\ldots,x_k$.
   b. V_i is the set of all vertices with subscript i.
   c. (x_i,y_j) is an edge if $i\neq j$ and $(x,y)\in E$. Note that within V_i there are no edges.
3. If G' has a clique with a vertex in each V_i, then G has a clique of size k. We leave the easy proof to the reader. □

Exercise 7.21.

1. Show that MULTI COLORED CLIQUE reduces to CLIQUE.
2. Show that the construction in the proof of theorem 7.20 works.

> SET COVER
> *Instance:* n and Sets $S_1,\ldots,S_m\subseteq\{1,\ldots,n\}$.
> *Question:* What is the smallest size of a subset of S_i's that covers all of the elements in $\{1,\ldots,n\}$?

Exercise 7.22. Prove there is a parameterized reduction from DOMINATING SET to SET COVER.

For the next exercise we will consider the following variant of the NODE-WEIGHTED STEINER TREE.

> STRONGLY CONNECTED STEINER SUBGRAPH PROBLEM (SCSS)
> *Instance:* A directed graph $G = (V, E)$, a set $U \subseteq V$, and $k \in \mathbb{N}$.
> *Question:* Is there a strongly connected subgraph of G with $\leq k$ vertices that contains every vertex of U? (A directed graph is strongly connected if there is a directed path between every pair of vertices.)

Exercise 7.23. Show that there is a parameterized reduction from MULTI COLORED CLIQUE to SCSS. Note that this shows SCSS is W[1]-hard.

7.9 DOMINATING SET

The reader should refer to section 2.5.2 for a definition of DOMINATING SET.

It is not obvious (and it is likely false) that DOMINATING SET is in W[1]. After guessing the k vertices for the dominating set D one then has to check whether *every* vertex is in D or adjacent to D. This takes $O(kn)$ steps. Why is DOMINATING SET (apparently) not in W[1] whereas CLIQUE is? Because CLIQUE is local: Once you guess a set C for the clique, you need only check whether every pair in C has an edge. We will later define W[2] and note (but not prove) that DOMINATING SET is W[2]-complete. For now we show the following:

Theorem 7.24. *DOMINATING SET is W[1]-hard.*

Proof. We reduce MULTICOLORED INDEPENDENT SET to DOMINATING SET.

1. Input $G = (V, E)$ with V partitioned as $V_1 \cup \cdots \cup V_k$.
2. Create a graph G', which will be formed by adding vertices and edges to G as follows:
   a. For each i (1) put an edge between every pair of vertices in V_i, (2) create two new vertices x_i, y_i, (3) for all $v \in V_i$ put in an edge from v to x_i and from v to y_i. Note that there is no edge between x_i and y_i (figure 7.1 makes it look like there is an edge between x_i and y_i; however, there is not.)
   b. For every edge $e = (u, v)$, where $u \in V_i$, $v \in V_j$, $i \neq j$, create a new node w_e. Put an edge between w_e and (1) every vertex of $V_i - \{u\}$ and (2) every vertex in $V_j - \{v\}$. (See figure 7.1 for a picture of G'.)

We will show that G has an independent set with one vertex from each V_i if and only if G' has a dominating set of size k.

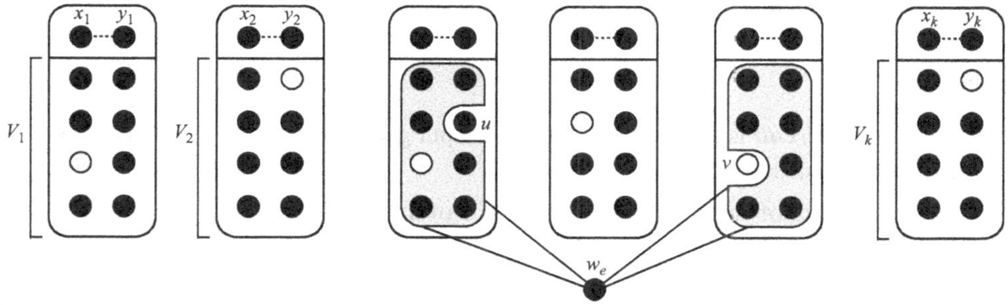

Figure 7.1
Reduction from MULTICOLORED INDEPENDENT SET to DOMINATING SET.

Assume G has an independent set I with one vertex from each V_i. We show that I is a dominating set for G'. Since there is one vertex in each V_i, $|I| = k$ and every vertex in $V_i \cup \{x_i, y_i\}$ is covered. Let $e = (u, v)$ and w_e be as in the construction. Since I is an independent set in G I cannot have both u and v. Assume $u \in V_i \cap \overline{I}$. The element $u_i \in V_i \cap I$ is not u and hence it is adjacent to w_e, so w_e is covered.

Assume that G' has a dominating set D of size k.

1. We first show that it has exactly one vertex in each V_i. $x_i \in V_i$ and all of its neighbors are in V_i, hence some element of $V_i \cup \{x_i\}$ is in D. Similarly, some element of $V_i \cup \{y_i\}$ is in D. If $x_i \in D$, then we need $y_i \in D$ to cover y_i. Then there would be two elements in $D \cap V_i$. Since $|D| = k$, there would be some j with $D \cap V_j = \emptyset$. Then neither x_j nor y_j would be covered.
2. We have that D has one vertex in each V_i. We now prove that D is an independent set in G. Assume, by way of contradiction, that there exists $v_i \in V_i \cap D$, $v_j \in V_j \cap D$, such that $(v_i, v_j) \in E$. Denote this edge e. Then the node w_e in G' is not covered, which contradicts D being a dominating set. □

Exercise 7.25. Show that CLIQUE reduces to the following problems.

1. SPARSEST CUT.
2. Connected DOMINATING SET (the dominating set has to be connected).
3. Independent DOMINATING SET (the dominating set has to be independent).

What happens if we restrict G to be planar? The problem is known to be in FPT. See Alber et al. [ABF$^+$02] for the most recent results and the history of prior results. In a nutshell: PLANAR DOMINATING SET was in $O(c^k n)$ for some c, and Alber et al. [ABF$^+$02] got it down to $O(c^{\sqrt{k}} n)$ for some c.

7.10 Circuit SAT and Weighted Circuit SAT

Definition 7.26.

1. A ***Boolean circuit*** is a circuit consisting of input gates, negation (~), AND (∧), OR (∨), and output gates.
2. An assignment of variables has ***weight k*** if k of the input are TRUE.

> Circuit SAT and Weighted Circuit SAT
> *Instance:* A Boolean circuit C and (for Weighted Circuit SAT) $k \in \mathbb{N}$.
> *Question:* Is there an assignment on the inputs of C (for Weighted Circuit SAT the assignment has weight k) such that the output is T?
> *Note:* This is a terrible name for the problem since the term "weighted" usually means that weights are given.

Theorem 7.27.

1. *Independent Set reduces to Weighted Circuit SAT. Hence Weighted Circuit SAT is W[1]-hard.*
2. *Dominating Set reduces to Weighted Circuit SAT. We will later see that this means it is unlikely that Weighted Circuit SAT is in W[1].*

Proof. 1) We reduce Independent Set to Weighted Circuit SAT. See the circuit labeled "independent set" in figure 7.2 for an example of the reduction.

1. Input $G = (V, E)$ and $k \in \mathbb{N}$.
2. We create a circuit C as follows.
   a. For each $v \in V$ we have an input.
   b. On the second level of the circuit we have the negation of all of the inputs.
   c. On the third level, for all $(u, v) \in E$, there is a gate that computes $\neg u \vee \neg v$.
   d. The fourth level is the AND of all of the gates on the third level.
3. It is easy to see that G has an Independent Set of size k if and only if there is an input of weight k that satisfies C.

2) We reduce Dominating Set to Weighted Circuit SAT. See the circuit labeled "dominating set" in figure 7.2 for an example of the reduction.

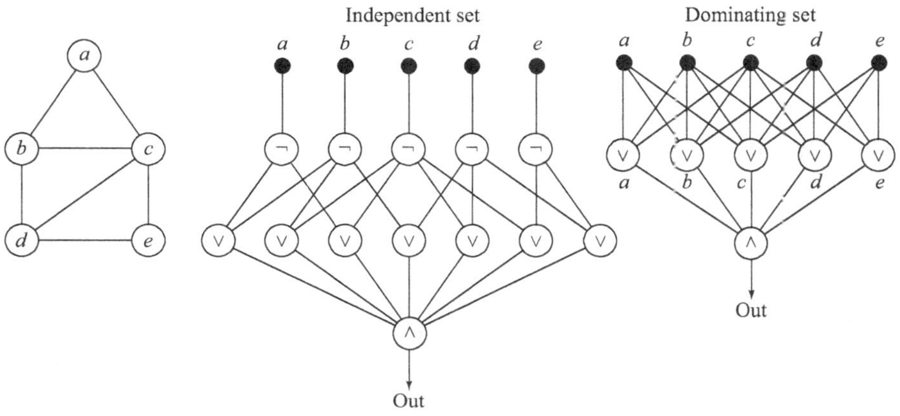

Figure 7.2
Circuits for INDEPENDENT SET and for DOMINATING SET.

1. Input $G = (V, E)$ and $k \in \mathbb{N}$.
2. We create a circuit C as follows.
   a. For each $v \in V$ we have an input.
   b. For every $v \in V$ we have a gate at the second level that computes the OR of (from the input level) v and all of the neighbors of v.
   c. The third level is the AND of all of the gates on the second level.
3. It is easy to see that G has a DOMINATING SET of size k if and only if there is an input of weight k that satisfies C. □

7.11 *W*-Hierarchy

Let us recap what we have.

1. The following problems are W[1]-complete and, hence, equivalent to each other: NONDET TM ACCEPTANCE, INDEPENDENT SET, CLIQUE, INDEPENDENT SET restricted to regular graphs, CLIQUE restricted to regular graphs, PARTIAL VERTEX COVER, MULTI COLORED CLIQUE, MULTICOLORED INDEPENDENT SET, SCS, RSCS, and FLOOD-IT ON TREES.
2. The following problems are W[1]-hard but do not seem to be in W[1]: DOMINATING SET, WEIGHTED CIRCUIT SAT.

Why is (say) CLIQUE in W[1] but DOMINATING SET does not appear to be? As noted earlier CLIQUE is local: Once you have a potential clique of k vertices, to check they are a clique you need only look at those k vertices. DOMINATING SET is global: Once you have a potential dominating set of k vertices, to check that they are a dominating set you need to look at all vertices.

With an eye toward formalizing the difference, we state the difference in terms of circuits. The different circuits we used for INDEPENDENT SET and DOMINATING SET (see figure 7.2) are instructive. Both circuits take a set of vertices (as a bit vector) and determine whether it satisfies the condition (INDEPENDENT SET or DOMINATING SET). But note the following:

1. In the INDEPENDENT SET-circuit, all paths from an input to the output pass through only one gate that has a large number of inputs (In figure 7.2 the last gate has 7 inputs which is the number of edges in the graph. An independent set circuit for a graph with e edges will have e inputs to the last gate. Note that e is at most $O(n^2)$).
2. In the DOMINATING SET-circuit, all paths from an input to the output pass through two gates that have a large number of inputs: (1) on the second level, gate x has $deg(x)+1$ inputs, and (2) the output node will have n (the number of vertices) inputs.

We will measure the complexity of a problem by the maximum number of big gates it has on a path from the input to the output.

Definition 7.28.

1. A ***large gate*** is a gate that has more than two inputs. We may make it more than some constant number of inputs if we are discussing a family of circuits.
2. The ***depth of a circuit*** is the maximum length of a path from an input to the output.
3. The ***weft of a circuit*** is the maximum number of large gates on a path from an input to the output.

> $C[t,d]$
> *Instance:* A circuit with weft at most t and depth at most d, and a parameter k.
> *Question:* Is there a satisfying input of weight k?

We define the W-hierarchy. Downey and Fellows [DF99] first defined the W-hierarchy; however, J. Buss and Islam [BI06] is an excellent modern reference with simplified proofs.

Definition 7.29. Let A be a parameterized problem. Let $t \in \mathbb{N}$.

1. $A \in$ ***W[t]*** if there is a constant d (which may be a function of k) and a reduction from A to $C[t,d]$. (This will usually be easy to show.)
2. A is ***W[t]-hard*** if, for all d, $C[t,d]$ reduces to A.

3. A is ***W[t]-complete*** if it is in W[t] and is W[t]-hard.
4. $A \in$ ***W[SAT]*** if A reduces to SAT. Hard and complete are defined in the usual way.
5. $A \in$ ***W[P]*** if A reduces to CIRCUIT SAT. Hard and complete are defined in the usual way.
6. $A \in$ ***XP*** if there exists functions f, g such that A can be solved in time $f(k) \cdot n^{g(k)}$. Hard and complete are defined in the usual way.

The following are known.

1. FPT problems are reducible to $C[0, O(f(k))]$ and thus in W[0].
2. We have defined W[1] in two ways, one using NONDET TM ACCEPTANCE and one using $C[1, d]$. These are equivalent.
3. One can also define W[i] just like the NONDET TM ACCEPTANCE definition of W[1] except we allow the Turing machine to have i tapes.
4. If $i \leq j$, then W[i] $\subseteq$ W[j] (this is obvious). This hierarchy is believed to be proper.
5. INDEPENDENT SET is W[1]-complete, and DOMINATING SET $\in$ W[2]-complete. Hence if DOMINATING SET reduces to INDEPENDENT SET, then W[1] = W[2], which, as noted in the last point, is thought to be unlikely.
6. There are many problems that are W[1]-complete. There are a few that are W[2]-complete. There seem to be very few problems that are in higher levels or even seem to be in higher levels.
7. Downey and Fellows [DF99] (see also [Nie06]) showed that the following problem is W[2]-complete.

HITTING SET
Instance: A hypergraph $H = (V, E)$ and a number $k \in \mathbb{N}$. (k will be the parameter.)
Question: Is there a set $V' \subseteq V$. $|V'| = k$, such that, for all $e \in E$, there exists $v \in V' \cap e$?
The set V' is called a ***hitting set***.

So we have

$$\text{FPT} = \text{W}[0] \subseteq \text{W}[1] \subseteq \text{W}[2] \subseteq \cdots \subseteq \text{W}[\text{SAT}] \subseteq \text{W}[\text{P}] \subseteq \text{XP}.$$

There is a (somewhat) natural problem in XP − FPT. We present it and give some points about the proof.

> DETERMINISTIC TM EMPTY STRING ACCEPTANCE (DTMESA)
> *Instance:* A Deterministic Turing Machine M, $n \in \mathbb{N}$ (in unary) and $k \in \mathbb{N}$ (the parameter).
> *Question:* Does M accept the empty string in time $\leq n^k$? Note that DETERMINISTIC TM EMPTY STRING ACCEPTANCE $\in$ XP.

Downey and Fellows [DF99] proved the following:

Theorem 7.30.

1. $XP - FPT \neq \emptyset$. *This is proven by a diagonalization and, hence, does not yield a natural set in* $XP - FPT$. *Let* $D \in XP - FPT$ *be this set.*
2. *DETERMINISTIC TM EMPTY STRING ACCEPTANCE is XP-complete under parameterized reductions. Hence D reduces to DETERMINISTIC TM EMPTY STRING ACCEPTANCE.*
3. *DETERMINISTIC TM EMPTY STRING ACCEPTANCE* $\in XP - FPT$. *This follows from items 1 and 2.*

Very few other XP-complete sets are known.

Note: Downey et al. [DFR98] have defined an alternative (but equivalent) definition of the W hierarchy that uses descriptive complexity theory.

For the next exercise we will consider the following variant of DOMINATING SET.

> CONNECTED DOMINATING SET
> *Instance:* A graph G and a number k.
> *Question:* Is there a dominating set X of size $\leq k$ such that X induces a connected graph?

Exercise 7.31.

1. Prove that there is a parameterized reduction from DOMINATING SET to CONNECTED DOMINATING SET. (Since DOMINATING SET is W[2]-complete, this shows that CONNECTED DOMINATING SET is W[2]-hard.)
2. Prove that CONNECTED DOMINATING SET $\in$ W[2] by creating an instance of WEIGHTED CIRCUIT SAT with weft two for it.
3. Prove that CONNECTED DOMINATING SET is W[2]-complete. (This follows from the first two parts.)

7.12 Lower Bounds on Approximations via W[1]-Hardness

We define polynomial time approximation schemes (PTAS) and efficient polynomial time approximation Schemes (EPTAS). (We will study PTAS a lot in chapter 9.) We then use the assumption W[1] $\neq$ FPT to show that some problems are unlikely to have an EPTAS. In chapters 8 and 9 we will use the assumption P $\neq$ NP to show that some problems are unlikely to have a PTAS.

Definition 7.32.

1. Let A be a min-problem (e.g., given a graph, return the size of the smallest vertex cover). A ***polynomial time approximation scheme*** (PTAS) for A is an algorithm that takes as input (x, ε) (x an instance of A and $\varepsilon > 0$) and outputs a solution that is $\leq (1+\varepsilon)A(x)$. The only running time constraint is that if we fix ε, the PTAS must run in time polynomial in the size of the remaining input. Note that this allows very bad running times in terms of ε. For example, a PTAS can run in time $n^{2^{2^{\varepsilon^{-1}}}}$ because, for any given value of ε, this is a polynomial running time.
2. Let A be a max-problem (e.g., given a graph, return the size of the largest clique). A ***polynomial time approximation scheme (PTAS)*** for A is an algorithm that takes as input (x, ε) (x an instance of A and $\varepsilon > 0$) and outputs a solution that is $\geq (1-\varepsilon)A(x)$. We have the same running time constraint as in part 1.

Taking a hint from our study of FPT we define the following.

Definition 7.33. Let A be a min-problem (a similar definition can be made for max-problems). An ***efficient polynomial time approximation scheme*** (EPTAS) for A is an algorithm that takes as input (x, ε) (x an instance of A and $\varepsilon > 0$) and outputs a solution that is $\leq (1+\varepsilon)A(x)$. The only running time constraint is that there is a computable function f so that the algorithm runs in time $f(\frac{1}{\varepsilon}) \cdot n^{O(1)}$. Note that this allows very bad running times in terms of ε. For example, a EPTAS can run in time $2^{2^{\varepsilon^{-1}}} n^2$ because, for any given value of ε, this is of the right form. One caveat: We will have a problem where sometimes there is no solution at all. In this case, an EPTAS will just output "no solution."

The following problem comes up in the human genome project. We use the formulation of Deng et al. [DLL+02].

Definition 7.34. If x, y are strings of the same length, then $d(x, y)$ is the number of places they differ on.

DISTINGUISHED SUBSTRING SELECTION

Instance: A set $B = \{b_1, \ldots, b_{n_1}\}$ of bad strings, a set $G = \{g_1, \ldots, g_{n_2}\}$ of good strings, and $L, k_b, k_g \in \mathbb{N}$. All of the strings in B are of length $\geq L$, and all the strings in G are of length L. All of the strings are over the same alphabet Σ, which we will assume is a constant. We will take $n = \max\{n_1, n_2\}$, and we assume that the strings in B are of length $\leq 2L$, so the input size is $O(nL)$.

Question: (Set Version) Is there a string s, $|s| = L$, such that the following occur:

1. For every $1 \leq i \leq n_1$, there is a substring t_i of b_i, $|t_i| = L$, such that $d(s, t_i) \leq k_b$. (So s is close to a substring of every bad string.)
2. For every $1 \leq i \leq n_2$, $d(s, g_i) \geq k_g$. (So s is far from every good string.)

Output: (Function Version) If no such string s exists then output NO; however if a string exists, output one of them.

Note: The terminology "good" and "bad" has its motivation in the application to designing genetic markers to distinguish the sequences of harmful germs, for which it is good for the markers to bind, from the human sequences, for which it is bad for the markers to bind. This terminology may have its origin in Lanctôt et al. [LLM$^+$03].

Deng et al. [DLL$^+$02] proved the following.

Theorem 7.35. *There is a PTAS for DISTINGUISHED SUBSTRING SELECTION. In particular, there is a function f and an algorithm A such that on an instance of DISTINGUISHED SUBSTRING SELECTION and an ε, A runs in time time $(nL)^{f(1/\varepsilon)}$ and outputs, if it exists, an s such that the following holds:*

1. *For every $1 \leq i \leq n_1$, there is a length-L substring t_i of b_i such that $d(s, t_i) \leq (1 + \varepsilon)k_b$.*
2. *For every $1 \leq i \leq n_2$, $d(s, g_i) \geq (1 - \varepsilon)k_g$.*

We have used the term DISTINGUISHED SUBSTRING SELECTION for both the function version and set version of the problem. We will now need to distinguish them.

Notation 7.36.

1. DISTINGUISHED SUBSTRING SELECTION-FUN is the function version of DISTINGUISHED SUBSTRING SELECTION.
2. DISTINGUISHED SUBSTRING SELECTION-SET-k_b, k_g is the parameterized version with k_b, k_g fixed.

We will now link approximating DISTINGUISHED SUBSTRING SELECTION-FUN to DISTINGUISHED SUBSTRING SELECTION-SET-k_b, k_g having an FPT. Our proof is essentially due to

Cesati and Trevisan [CT97] who proved a general theorem relating EPTASs and FPT (which you will prove in the exercises after the next theorem).

Theorem 7.37. *If* DISTINGUISHED SUBSTRING SELECTION-FUN *has an* EPTAS, *then the* DISTINGUISHED SUBSTRING SELECTION-SET-k_b, k_g *is in* FPT.

Proof. Assume there is an EPTAS for DISTINGUISHED SUBSTRING SELECTION-FUN. It runs in time $f(1/\varepsilon) \cdot (nL)^{O(1)}$.

The following is an FPT algorithm for DISTINGUISHED SUBSTRING SELECTION-SET-k_b, k_g. Fix k_b and k_g and let $k = \max\{k_b, k_g\}$.

1. Input an instance I of DISTINGUISHED SUBSTRING SELECTION-SET-k_b, k_g. We take I to be the entire instance including k_b, k_g.
2. Run the EPTAS on I with $\varepsilon = \frac{1}{2k}$. Note that this takes time $f(2k)(nL)^{O(1)}$.
3. (This is commentary and is not part of the algorithm.)
   If the EPTAS outputs a string s, then the following holds:
   a. For every $1 \le i \le n_1$ there is a length-L substring t_i of b_i such that $d(s, t_i) \le (1 + \frac{1}{2k})k_b = k_b + \frac{k_b}{2k}$. Since $d(s, t_i) \in \mathbb{N}$, we have $d(s, t_i) \le k_b$.
   b. For every $1 \le i \le n_2$, $d(s, g_i) \ge (1 - \frac{1}{2k})k_g = k_g - \frac{k_g}{2k}$. Since $d(s, t_i) \in \mathbb{N}$, we have $d(s, g_i) \ge k_g$.
   c. Because of the above two points, $I \in$ DISTINGUISHED SUBSTRING SELECTION-SET-k_b, k_g.

   If the EPTAS outputs "There is no such string," then clearly $I \notin$ DISTINGUISHED SUBSTRING SELECTION-SET-k_b, k_g.
4. If the EPTAS returns a string, then output YES $I \in$ DISTINGUISHED SUBSTRING SELECTION-SET-k_b, k_g.
   If the EPTAS does not return a string, then output NO, $I \notin$ DISTINGUISHED SUBSTRING SELECTION-SET-k_b, k_g.

From the comments made in the algorithm, we have that it is correct and works in time $f(k)(nL)^{O(1)}$). Hence DISTINGUISHED SUBSTRING SELECTION-SET-k_b, k_g is FPT. □

The above theorem is only interesting if DISTINGUISHED SUBSTRING SELECTION-SET-k_b, k_g is W[1]-hard. It is! Gramm et al. [GGN06] proved the following, which we will not prove.

Theorem 7.38.

1. *DISTINGUISHED SUBSTRING SELECTION-SET-k_b, k_g is W[1]-hard.*
2. *If there is an EPTAS for DISTINGUISHED SUBSTRING SELECTION, then W[1] = FPT. (This follows from part 1.)*

Exercise 7.39.

1. Show that, if there is an EPTAS for DOMINATING SET, then FPT = W[1].
2. Formulate a general theorem that links EPTASs and FPT.

The general theorem that you will prove in exercise 7.39 and that Cesati and Trevisan [CT97] proved has very few applications. See the paper of Cesati and Trevisan for one more.

7.13 GRID TILING

All of the results in this section are from Cygan et al. [CFK+15].

We introduce two parameterized grid tiling problems and obtain lower bounds on them. These problems are contrived; however, by reductions, we will use them to get lower bounds on other parameterized problems that we do care about.

7.13.1 GRID TILING

Definition 7.40. Let G be a grid.

1. Two cells are ***up-down neighbors*** if one is directly above the other.
2. Two cells are ***left-right neighbors*** if one is directly to the right of the other.

GRID TILING PROBLEM (GRID)
Instance: A $k \times k$ grid, an $n \in \mathbb{N}$, and each cell $S(i,j)$ in the grid has a subset of $\{1,\ldots,n\} \times \{1,\ldots,n\}$. (We will use $S(i,j)$ to denote the set of ordered pairs.)
Question: Is there a set of ordered pairs

$$\{p_{i,j} = (x_{i,j}, y_{i,j}) \in S_{i,j} : 1 \leq i,j \leq n\}$$

such that the following happens?

- For all $1 \leq j \leq n$, $1 \leq i \leq n-1$, $y_{i,j} = y_{i+1,j}$ (so all of the up-down neighbors have the same second coordinate).
- For all $1 \leq i \leq n$, $1 \leq j \leq n-1$, $x_{i,j} = x_{i,j+1}$ (so all of the left-right neighbors have the same first coordinate).

Such a way to pick out ordered pairs is called ***a solution***.
Note: Grid Numbering: The cell $S(1,1)$ is the bottom left cell. THE k-GRID TILING LE PROBLEM.
Note: Example 7.41 gives an instance of GRID and a solution.
Note: As presented, this is not a parameterized problem; however, it can easily be made into one. k is the parameter.

Example 7.41. An instance and solution for GRID with $k=3$ and $n=100$.

(2, 13) **(7, 8)** (8, 12)	**(7, 11)** (8, 8)	**(7, 1)** (11, 2)
(1, 12) **(9, 8)** (10, 11) (20, 80)	(5, 100) (6, 99) **(9, 11)**	**(9, 1)** (41, 8)
(1, 1) **(21, 8)**	**(21, 11)** (15, 37)	**(21, 1)**

Theorem 7.42.

1. *There is a k-linear FPT reduction from CLIQUE to GRID.*
2. *GRID is W[1]-hard. (This follows from part 1.)*
3. *Let f be any computable function. Assuming ETH, GRID requires $f(k) \cdot n^{\Omega(k)}$ (This follows from part 1 and exercise 7.57.)*

Proof. Here is the reduction:

1. Input (G, k). Let $G=(V, E)$. We assume $V=\{1, \ldots, n\}$. The k-parameter for the grid problem will be k, and the n will be n.
2. For $1 \leq i, j \leq k$, we define the set $S(i,j)$:
   a. For $1 \leq i \leq k$, $S(i, i)=\{(a, a) \mid 1 \leq a \leq n\}$. (So all of the diagonal cells have the same set of ordered pairs.)
   b. For $1 \leq i < j \leq k$, $S(i,j)=\{(a, b) \mid \{a, b\} \in E\}$. (So all of the off-diagonal cells have the same set of ordered pairs. Note that if a cell has (a, b), then it also has (b, a).)

If G has k-clique $\{v_1, \ldots, v_k\}$, then there is a solution to the grid problem:

1. For $1 \leq i \leq k$, pick (v_i, v_i) from $S(i, i)$.
2. For $1 \leq i < j \leq n$, pick (v_i, v_j) out of $S(i,j)$ and $S(j, k)$. Note that since $\{v_1, \ldots, v_k\}$ is a clique, $(v_i, v_j) \in E$, so $(v_i, v_j) \in E$.

We leave the following as an exercise: If the grid has a solution, then G has a clique of size k. □

We will now use the lower bounds on GRID to get lower bounds on a parameterized version of list coloring.

> LIST COLORING and PLANAR LIST COLORING
> *Instance:* Graph $G = (V, E)$ and, for every $v \in V$, a subset L_v of colors. We take the colors to be $\{1, \ldots, n\}$ and note that n is *not* the number of vertices.
> *Instance:* PLANAR LIST COLORING is LIST COLORING restricted to planar graphs.
> *Instance:* Parameterized PLANAR LIST COLORING will also have as input a parameter k, and the graph is restricted to treewidth $\leq k$. We will still call this problem PLANAR LIST COLORING (this is not standard).
> *Question:* Is there a proper coloring of G where vertex v is colored by some color in L_v?

Theorem 7.43.

1. *There is a k-linear FPT reduction from GRID to PLANAR LIST COLORING.*
2. *PLANAR LIST COLORING is W[1]-hard. (This follows from part 1.)*
3. *Let f be any computable function. Assuming ETH, PLANAR LIST COLORING requires $f(k) \cdot n^{\Omega(k)}$. (This follows from part 1 and exercise 7.57.)*

Proof. Here is the reduction:

1. Input: k, n, a $k \times k$ grid of cells $S(i,j)$ that contain a set of elements of $\{1, \ldots, n\} \times \{1, \ldots, n\}$. We will call this instance of GRID, $\mathcal{G}$.
2. For every $a, a' \in \{1, \ldots, n\}$ such that $a < a'$, and every $b, b' \in \{1, \ldots, n\}$ (no restriction), create a vertex v with $L_v = \{(a, b), (a', b')\}$. Let X be the set of these $\binom{n}{2}n^2$ vertices. We will need several copies of X, but we do not subscript it to avoid too much notation.
3. For every $1 \leq i \leq j \leq n$, create node $v_{i,j}$ with $L_{v(i,j)} = S(i,j)$.
4. We now put in the horizontal edges. For every $1 \leq i \leq k-1$ and $1 \leq j \leq k$, take a copy of X. Put an edge between (1) $v_{i,j}$ and every vertex in X, and (2) $v_{i+1,j}$ and every vertex in X.
5. We now put in the vertical edges. For every $1 \leq i \leq k$ and $1 \leq j \leq k-1$, take a copy of X. Put an edge between (1) $v_{i,j}$ and every vertex in X, and (2) $v_{i,j+1}$ and every vertex in X.
6. Call the resulting graph together with the lists of colors (G, L).

Exercise 7.44 completes the proof. □

Exercise 7.44. This exercise refers to theorem 7.43

1. Show that $\mathcal{G}$ has a solution if and only if (G, L) has a list coloring.
2. Show that G has treewidth $\leq k$.

7.13.2 Grid tiling with $\leq$

We now look at a variant of grid tiling that will help us prove a lower bound on the scattered set problem.

> THE GRID TILING LE PROBLEM (GRID-LE)
> *Instance:* A $k \times k$ grid, an $n \in \mathbb{N}$, and each cell $S(i,j)$ in the grid has a subset of $\{1, \ldots, n\} \times \{1, \ldots, n\}$. (We will use $S(i,j)$ to denote the set of ordered pairs.)
> *Question:* Is there a set of ordered pairs
>
> $$\{p_{i,j} = (x_{i,j}, y_{i,j}) \in S_{i,j} : 1 \leq i,j \leq n\}$$
>
> such that the following happens?
>
> - For all $1 \leq j \leq n$, $1 \leq i \leq n-1$, $y_{i,j} \leq y_{i+1,j}$ (so all of the up-down neighbors have the second coordinate monotone increasing as you go up).
> - For all $1 \leq i \leq n$, $1 \leq j \leq n-1$, $x_{i,j} \leq x_{i,j+1}$ (so all of the left-right neighbors have the first coordinate monotone increasing as you go right).
>
> Such a way to pick out ordered pairs is called ***a solution***.
> *Note:* Grid Numbering: The cell $S(1,1)$ is the bottom left cell.
> *Note:* Example 7.45 gives an instance of GRID and a solution.

Example 7.45. An instance and solution for GRID-LE with $k = 3$ and $n = 100$.

(2, 13) **(7, 9)** (8, 12)	**(11, 27)** (8, 8)	**(87, 17)** (11, 2)
(1, 12) **(9, 8)** (10, 11) (20, 80)	(5, 100) (6, 99) **(9, 15)**	**(84, 17)** (41, 8)
(1, 1) **(21, 8)**	**(31, 11)** (15, 37)	**(42, 1)**

Theorem 7.46.

1. *There is a k-linear FPT reduction from GRID to GRID-LE.*
2. *GRID-LE is W[1]-hard. (This follows from part 1.)*
3. *Let f be any computable function. Assuming ETH, GRID-LE requires $f(k) \cdot n^{\Omega(k)}$. (This follows from part 1 and exercise 7.57.)*

$S^I_{4i-3,4j-3}$: $(iN - z, jN + z)$	$S^J_{4i-3,4j-2}$: $(iN + \alpha, jN + z)$	$S^I_{4i-3,4j-1}$: $(iN - \alpha, jN + z)$	$S^J_{4i-3,4j}$: $(iN + z, jN + z)$
$S^I_{4i-2,4j-3}$: $(iN - z, jN + b)$	$S^J_{4i-2,4j-2}$: $((i+1)N, (j+1)N)$	$S^I_{4i-2,4j-1}$: $(iN, (j+1)N)$	$S^J_{4i-2,4j}$: $(iN + z, (j+1)N + b)$
$S^I_{4i-1,4j-3}$: $(iN - z, jN - b)$	$S^J_{4i-1,4j-2}$: $((i+1)N, jN)$	$S^I_{4i-1,4j-1}$: (iN, jN)	$S^J_{4i-1,4j}$: $(iN + z, (j+1)N - b)$
$S^I_{4i,4j-3}$: $(iN - z, jN - z)$	$S^J_{4i,4j-2}$: $((i+1)N + \alpha, jN - z)$	$S^I_{4i,4j-1}$: $((i+1)N - \alpha, jN - z)$	$S^J_{4i,4j}$: $(iN + z, jN - z)$

Figure 7.3
Grid tiling reduction

Proof sketch. Here is the reduction:

Takes the same input as grid tiling but instead requires that the first coordinate of $p_{i,j} \leq$ the first coordinate of $p_{i+1,j}$. Similarly, the second coordinate of $p_{i,j} \leq$ the second coordinate of $p_{i,j+1}$. We can prove this problem is W[1]-hard and imply no $f(k) \cdot n^{o(k)}$ algorithm by reduction from grid tiling. We use the gadget in figure 7.3 to blow up each tile in our original instance into a grid of four by four tiles. □

> k-SCATTERED SET (k-SCAT SET)
> *Instance:* A graph G and a number d.
> *Question:* Are there k vertices with pairwise distances $\geq d$?

If $d = 2$, then the k-SCAT SET problem is just INDEPENDENT SET$_k$, which we already know is W[1]-hard (though we do not know if ETH yields an $f(k) \cdot n^{\Omega(k)}$ lower bound). However, PLANAR INDEPENDENT SET is FPT. What about Planar k-SCAT SET? We state what is known:

Theorem 7.47.

1. *There is a k-linear FPT reduction from GRID-LE to k-SCAT SET.*
2. *k-SCAT SET is W[1]-hard. (This follows from part 1.)*
3. *Let f be any computable function. Assume ETH. k-SCAT SET requires $f(k) \cdot n^{\Omega(k)}$. (This follows from part 1 and exercise 7.57.)*

7.13.3 The k-Unit Disk Graphs Problem

> k-UNIT DISK GRAPH
> *Instance:* A set of points P in the plane.
> *Question:* Can you draw k unit disks centered on $p \in P$ without the disks intersecting?

Theorem 7.48.

1. *There is a reduction from GRID-LE to UDG$_k$ such that an instance of GRID-LE with parameter k is mapped to an instance of UDG$_k$ with parameter k^2.*
2. *UDG$_k$ is W[1]-hard (this follows from part 1).*
3. *If there is an $f(k) \cdot n^{o(\sqrt{k})}$ algorithm for UDG$_k$ then there is an $f(k^2) \cdot n^{o(k)}$ algorithm for GRID-LE (which is thought to be unlikely).*

Proof sketch. We show that GRID-LE reduces to UDG$_k$. The parameter will go from k to k^2.

1. Input a $k \times k$ grid where each cell $S(i,j)$ has a subset of $\{1, \ldots, n\} \times \{1, \ldots, n\}$.
2. We create an instance of UDG$_{k^2}$ as follows.
   a. For every $S_{i,j}$ take all the ordered pairs in $S_{i,j}$ and arrange them in a grid in the obvious way.
   b. Take each of these grids of points and arrange them into a $k \times k$ grid.
   c. Pick a distance between the grids-of-points carefully so that the construction works.

See figure 7.4 for an example of the reduction.

□

$S[1,3]$: (1,1) (2,5) (3,3)	$S[2,3]$: (3,2) (2,3)	$S[3,3]$: (5,4) (3,4)
$S[1,2]$: (5,1) (1,4) (5,3)	$S[2,2]$: (3,1) (2,2)	$S[3,2]$: (1,1) (2,3)
$S[1,1]$: (1,1) (3,1) (2,4)	$S[2,1]$: (2,2) (1,4)	$S[3,1]$: (1,3) (2,3) (3,3)

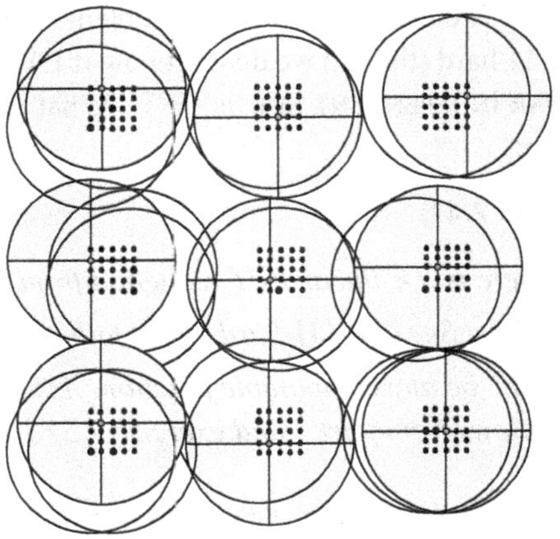

Cygan, Fomin, Kowalik, Lokshtanov, Marx, Pillipczuk, Pilipczuk, Saurabh 2015

Figure 7.4
An example of the reduction from GRID-LE to UDG_k.

7.14 Further Results

7.14.1 Another Look at DOMINATING SET

We have stated that DOMINATING SET is W[2]-complete and, hence, unlikely to be in FPT. However, using ETH and SETH, one can obtain sharper bounds on the parameterized complexity of DOMINATING SET.

Let $k \in \mathbb{N}$. Let DOMINATING SET_k be the problem of, given a graph G, is there a dominating set of size k. Clearly this problem is in time $O(n^{k+1})$. Eisenbrand and Grandoni [EG04] have obtained slightly better algorithms. We state two known lower bounds. They are probably folklore since our only source is a workshop on fine-grained complexity held by the Max Planck Institute in 2019 [Unk19].

Theorem 7.49.

1. *Assume ETH. There exists $\delta > 0$ such that, for large k, DOMINATING SET_k requires time $\Omega(n^{\delta k})$.*
2. *Assume SETH. Let $k \geq 3$ and $\varepsilon > 0$. DOMINATING SET_k requires time $\Omega(n^{k-\varepsilon})$.*

Those same notes leave the following as an exercise:

Exercise 7.50. Assume ETH. Show that SUBSET SUM cannot be solved in time $2^{o(n)}$.

7.14.2 Restrictions on Graphs

Some graph problems are in FPT if the graphs are restricted. Courcelle [Cou90] and independently Borie et al. [BPT92] showed the following: Let $\mathcal{P}$ be some graph property that is definable in monadic second order logic. Let $k \in \mathbb{N}$. There is a linear time algorithm for $\mathcal{P}$ restricted to graphs of treewidth $\leq k$. (See Flum and Grohe [FG06] for a complete treatment of Courcelle's theorem and its applications.) From this theorem the following problems are FPT with the parameter being the treewidth.

1. Given a boolean circuit, is it satisfiable?
2. Given a graph G and a number k, is G k-colorable?
3. Given a graph G and a number k, does G give an independent set of size k?
4. Given a graph G and a number k, is the crossing number of $G \leq k$? (This one has parameter $k +$ treewidth.)

These results raised the question of whether bounding the clique width can also be used to put a problem into FPT.

Let f be any computable function. Let t bound the clique number. Fomin et al. [FGLS10] have shown that if you assume ETH, then several graph problems restricted to graphs of clique width $\leq t$ cannot be solved in time $f(t) \cdot n^{o(t)}$.

7.14.3 Perfect Codes: A Problem with an Interesting History

> Perfect Codes
> *Instance:* A graph $G = (V, E)$ and $k \in \mathbb{N}$.
> *Question:* Is there a $V' \subseteq V$, $|V'| = k$, such that, for each vertex $v \in V$, there is exactly one $v' \in V'$ that is either v or adjacent to v.

The problem Perfect Codes has an interesting history.

Theorem 7.51.

1. *(Downey and Fellows [DF99]) Independent Set reduces to Perfect Codes, hence Perfect Codes is W[1]-hard.*
2. *(Downey and Fellows [DF99]) Perfect Codes ∈ W[2].*
3. *(Cesati [Ces03]) Perfect Codes reduces to Nondet TM Acceptance and hence Perfect Codes ∈ W[1]. Therefore Perfect Codes is W[1]-complete.*

Downey and Fellows conjectured that Perfect Codes is intermediary: in W[2] − W[1] but not W[2]-complete. Hence it was a surprise when Cesati showed that Perfect Codes is W[1]-complete. It was interesting that Cesati showed PERFECT CODES $\in$ W[1] by a reduction.

We briefly discuss why the problem is called PERFECT CODE. If x, y are strings of the same length, then $d(x,y)$ is the number of bits they differ on. Let $G_n = (V, E)$ be the graph with $V = \{0,1\}^n$ and $E = \{(x,y) \mid d(x,y) = 1\}$.

1. G_3 has a perfect code of size 2: Use the vertices $\{000, 111\}$.
2. For all n such that $n+1$ is a power of 2, G_n has a perfect code of size $\lg n + 1$.
3. Perfect codes are useful for error-correcting codes.
4. The literature on error correcting code is vast.

7.14.4 Consequence of ETH for Parameterized Complexity

We show that, assuming ETH, we can obtain better lower bounds for both problems within FPT and outside of FPT.

ETH Implies Lower Bounds on Parameterized Problems Recall from theorem 7.1 that VERTEX COVER is in time $O(kn + k^2 2^k)$. Can we lessen the dependence on k by getting, for example, a $O(2^{k^{0.9}} n^3)$ algorithm (this would be worse then $O(kn + k^2 2^k)$ when k is small but might be better when k is large). What about other parameterized problems? Cai and Juedes [CJ03] showed that such an algorithm would imply that ETH is false:

Theorem 7.52. *Assume ETH. Let $L \in \mathbb{N}$. Then VERTEX COVER, DOMINATING SET, and DIRECTED HAMILTONIAN CYCLE cannot be solved in time $2^{o(k)} n^L$.*

Proof. We prove this for VERTEX COVER. The rest are similar.

By theorem 6.11, assuming ETH, VERTEX COVER requires $2^{\Omega(n)}$ time. If VERTEX COVER could be done in $2^{o(k)} n^L$ time, then, since $k \le n$, this would yield a $2^{o(n)} n^L$ algorithm for VERTEX COVER. This violates the lower bound of $2^{\Omega(n)}$. □

Exercise 7.53. Prove the rest of theorem 7.52.

Exercise 7.54. Obtain results similar to theorem 7.52 for PLANAR VERTEX COVER, PLANAR DOMINATING SET, PLANAR DIRECTED HAMILTONIAN CYCLE.

We note upper bounds (i.e., algorithms) for PLANAR DOMINATING SET and PLANAR VERTEX COVER and leave it to the reader to contrast them with the lower bounds from exercise 7.54.

1. Alber et al. [ABF$^+$02] show that PLANAR DOMINATING SET set can be solved in $2^{O(\sqrt{k})} n^{O(1)}$ time.
2. Demaine et al. [DFHT05] show that PLANAR k-VERTEX COVER can be solved in $2^{O(\sqrt{k})} n^{O(1)}$ time.

There are other planar FPT problems for which, assuming ETH, we have matching upper and lower bounds.

ETH Implies $f(k) \cdot n^{\Omega(k)}$ Lower Bounds Recall that, by theorem 7.16, assuming FPT $\neq$ W[1], CLIQUE and INDEPENDENT SET are not FPT. More precisely, there is no function f such that CLIQUE can be solved in time $f(k) \cdot n^{O(1)}$. What about (say) $f(k) \cdot n^{\sqrt{k}}$? What about other parameterized problems?

Theorem 7.55. *Assume ETH. Let $f(k)$ be any computable function. There is no algorithm for CLIQUE or INDEPENDENT SET that takes time $f(k) \cdot n^{o(k)}$ time.*

Proof. By theorem 6.11, assuming ETH, 3-COLORING requires $2^{\Omega(n)}$ time.

Assume CLIQUE is solvable in $f(k) \cdot n^{o(k)}$ time. Then there exists a monotone increasing and unbounded s such that CLIQUE is solvable in time $f(k) \cdot n^{k/s(k)}$. We give an algorithm for 3-COLORING that first does a reduction from 3-COLORING to CLIQUE and then runs the CLIQUE algorithm.

We can assume that $f(k)$ is monotone increasing and unbounded. This will come up toward the end.

1. Input $G=(V,E)$, a graph. We will derive the value of k that we need later. We impose one condition now: k divides n for notational convenience.
2. Split V into k groups $V_1, \ldots, V_k$ of roughly n/k vertices each.
3. For each $1 \leq i \leq k$, find all valid 3-colorings of V_i. For each coloring, we have a new vertex. There are at most $3^{|V_i|} \leq 3^{\lceil n/k \rceil}$ vertices per i, so at most $k \cdot 3^{\lceil n/k \rceil}$ new vertices. These new vertices will be our vertex set V'. Now we analyze the time for this step. There are k V_i's to check. For each one, we look at the $3^{\lceil n/k \rceil} = O(3^{n/k})$ possible 3-colorings. We check each possible 3-coloring to see if it really is a 3-coloring, which takes $O((n/k)^2)$ time. Hence the total time is

$$O\left(k \cdot (n/k)^2 \cdot 3^{n/k}\right) = O\left(\frac{n^2}{k} \cdot 3^{n/k}\right) = O\left(n^2 \cdot 3^{n/k}\right).$$

4. If two vertices in V' correspond to the same V_i, there is no edge between them.
5. For all $1 \leq i < j \leq \frac{n}{k}$, for all $x \in V_i$, $y \in V_j$, we do the following: If the coloring $x \cup y$ of $V_i \cup V_j$ is valid, then put an edge between x and y; otherwise, do not. We analyze how long this step takes. There are $\binom{n/k}{2} \leq \frac{n^2}{2k^2}$ pairs (i,j). For each pair, there are $3^{n/k} \cdot 3^{n/k} = 3^{2n/k}$ pairs (x,y). For each pair, it takes $O((2n/k)^2) = O(n^2/k^2)$ time to check whether $x \cup y$ is a valid coloring. Hence the total time for this step is

$$O\left(\frac{n^2}{2k^2} \cdot 3^{2n/k} \cdot \frac{n^2}{k^2}\right) = O\left(\frac{n^4}{k^4} \cdot 3^{2n/k}\right) = O\left(n^4 \cdot 3^{2n/k}\right).$$

6. Let E' be the set of edges, so $G' = (V', E')$ is the new graph.
7. (This is not a step; it is a comment.)
   So far the algorithm has taken time $O(n^4 \cdot 3^{2n/k})$ and produced a graph on $\le k \cdot 3^{n/k}$ vertices. It is easy to see that G has a 3-coloring if and only if G' has a k-clique.
8. Run the CLIQUE algorithm on (G', k). This takes time

$$\le f(k) \cdot \left(k \cdot 3^{n/k}\right)^{k/s(k)} = f(k) \cdot k^{k/s(k)} \cdot 3^{n/s(k)}.$$

   If the algorithm returns YES, then output YES. If the algorithm returns NO, then output NO.

We analyze the running time and pick k. The running time is

$$O\left(n^4 \cdot 3^{3n/k} + f(k) \cdot k^{k/s(k)} \cdot 3^{n/s(k)}\right) = n^4 \cdot 2^{O(n/k)} + f(k) \cdot k^{k/s(k)} \cdot 2^{O(n/s(k))}.$$

Set $k(n)$ to be the largest value of k such that $f(k) \le n$ and $k^{k/s(k)} \le n$. Because $f(k)$ and $s(k)$ are monotone increasing and unbounded, $k(n)$ is also monotone increasing and unbounded. Also note that $k(n) < n$, so it makes sense for its role in the algorithm. The running time of our 3-coloring algorithm is then

$$n^4 \cdot 2^{O(n/k)} + f(k) \cdot k^{k/s(k)} \cdot 2^{O(n/s(k))} \le n^3 \cdot 2^{O(n/k(n))} + n^2 \cdot 2^{O(n/s(k(n)))} = 2^{o(n)},$$

which contradicts the lower bound on 3-COLORING that comes from ETH. □

We can now use CLIQUE and a certain kind of reduction to get, assuming ETH, $f(k) \cdot n^{\Omega(k)}$ lower bounds.

Definition 7.56. Let A and B be parameterized problems. A ***k-linear FPT reduction from A to B*** is an FPT reduction such that when (x, k) is mapped to (y, k'), $k' = O(k)$.

Exercise 7.57. Assume ETH. Let f be a computable function.

1. Let A be a parameterized problem with parameter k. Assume that CLIQUE reduces to A with a k-linear reduction (so A is W[1]-hard). Then A requires $f(k) \cdot n^{\Omega(k)}$ time.
2. Assume ETH. Let f be any computable function. Then the following problems have running times $\ge f(k) \cdot n^{\Omega(k)}$: MULTI COLORED CLIQUE, MULTICOLORED INDEPENDENT SET, DOMINATING SET, SET COVER, and PARTIAL VERTEX COVER.

8

Basic Lower Bounds on Approximability via PCP and Gap Reductions

8.1 Introduction

In this chapter we will show that some problems are NP-hard to approximate.

When Cook and Levin showed SAT is NP-complete, they *could not* take some known NP-complete set A and show $A \leq_p$ SAT since *there were not known NP-complete sets!* SAT was the first one. We are initially in the same position with regard to hardness of approximation. To show a problem is hard to approximate, we cannot use a reduction. We need a basic problem (actually several basic problems) analogous to SAT for NP-completeness, that we have reason to believe is hard to approximate.

Chapter Summary

1. *We define approximation algorithms of a variety of types and give examples.*
2. *We define an appropriate type of reduction needed to prove that if f is hard to approximate, then g is hard to approximate.*
3. *We describe the PCP machinery and state theorems about it (without proof).*
4. *We use the PCP machinery and the reductions from item 2 to show that several problems are are NP-hard to approximate. We also state (but do not prove) that a few other problems are NP-hard to approximate.*

Convention 8.1. When working with a function problem, we will use the notation OPT(x) for the optimal value. For example, we will use OPT(x) instead of VERTEX COVER(x). When we use OPT, the problem at hand is understood. If we are dealing with two different problems we may use subscripts like $\text{OPT}_{\text{VERTEX COVER}}$.

We use TSP and MAX 3SAT (to be defined) as running examples. recall that TSP is the following:

- Input: a weighted graph G and a number k. The weights are nonnegative integers.
- Determine whether there is a Hamiltonian cycle of weight $\leq k$.

For most of this book we have looked at *decision problems* where every instance has a *yes* or *no* answer. For example, TSP is a YES-NO question.

In the real world TSP is *not* a decision problem; indeed, in the real world one wants to *find* the optimal (minimum weight) cycle. This is the *function* version of TSP. We touched on this distinction in chapter 0 and concluded (correctly) that, with regard to polynomial time, the decision problem and the function problem are equivalent. But let's get back to the real world. One way to cope with a problem being NP-hard is to approximate it. This concept only makes sense if we are talking about a *function*, not a *set*. In this chapter we will look at functions that are naturally associated with NP-complete problems and show that they are NP-hard to approximate. In this chapter we will show how to use *gap reductions* to get lower bounds on approximations contingent on $\mathrm{P} \neq \mathrm{NP}$.

We now define MAX 3SAT, which will be one of our basic problems.

> MAX 3SAT
> *Instance:* A 3CNF formula φ.
> *Question:* What is the max number of clauses that can be satisfied simultaneously? (One might also ask for the assignment that achieves this max.)

Definition 8.2. An ***optimization problem*** consists of the following:

- The set of instances of the problem (e.g., the set of weighted graphs for TSP, the set of 3CNF formulas for MAX 3SAT).
- For each instance: the set of possible solutions (e.g., the set of Hamiltonian cycles for TSP, the set of truth assignments for MAX 3SAT).
- For each solution: a nonnegative cost or benefit (e.g., for TSP the cost is the sum of weights on the Hamiltonian Cycle, for MAX 3SAT the benefit is the number of clauses satisfied).
- An objective: either min or max (e.g., min cost of a Hamiltonian cycle for TSP, max the number of clauses that are satisfied for MAX 3SAT).

The goal of an optimization problem is to find a solution that achieves the objective: Either minimize a cost or maximize a benefit.

Now we can define an NP-optimization problem. The class of all NP-optimization problems is called NPO; it is the optimization analog of NP.

Definition 8.3. An ***NP-optimization problem*** is an optimization problem with the following additional requirements:

- All instances and solutions can be recognized in polynomial time (e.g., (1) TSP: You can tell whether a proposed cycle is Hamiltonian, (2) MAX 3SAT: You can tell whether a proposed string is a truth assignment of length n).
- All solutions are of length polynomial in the length of the instance which they solve (e.g., (1) TSP: A Hamiltonian cycle is clearly of length polynomial in the size of the graph—it's actually shorter, (2) MAX 3SAT: An assignment is clearly of length polynomial in the size of the formula—it's actually shorter).
- The cost or benefit of a solution can be computed in polynomial time (e.g., (1) TSP: Given a Hamiltonian cycle in a weighted graph, one can easily find the weight of the cycle, and (2) given an assignment for a formula, one can easily find the number of clauses it satisfies).

We can convert any NPO problem into an analogous decision problem in NP. For a min problem, we ask "Is $\mathrm{OPT}(x) \leq q$?" and for a maximization problem, we ask "Is $\mathrm{OPT}(x) \geq q$?" The optimal solution can serve as a short easily verified certificate of a "yes" answer, and so these analogous decision problems are in NP.

This means that NPO is, in some sense, a generalization of NP problems.

Convention 8.4. For the rest of this chapter, *problem* means *NPO problem*. When A and B are mentioned they are NPO problems. We will often say whether A is a min-problem or a max-problem.

Note that when we speak of solving an NPO problem we only mean finding $\mathrm{OPT}(x)$, which is the cost or benefit of the optimal solution. So we are not quite in the real world: For us a solution to the TSP problem is just to say how much the min Ham cycle costs, not to find it. However, this will suit our purposes.

8.2 Approximation Algorithms

Let's say you are trying to find a polynomial-time algorithm that will return the cost of the optimal TSP solution. You do not succeed; however, you get a polynomial-time algorithm that returns a number that is at most twice the optimal. Is that good? Can you prove that no algorithm is better unless P = NP? Before asking these questions, we need to define our terms.

Definition 8.5. In the two definitions below, ALGORITHM is a polynomial-time algorithm and $c \geq 1$ is a constant. (We will later generalize to the case where c is a function.)

- Let A be a min-problem. ALGORITHM is a ***c-approximation algorithm for A*** if, for all valid instances x,

$$\text{ALGORITHM}(x) \leq c \times \mathrm{OPT}(x).$$

- Let A be a max-problem. ALGORITHM is a ***c-approximation algorithm for A*** if, for all valid instances x,

$$\mathrm{OPT}(x) \leq c \times \text{ALGORITHM}(x).$$

Some caveats and conventions.

1. The definition of c-approximation for a min-problem makes sense. For example, if you have an algorithm for approximate Euclidean TSP (We do!) that returns a number that is $\leq \frac{3}{2}\mathrm{OPT}$, that is called a $\frac{3}{2}$-approximation algorithm.
2. The definition of c-approximation for a max-problem is awkward. For example, if you have an algorithm for approximate MAX 3SAT that returns a number that is $\geq \frac{7}{8}\mathrm{OPT}$ (We do!), that is called an $\frac{8}{7}$-approximation algorithm. Really!
3. We personally do not like this definition. We will use it for min problems. For max problems we will either avoid it or make sure to remind the reader about what is going on.
4. Why is the definition the way it is? Because this way in both max and min cases we seek c-approximation algorithms where $c \geq 1$. We will later see that this makes the definition of ***polynomial time approximation schemes (PTAS)*** (to be defined later) smooth.

In some sense, an approximation algorithm is doing pretty well if it is a c-approximation algorithm with some constant value c. But sometimes we can do even better! There are cases where, for all $\varepsilon > 0$, there is a $(1+\varepsilon)$-approximation.

Definition 8.6.

1. Let A be a min-problem. A *PTAS* for A is an algorithm that takes as input (x, ε) (x an instance of A and $\varepsilon > 0$) and outputs a solution that is $\leq (1+\varepsilon)\mathrm{OPT}(x)$. The only running time constraint is that if we fix ε, the PTAS must run in time polynomial in the size of the remaining input. Note that this allows very bad running times in terms of ε. For example, a PTAS can run in time n^{1/ε^2} because for any given value of ε, this is a polynomial running time.
2. Let A be a max-problem. A PTAS for A is an algorithm that takes as input (x, ε) (x an instance of A and $\varepsilon > 0$) and outputs a solution that is $\geq (1-\varepsilon)\mathrm{OPT}(x)$. We have the same running time constraint as in part 1.

We define several complexity classes:

Definition 8.7.

1. The class PTAS is the set of all problems for which a PTAS exists. We use the term PTAS for both the type of an approximation algorithm and the set of all problems that have that type of approximation algorithm.

2. Let A be a min-problem. $A \in$ APX if there is a constant $c \geq 1$ and an algorithm M such that $M(x)$ is $\leq c \times \text{OPT}(x)$. (This is just a c-approximation; however, we phrase it this way so that you will see the other classes are variants of it.)
3. Let A be a max-problem. $A \in$ APX if there is a constant $c \geq 1$ and an algorithm M such that $M(x)$ is $\geq \frac{1}{c} \times \text{OPT}(x)$.
4. Let A be a min-problem. $A \in$ Log-APX if there is a constant $c > 0$ and an algorithm M such that $M(x)$ is $\leq c \times \log |x| \times \text{OPT}(x)$.
5. Let A be a max-problem. $A \in$ Log-APX if there is a constant c (it does not need to be ≥ 1) and an algorithm M such that $M(x)$ is $\geq \frac{1}{c \log |x|} \times \text{OPT}(x)$.
6. Let A be a min-problem. $A \in$ Poly-APX if there is a polynomial p and an algorithm M such that $M(x)$ is $\leq p(|x|) \times \text{OPT}(x)$.
7. Let A be a max-problem. $A \in$ Poly-APX if there is a polynomial p and an algorithm M such that $M(x)$ is $\geq \frac{1}{p(|x|)} \times \text{OPT}(x)$.

The following examples are known.

Example 8.8. All of our examples are variants of TSP.

1. The ***metric TSP*** problem is the TSP problem restricted to weighted graphs that are symmetric and satisfy the triangle inequality: $w(x,y) + w(y,z) \geq w(x,z)$. There is an algorithm discovered independently by Christofides [Chr22] (in 1976 unpublished, finally published in 2022, which is the reference) and Serdyukov [Ser78] (in 1978) that gives a $\frac{3}{2}$-approximation to the metric TSP problem. Hence the metric TSP problem is in APX. This algorithm is well known. It is called ***the Christofides-Serdyukov algorithm***.
2. Karlin, Klein, and Oveis-Gharan [KKG21], in 2021, obtained the first improvement over the $\frac{3}{2}$-approx. They showed that there is a $(\frac{3}{2} - \varepsilon)$-approximation to the metric TSP problem where $\varepsilon > 10^{-36}$. This does not improve the class that metric TSP is in—it is still in APX— but it is interesting that one can do better than $\frac{3}{2}$, which was, until this result, a plausible limit on approximation.
3. The ***Euclidean TSP*** problem is the TSP problem when the graph is a set of points in the plane and the weights are the Euclidean distances. Arora [Aro98] and Mitchell [Mit99], in 1998, independently showed a PTAS for the Euclidean TSP problem. Both of their algorithms will, on input n points in the plane (which defines the weighted graph) and ε, produce a $(1+\varepsilon)$-approximation in time $O(n(\log n)^{O(1/\varepsilon)})$.
4. Arora and Mitchell actually have an algorithm that works on n points in $\mathbb{R}^d$ that runs in time $O(n(\log n)^{O(\sqrt{d}/\varepsilon)^{d-1}})$.

8.3 Basic Hard-to-Approximate Problems

The following are basic hard-to-approximate problems. We include both the upper and the lower bounds. All of the lower bounds are under the assumption $\mathsf{P} \neq \mathsf{NP}$.

1. TSP $\notin$ Poly-APX.
2. CLIQUE $\in$ Poly-APX $-$ Log-APX.
3. SET COVER $\in$ Log-APX $-$ APX.
4. MAX 3SAT $\in$ APX $-$ PTAS.

From the results listed above we have the following.

Theorem 8.9. *If P ≠ NP then*

$$\mathsf{PTAS} \subset \mathsf{APX} \subset \mathsf{Log\text{-}APX} \subset \mathsf{Poly\text{-}APX}.$$

We will discuss all four of the results in the order given above. The lower bound on approximating TSP is elementary, and we can present it in its entirety. The lower bound on approximating CLIQUE and MAX 3SAT use the PCP machinery, and hence, we will have a section on PCP. The lower bound on approximating SET COVER uses a different machinery that is a close cousin to the PCP machinery; however, we will discuss it briefly.

8.4 Lower Bounds on Approximating TSP

Recall that, by example 8.8, the metric TSP problem (where $w(a,b) + w(b,c) \geq w(a,c)$) is in APX. What about TSP problems without that condition? We show that if TSP $\in$ Poly-APX, then $\mathsf{P} = \mathsf{NP}$. Let OPT(H) be the cost of the min Hamiltonian cycle in weighted graph H. Informally, we will map instances G of HAMILTONIAN CYCLE to instances G' of TSP such that:

If $G \in$ HAMILTONIAN CYCLE, then OPT(G') is small.
If $G \notin$ HAMILTONIAN CYCLE, then OPT(G') is large.

We will then use the alleged approximation algorithm for TSP to determine which is the case. This is called a ***gap reduction*** because of the large gap between the costs of the optimal routes.

Theorem 8.10. *If TSP ∈ Poly-APX, then P = NP.*

Proof. We assume that TSP $\in$ Poly-APX and show that HAMILTONIAN CYCLE $\in$ P. Since HAMILTONIAN CYCLE is NP-complete, this will show $\mathsf{P} = \mathsf{NP}$.

To avoid notational clutter, we call the algorithm for TSP $\in$ Poly-APX ***the approx algorithm***. Let $p(n)$ be such that the algorithm returns a number $\leq p(n)\mathrm{OPT}(G)$.

Let $c(n)$ be a polynomial to be named later. As part of our reduction we will map instances G of Hamiltonian Cycle to instances G' of TSP such that:

If $G \in$ Hamiltonian Cycle, then $\mathrm{OPT}(G') = n$.
If $G \notin$ Hamiltonian Cycle, then $\mathrm{OPT}(G') \geq c(n)$.

Here is an algorithm for Hamiltonian Cycle.

1. Input $G = (V, E)$, an unweighted graph.
2. Create an instance G' of TSP, using the same vertex set as G used, as follows: (1) if $e \notin E$ then give e weight $c(n)$, and (2) if $e \in E$ then give e weight 1. Note that G' is a complete weighted graph.
3. (This is a comment, not part of the algorithm.)
   a. If $G \in$ Hamiltonian Cycle, then $\mathrm{OPT}(G') \leq n$ since you can just use the Hamiltonian cycle.
   b. If $G \notin$ Hamiltonian Cycle, then $\mathrm{OPT}(G') \geq c(n)$ since any cycle in G' will have to use at least one edge of cost $c(n)$ (actually $\mathrm{OPT}(G') \geq c(n) + n - 1$, but this is not needed).
4. Run the approx algorithm on G'.
5. (This is a comment, not part of the algorithm.)
   a. If $G \in$ Hamiltonian Cycle, then the approx alg run on G' returns a route of size $\leq np(n)$.
   b. If $G \notin$ Hamiltonian Cycle, then the approx alg run on G' returns a route of size $\geq c(n)$.

   To ensure these cases do not overlap, we pick $c(n) > np(n)$.
6. If the approx alg outputs a number $\leq np(n)$, then output Yes. If the approx alg outputs a number $> c(n)$, then output No. By the commentary in the algorithm, no other case will occur. □

The key to the proof of theorem 8.10 was creating an instance of TSP that either had a very small or very large solution, called a ***gap***. This is a paradigm for most lower bounds on approximation.

8.5 Gap Lemmas

In this section we prove two easy lemmas that show how to use a reduction that causes a gap (like the one in theorem 8.10) to obtain a lower bound on approximation algorithms. This first lemma is for max-problems, and the second one is for min-problems. The proofs are similar, and hence, the proof of the second one is omitted.

Convention 8.11. We will often use the notation $|y|$. This is the size of y; however, we will use *size* in a different way for different inputs.

1. If y is a string, then $|y|$ is the length of y.
2. If y is a graph, then $|y|$ is the number of vertices.
3. If y is a 3CNF formula, then $|y|$ might be either the number of variables or the number of clauses depending on our application.

Definition 8.12. Let g be a max-problem (e.g., CLIQUE). Let $a(n)$ and $b(n)$ be functions from $\mathbb{N}$ to $\mathbb{N}$ such that $\frac{b(n)}{a(n)} < 1$. Then $\text{GAP}(g, a(n), b(n))$ is the following problem.

> $\text{GAP}(g, a(n), b(n))$
> *Instance:* y for which you are promised that either $g(y) \geq a(|y|)$ or $g(y) \leq b(|y|)$.
> *Question:* Is $g(y) \geq a(|y|)$?

Lemma 8.13. Let A be an NP-hard set. Let g be a max-problem. Let $a(n)$ and $b(n)$ be functions from $\mathbb{N}$ to $\mathbb{N}$ such that (1) $\frac{b(n)}{a(n)} < 1$, and (2) b is computable in time polynomial in n. Assume that there exists a polynomial time reduction that maps x to y such that the following occurs:

- If $x \in A$, then $g(y) \geq a(|y|)$.
- If $x \notin A$, then $g(y) \leq b(|y|)$.

Then:

1. $\text{GAP}(g, a(n), b(n))$ is NP-hard (this follows from the premise).
2. If there is an approximation algorithm for g that, on input y, returns a number $> \frac{b(|y|)}{a(|y|)} g(y)$, then P = NP.

Proof. We just prove part 2.

We use the reduction and the approximation algorithm to obtain $A \in$ P. Since A is NP-hard, we obtain P = NP.

Algorithm for A

1. Input x.
2. Run the reduction on x to get y.
3. Run the approximation algorithm on y.

4. (This is a comment and not part of the algorithm.)
   $x \in A \rightarrow g(y) \geq a(|y|) \rightarrow$ approx on y returns $> \frac{b(|y|)}{a(|y|)} a(|y|) = b(|y|)$.
   $x \notin A \rightarrow g(y) \leq b(|y|) \rightarrow$ approx on y returns $\leq b(|y|)$.
5. If the approx returns a number $> b(|y|)$, then output YES. Otherwise output NO. (This is the step where we need $b(|y|)$ to be computable in time polynomial in $|y|$.)

□

We now look at min-problems.

We use the same name, GAP$(g, a(n), b(n))$, for the following problem.

Definition 8.14. Let g be a min-problem (e.g., TSP). Let $a(n)$ and $b(n)$ be functions from $\mathbb{N}$ to $\mathbb{N}$ such that $\frac{b(n)}{a(n)} > 1$. Then GAP$(g, a(n), b(n))$ is the following problem.

We use the same notation, GAP$(g, a(n), b(n))$, for min-problems as we did for max-problems. When we use these lemmas, the meaning will be clear from context.

> GAP$(g, a(n), b(n))$
> *Instance:* y for which you are promised that either $g(y) \leq a(|y|)$ or $g(y) \geq b(|y|)$.
> *Question:* Is $g(y) \leq a(|y|)$?

We now state a lemma that is useful for obtaining lower bounds on approximation min-problems. The proof is similar to that of lemma 8.13 and, hence, is omitted.

Lemma 8.15. Let A be an NP-complete set. Let g be a min-problem. Let $a(n)$ and $b(n)$ be functions from $\mathbb{N}$ to $\mathbb{N}$ such that (1) $\frac{b(n)}{a(n)} > 1$, and (2) b is computable in time polynomial in n. Assume there exists a polynomial time reduction that maps x to y such that the following occurs:

- If $x \in A$ then $g(y) \leq a(|y|)$.
- If $x \notin A$ then $g(y) \geq b(|y|)$.

Then:

1. GAP$(g, a(n), b(n))$ is NP-hard (this follows from the premise).
2. If there is an approximation algorithm for g that, on input y, returns a number $< \frac{b(|y|)}{a(|y|)} g(y)$, then P $=$ NP.

Definition 8.16. We will refer to reductions like the ones in lemma 8.13 and 8.15 as ***gap reductions with ratio*** $\frac{b(n)}{a(n)}$.

8.6 PCP Machinery

In this section we discuss a characterization of NP in terms of ***probabilistically checkable proofs***. This characterization has a hard (and long) proof that we will omit. However, once we have the characterization, we will use it to construct gap reductions that will show some approximation problems are NP-hard.

Recall the following notation and definition.

Notation 8.17. Let $\exists^p y$ mean that there exists y such that $|y|$ is polynomial in $|x|$, where x is understood. Let $\forall^p y$ mean for all y such that $|y|$ is polynomial in $|x|$, where x is understood.

Definition 8.18. $A \in$ NP if there exists a polynomial predicate B such that

$$A = \{x \mid \exists^p y : B(x, y)\}.$$

We want to rewrite this and modify it.

Definition 8.19.

1. An ***oracle turing machine–bit access*** (henceforth OTM-BA) is an oracle Turing machine where (1) the oracle is a string of bits and (2) the requests for the bits is made by writing down the address of the bit. By convention, if the string is s long and a query is made for bit $t > s$, then the answer is NO.
2. We denote an oracle Turing machine by $M^{()}$. If $M^{()}$ is an oracle Turing machine and y is the string being used for the oracle, and x is an input, we denote the computation of $M^{()}$ on x with oracle y by $M^y(x)$.
3. A ***polynomial OTM-BA (POTM-BA)*** is an OTM-BA that runs in polynomial time. Note that a POTM-BA can use an oracle string of length 2^{poly} since it can write down that it wants bit position (say) 2^{n^2} with n^2 bits.
4. We give two equivalent definitions of a ***randomized POTM-BA (RPOTM-BA)***. One is intuitive and the other is better for proofs.
   a. An ***RPOTM-BA*** is a POTM-BA that is allowed to flip coins. So there will be times where, rather than do STEP A, it will do STEP A with probability (say) 1/3 and STEP B with probability 2/3. Hence we cannot say *The machine accepts x using oracle bit string y*, but we can say *The machine will accept x using oracle bit string y with probability* ≥ 0.65.
   b. Note that for a RPOTM-BA computation, many coins are flipped and are used. We can instead think of the string of coin flips as being part of the input, and then asking what fraction of the inputs accept. Formally, an ***RPOTM-BA*** is a POTM-BA that has 2 inputs x, τ. We will be concerned with the fraction of τ's ($|\tau|$ will be a function of $|x|$) for which $M^y(x, \tau)$ accepts. We will refer to this as *the probability*

that x with oracle bit string y is accepted since we think in terms of the string τ being chosen at random. We will refer to τ as a string of coin flips.

The following is an alternative definition of NP.

Definition 8.20. $A \in \mathsf{NP}$ if there exists a POTM-BA $M^{()}$ such that

$x \in A \Rightarrow \exists^p y : M^y(x) = 1$
$x \notin A \Rightarrow \forall^p y : M^y(x) \neq 1.$

If $x \in A$, then we think of y as being the EVIDENCE that $x \in A$. This evidence is short (only $p(|x|)$ long) and checkable in polynomial time. Note that the computation of $M^y(x)$ may certainly use all of the bits of y. What if we (1) restrict the number of bits of the oracle that the computation can look at and (2) use an RPOTM-BA?

Definition 8.21. Let $q(n)$ and $r(n)$ be monotone increasing functions from $\mathbb{N}$ to $\mathbb{N}$. An ***$r(n)$-random $q(n)$-query RPOTM-BA $M^{()}$*** is an RPOTM-BA where, for all y and for all x of length n, $M^y(x)$ flips $r(n)$ coins and makes $q(n)$ queries.

Definition 8.22. Let $r(n)$ and $q(n)$ be monotone increasing functions from $\mathbb{N}$ to $\mathbb{N}$ and $\varepsilon(n)$ be a monotone decreasing function from $\mathbb{N}$ to $[0, 1)$. $A \in \mathsf{PCP}(r(n), q(n), \varepsilon(n))$ if there exists an $r(n)$-random, $q(n)$-query RPOTM-BA $M^{()}$ such that, for all n, for all $x \in \{0, 1\}^n$, the following holds.

1. If $x \in A$, then there exists y such that, for all τ with $|\tau| = r(n)$, $M^y(x, \tau)$ accepts. In other words, the probability of acceptance is 1.
2. If $x \notin A$, then for all y at most $\varepsilon(n)$ of the τ's with $|\tau| = r(n)$ make $M^y(x, \tau)$ accept. In other words, the probability of acceptance is $\leq \varepsilon(n)$.

We are only going to be concerned with $r(n) = O(\log n)$ and $q(n) = O(1)$ or $O(\log n)$. We will see below that we can assume $|y| = 2^{q(n)+r(n)}$, which is polynomial in n.

Note: The queries are made adaptively. This means that the second question asked might depend on the answer to the first. Hence if $M^y(x, \tau)$ asks $q(n)$ questions, then the total number of bit positions of the oracle string that are relevant for a fixed (x, τ) is $2^{q(n)} - 1$. Since there are $2^{r(n)}$ values of τ, there are a total of $2^{q(n)+r(n)} - 1 \leq 2^{q(n)+r(n)}$ bit positions of the oracle string that are relevant for fixed x. Hence we can take $|y| = 2^{q(n)+r(n)}$.

Example 8.23.

1. $\mathsf{SAT} \in \mathsf{PCP}(0, n, 0)$. The y value is the satisfying assignment. $M^{()}$ makes all n queries and does not use random bits.
2. If φ is a formula, let C be the number of clauses in it. $\mathsf{3SAT} \in \mathsf{PCP}(\lg(C), 3, \frac{C-1}{C})$. The y value is the satisfying assignment. M picks a random clause and queries the three truth assignments. If they satisfy the clause, output YES; else NO. If $\varphi \in \mathsf{3SAT}$, then the

algorithm will return YES. If $\varphi \notin$ 3SAT, then the worst case is if y satisfies all but one of the clauses, and hence, the probability of error is $\leq \frac{C-1}{C}$.

3. Let $L \in \mathbb{N}$. 3SAT $\in$ PCP$(L\lg(C) + O(1), 3L, \frac{C-L}{C})$. Iterate the proof in part 2 L times.

Arora et al. [ALM$^+$98], building on the work of Arora et al. [AS98], proved the following theorem. This theorem is the cornerstone of lower bounds for approximations and is often called ***the PCPtheorem***.

We omit the proof, which is difficult.

Theorem 8.24.

1. *SAT* $\in$ *PCP*$(O(\log n), O(1), \frac{1}{2})$.
2. *For all constants* $0 < \varepsilon < 1$, *SAT* $\in$ *PCP*$(O(\log n), O(1), \varepsilon)$. *(This is easily obtained by iterating the protocol from part 1.)*
3. *SAT* $\in$ *PCP*$(O(\log^2 n), O(\log n), \frac{1}{n})$. *(This is easily obtained by iterating the protocol from part 1.)*

The result SAT $\in$ PCP$(O(\log^2 n), O(\log n), \frac{1}{n})$ is *not* good enough for proving problems hard to approximate. Ajtai–Komlós–Szemerédi [AKS87] and Impagliazzo and Zuckerman [IZ89] improved it by using a technique to reuse random bits. The technique involves doing a random walk on an expander graph. The next theorem is *not* in their papers; however, one can obtain it from their papers. See V. Vazirani [Vaz01] (theorem 29.18).

Theorem 8.25.

1. *SAT* $\in$ *PCP*$(O(\log n), O(\log n), \frac{1}{n})$.
2. *If* $A \in$ *NP, then there exists* $c, d \in \mathbb{N}$ *such that* $A \in$ *PCP*$(c\log n, d\log n, \frac{1}{n})$. *(This follows from part 1.)*

8.7 Clique Is Hard to Approximate

Arora et al. [ALM$^+$98], building on the work of Feige et al. [FGL$^+$96], proved that CLIQUE is hard to approximate. We will recognize the proof as a gap reduction.

Notation 8.26. If G is a graph, then OPT(G) is the size of the largest clique in G.

We state both the upper and lower bound in this proof; however, we only prove the lower bound.

Note that approximating CLIQUE is trivially in Poly-APX (just output N, the number of vertices). The following theorem improved on this trivial algorithm; however, it does not put CLIQUE in a better approximation class. We will see later that, assuming P $\neq$ NP, approximating clique cannot be put into a better approximation class.

In the following theorem, the size of a graph is the number of vertices. We use N for the number of vertices since n will be the length of a string in an NP-set A.

Theorem 8.27.

1. *(Feige [Fei04]) There is an algorithm that, on input graph G with N vertices, outputs a clique of size*
$$\Omega\left(\frac{\log^3 N}{N(\log\log N)^2}\right) OPT(G).$$
2. *Let A be an NP-complete problem. Let c, d be such that* $A \in \mathrm{PCP}(c\lg n, d\lg n, \frac{1}{n})$ *(such a c, d exist by theorem 8.25). There is a reduction that maps* $x \in \Sigma^*$ $(|x| = n)$ *to a graph G on* $N = n^{c+d}$ *vertices such that:*
   - *If* $x \in A$ *then* $OPT(G) \geq n^c = N^{c/(c+d)}$.
   - *If* $x \notin A$ *then* $OPT(G) \leq n^{c-1} = N^{(c-1)/(c+d)}$.
3. $\text{GAP}(\text{CLIQUE}, N^{c/(c+d)}, N^{(c-1)/(c+d)})$ *is NP-hard. (This follows from part 2 and lemma 8.13.)*
4. *If there is an approximation algorithm for* CLIQUE *that, on input G, where G has N vertices, returns a number* $> \frac{1}{N^{1/(c+d)}} OPT(G)$, *then P = NP. (This follows from part 2 and lemma 8.13.)*
5. *Assuming P ≠ NP,* CLIQUE ∈ Poly-APX − Log-APX. *(This follows from parts 1 and 4)*

Proof. We just prove part 2.

Let A, c, d be as in the statement of the theorem. To avoid notational clutter, we say "Run the PCP on (x, τ)" rather than give the RPOTM-BA for A a name.

1. Input x of length n.
2. Form a graph $G = (V, E)$ as follows:
   a. $V = \{0, 1\}^{c\lg n + d\lg n}$. Note that there are $N = n^{c+d}$ vertices.
   b. This step will help us determine the edges. For each vertex, write it as $\tau\sigma$, where $|\tau| = c\lg n$ and $|\sigma| = d\lg n$. Run the PCP on (x, τ) and answer the ith query made with the ith bit of σ. Keep track of which queries were made, what the answers were, and if the computation was accepted.
   c. We now determine the edges. Let $\tau\sigma$ and $\tau'\sigma'$ be two vertices. We know both the queries and the answers made when running PCP(x, τ) using σ for the answers, and running PCP(x, τ') using σ' for the answers. Connect the two vertices $\tau\sigma$ and $\tau'\sigma'$ if (1) both represent computations that accept, (2) $\tau \neq \tau'$, and (3) the answers to queries do not contradict.
3. Output the graph.

Note the following:

1. If $x \in A$, then there exists a consistent way to answer the bit-queries such that, for all $\tau \in \{0, 1\}^{c \lg n}$, the PCP on (x, τ) accepts. Hence $\mathrm{OPT}(G) \geq 2^{c \lg n} = n^c = N^{c/(c+d)}$.
2. If $x \notin A$, then any consistent way to answer the bit-queries will make $\leq \frac{1}{n}$ of the $\tau \in \{0, 1\}^{c \lg n}$ accept. Hence $\mathrm{OPT}(G) \leq n^{c-1} = N^{(c-1)/(c+d)}$.

□

Note: Theorem 8.27 showed that, for $\delta = \frac{1}{c+d}$, if there is an algorithm that returns a number $> \frac{1}{n^\delta}\mathrm{OPT}(G)$ then P = NP. Better results are known:

1. Hastad [Has99] showed that, for all $0 \leq \delta < 1$, if there is an algorithm that returns a number $\geq \frac{1}{n^\delta}\mathrm{OPT}(G)$ then ZPP = NP.
2. Zuckerman [Zuc07] showed that, for all $0 \leq \delta < 1$, if there is an algorithm that returns a number $\geq \frac{1}{n^\delta}\mathrm{OPT}(G)$ then P = NP.

8.8 Set Cover Is Hard to Approximate

The definition of Set Cover is in section 7.8.2.

Notation 8.28. An algorithm ***approximates Set Cover within a factor of $f(n)$*** if it outputs a number that is $\leq f(n)\mathrm{OPT}$.

The following are known.

Theorem 8.29.

1. *Chvatal [Chv79] showed that a simple greedy algorithm approximates Set Cover within a factor of* $\ln(n)$.
2. *Slavík [Sla97] gave a slight improvement by replacing the* $\ln n$ *with*
   $\ln n - \ln(\ln n) - O(1)$.
3. *Lund and Yannakakis [LY94] showed that, for any* $0 < c < 1/4$, *if Set Cover can be approximated within a factor of* $c \lg n$ *then NP ⊆ DTIME*$(n^{polylog(n)})$.
4. *Dinur and Steurer [DS13] showed that if there exists* $\varepsilon > 0$ *such that Set Cover can be approximated within a factor of* $(1 - \varepsilon)\ln n$, *then P = NP.*
5. *There were many intermediary results between Lund and Yannakakis [LY94] and Dinur and Steurer [DS13]. Melder [Mel21] presents a guide to the papers needed to obtain the result and how they fit together. This guide is summarized in in figure 8.1.*

Note: How does m, the number of sets, impact these results? The lower bound proofs apply for m very small, like $n^{0.0001}$. Hence, when we later do reductions of Set Cover to other problems, we can assume m is small.

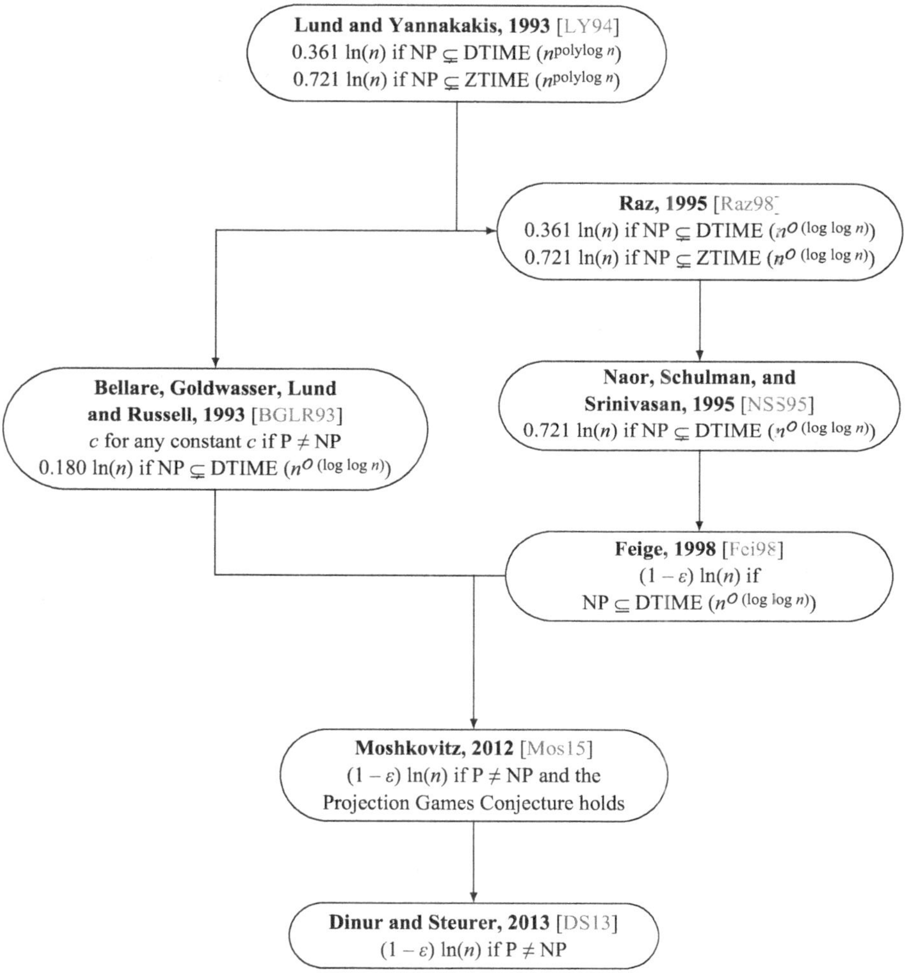

Figure 8.1
The history of SET COVER lower bounds.

The lower bound papers do not use PCPs. They instead use a close cousin: 2-prover-1-round interactive proof systems.

For the next exercise we will consider the following problem, which can be seen as the dual of SET COVER.

> MAX COVERAGE
> *Instance:* n, Sets $S_1, \ldots, S_m \subseteq \{1, \ldots, n\}$, and k.
> *Question:* What is the largest size of a set $X \subseteq \{1, \ldots, n\}$ such that there exists a set of k S_i's that contain the elements of X?

Exercise 8.30. Let $\varepsilon > 0$. Show that if there is an approximation algorithm for MAX COVERAGE that returns $(1 - \frac{1}{e} + \varepsilon)$OPT, then P = NP.
Hint: Use theorem 8.29.4 and a gap reduction.

8.9 MAX 3SAT Is Hard to Approximate

Arora et. al [ALM+98] proved that MAX 3SAT is hard to approximate. We will recognize the proof as a gap reduction.

Notation 8.31. If φ is a 3CNF formula (so every clause has ≤ 3 literals), then OPT(φ) is the max number of clauses that can be satisfied simultaneously.

We state both the upper and lower bound in this proof. We prove one of the upper bounds; however, our main interest is in the lower bound.

Notation 8.32. Let $q \in \mathbb{N}$. Then $C(q)$ is the maximum number of clauses in a 3CNF formula on 2^q variables. Note that $C(q) = O(2^{3q})$. Since q is constant, $C(q)$ is constant.

In the following theorem, the size of a 3CNF formula is the number of clauses.

Theorem 8.33.

1. *Restrict MAX 3SAT to formulas that have* exactly *three literals per clause. There is an algorithm that, given such a φ, returns a number that is $\geq 0.875 OPT(\varphi)$.*
2. *Karloff and Zwick [KZ97] have a randomized polynomial time algorithm for MAX 3SAT (note that clauses can have 1, 2, or 3 literals) that, on input φ, does the following: (1) if $\varphi \in SAT$ returns an assignment that satisfies $\geq 0.875 OPT(\varphi)$ of the clauses, and (2) if $\varphi \notin SAT$ then there is good evidence that the algorithm still returns an assignment that satisfies $\geq 0.875 OPT(\varphi)$ clauses.*

3. *Let A be NP-complete. Let $c, q \in \mathbb{N}$ such that $A \in PCP(c \lg n, q, 0.25)$ (such a c, q exist by theorem 8.24). There is a reduction that maps $x \in \Sigma^*$ to a 3CNF formula φ such that:*
   a. *If $x \in A$, then $OPT(\varphi) = |\varphi|$. ($\varphi \in 3SAT$ so all $|\varphi|$ clauses are satisfied.)*
   b. *If $x \notin A$, then $OPT(\varphi) \leq \left(1 - \frac{3}{4C(q)}\right)|\varphi|$.*
   c. *The output φ has exactly three literals per clause.*
4. *GAP(MAX 3SAT, m, $\left(1 - \frac{3}{4C(q)}\right)m$) is NP-hard (where m is the number of clauses). This holds even if φ is restricted to having exactly three literals per clause. (This follows from part 3 and lemma 8.13.)*
5. *If there is an approximation algorithm for MAX 3SAT that, on input φ, returns a number $> \left(1 - \frac{3}{4C(q)}\right)OPT(\varphi)$, then P = NP. (This follows from part 3 and lemma 8.13.)*
6. *Assuming P ≠ NP, MAX 3SAT ∈ APX − PTAS. (The problem that separates them is MAX 3SAT restricted to formulas that have exactly three literals per clause. This follows from parts 1 and 5.)*

Proof sketch. We just prove parts 1 and 3.

1) We first give a randomized algorithm: Assign each variable to TRUE or FALSE at random. The probability of a particular clause being satisfied is $\frac{7}{8} = 0.875$, so by linearity of expectation, we expect $\frac{7}{8}$ of the clauses to be satisfied. This gives a randomized algorithm that outputs a number $\geq 0.875\mathrm{OPT}(\varphi)$. This algorithm can be derandomized using the method of conditional probabilities. Details can be found in either V. Vazirani's book [Vaz01] or Shmoys-Williamson's book [WS11].

3) Let A, c, q be as in the statement of the theorem. To avoid notational clutter, we say "Run the PCP on (x, τ)" rather than give the RPOTM-BA for A a name.

1. Input x.
2. Form a 3CNF formula ψ as follows:
   a. The PCP for A can only make $2^{q+c\lg n} = 2^q n^c$ possible bit-queries. There are $2^q n^d$ variables, one for each possible bit-query.
   b. For every $\tau \in \{0,1\}^{c\lg n}$ do the following. For every $\sigma \in \{0,1\}^q$ run the PCP(x, τ) using σ for the query answers. Keep track of which ones accepted and which ones rejected. From this information, form a formula on $\leq 2^q$ variables that is TRUE if and only if the PCP accepts with those answers (using τ for the coin flips). Constructing the formula takes polynomial time since q (and hence 2^q) is constant. Convert this formula to 3CNF (this easily takes polynomial time). Call the result ψ_τ. Note that ψ_τ has between 1 and $C(q)$ clauses.
   c. ψ is the AND of the n^c formulas from the last step. Note that ψ is in 3CNF form.

Note the following.

1. Assume $x \in A$. Then there exists a consistent way to answer the bit-queries such that, for all $\tau \in \{0,1\}^{c \lg n}$, the PCP on (x, τ) accepts. Hence every ψ_τ can be satisfied simultaneously. Therefore the fraction of clauses of ψ that can be satisfied is 1. Hence $\mathrm{OPT}(\varphi) = |\varphi|$.
2. Assume $x \notin A$. Then every consistent way to answer the bit-queries will make $\le \frac{1}{4}$ of the $\tau \in \{0,1\}^{c \lg n}$ accept. We need to estimate the fraction of clauses of ψ that are satisfied. This fraction is maximized when the following occurs: (1) all n^c of the ψ_τ have $C(q)$ clauses, and (2) there is an assignment that satisfies all $C(q)$ clauses in $1/4$ of the ψ_τ, and $C(q)-1$ clauses in $3/4$ of the ψ_τ. Hence the fraction of clauses satisfied is
$$\frac{(n^c/4)C(q) + (3n^c/4)(C(q)-1)}{n^c C(q)} = \frac{n^c C(q) - (3n^c/4)}{n^c C(q)} = 1 - \frac{3}{4C(q)}.$$
Hence $\mathrm{OPT}(\varphi) \le \left(1 - \frac{3}{4C(q)}\right)|\varphi|$.

□

Theorem 8.33 shows that if there is an algorithm that returns a number $>(1 - \frac{3}{4C(q)})\mathrm{OPT}(\varphi)$ then P = NP. This is sufficient to show MAX 3SAT $\notin$ PTAS but still leaves open the question of whether the best known approximation, $0.875\mathrm{OPT}(\varphi)$, is optimal. It is. Hastad [Has01] proved the following:

Theorem 8.34.

1. *Let $0 < \varepsilon < 1$. Let our formulas have $4m$ clauses. Then*
$$\text{GAP}(\text{MAX 3SAT}, 4(1-\varepsilon)m, 3.5(1+\varepsilon)m)$$
is NP-hard.
2. *Let $0 < \varepsilon < 1$. Let our formulas have m clauses. Then*
$$\text{GAP}(\text{MAX 3SAT}, (1-\varepsilon)m, 0.875(1+\varepsilon)m)$$
is NP-hard. (This follows from part 1.) This result holds when MAX 3SAT is restricted to having exactly three literals per clause. Hence this result is a lower bound that matches the upper bound in theorem 8.33.1.
3. *Let $0 < \delta < 1$. If there is an approximation algorithm for MAX 3SAT that, on input φ, returns a number $>(0.875 + \delta)OPT(\varphi)$, then P = NP. (This follows from part 2.)*

In section 8.10 we will use theorem 8.34 to obtain lower bounds on how well one can approximate VERTEX COVER. In chapter 9 we will use theorem 8.34 to obtain many lower bounds on approximation.

8.10 Vertex Cover Is Hard to Approximate

We show a lower bound on how well Vertex Cover can be approximated by using a reduction from the GAP version of Max 3SAT that was discussed in theorem 8.34 to a GAP version of Vertex Cover.

Theorem 8.35. *We consider Vertex Cover. There is an algorithm that will, given a graph G, output a number that is* $\leq 2OPT(G)$.

Proof sketch. Here is the algorithm.

1. Input $G=(V,E)$.
2. Let $M=\emptyset$. Add edges to M so that no two edges in M share a vertex, until you can add no more. This gives a maximal matching.
3. Output U, the set of endpoints of the edges in M.

U is a vertex cover since, if $e=(a,b)$ has neither endpoint in U, then M was not maximal as it could have added e.

We leave it to the reader to show that $\text{OPT} \geq |M| \geq \frac{|U|}{2}$.

□

The following theorem seems to be new; however, it is weaker than the best known results. We prove a lower bound on approximating Vertex Cover; however, we give enough information in the proof so that the reader can obtain lower bounds on approximating Independent Set and Clique.

Theorem 8.36. *We are considering the problem Vertex Cover. Let* $\delta > 0$. *If there is an algorithm that, on input a graph G, returns a number* $\leq (\alpha - \delta)OPT(G)$ *where* $\alpha = 1.107$, *then P = NP.*

Proof sketch. We assume that there is an algorithm that, given a graph G, returns a number $\leq(\alpha-\delta)\text{OPT}_{\text{Vertex Cover}}(G)$. To avoid notational clutter, we will call it ***the algorithm***. We will prove this with α and only at the end see what α has to be to make the proof work.

Let ε be a parameter to be named later. It will depend on δ and α. By theorem 8.34

$$\text{GAP}(\text{Max 3SAT}, (1-\varepsilon)m, 0.875(1+\varepsilon)m)$$

is NP-hard. We present a polynomial-time algorithm for this GAP problem that uses the approximation algorithm for Vertex Cover.

We will use the fact that, if G has n vertices, then $\text{OPT}_{\text{Independent Set}}(G) = n - \text{OPT}_{\text{Vertex Cover}}(G)$.

1. Input $\varphi = C_1 \wedge \cdots \wedge C_m$, a formula in 3CNF where every clause has exactly three literals. We are promised that either

a. $\mathrm{OPT}_{\text{MAX 3SAT}}(\varphi) \geq (1-\varepsilon)m$, or

b. $\mathrm{OPT}_{\text{MAX 3SAT}}(\varphi) \leq (0.875+\varepsilon)m$.

2. Create a graph G as follows.

   a. For each C_i we have a vertex for each literal. Hence G has $3m$ vertices.

   b. Put an edge between every pair of vertices in the same C_i. Put an edge between vertices from different C_i's if they contradict each other.

3. (This is commentary, not part of the algorithm.) The reader should be able to work out that:

   If $\mathrm{OPT}_{\text{MAX 3SAT}}(\varphi) \geq (1-\varepsilon)m$, then $\mathrm{OPT}_{\text{VERTEX COVER}}(G) =\leq 3m-(1-\varepsilon)m = (2+\varepsilon)m$

   If $\mathrm{OPT}_{\text{MAX 3SAT}}(\varphi) \leq (0.875+\varepsilon)m$, then $\mathrm{OPT}_{\text{VERTEX COVER}}(G) \geq (2.125-\varepsilon)m$

4. Run the algorithm on G to produce number z.

5. (This is commentary, not part of the algorithm.)
   Note the following two cases.

   a. $\mathrm{OPT}_{\text{MAX 3SAT}}(\varphi) \geq (1-\varepsilon)m$ so $\mathrm{OPT}_{\text{VERTEX COVER}}(G) \leq (2+\varepsilon)m$. Hence $z \leq (\alpha - \delta)\ (2+\varepsilon)m$.

   b. $\mathrm{OPT}_{\text{MAX 3SAT}}(\varphi) \leq (0.875+\varepsilon)m$ so $\mathrm{OPT}_{\text{VERTEX COVER}}(G) \geq (2.125-\varepsilon)m$. Hence $z \geq (2.125-\varepsilon)m$.

   To make these two cases disjoint we need

$$(\alpha-\delta)(2+\varepsilon) < (2.124-\varepsilon)$$

$$\alpha-\delta < \frac{2.214-\varepsilon}{2+\varepsilon}.$$

   We want to make α as large as possible so take $\alpha = \frac{2.214}{2} = 1.107$. Given δ we can then pick ε such that the inequality above holds.

6. If $z \leq (\alpha-\delta)(2+\varepsilon)m$, then output $\mathrm{OPT}(\varphi) \geq (1-\varepsilon)m$.

   If $z \geq (2.125-\varepsilon)$, then output $\mathrm{OPT}(\varphi) \leq (0.875+\varepsilon)m$.

□

Exercise 8.37. In the proofs of theorems 8.27 and 8.33 we used the GAP lemmas (lemmas 8.13, 8.15) which have as a premise a reduction from an *NP-hard set A* to a GAP problem. In the proof of theorem 8.36 we directly reduced a GAP problem to a GAP problem. Devise and prove gap lemmas that use a reduction from a GAP problem to a GAP problem.

Are better lower bounds for approximating VERTEX COVER known? Yes. We state two results.

Theorem 8.38. *Let $\delta > 0$.*

1. *(Hastad [Has01]) If there is an algorithm that, on input a graph G, returns a number $\le (1.166\ldots - \delta)OPT(G)$, then P = NP (the number is actually $\frac{7}{6}$).*
2. *(Dinur & Safra [DS05]) If there is an algorithm that, on input a graph G, returns a number $\le (1.3606\ldots - \delta)OPT(G)$, then P = NP (the number is actually $10\sqrt{5} - 21$).*
3. *The 2-to-2 unique games conjecture was proven in a sequence of papers [KMS17, KMS18, DKK$^+$18a, DKK$^+$18b]. See [Kho19] for an overview and how the conjecture implies the following. If there is an algorithm that, on input a graph G, returns a number $\le (1.414\ldots - \delta)$, then P = NP (the number is actually $\sqrt{2}$).*

Are better lower bounds for approximating VERTEX COVER known? There are no NP-hardness results known. However, in chapter 10 we will state the *unique games conjecture* from which several lower bounds can be proven, including a lower bound of $2 - \delta$ for approximating VERTEX COVER.

8.11 Projects

The lower bounds in this chapter used the PCP machinery as a black box. The lower bounds on SET COVER used other hard theorems as a black box. This chapter could have been titled *Hardness Results on Approximation Made Easy By Leaving Out the Hard Parts.*

One can lean into this mentality and have more write-ups along those lines.

Project 8.39.

1. *Take the theorems in note 8.7 and write them up leaving out the hard parts.*
2. *Take theorem 8.34 and write it up leaving out the hard parts.*

Or one might try to get easier proofs.

Project 8.40. *Obtain easier proofs of lower bounds on approximation by either finding a direct proof that avoids the PCP machinery or making the PCP machinery easier.*

9

Inapproximability

9.1 Introduction

In chapter 8 we stated the following:

Theorem 9.1. *Assuming P $\neq$ NP:*

1. *TSP $\notin$ Poly-APX.*
2. *CLIQUE $\in$ Poly-APX $-$ Log-APX.*
3. *SET COVER $\in$ Log-APX $-$ APX.*
4. *MAX 3SAT $\in$ APX $-$ PTAS.*

In this chapter we will use these results to show that problems are hard to approximate.

Chapter Summary

1. *Assume P $\neq$ NP. We use reductions from MAX 3SAT to show that several problems are not in PTAS.*
2. *Assume P $\neq$ NP. We use reductions from SET COVER to show that several problems are not in APX.*
3. *We briefly discuss Log-APX and Poly-APX; however, there are very few results about them.*
4. *We briefly discuss the class NPO, which captures some other aspects of approximation.*

Convention 9.2. For the rest of this chapter, "problem" means ***NPO problem***. When A and B are mentioned, they are NPO problems. We will often say whether A is a min problem or a max problem.

Notation 9.3. Recall that, if A is an NPO, then the cost (if it's a min problem) or benefit (if it's a max problem) of a solution can be computed in polynomial time. If the problem is understood, x is an instance, and y is a solution, then benefit(x, y) is the benefit and cost(x, y) is the cost. For example, if the problem is CLIQUE, then the instance is a graph G, the solution is a clique C, and benefit(G, C) is the size of C.

Everything in sections 9.2, 9.3, 9.4, and 9.5 is by Papadimitriou and Yannakakis [PY91] with a caveat. They had a very different outlook since theorem 8.33 was not known when they wrote their paper. For that reason they could not have proven a statement like

If VERTEX COVER with maximum degree 4 has a PTAS, then P = NP.

However, they had the reductions in place so that once approximating MAX 3SAT was shown to be NP-hard, that kind of result followed.

9.2 Lower Bounds on Approximability

Papadimitriou and Yannakakis [PY91] proved the following.

Theorem 9.4. *If MAX 3SAT ∈ PTAS, then every problem in APX has a PTAS.*

Note that MAX 3SAT ∈ APX. Papadimitriou and Yannakakis called problems like MAX 3SAT ***APX-complete*** (we define this later). They then showed many problems were APX-complete.

We contrast what they could conclude with what we can conclude. Let X be a problem in APX that we wish to show is likely not in PTAS.

1. They showed that there is an approximation preserving reduction (to be defined) from MAX 3SAT to X. Hence, *if X has a PTAS, then PTAS = APX*. Therefore X is unlikely to have a PTAS.
2. We use the same reductions they did; however, armed with theorem 8.33, we can say, *if X has a PTAS, then P = NP.*

We will use their reductions; however, we will not need the notion of APX-complete since we have theorem 8.33. We may still use the terminology for convenience.

We define the appropriate reductions, which are more complicated than the reduction $\leq_p$.

Definition 9.5. Let A and B be 2 optimization problems. An ***approximation preserving reduction (APR)*** from A to B is a pair of polynomial-time functions (f, g) such that the following holds:

- If x is an instance of A, then $f(x)$ is an instance of B.
- If y is a solution for $f(x)$, then $g(x, y)$ is a solution for x. We will later define types of reductions where a good solution for B maps to a good solution for A, for some notion of "good." Since we are only interested in good (whatever that might mean) solutions, we may restrict g to solutions y that do not have an obvious improvement. For example, if the problem is MAXSAT (maximize the number of clauses

satisfied), we may assume that any variable that appears without negations is set to TRUE.

- We usually do not use the notation f and g. Instead (1) x will be an instance of A, x' will be the instance of B that x maps to ($f(x)=x'$), and (2) y' will be the solution to x', and y will be the solution of x that y' maps to ($g(y')=y$).
- Note that we have not specified how this reduction preserves approximation. And we won't. The name *approximation preserving* is not quite right. We will define reductions that begin "Blah is an approximation preserving reduction that also has property Blah Blah."

The idea here is that f transforms instances of A into instances of B while g transforms solutions for B instances into (in some sense equally good) solutions for the A instance.

Let's refine the above idea further:

Definition 9.6.

1. A ***PTAS reduction*** is an approximability preserving reduction satisfying the following additional constraint:

   For any $\varepsilon > 0$, there is a $\delta = \delta(\varepsilon) > 0$ such that, if y' is a $(1+\delta(\varepsilon))$-approximation to B, then y is a $(1+\varepsilon)$-approximation to A. Note that here we allow the f and g functions to depend on ε. We will assume that the function δ is monotone increasing and goes to infinity.
2. A ***strict reduction*** is an APX reduction where $\delta(\varepsilon)=\varepsilon$. This is a very strong concept of reduction.

Definition 9.7.

1. An ***APX reduction*** is a PTAS-reduction for which the δ function is linear in ε. This is a convenient reduction to use because if $B \in O(f)$-APX, then $A \in O(f)$-APX.

Papadimitriou and Yannakakis defined APX-hard and APX-complete. We will give a different definition that is equivalent and makes more sense given that we have theorem 8.33.

Their definition:

Definition 9.8.

1. A problem B is ***APX-hard*** if, for all $A \in$ APX, there is an APX reduction from A to B.
2. A problem B is ***APX-complete*** if B is APX-hard and $B \in$ APX.

Our definition:

Definition 9.9.

1. A problem B is ***APX-hard*** if there is an APX reduction from MAX 3SAT to B.
2. A problem B is ***APX-complete*** if B is APX-hard and $B \in$ APX.

Henceforth we use our definition of APX-complete.

Note: The equivalence of these two definitions is not obvious. It depends on the theorem (due to Papadimitriou and Yannakakis) that MAX 3SAT is APX-complete using definition 9.8. We will not prove their theorem, nor do we need the equivalence of the two definitions of APX-hard and APX-complete.

The following follows from theorem 8.33 and our definition of APX-complete.

Theorem 9.10. *If B is APX-complete, then:*

1. $B \in$ *APX.*
2. *If* $B \in$ *PTAS, then* P $=$ NP.

Exercise 9.11. Assume there is a PTAS reduction from A to B.

1. Prove that, if $B \in$ PTAS, then $A \in$ PTAS. We will use the contrapositive form: If $A \notin$ PTAS, then $B \notin$ PTAS.
2. Prove that, if $B \in$ APX, then $A \in$ APX. We will use the contrapositive form: If $A \notin$ APX, then $B \notin$ APX.
3. Assume A reduces to B with a strict reduction. Also assume that A and B are both MAX-problems (a similar theorem holds if they are Min-problems). Let $\varepsilon > 0$. Prove that, if B has a $(1+\varepsilon)$-approximation, then A has a $(1+\varepsilon)$-approximation. We will use the contrapositive form: If A does not have a $(1+\varepsilon)$-approximation, then B does not have a $(1+\varepsilon)$-approximation.

PTAS reductions are all a bit awkward to work with directly because all of the approximability results use a multiplicative factor. In practice, people use L-reductions.

The reader is advised to look at convention 8.1 again for the use of the notation OPT.

Definition 9.12. An ***L-reduction*** from A to B is an approximation preserving reduction which satisfies the following two properties:

- $\mathrm{OPT}_B(x') = O(\mathrm{OPT}_A(x))$.
- If A is a min problem, then

$$|\mathrm{cost}_A(y) - \mathrm{OPT}_A(x)| = O(|\mathrm{cost}_B(y') - \mathrm{OPT}_B(x')|).$$

- If A is a max problem, then

$$|\mathrm{benefit}_A(y) - \mathrm{OPT}_A(x)| = O(|\mathrm{benefit}_B(y') - \mathrm{OPT}_B(x')|).$$

We denote this reduction by $A \leq_L B$. (The L stands for "linear".)

L-reductions are stronger than APX reductions (as we will show) so the existence of an L-reduction implies the existence of an APX reduction which implies the existence of a PTAS reduction. Note that L-reductions are not stronger than strict reductions.

Theorem 9.13. *If $A \leq_L B$, then that same reduction is an APX reduction.*

Proof. Let's prove that the L-reduction is an APX reduction. We will do this in the minimization case.

First, $\mathrm{OPT}_B(x') = O(\mathrm{OPT}_A(x))$, so there exists a constant $\alpha > 0$ such that

$$\mathrm{OPT}_B(x') \leq \alpha \mathrm{OPT}_A(x).$$

Second, $|\mathrm{cost}_A(y) - \mathrm{OPT}_A(x)| = O(|\mathrm{cost}_B(y') - \mathrm{OPT}_B(x')|)$, so because this is a minimization problem, there exists a positive constant β such that

$$\mathrm{cost}_A(y) - \mathrm{OPT}_A(x) \leq \beta(\mathrm{cost}_B(y') - \mathrm{OPT}_B(x')).$$

To show that the L-reduction is an APX reduction we need to show that there is a constant γ such that, if

$$\mathrm{cost}_B(x') \leq (1 + \gamma\varepsilon)\mathrm{OPT}_B(x'),$$

then

$$\mathrm{cost}_A(x) \leq (1 + \varepsilon)\mathrm{OPT}_A(x).$$

We will derive γ. Let $\delta = \gamma\varepsilon$.

Assume that $\mathrm{cost}_B(y') \leq (1 + \delta)\mathrm{OPT}_B(x')$. We want to obtain

$$\mathrm{cost}_A(y) \leq (1 + \varepsilon)\mathrm{OPT}_A(x).$$

We know that

$$\mathrm{cost}_A(y) \leq \mathrm{OPT}_A(x) + \beta\mathrm{cost}_B(y') - \beta\mathrm{OPT}_B(x').$$

We factor out $\mathrm{OPT}_A(x)$ to get

$$\mathrm{cost}_A(y) \leq \mathrm{OPT}_A(x)\left(1 + \frac{\beta\mathrm{cost}_B(y')}{\mathrm{OPT}_A(x)} - \frac{\beta\mathrm{OPT}_B(x')}{\mathrm{OPT}_A(x)}\right).$$

Because

$$\frac{1}{\mathrm{OPT}_A(x)} \leq \frac{\alpha}{\mathrm{OPT}_B(x')},$$

we have

$$\mathrm{cost}_A(y) \leq \mathrm{OPT}_A(x)\left(1 + \frac{\alpha\beta\mathrm{cost}_B(y')}{\mathrm{OPT}_B(x')} - \frac{\alpha\beta\mathrm{OPT}_B(x')}{\mathrm{OPT}_B(x')}\right).$$

Hence,

$$\mathrm{cost}_A(y) \leq \mathrm{OPT}_A(x)\left(1 + \frac{\alpha\beta\mathrm{cost}_B(y')}{\mathrm{OPT}_B(x')} - \alpha\beta\right).$$

Because $\text{cost}_B(y') \leq (1+\delta)\text{OPT}_B(x')$, we have $\frac{\text{cost}_B(y')}{\text{OPT}_B(x')} \leq 1+\delta$, so

$$\text{cost}_A(y) \leq OPT_A(x)(1+\alpha\beta(1+\delta)-\alpha\beta) = \text{OPT}_A(x)(1+\alpha\beta\delta).$$

Hence,

$$\text{cost}_A(y) \leq \text{OPT}_A(x)(1+\alpha\beta\gamma\varepsilon).$$

It suffices to take $\gamma = \frac{1}{\alpha\beta}$. □

Exercise 9.14.

1. Show that if $A \leq_L B$ and $B \leq_L C$, then $A \leq_L C$.
2. Show that if $A \leq_L B$ with a strict reduction and $B \leq_L C$ with a strict reduction, then $A \leq_L C$ with a strict reduction.

9.3 Max 3SAT and Its Variants

The results in this section are essentially due to Papadimitriou and Yannakakis [PY91]. We define several SAT problems. We include Max 3SAT since the other SAT problems are variants of it.

> Max 3SAT
> *Instance:* A formula φ where every clause has ≤ 3 literals.
> *Output:* The max number of clauses that can be satisfied by an assignment.

> Max 3SAT-a
> *Instance:* A formula φ where every clause has ≤ 3 literals and every variable occurs $\leq a$ times.
> *Output:* The max number of clauses that can be satisfied by an assignment.
> *Note:* This problem is not interesting in its own right; however, it will be used to show that several problems on graphs of bounded degree are hard to approximate.

> Max 2SAT
> *Instance:* A formula φ where every clause has ≤ 2 literals
> *Output:* The max number of clauses that can be satisfied by an assignment.

> MAX NAE 3SAT
> *Instance:* A formula φ where every clause has ≤ 3 literals.
> *Output:* The max number of clauses that can be satisfied by an assignment with the extra condition that no clause has all of its literals TRUE.
> *Note:* NAE stands for not-all-equal.
> *Note:* This problem is not interesting in its own right; however, it will be used to show that MAX CUT is hard to approximate.

We will need the following exercise.

Exercise 9.15. Let φ be a formula with C clauses where every clause has ≤ 3 literals. Prove that the maximum number of clauses that can be satisfied is $\geq \Omega(C)$ (and hence clearly $\Theta(C)$).
Hint: Use theorem 8.33.1.

9.4 Reductions from MAX 3SAT and Its Variants

We first show a reduction from MAX 3SAT to MAX 3SAT-3 that is *not* a PTAS reduction to motivate that we need a more sophisticated reduction.

1. Input $\varphi(x_1, \ldots, x_n)$. Assume φ has m clauses.
2. For each variable x that occurs ≥ 4 times, do the following:
   a. Let k be the number of times x occurs. Introduce new variables $z_1, \ldots, z_k$.
   b. Replace the k occurrences of x with $z_1, \ldots, z_k$.
   c. Add the clauses $(z_1 \rightarrow z_2)$, $(z_2 \rightarrow z_3)$, ..., $(z_{L-1} \rightarrow z_k)$, $(z_k \rightarrow z_1)$. These clauses are an attempt to force all of the z_i to have the same truth value. If this was a decision-problem reduction, then the attempt would succeed. (Formally we would use $z_1 \vee \neg z_2$ for $z_1 \rightarrow z_2$.)

   Let φ' be the new formula.

Clearly every variable occurs ≤ 3 times. Clearly $\varphi \in 3\text{SAT}$ if and only if $\varphi' \in 3\text{SAT}$. But we need more. We need to be able to take an assignment that satisfies many clauses of φ' and map it to an assignment that satisfies many clauses of φ.

We give an example to show why the reduction does not work.

$$\varphi(x) = (x \vee x \vee x) \wedge \cdots \wedge (x \vee x \vee x) \wedge \cdots \wedge (\neg x \vee \neg x \vee \neg x) \wedge \cdots \wedge (\neg x \vee \neg x \vee \neg x),$$

where there are m $(x \vee x \vee x)$ clauses and m $(\neg x \vee \neg x \vee \neg x)$ clauses. Note that the MAX 3SAT value is m.

Formula φ' will be

$$(z_1 \vee z_2 \vee z_3) \wedge \cdots \wedge (z_{3m-2} \vee z_{3m-1} \vee z_{3m}) \wedge$$
$$(\neg z_{3m+1} \vee \neg z_{3m+2} \vee \neg z_{3m+3}) \wedge \cdots \wedge (\neg z_{6m-2} \vee \neg z_{6m-1} \vee \neg z_{6m}) \wedge$$
$$(z_1 \rightarrow z_2) \wedge \cdots \wedge (z_{6m-1} \rightarrow z_{6m}) \wedge (z_{6m} \rightarrow z_1).$$

If we set $z_1, \ldots, z_{3m}$ to TRUE and $z_{3m+1}, \ldots, z_{6m}$ to FALSE, then we satisfy every single clause except $z_{3m} \rightarrow z_{3m+1}$. That's $2m-1$ clauses. Hence φ' has a satisfying assignment that satisfies $\frac{2m-1}{2m}$ of its clauses. The max fraction of clauses that can be satisfied by the original φ is $\frac{1}{2}$. Thats a big difference. More to the point, there is no useful way to take this assignment for φ' and map it to an assignment for φ that satisfies many clauses. Hence if we want to reduce MAX 3SAT to MAX 3SAT-3, we will need to use a more intricate reduction.

In the failed reduction, we used a ***cycle*** to connect the different variables that are supposed to all have the same truth value. In the correct reduction, we will use a more complicated graph.

Recall that a graph has ***maximum degree d*** if its vertices all have degree at most d.

Definition 9.16. Let $d \in \mathbb{N}$. A ***d-regular expander graph*** is a graph $G = (V, E)$ where (1) every vertex has degree d and (2) for every partition $V = V_1 \cup V_2$ the number of edges from vertices in V_1 to vertices in V_2 is $\geq \min(|V_1|, |V_2|)$.

Note: There are many different notions of expander graph that are *not* equivalent. Hence you may come across a different definition in the literature. They all involve graphs that somehow combine high connectivity with a small number of edges. We will only use the definition above.

The following is well known, and also in the paper by Papadimitriou and Yannakakis [PY91, page 432].

Exercise 9.17. Prove that, for all $k \equiv 0 \pmod 2$, there exists a 3-expander graph on k vertices.

Theorem 9.18.

1. *MAX 3SAT* $\leq_L$ *MAX 3SAT-7.*
2. *MAX 3SAT-7* $\leq_L$ *MAX 3SAT-3.*
3. *MAX 3SAT* $\leq_L$ *MAX 3SAT-3 (this follows from the parts 1 and 2, and exercise 9.14).*
4. *MAX 3SAT-3 is APX-complete (this follows from part 3 and theorem 8.33.1).*
5. *Assume* $\mathrm{P} \neq \mathrm{NP}$. *Let* $a \geq 3$. *Then MAX 3SAT-a* $\notin$ *PTAS (this follows from part 4 and theorem 9.10.2).*

Proof sketch.

1) Here is the reduction:

1. Input $\varphi(x_1, \ldots, x_n)$. We will assume every variable occurs an even number of times. The modifications needed if a variable occurs an odd number of times are left to the reader.
2. For each variable x that occurs ≥ 8 times, do the following:
   a. Let k be the number of times x occurs. Introduce new variables $z_1, \ldots, z_k$.
   b. Replace the k occurrences of x with $z_1, \ldots, z_k$.
   c. Let G be a 3-expander graph on k vertices $\{1, \ldots, k\}$ (which exists by exercise 9.17). For every edge $\{i, j\}$, add the clauses $(z_i \rightarrow z_j)$ and $(z_j \rightarrow z_i)$. (Formally, we would use $z_i \vee \neg z_j$ for $z_i \rightarrow z_j$.) Note that z_i will occur at most 7 times in φ': (1) once in the place it replaces x in the original formula, (2) 3 times in clauses of the form $z_i \vee \neg z_j$, and (3) 3 times in clauses of the form $z_j \vee \neg z_i$. The last two bounds come from G having maximum degree 3.

   Let the new formula be φ'.

How many times does a variable z occur in φ'? If z occurs $k \leq 7$ times in φ, then it will occur $k \leq 7$ times in φ'. If z occurred ≥ 8 times in φ, then it will occur 7 times in φ' as noted above.

We show how to go from an assignment for φ' to an assignment for φ. Let $\vec{b}'$ be an assignment for φ'. Let x be a variable in φ that occurred ≥ 8 times in φ. We associated with x a set of variables that we want to all be assigned the same truth value. Let Z_{TRUE} be the subset of those variables that are assigned TRUE, and Z_{FALSE} be the subset of those variables that are assigned FALSE. We will show that *more* clauses of φ' can be satisfied by assigning all of them the same truth value. Recall that we have many clauses relating the variables associated with x via an expander graph.

Assume $|Z_{\text{TRUE}}| > |Z_{\text{FALSE}}|$ (the other case is similar). The expander graph has $\geq |Z_{\text{TRUE}}|$ edges from Z_{TRUE} to Z_{FALSE}. For every edge (i,j) where $i \in Z_{\text{TRUE}}$ and $j \in Z_{\text{FALSE}}$, there are *two* clauses, $z_i \rightarrow z_j$ and $z_j \rightarrow z_i$. Hence there are $\geq 2|Z_{\text{TRUE}}|$ clauses that connect a variable from Z_{TRUE} to a variable from Z_{FALSE}. If we change the assignment of everything in Z_{FALSE} to TRUE, then we make $\geq 2|Z_{\text{TRUE}}|$ clauses TRUE. How many clauses are now FALSE? Each variable in Z_{TRUE} appeared at most once in the non-expander-part of the formula. Hence $\leq |Z_{\text{TRUE}}|$ of the clauses are now FALSE. So the net gain is $\geq |Z_{\text{TRUE}}| > 0$. Recall that we began with a variable x in φ which was associated with $Z_{\text{FALSE}} \cup Z_{\text{TRUE}}$ with $|Z_{\text{TRUE}}| > |Z_{\text{FALSE}}|$. We now know that to maximize the number of clauses satisfied, $Z_{\text{FALSE}} = \emptyset$. Hence all of the variables associated with x are assigned the same value. *This is the reason we use expander graphs rather than cycles!*

We recap and get back to our issue. We can assume that $\vec{b}'$ assigns variables as we intended. Hence it is easy to map to $\vec{b}$, which only assigns the original variables of φ in the obvious way. We now show that we have an L-reduction.

The expander graph for x_i has k_i vertices and $3k_i/2$ edges. Each edge corresponds to 2 clauses that will both be set to TRUE. Hence

$$\text{benefit}(\varphi', \vec{b}') = \text{benefit}(\varphi, \vec{b}) + 3\sum_{i=1}^{n} k_i.$$

Hence

$$\text{OPT}(\varphi') = \text{OPT}(\varphi) + 3\sum_{i=1}^{n} k_i.$$

If we subtract we get

$$|\text{OPT}(\varphi') - \text{benefit}(\varphi', \vec{b}')| = |\text{OPT}(\varphi) - \text{benefit}(\varphi, \vec{b})|.$$

Hence we have the second condition for being an L-reduction. Note that φ' has $m + O(\sum_{i=1}^{n} k_i) = O(m)$ clauses and φ has $O(m)$ clauses. Hence, by exercise 9.15, $\text{OPT}(\varphi) = O(\text{OPT}(\varphi'))$. □

Exercise 9.19. Prove theorem 9.18.2.

Tovey [Tov84] proved a related result, though we will not need it or prove it:

Theorem 9.20. *Let* MAX *E3SAT-E5 be the function that takes a formula where (1) every clause has* exactly *3 literals, and (2) every variable occurs* exactly *5 times, and returns (as usual) the assignment that maximizes the number of clauses satisfied. Then* MAX *3SAT* $\le_L$ MAX *E3SAT-E5.*

9.5 Formulas and Graphs

In this section, we start with MAX 3SAT, which we already know has no PTAS (from theorem 8.33) and form a chain of reductions through several graph and formula problems. It is of interest that we start with a formula problem, then get some graph problems, then get back to formulas, then graphs again.

Definition 9.21.

1. INDEPENDENT SET-B-a is the function that, given a graph G with maximum degree a, returns the size of the maximum independent set of G.
2. VERTEX COVER-B-a is the function that, given a graph G with maximum degree a, returns the size of the minimum vertex cover of G.
3. DOMINATING SET-B-a is the function that, given a graph G with maximum degree a, returns the size of the minimum dominating set of G.

Theorem 9.22. *Within this theorem a statement of the form $X \notin$ PTAS is contingent on $P \neq NP$.*

1. *MAX 3SAT-3 $\leq_L$ INDEPENDENT SET-B-4 (the reduction is strict).*
2. *For all $\Delta \geq 4$, INDEPENDENT SET-B-Δ is APX-complete. This follows from part 1 and the result of Halldórsson-Radhakrishnan [HR97] that INDEPENDENT SET-B-Δ has a $\frac{\Delta+2}{3}$-approximation via a greedy algorithm. (Recall that this means there is an algorithm that returns an independent set of size $\geq \frac{3}{\Delta+2} OPT$.)*
3. *INDEPENDENT SET-B-4 $\leq_L$ VERTEX COVER-B-4 (the reduction is strict).*
4. *For all $\Delta \geq 4$, VERTEX COVER-B-Δ and VERTEX COVER are both APX-complete. This follows from part 3 and the 2-approximation for VERTEX COVER (unbounded degree) that we showed in theorem 8.35.*
5. *VERTEX COVER-B-4 $\leq_L$ DOMINATING SET-B-4 (the reduction is strict).*
6. *For all $\Delta \geq 4$, DOMINATING SET-B-Δ is APX-complete. This follows from part 5 and the folklore result that DOMINATING SET-B-Δ has an $O(\log \Delta)$-approximation by a greedy algorithm.*
7. *INDEPENDENT SET-B-4 $\leq_L$ MAX 2SAT.*
8. *MAX 2SAT is APX-complete. This follows from part 7 and a simple approximation algorithm for MAX 2SAT similar to the proof of theorem 8.33.1.*
9. *MAX 2SAT $\leq_L$ MAX NAE 3SAT (this reduction is strict).*
10. *MAX NAE 3SAT $\in$ APX $-$ PTAS. The lower bound follows from part 9. We leave as an exercise to show that MAX NAE 3SAT $\in$ APX.*
11. *MAX NAE 3SAT $\leq_L$ MAX CUT. (MAX CUT was defined in section 2.7.)*
12. *MAX CUT $\in$ APX $-$ PTAS. Part 11 shows the lower bound. There is a simple randomized algorithm that returns a cut that is $\geq 0.5OPT$: for each vertex v flip a fair coin to decide which half of the partition the vertex v goes into. By the method of conditional probabilities the algorithm can be derandomized.*

Proof. 1) We give the reduction from MAX 3SAT-3 to INDEPENDENT SET-B-4.

1. Input a formula φ where every clause has ≤ 3 literals and every variable appears ≤ 3 times.
2. We construct a graph G.
   a. For every occurrence of a literal there is a vertex.
   b. For every clause of the form $(L_1 \vee L_2 \vee L_3)$ where the L_i's are literals, put in edges between each pair of L_i's. For every clause of the form $(L_1 \vee L_2)$ where the L_i's are literals, put in edges between L_1 and L_2.

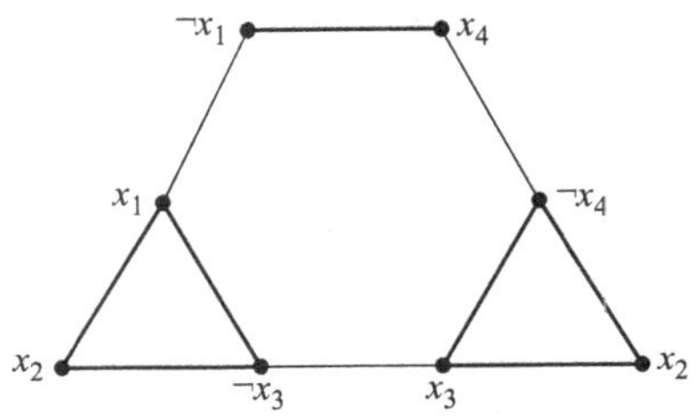

Figure 9.1
L-Reduction from MAX 3SAT-3 to INDEPENDENT SET-B-4.

c. For all variables x, if x is in one clause and $\bar{x}$ is in another clause, put an edge between them.

See figure 9.1.

Let v be a vertex. It corresponds to variable x in clause C. There will be two edges from v to the other vertices in clause C. There will be ≤ 2 edges to other clauses since they go to vertices associated with occurrences of $\neg x$, and there are at most two of those. Hence the degree of v is ≤ 4. The same reasoning holds when v is associated with $\neg x$.

We leave it to the reader to prove that this is a strict reduction.

3) If $G = (V, E)$ is a graph and U is an independent set, then $V - U$ is a vertex cover. Hence to show INDEPENDENT SET-B-4 $\leq_L$ VERTEX COVER-B-4 do the following: (1) just map G to G, and (2) map a vertex cover for G to its complement to get an independent set for G. The proof that this is a strict L-reduction is trivial.

5) VERTEX COVER-B-4 $\leq_L$ DOMINATING SET-B-4 by the following reduction: Map a graph G to the graph obtained by, for every edge (u, v), adding a vertex x that has an edge to u and to v. No other edges use x. We map a dominating set of G' to a vertex cover of G as follows:

1. Input G' and a dominating set D' for G'.
2. If one of the new vertices x, adjacent to both u, v, is in D, then either (1) one of u, v is in D so x can be removed, or (2) neither of u, v is in D, so remove x and add (say) u.
3. The new set (which might not be a dominating set for G') is a vertex cover U for G, so output it.

See figure 9.2.

Clearly $\text{benefit}(D) = \text{benefit}(U)$, so $\text{OPT}(D) = \text{OPT}(U)$. So we have a strict reduction.

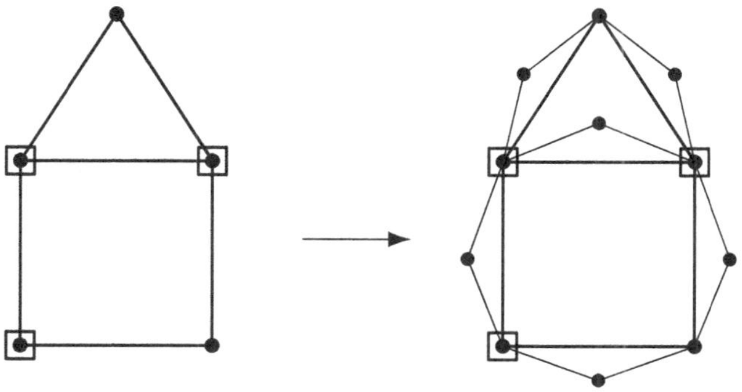

Figure 9.2
L-Reduction from VERTEX COVER-B-4 to DOMINATING SET-B-4.

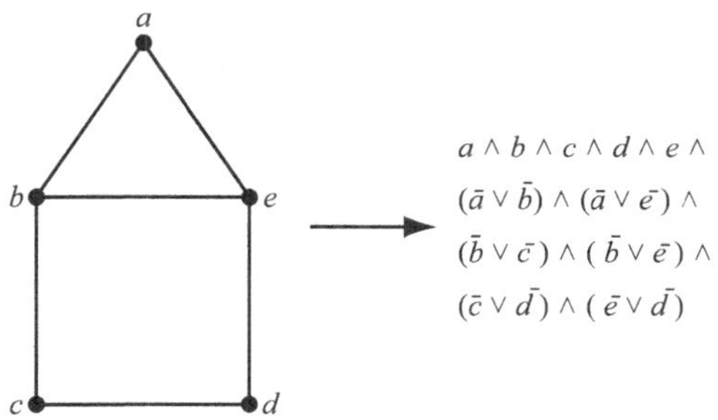

Figure 9.3
L-Reduction from INDEPENDENT SET-B-4 to MAX 2SAT.

7) INDEPENDENT SET-B-4 $\leq_L$ MAX 2SAT by the following reduction from graphs to formulas.

1. Input $G = (V, E)$, a graph of maximum degree 4.
2. Form a formula φ as follows:
   a. For every vertex v we have clause $\{v\}$.
   b. For every edge (u, v) we have clause $\{\overline{u} \vee \overline{v}\}$.
   (See figure 9.3.)

We show how to map an assignment for φ to an independent set of G. Recall from the definition of reduction that we need only consider assignments that cannot be improved in an obvious way.

If the assignment makes some 2-clause $(\overline{u} \vee \overline{v})$ FALSE, then change the assignment to make u FALSE. The number of clauses satisfied will not decrease. Hence we will only consider assignments where all of the 2-clauses are satisfied. Given such an assignment, we map it to the set of vertices associated with variables that are set to TRUE. Because of the change we made to the assignment, this will correspond to an independent set.

We now show that this is an L-reduction. Any assignment for φ can be improved (or at least not made worse) by the procedure described above: Make every clause $(\overline{u} \vee \overline{u})$ TRUE. The remaining 1-clauses that are set to TRUE must correspond to a maximum independent set in G. Hence we can start with an assignment $\vec{b}$ for φ that makes every 2-clause TRUE. Map it to the set of vertices U corresponding to the 1-clauses being TRUE. Clearly those vertices form an independent set. Hence benefit$(\vec{b}) = $ benefit$(U) + |E|$. Hence the second condition for being an L-reduction is satisfied.

Since G has maximum degree 4, $|E| = O(|V|)$ and $\text{OPT}(G) = \Theta(|V|)$ (by a greedy algorithm). Since φ has $|V|$ 1-clauses, $\text{OPT}(\varphi) = \Theta(|V|)$. Since $\text{OPT}(G)$ and $\text{OPT}(\varphi)$ are both $\Theta(|V|)$, the first condition of being an L-reduction, $\text{OPT}(\varphi) = O(\text{OPT}(G))$, is satisfied.

9) MAX 2SAT $\leq_L$ MAX NAE 3SAT by the following reduction from formulas to formulas.

1. Input φ, a formula where every clause has ≤ 2 literals.
2. Let z be a new variable.
3. We form a formula φ' by adding $\vee z$ to all 2-clauses, and $\vee z \vee z$ to all 1-clauses.

Let $\vec{b}'$ be an assignment for φ' where m of the clauses are satisfied but no clause has all three literals set the same. If z is set to TRUE, then the m satisfied clauses all have at least one literal set to FALSE. Hence if we flip the truth value of all the variables (note z is now FALSE) we still get an assignment where there are m clauses set to TRUE, and none of them have all literals set the same. We need only consider such assignments.

Since z is set to FALSE, the assignment, not including z, is an assignment for φ that makes m clauses TRUE. Hence benefit$(\varphi', \vec{b}') = $ benefit$(\varphi, \vec{b})$, so we have a strict reduction.

11) MAX NAE 3SAT $\leq_L$ MAX CUT by the following reduction from formulas to graphs.

1. Input φ, a formula in 3CNF form. It has n variables $x_1, \ldots, x_n$. x_i appears k_i times.
2. We form the graph G as follows (G will actually be a multigraph).
   a. For every variable x, let both x and $\overline{x}$ be vertices, and put k edges between them where k is the number of occurrences of x (see figure 9.4, left side).
   b. For every 3-clause $L_1 \vee L_2 \vee L_3$ put edges between L_1, L_2, and L_1, L_3, and L_2, L_3. Similar for 2-clauses (see figure 9.4, right side). For 1-clauses no edges are added.

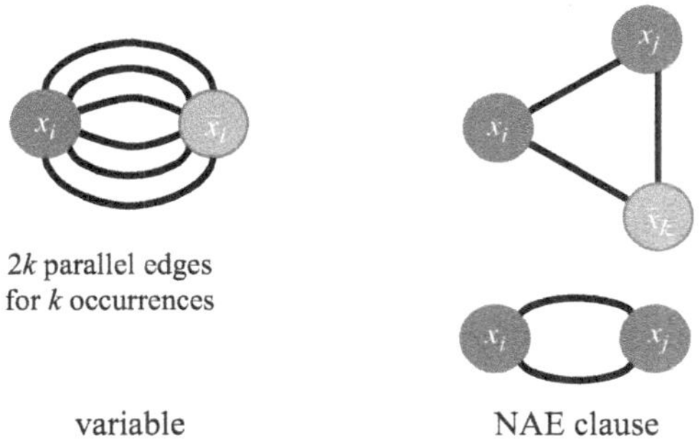

Figure 9.4
L-Reduction from MAX NAE 3SAT to MAX CUT.

We need to show how to map a partition of vertices (V_1, V_2) to an assignment $\vec{b}$. It is easy to see the only partitions worth considering have, for every variable x, x and $\neg x$ in different parts. Hence we can map (V_1, V_2) to the assignment that sets every literal in V_1 to TRUE. (We could have used V_2 but it won't matter which since MAX NAE 3SAT treats TRUE and FALSE equally.)

We leave it to the reader to finish this proof by comparing OPT(G) to OPT(φ). □

More is known about MAX CUT.

Note: We assume P $\neq$ NP in this note.

1. Alimonti and Kann [AK00] showed that VERTEX COVER-B-3, INDEPENDENT SET-B-3, and DOMINATING SET-B-3 have no PTAS.
2. Hastad [Has01, theorem 8.2] and Trevisan et al. [TSSW00, theorem 4.4] showed that MAX CUT does not have a $(\frac{17}{16} + \varepsilon)$-approximation (meaning that there is no algorithm that returns $\geq (\frac{16}{17} + \varepsilon)$OPT). In chapter 10 we will revisit MAX CUT assuming the unique game conjecture.

Note: Berman and Karpinski [BK99] have obtained the following concrete numbers:

1. VERTEX COVER-B-3 has no poly time $(\frac{145}{144} - \varepsilon)$-approximation.
2. VERTEX COVER-B-4 has no poly time $(\frac{79}{78} - \varepsilon)$-approximation.
3. VERTEX COVER-B-5 has no poly time $(\frac{74}{73} - \varepsilon)$-approximation.
4. INDEPENDENT SET-B-3 has no poly time $(\frac{140}{139} - \varepsilon)$-approximation.
5. INDEPENDENT SET-B-4 has no poly time $(\frac{74}{73} - \varepsilon)$-approximation.
6. INDEPENDENT SET-B-5 has no poly time $(\frac{68}{67} - \varepsilon)$-approximation.

Here are open problems:

Open Problem 9.23.

1. *Obtain concrete lower bounds for* Independent Set-*B*-6*,* Independent Set-*B*-7*, etc.*
2. *Close the gap between the lower bounds presented in note 9.5 and the upper bounds discussed in theorem 9.22.*
3. *Obtain concrete lower bounds for* Dominating Set-*B*-3*,* Dominating Set-*B*-4*, etc.*

For the next exercise we will consider the following variant of Set Cover.

> Exact Set Cover
> *Instance:* n and sets $S_1, \ldots, S_m \subseteq \{1, \ldots, n\}$.
> *Output:* The smallest size of a subset of S_i's that covers all of the elements in $\{1, \ldots, n\}$ with every element of $\{1, \ldots, n\}$ in exactly one of the chosen S_i's.

Exercise 9.24.

1. Show that Max Cut $\leq_L$ Exact Set Cover. What can you conclude from this in terms of approximations for Exact Set Cover?
2. Show that the problem of just finding a feasible solution for Exact Set Cover is NP-hard. What can you conclude from this in terms of approximations for Exact Set Cover?

9.6 Other Complexity Classes for Approximations

APX-completeness is not the be-all and end-all of complexity in approximations. There are other classes both above it and below it in complexity.

9.6.1 Log-APX-Completeness

Everything in this section is due to Escoffier and Paschos [EP06].

Combining exercise 9.14 and 8.29.4 we easily have the following.

Theorem 9.25. *Let $A_1, \ldots, A_m$ be such that* Set Cover $\leq_L A_1 \leq_L \cdots \leq_L A_m$*. Assume* P $\neq$ NP*. There is a c such that A_m does not have a $c \ln n$-approximation.*

Chlebík and Chlebíková [CC08] showed the following.

Theorem 9.26.

1. *There is a* $(\ln \Delta + 2)$*-approximation for* DOMINATING SET *where* Δ *is the maximum degree.*
2. *There is a* $(\ln n + 2)$*-approximation for* DOMINATING SET *(no bound on the degree).*
3. *Assume* $P \neq NP$*. For all* $c < 1$ *there is no* $c \ln \Delta$*-approximation for* DOMINATING SET *restricted to graphs of maximum degree* Δ*.*
4. *Assume* $P \neq NP$*. For all* $c < 1$ *there is no* $c \ln n$*-approximation for* DOMINATING SET *(no bound on the degree).*

Proof sketch. 1,2) A greedy algorithm where you always take the vertex of max degree works for both parts 1 and 2.

3,4) We prove part 4. Part 3 is similar.

We show SET COVER $\leq_L$ DOMINATING SET.

1. Input n and $S_1, \ldots, S_m \subseteq \{1, \ldots, n\}$. Let $U = \{1, \ldots, n\}$.
2. Form a graph $G = (V, E)$ as follows:
   a. There is a vertex for every element of $\{1, \ldots, n\}$ and for every S_i. Hence there are $n + m$ vertices. We call the vertices associated with $\{1, \ldots, n\}$ the ***U vertices***, and the vertices associated with the $S_1, \ldots, S_m$, the ***S vertices***.
   b. All vertices representing the S_i are connected. This forms a clique of size n.
   c. For all $i \in \{1, \ldots, n\}$ and all $j \in \{1, \ldots, m\}$ connect i to j if $i \in S_j$.
   d. Note that $\Delta \leq n$.

Let D be a dominating set. If there are any U vertices in D, then they can be replaced by the S vertex they connect to. Hence we can assume that every dominating set consists only of S vertices.

We map a dominating set to the S-sets that its S vertices correspond to. The size of the dominating set is exactly the size of a covering. Hence this is a strict reduction. □

Exercise 9.27. Show that DOMINATING SET $\leq_L$ SET COVER.

We now show another inapproximability result. The ***15 puzzle*** [Wika] is a classic puzzle from the 1870s. It can be generalized to the $n^2 - 1$ puzzle. Ratner and Warmuth [RW90] showed that solving the $n^2 - 1$ puzzle is NP-hard. It is not known if the function version (trying to minimize the number of moves) is in APX.

Călinescu et al. [CDP08] worked on MOTION PLANNING, which is a generalization of the $n^2 - 1$ puzzle:

Motion Planning *Instance:* A graph $G = (V, E)$ with the vertex set split into 2 (possibly overlapping) sets V_1, V_2 of the same size. The elements of V_1 are called ***tokens*** and each one has a ***robot*** on it. The elements of V_2 are called ***targets***. *Output:* A ***move*** is when a robot goes on a path with no other robots on it. Note that a robot may go quite far in one move. We want a final configuration where all the robots are on the vertices in V_2 (only one robot can fit on a vertex). Output the minimum number of moves needed to do this.

Theorem 9.28. *Set Cover $\leq_L$ Motion Planning. Hence there exists a constant c such that, assuming P $\neq$ NP, Motion Planning is not $c \ln n$-approximable.*

Proof. We give a reduction Set Cover $\leq_L$ Motion Planning.

1. Input is an n and $S_1, \ldots, S_m \subseteq \{1, \ldots, n\}$.
2. Form a graph (shown in figure 9.5) with:
   a. C is the set of m vertices, one for each S_j.
   b. B is the set of n vertices, one for each element of $\{1, \ldots, n\}$. Put an edge between elements of B and C if the element associated with B is in the set associated with C.
   c. A is a set of m vertices that form a path graph. The rightmost element of A has edges to all vertices in B.
3. Robots are put on the elements of $A \cup B$ and the target is $B \cup C$.

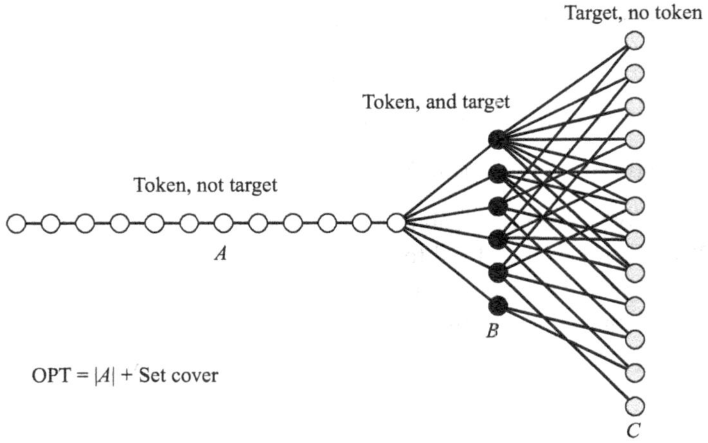

Figure 9.5
L-Reduction from Set Cover to Motion Planning.

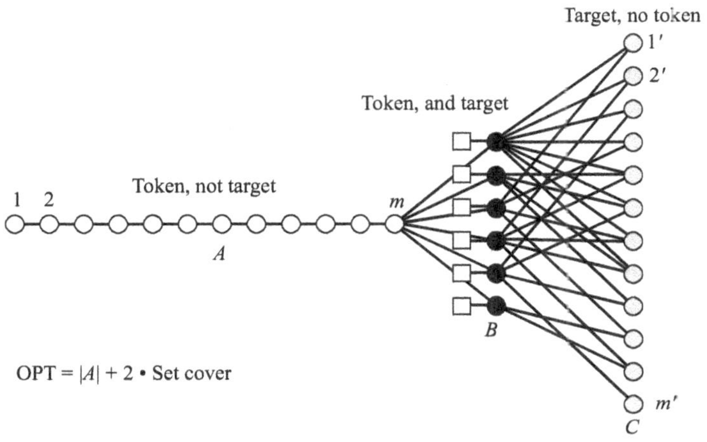

Figure 9.6
L-Reduction from Set Cover to Motion Planning modified.

The optimal solution takes a number of moves equal to $|A|$ plus the size of a minimal set cover. This is because we need one move to fill every location in C, which requires exactly $|A|$ moves. Furthermore, we need one move to cover every location vacated by a robot in B. But the number of robots in B that must move is exactly the size of the minimum set cover.

Thus, the conditions of the L-reduction are satisfied, and this problem is NP-complete. If two robots are not allowed to occupy the same vertex at once, figure 9.6 shows a small modification that can be used to get around this problem. The optimal solution in this case is $|A|$ plus twice the size of a minimal set cover. □

Node-Weighted Steiner Tree (NWST)
Instance: A graph $G=(V,E)$ with weights on its nodes, a set $T \subseteq V$ marked as terminal, and a node $r \in V$.
Output: A set of vertices S such that (1) the graph induced by $T \cup S$ connects all the terminal nodes to r and (2) S has the minimum sum of weights of vertices of any set that satisfies (1).

Theorem 9.29.

1. *There is an $O(\log n)$-approximation for NWST. This was proven by Moss and Ranbani [MR07]. We omit the proof.*
2. *There is an L-reduction from Set Cover to Node-Weighted Steiner Tree. Hence there is a constant c such that, assuming P $\neq$ NP, there is no $c \ln n$-approx for NWST. (This follows from the reduction and theorem 8.29.4).*

Proof. We give the reduction:

1. Input an instance of SET COVER: n and $S_1, \ldots, S_m \subseteq \{1, \ldots, n\}$.
2. Construct a graph as follows:
   a. For every $i \in \{1, \ldots, n\}$ there is a node of weight 0. These are the terminal nodes. For every set S_j there is a node of weight 1.
   b. There is a node r that has an edge to each S_j.
   c. If $i \in S_j$, then there is an edge between i and S_j.

See figure 9.7 for an example.

It is easy to see that there is a cost d solution to NODE-WEIGHTED STEINER TREE if and only if there is a set cover of size d. □

Note: One can define the ***edge weighted Steiner tree (EWST)*** problem. It is NP-complete. Bykra et al. [BGRS13] showed there is a 1.39-approximation for EWST. Hence, assuming $\mathsf{P} \neq \mathsf{NP}$, there is no reduction from SET COVER to EWST.

> GROUP STEINER TREE (GST)
> *Instance:* A graph $G = (V, E)$ with weights on its edges, sets $V_1, \ldots, V_k \subseteq V$, and a node $r \in V$.
> *Output:* A set of nodes S of such that (1) the graph induced by S connects some vertex of each V_i and (2) the graph induced by S has the minimum weights of the edges among all S that satisfy (1).
> *Note:* The graph induced will be a tree.

Theorem 9.30.

1. *The GROUP STEINER TREE problem has (1) a randomized polynomial-time algorithm that gives an $O((\log^2 n \log\log n \log k)$-approximation for general graphs, and (2) an $O(\log n \log\log k)$-approximation for trees. These were obtained by Garg et al. [GKR00]. We omit the proofs.*
2. *SET COVER $\leq_L$ GROUP STEINER TREE. We can restrict GROUP STEINER TREE to trees where every edge has weight 0 or 1. This reduction is exact.*
3. *If there is a $(1 - o(1)) \ln n$-approx for GROUP STEINER TREE, then $\mathsf{P} = \mathsf{NP}$. (This follows from parts 1 and 2 and theorem 8.29.4.)*

Proof. We prove part 2 by giving the reduction.

1. Input an instance of SET COVER: n and $S_1, \ldots, S_m \subseteq \{1, \ldots, n\}$.
2. Construct a graph as follows:

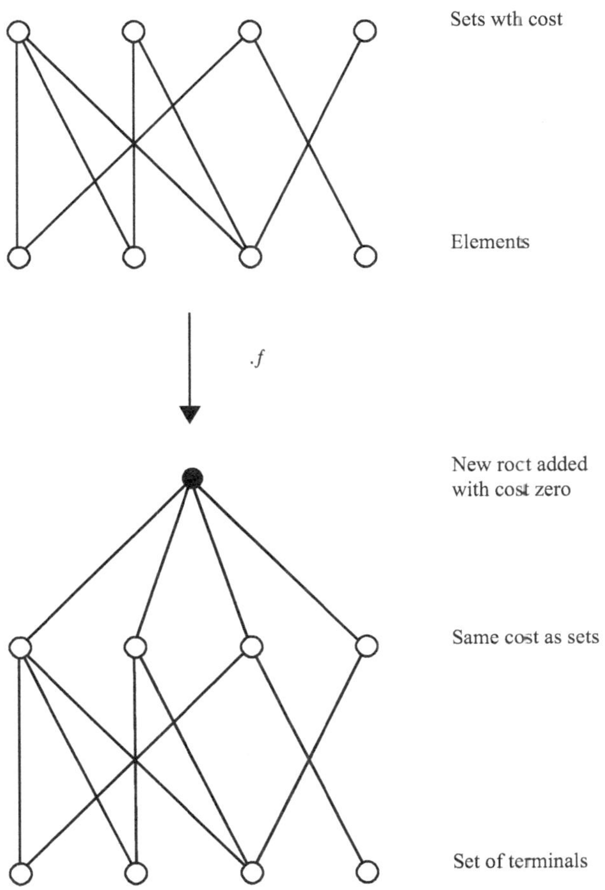

Figure 9.7
APX-Reduction from Set Cover to NWST.

a. The root is r. It has children $S_1, \ldots, S_m$. The weights of the edges from r to each S_j are 1.
b. Each S_j has children labeled with the elements of S_j. These will be the leaves. (So you will have several leaves with the same label.) The edges from the i to S_j have weight 0.
c. For $1 \le i \le n$, V_i is the set of all leaves labeled i.

We leave it to the reader to show this is an L-reduction.

We do an example. Let $n = 8$ and
$S_1 = \{1, 3, 8\}$.

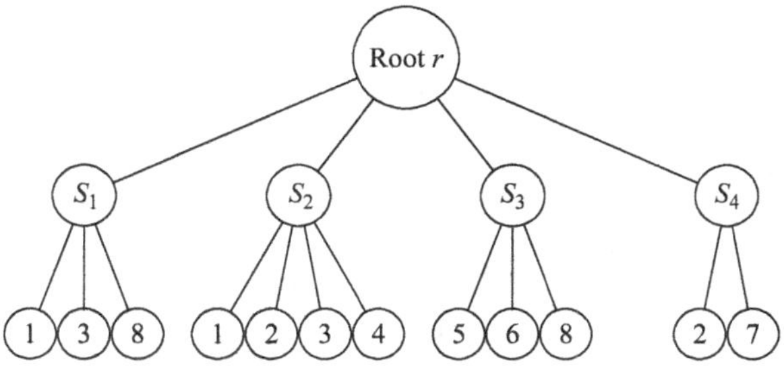

Figure 9.8
L-Reduction from SET COVER to GST.

$S_2 = \{1, 2, 3, 4\}$.
$S_3 = \{5, 6, 8\}$.
$S_4 = \{2, 7\}$.

The resulting graph is in figure 9.8. The weights on the edges are as follows: (1) all of the edges from r to an S_i have weight 1, and (2) all of the edges from S_i to one of its children have weight 0. □

9.6.2 Poly-APX-Complete

Bazgan et al. [BEP05] defined Poly-APX, which is a framework for saying that problems cannot be approximated any better than within a polynomial factor. Two problems in this class (each of which can easily be reduced to the other) are finding the largest clique and the largest independent set. There are very few other problems in this class.

9.6.3 EXP-APX-Complete

Escoffier and Paschos [EP06] defined EXP-APX, which is a framework for saying that problems cannot be approximated any better than within an exponential factor. EXP-APX-complete problems occur when there are numbers in the problem that can have large size, increasing the difficulty of approximation. The most well-known problem in this class is the non-metric traveling salesman problem, where edges can be exponential in the size of the input. We do not know of any other natural problems that are in this class. There are very few other problems in this class.

9.6.4 NPO-Complete

Crescenzi et al. [CKST99] defined the class NPO and NPO-complete to capture other aspects of approximation (and lack thereof). Examples of problems that are NPO-complete:

- MAX-WEIGHTED-SAT: Given a CNF formula φ, find a satisfying assignment that maximizes the sum of the weights of the variables set to true. That sum is the output.
- MIN-WEIGHTED-SAT: Given a CNF formula φ, find a satisfying assignment that minimizes the sum of the weights of the variables set to true. That sum is the output.
- Both the min and max version of 0-1 PROGRAMMING (defined in example 0.13).

Jonsson [Jon98] defined the class NPOPB and NPOPB-complete. This class contains solutions that are recognizable in polynomial time, whose costs are also computable in polynomial time. (The PB in the title stands for "polynomially bounded.")

However, computing even a valid solution is NP-complete. The distinction between NPOPB-complete and NPO-complete is similar to that of between Poly-APX-complete and EXP-APX-complete respectively; it's usually due to the presence of large numbers.

NPOPB-complete problems are difficult to approximate polynomially even when the solutions are trivial. Examples of problems in this class include:

- minimum independent dominating set
- shortest Turing-machine computation
- longest induced path
- longest path with "forbidden pairs" (pairs of vertices such that the path cannot cross both)

9.7 Further Results

9.7.1 EDGE MATCHING

We can define optimization versions of edge-matching puzzles from section 5.5.1.

EDGE MATCHING APPROXIMATION
Instance: A multiset of n unit squares with the edges colored, and a target rectangle T of area n.
Question: There are two objectives:

1. How many of the n squares can you pack into T such that all tiles sharing an edge have matching colors?
2. If you pack all n squares into T, how many of the edge pairs can match?

Bosboom et al. [BDD+17] showed that both versions of EDGE MATCHING APPROXIMATION are NP-hard to approximate within a factor of 0.9999999702, even when the target rectangle T has dimensions $1 \times n$. On the other hand, both versions of EDGE MATCHING APPROXIMATION have easy $\frac{1}{2}$-approximations (at least in the $1 \times n$ case), so these results are tight up to constant factors.

9.7.2 MAX FEASIBLE LINEAR SYSTEM

> MAX FEASIBLE LINEAR SYSTEM
> *Instance:* A system of linear equations with coefficient in $\mathbb{Z}$: $A \cdot \vec{x} = \vec{b}$.
> *Output:* A maximum-size subset of the equations that has a solution over $\mathbb{Q}$.

MAX FEASIBLE LINEAR SYSTEM *looks* easy since, given matrix A and vector $\vec{b}$ one can, in polynomial time, do the following (by Gaussian elimination):

- Determine whether there is a solution, and if so then find one, and if not, then produce a certificate of infeasibility.
- If there is no solution, then find an $\vec{x}$ such that $A\vec{x}$ is close to $\vec{b}$. More precisely $\vec{x}$ is such that, $A\vec{x} - \vec{b}$ has the *least mean squared error*.

Nevertheless, Amaldi and Kann [AK95] showed the following:

- The natural decision formulation of MAX FEASIBLE LINEAR SYSTEM is NP-hard.
- Many variants and restrictions of MAX FEASIBLE LINEAR SYSTEM are NP-hard.
- Assume P $\neq$ NP. Many variants of MAX FEASIBLE LINEAR SYSTEM are hard to approximate. The hardness varies with the variant. Some are in APX but not PTAS, and some are harder to approximate than that.

9.7.3 SHORTEST VECTOR PROBLEM and Its Variants

We defined lattices, norms, and SHORTEST VECTOR PROBLEM in section 6.7.1. We recall the definition of SHORTEST VECTOR PROBLEM for the reader's convenience.

In the next problem let $p \in [1, \infty]$.

> SHORTEST VECTOR PROBLEM$_p$ (SVP$_p$) and the Approximate Versions
> *Instance:* A lattice $\mathcal{L}$ specified by a basis.
> *Output:* A shortest vector in that basis using the p-norm. (For definition of p-norm see section 6.7.1.)

Output: Let $f(n)$ be a function (think of it as increasing). A vector that is $\leq f(n)\mathrm{SVP}_p(L)$. (Such a vector is called an ***f*(*n*)-approximation for SVP**$_p$.)

Note: Approximate Shortest Vector Problem is an important problem for cryptography. Many crypto systems are based on factoring or discrete log being hard; however, both of these problems can be solved quickly using a quantum computer. To prepare for the day when quantum computers (or perhaps some other method) can factor or solve discrete log quickly, people have devised crypto systems based on approximate Shortest Vector Problem. It is believed that approximate Shortest Vector Problem cannot be solved by a quantum computer.

Lenstra et al. [LLL82] (see Regev and Kaplan [RK04, remark 8]) showed that there is a $(4/3)^n$-approximation to SVP_2. They used it to obtain an algorithm to factor polynomials. Schnorr [Sch87] improved this to a $2^{n(\log\log n)^2/\log n}$-approximation. Combining this with results by Ajtai et al. [AKS01] leads to a $2^{n(\log\log n)/\log n}$-approximation (See corollary 15 of [AKS01] and plug in $k = O(\log n)$.) Theorem 9.31 gives a lower bounds on approximation. Unfortunately the upper and lower bounds are far apart.

H. Bennett [Ben23] has a survey that covers both the fine-grained complexity of Shortest Vector Problem (e.g., uses assumptions like ETH) and the computational complexity of Shortest Vector Problem (e.g., uses assumptions like $\mathsf{P} \neq \mathsf{NP}$). We state some of the lower bounds and also refer the reader to the survey or the papers cited for earlier results on this topic.

Theorem 9.31.

1. *(Boas [vEB81]) SVP_∞ is NP-hard.*
2. *(Ajtai [Ajt98]) SVP_2 is NP-hard under randomized reductions.*
3. *(Regev and Rosen [RR06]) SVP_1 is NP-hard under randomized reductions.*
4. *(Micciancio [Mic00]) Let $p \in [1,\infty) \cup \{\infty\}$ and $\varepsilon > 0$. $(2^{p/2} - \varepsilon)$-approximating SVP_p is NP-hard under randomized reductions.*
5. *(Khot [Kho05]) Assume $NP \not\subseteq RP$. Let $p \in (1,\infty)$. There is no poly-time algorithm for constant factor approximate SVP_p.*
6. *(Regev and Rosen [RR06]) Assume $NP \not\subseteq RP$. There is no poly-time algorithm for constant factor approximate SVP_1.*
7. *(Haviv and Regev [HR12] Assume $NP \not\subseteq RTIME(2^{polylog(n)})$. Let $p \in [1,\infty)$. Let $\varepsilon > 0$. There is no poly-time algorithm for $2^{(\log n)^{1-\varepsilon}}$-factor approximation for SVP_p.*
8. *(Micciancio [Mic12]) This paper presented new proofs of the results of Khot [Kho05] and Haviv and Regev [HR12] that, while still using random reductions, seem more*

likely to be able to be derandomized. If the reductions can be derandomized, then (1) Khot's result can be improved to use the hardness assumption $\mathsf{P} \neq \mathsf{NP}$, *and (2) Haviv and Regev's result can be improved to use the hardness assumption* $\mathsf{NP} \not\subseteq \mathsf{DTIME}(2^{polylog(n)})$.

9. *(H. Bennett and Peikert [BP22]) Let* $p \in [1, \infty)$. SVP_p *is NP-hard under randomized reductions. This proof has some aspects to it that make derandomizing it plausible. If this is shown, then the hardness assumption of* $\mathsf{NP} \not\subseteq \mathsf{RP}$ *can be changed to* $\mathsf{P} \neq \mathsf{NP}$.

9.7.4 Online Nearest Neighbor

A useful problem in data structures is to store a set of points A (in some space) so that, given a point x (not in A), you can determine the point in A that is closest to x. You may be allowed to preprocess the points.

> ONLINE NEAREST NEIGHBOR (ONNN_p and γ-ONNN_p):
> *Instance:* (To Preprocess) A set of points A in $\mathbb{R}^d$. We will assume there are n points.
> *Instance:* A query point x.
> *Output:* (ONNN_p) A point $y \in A$ that is closest to x in the p-norm.
> *Output:* (γ-ONNN_p where $\gamma > 1$) We will call the distance to the closest point OPT. A point $y \in A$ such that $\|x - y\|_p \leq \gamma \text{OPT}$. (We will also allow distances other than p-norms such as edit distance and Hamming distance.)

1. If you do no preprocessing and, given x, compute its distance to every point in A, this takes $O(n)$ time (assuming that computing a distance takes $O(1)$ time). This is considered a lot of time for data structures since (1) n is large and (2) computing a distance is costly even if it $O(1)$.
2. Assume you knew ahead of time the set of query points. You could, in the preprocessing stage, determine for each query point which point of A it is closest to. This would yield query time $O(1)$ but (1) a lot of time for preprocessing and (2) a lot of space for the data structure.

Is there a way to get both quick preprocessing and quick query times? What if you settle for an approximation? Assuming SETH there are lower bounds on the trade-off.

Theorem 9.32.

1. *(Rubinstein [Rub18]) Let* $p \in \{1, 2\}$ *Assume SETH. Let* $\delta, c > 0$. *There exists* $\varepsilon = \varepsilon(\delta, c)$ *such that no algorithm for* ONNN_p *has (1) preprocessing time* $O(n^c)$, *(2) query time* $O(n^{1-\delta})$ *and solves* $(1+\varepsilon)$-ONNN_p. *(The result also holds for edit-distance and Hamming-distance.)*

2. *(Ko and Song [KS20]) Assume SETH. Let $\delta, c > 0$. There exists $\varepsilon \in \{0, 1)$ such that no algorithm for* ONNN$_p$ *has (1) preprocessing time polynomial in n, (2) query time* $O(n^{1-\delta})$ *and solves* $(1+\varepsilon)$*-*ONNN$_p$.

9.7.5 APX-Intermediate Problems

Let A be a min problem such that there exists an algorithm that returns a number $\leq OPT_A(x) + 1$. This is in APX but seems better than that, so perhaps not APX-complete.

We present two problems with the following properties: (1) there is a better-than-APX approximation algorithm, as above; (2) there is currently no PTAS for them; and (3) if the problem is APX-complete, then the polynomial hierarchy collapses.

For both problems the third point is due to Crescenzi et al. [CKST99], and for the first point we will provide a reference.

BIN PACKING
Instance: A finite set of positive rationals U.
Output: The minimum number of bins needed to partition U into sets whose sum is ≤ 1?
Note: Rothvoß [Rot13a] showed there is an $\text{OPT} + O(\log(\text{OPT}) \log\log(\text{OPT})$-approximation.

EDGE CHROMATIC NUMBER
Instance: A graph G.
Output: The minimum number of colors needed to color the edges so that no two incident edges are the same color.
Note: By Vizing's theorem [Viz64], a graph with maximum degree Δ has a $\Delta + 1$-edge coloring. Hence we can always get an answer $\leq \text{OPT}(G) + 1$. (Misra and Gries [MG92] obtained a constructive proof of Vizing's theorem that yields a polynomial-time algorithm that finds the $\Delta + 1$-coloring. This is not needed for our purpose of just finding the edge-chromatic number.)

10

Unique Games Conjecture

10.1 Introduction

As the title indicates, this is a chapter on *the unique games conjecture*. This is a deep field of study that interacts with many branches of mathematics. Hence we will only be able to scratch the surface.

Chapter Summary

1. *We motivate and state the* UNIQUE GAMES CONJECTURE.
2. *We state consequence of the* UNIQUE GAMES CONJECTURE. *These consequences are mostly lower bounds on approximation that are better than those obtained by assuming* $P \neq NP$.
3. *We discuss results obtained because of the study of the* UNIQUE GAMES CONJECTURE. *These mostly have to do with limits on particular types of algorithms.*
4. *We discuss both reasons to believe, and reasons to not believe, the* UNIQUE GAMES CONJECTURE.

10.2 Our Failure to Obtain Matching Upper and Lower Bounds on Approximation

In chapters 8 and 9 we stated (and in some cases proved) upper and lower bounds on many approximation problems. The lower bounds assumed the usual hardness assumption $P \neq NP$ and used the PCP machinery, or reductions. We recall two sets of results.

MAX 3SAT Let OPT be the max number of clauses that can be satisfied. We list upper and lower bounds on approximating MAX 3SAT.

1. (Folklore) There is a polynomial-time algorithm that will, given a 3CNF formula φ, find an assignment that will satisfy $\frac{7}{8}\text{OPT}(\varphi)$. We presented this in theorem 8.33. The proof was easy.

2. (Arora et al. [ALM$^+$98]) Assume P $\neq$ NP. There exists $\varepsilon > 0$ such that there is no polynomial-time algorithm that will, given a 3CNF formula φ, find an assignment that satisfies $(1-\varepsilon)\text{OPT}(\varphi)$ clauses. We presented this in theorem 8.33. It was easy an easy proof *given* the PCP theorem (theorem 8.24).
3. (Hastad [Has01]) Assume P $\neq$ NP. Let $\delta > 0$. There is no polynomial-time algorithm that will, given a 3CNF formula φ, find an assignment that satisfies $(\frac{7}{8}+\delta)\text{OPT}(\varphi)$ clauses. We stated this as theorem 8.34 but did not prove it. The proof is difficult and requires (a) a different kind of proof system which we will discuss in section 10.3 and (b) hard math as evidenced by the terms "fourier transforms" and "long codes."

To summarize: Assuming P $\neq$ NP, the upper and lower bounds on approximating MAX 3SAT *match*.

VERTEX COVER Let OPT be the min size of a vertex cover. We list upper and lower bounds on approximating VERTEX COVER.

1. (Theorem 8.35) There is a polynomial-time algorithm that will, given a graph G, find a vertex cover of size $\leq 2\text{OPT}(G)$.
2. The 2-to-2 Unique Games Conjecture (this is complicated—we won't be defining it) was proven in a sequence of papers [KMS17, KMS18, DKK$^+$18a, DKK$^+$18b]. See [Kho19] for an overview and how the conjecture implies the following: If there is an algorithm that, on input a graph G, returns a number $\leq(1.414\ldots-\delta)$ then P $=$ NP (the number is actually $\sqrt{2}$).

To summarize: Even assuming P $\neq$ NP, and using hard math, the upper and lower bounds on approximating VERTEX COVER *do not match*.

Upshot Assume P $\neq$ NP. For MAX 3SAT we obtain matching bounds for approximation; however for VERTEX COVER we can only obtain non-matching bounds.

So back to VERTEX COVER. The bounds do not match. So ... now what do we do?

- Try to get a better approximation algorithm for VERTEX COVER. The result that gives $\leq 2\text{OPT}(G)$ is very old, and there have been no improvements on it. So this is unlikely. The consensus of the theory community is that $\leq 2\text{OPT}(G)$ is the best polynomial-time approximation possible.
- Try to use even harder math or more finely tuned prover-systems to get better lower bounds. Khot [Kho10] gives reasons why this may not be possible.

So *now* what do we do? In section 10.3 we will describe a game that was used to obtain better lower bounds than PCP for many problems; however, most of those lower bounds did not match the upper bounds. Then, in section 10.4, we will tweak that game to come up with another one that *seems* NP-hard. The assumption that it *is* NP-hard will be dubbed.

The Unique Games Conjecture.

From that conjecture one can obtain (1) *matching* upper and lower bounds for approximating VERTEX COVER and several other problems for which assuming $P \neq NP$ did not seem to suffice, (2) *better* lower bounds for some problems than can be obtained from assuming $P \neq NP$, and (3) *better* lower bounds for some problems if you assume variants of the unique games conjecture.

Warning: Math games are not fun games. See two blogs of Gasarch [Gas09a, Gas09b] for more on this.

Most of this chapter will be about the unique games conjecture and what it implies for lower bounds on approximation. However, we will also discuss a surprising application it has to integrality gaps.

10.3 2-Prover-1-Round Game

Arora et al. [ABSS97] defined the following problem as a tool to prove better lower bounds on approximation.

LABEL COVER (APPROXIMATE LABEL COVER)
Instance:

1. A bipartite graph (V, W, E). (For the approximate version there is also an input $0 < \delta < 1$.)
2. $a, b \in \mathbb{N}$ with $a \geq b$. (Note that the only restriction on a, b is $a \geq b$. This is one of the things we will tweak.)
3. For each edge $e \in E$ a surjection $f_e : [a] \to [b]$. (Note that the only restriction on f_e is that it be surjective. This is one of the things we will tweak.)

Question: Is there a way to label every $v \in V$ with an element $\ell(v) \in [a]$, and every $w \in W$ with an element $\ell(w) \in [b]$, such that, for every $e = (v, w), f_e(\ell(v)) = \ell(w)$? (For the approximate version we want to know whether we can satisfy the fraction δ of the constraints.) Such a labeling is called a ***label covering***.

Note: This problem is also called a "2-Prover-1-Round Game." We will see why in definition 10.2.

Exercise 10.1. Show that LABEL COVER is NP-complete.

We now reinterpret the problem as a game where two provers are trying to convince one verifier that a graph has a label covering.

Definition 10.2. The ***2-prover-1-round game*** goes as follows. The board consists of the following:

1. A bipartite graph (V, W, E).
2. $a, b \in \mathbb{N}$ with $a \geq b$.
3. For each edge $e \in E$ a surjection $f_e : [a] \rightarrow [b]$.

The players are (1) the verifier, and (2) a pair of provers P_1, P_2. The game is played as follows:

- The Verifier randomly picks a $(u, w) \in E$ and sends u to P_1 and w to P_2. P_1 and P_2 cannot communicate.
- P_1 outputs an $\ell(u) \in [a]$, P_2 outputs a $\ell(w) \in [b]$.
- If $f_e(\ell(u)) = \ell(w)$ then the provers win. If not then the verifier wins.

Clearly if there is a label covering, then the provers can win. Also note that if δ of the edges can be satisfied, then there is a strategy for the provers to win δ of the time.

You can view the game as the provers wanting to convince the verifier that there is a label covering. The more rounds of the game they play the harder it will be to fool the verifier. How fast does the error decrease? Raz [Raz98] showed that the error decreases exponentially. This is a very hard and deep theorem that is the key to many other results. It is called ***Raz's parallel repetition theorem***. The following can be proven from the PCP theorem (theorem 8.24) and Raz's parallel repetition theorem.

Definition 10.3. Let U be an instance of the label cover problem. If δ is the largest fraction of edges that can be satisfied then $\mathrm{OPT}(U) = \delta$.

Theorem 10.4. *Let $0 < \delta < 1$. Then there exists a constant C such that the following happens: Restrict the inputs U to label cover where $a = \Theta((1/\delta)^C)$. It is NP-hard to distinguish between the following cases:*

- *$OPT(U) = 1$ (so there is a label covering)*
- *$OPT(U) \leq \delta$ (so there is no way to cover more than δ of the edges)*

Intuitively it is hard to distinguish the cases where there is a label covering from the case where its hard to even approximate a label covering.

This theorem was a key ingredient in getting lower bounds for approximation by Bellare et al. [BGS98] (MAX 3SAT, MAX CUT, VERTEX COVER, Clique, Chromatic Number), Dinur et al. [DGKR05] (Hypergraph VERTEX COVER), Dinur and Safra [DS02] (VERTEX COVER), Khot [Kho19] (VERTEX COVER), Dinur and Steurer [DS13] (Set Cover), Guruswami et al. [GHS02] (c-coloring a 2-colorable 4-uniform hypergraph), Hastad [Has99] (Clique), Hastad

[Has01] (Max 3SAT, Vertex Cover), Khanna et al. [KLS00], (4-coloring 3-colorable graphs), Khot [Kho01] (Chromatic Number, Clique) and others. These results offer some good news and some bad news:

- *Good news:* In all cases the lower bounds obtained involved concrete numbers (e.g., Vertex Cover cannot be approximated any better than $1.41\text{OPT}(G)$), in contrast to proofs where the constant was buried in the details of the PCP theorem (e.g., theorems 8.27 and 8.33).
- *Bad news:* In most cases these improved lower bounds still were not tight. (Exceptions: Dinur and Steurer's lower bound on set cover is tight, and Hastad's lower bound on Max 3SAT is tight.)
- *Worse news:* Khot [Kho10] gives reasons why the lower bounds obtained by using theorem 10.4 cannot be improved.

So again... what do we do? We will tweak the definition of 2-prover-1-round games in the next section.

10.4 Unique Games Conjecture

We define a tweak on the Label Cover problem.

Unique Label Cover
Instance:

1. A bipartite graph (V, W, E). (For the approximate version also $0 < \delta < 1$.)
2. $a \in \mathbb{N}$ (this is one of the tweaks—for Label Cover we needed $a, b \in \mathbb{N}$).
3. For each edge $e \in E$ a bijection $f_e : [a] \to [a]$. (This is one of the tweaks—for Label Cover was a surjection from $[a]$ to $[b]$.)

Question: (This is the same as Label Cover.) Is there a way to label every $v \in V$ with an element $\ell(v) \in [a]$, and every $w \in W$ with an element $\ell(w) \in [b]$, such that, for every $e = (v, w)$, $f_e(\ell(v)) = \ell(w)$? (For the approximate version we want to know whether we can satisfy the fraction δ of the constraints.) Such a labeling is called a ***label covering***.

Note: This problem is also called a ***unique 2-prover-1-round game***. We will see why in definition 10.5.

Definition 10.5. We define the ***unique 2-prover-1-round game***. The board consists of the following:

1. A bipartite graph (V, W, E).
2. $a \in \mathbb{N}$.
3. For each edge $e \in E$ a bijection $f_e : [a] \to [a]$.

The players are (1) the verifier, and (2) a pair of provers P_1, P_2. The game is played as follows

- The verifier randomly picks a $(u, w) \in E$ and sends u to P_1 and w to P_2. P_1 and P_2 cannot communicate.
- P_1 outputs an $\ell(u) \in [a]$, P_2 outputs an $\ell(w) \in [a]$.
- If $f_e(\ell(u)) = \ell(w)$, then the provers win. If not, then the verifier wins.

Clearly if there is a label covering, then the provers can win. Also note that if δ of the edges can be satisfied, then there is a strategy for the provers to win δ of the time.

You can view the game as the provers wanting to convince the verifier that there is a label covering.

Definition 10.6. Let U be an instance of the unique 2-prover 1-round game. If δ is the largest fraction of edges that can be satisfied, then $\mathrm{OPT}(U) = \delta$.

The following exercise may surprise you.

Exercise 10.7. Let $0 < \delta < 1$. The following two cases can be distinguished in polynomial time:

- $\mathrm{OPT}(U) = 1$ (so there is a label covering)
- $\mathrm{OPT}(U) \leq \delta$ (so there is no way to cover more than δ of the edges)

The only possible way to get a hard problem out of unique games is to allow 2-sided error.

UNIQUE GAME CONJECTURE(UGC)

Conjecture 10.8. *Let $0 < \varepsilon, \delta < 1$. Then there exists a constant C such that the following happens. Restrict the inputs U to instances where $a = C$. It is NP-hard to distinguish between the following cases:*

- *$OPT(U) \geq 1 - \varepsilon$ (so there is a fairly good approximate label covering)*
- *$OPT(U) \leq \delta$ (so there is no way to cover more than δ of the edges)*

In the next section we will list many consequences of the unique games conjecture.

10.5 Assuming UGC . . .

In this section we list consequences of the unique games conjecture for lower bounds on approximation. In each subsection we define the problems in our list that have not been defined elsewhere in this book. We sometimes even define a problem that we have defined earlier since we need to say something about it that was not mentioned earlier.

The tables in this chapter abbreviate *Approximation Factor* by *AF*, and *Folklore* by *FL*.

The following notation comes up in several definitions.

Definition 10.9. Let $G=(V,E)$ be a graph. Let $V_1, V_2 \subseteq V$. Then $E(V_1, V_2)$ is the set of edges that have one endpoint in V_1 and the other in V_2.

We divide the problems we look at into three categories. All three categories (1) have to do with lower bounds on approximation and (2) yield better lower bounds when using UNIQUE GAMES CONJECTURE as a hardness assumption than just assuming $\mathsf{P} \neq \mathsf{NP}$.

The three categories are as follows:

- Problems where UNIQUE GAMES CONJECTURE yields matching upper and lower bounds on approximations.
- Problems where UNIQUE GAMES CONJECTURE yields better lower bounds than assuming $\mathsf{P} \neq \mathsf{NP}$; however, the upper and lower bounds still do not match.
- Problems where hardness assumption extensions of UNIQUE GAMES CONJECTURE yields better lower bounds than hardness assumption $\mathsf{P} \neq \mathsf{NP}$. We denote these extensions UNIQUE GAMES CONJECTURE+.

In the tables below we allow randomized algorithms, with high probability of success, in the column *Best Approx*.

Note: In the next three sections we will use definition 8.5 of α-approximation. Reread that definition and the paragraph that follows it to see why the standard notation is awkward.

10.5.1 Problems Where UGC Implies Optimal Lower Bounds

> VERTEX COVER ON k-HYPERGRAPHS
> *Instance:* k-Hypergraph $G=(V,E)$. Note that $E \subseteq \binom{V}{k}$.
> *Output:* The size of the smallest set $V' \subseteq V$ such that every edge $E=\{v_1, \ldots, v_k\}$ has some element of V' in it.

MAX CUT
Instance: A graph $G = (V, E)$.
Output: The largest value $|E(V_1, V_2)|$ can have where (V_1, V_2) ranges over all partitions of V.
Note: Goemans and Williamson [GW95] have a randomized polynomial-time algorithm for MAX CUT that returns a value $\geq \beta$OPT where

$$\beta = \min_{0 \leq \theta \leq \pi} \frac{2}{\pi} \frac{\theta}{1 - \cos\theta} \sim 0.87856.$$

Let $\alpha_{MC} = \frac{1}{\beta}$. Using definition 8.5, α_{MC} is the approximation ratio.

MAX 2SAT
Instance: A 2CNF formula $\varphi = C_1 \wedge \cdots \wedge C_k$.
Output: The largest fraction of clauses that an assignment can satisfy.
Note: Lewin et al. [LLZ02] have a randomized polynomial-time algorithm for MAX 2SAT that returns a value $\geq \beta$OPT where β is hard to describe but is around 0.94. Let $\alpha_{LLZ} = \frac{1}{\beta}$. Using definition 8.5, α_{LLZ} is the approximation ratio.

MAX ACYCLIC SUBGRAPH (MAS)
Instance: A directed graph $G = (V, E)$.
Output: The size of the largest acyclic subgraph.

Problem	Best Approx	UGC	P $\neq$ NP
VERTEX COVER	2	$2 - \varepsilon$ [KR08, BK09]	$1.414 - \delta$ See theorem 8.38.3 for refs.
VERTEX COVER on k-uniform hypergraphs ($k \geq 3$)	k	$k - \varepsilon$ [KR08, BK10]	$k - 1 - \varepsilon$ [DGKR05]
MAX CUT	α_{MC}[GW95]	$\alpha_{MC} - \varepsilon$ [KKMO07] [OW08, KO09]	$17/16 - \varepsilon$ [Has01]
MAX 2SAT	α_{LLZ} [LLZ02]	$\alpha_{LLZ} - \varepsilon$ [Aus07]	APX-hard [PY91]
MAS	2 [New00]	$2 - \varepsilon$ [GHM$^+$11]	$66/65 - \varepsilon$ [New00]

10.5.2 Problems Where UGC Implies Good but Not-Optimal Lower Bounds

FEEDBACK ARC SET
Instance: A directed graph $G=(V,E)$.
Output: The size of the smallest set of edges that contains at least one edge from every directed cycle.

Definition 10.10. Let $G=(V,E)$ be a graph and $V' \subseteq V$. The ***sparsity of V'***, denoted $S(V')$, is

$$\frac{|E(V',V-V')|}{\min\{|V'|,|V-V'|\}}.$$

SPARSEST CUT (SC). Also called MIN CUT.
Instance: A graph $G=(V,E)$.
Output: The minimum possible sparsity of a set of vertices of G.

MIN-2SAT-DELETION
Instance: A 2CNF formula $\varphi = C_1 \wedge \cdots \wedge C_k$, where no clause is of the form $\bar{x} \vee \bar{y}$.
Output: The max number of clauses that can be satisfied. (We give results for this version.)
Note: An equivalent formulation, which is where the name comes from, is as follows: Given a 2CNF formula φ (no restrictions), what is the min number of clauses that need to be deleted from φ such that the remaining formula is satisfiable? It is equivalent in that the answer to one of them is k minus the answer to the other one.

MINUNCUT
Instance: A Graph $G=(V,E)$.
Output: A set $V' \subseteq V$ that minimizes $|E(V',V')| + |E(V-V',V-V')|$.

Problem	Best app	UGC	P ≠ NP
FBACK ARC SET	$O(\log n \log\log n)$ [Sey95] [ENSS98]	$\omega(1)$ [GHM+11]	1.36 (Red from VC)
SPARSEST CUT	$O(\sqrt{\log n})$ [ALN05] [ARV09]	$\omega(1)$ [CKK+06]	NONE
MIN-2SAT-DEL	$O(\sqrt{\log n})$ [ACMM05]	$\omega(1)$ [Kho02] $\omega(1)$ [CKK+06]	APX-hard (FL)
MINUNCUT	$O(\sqrt{\log n})$ [ACMM05]	$\omega(1)$ [Kho02]	APX-complete (FL)

10.5.3 Problems Where UGC Is Used to Obtain Lower Bounds

We present several problems where, by assuming an extension of UNIQUE GAMES CONJECTURE, better lower bounds on approximation can be found. For details on those extensions, see the survey by Khot [Kho10].

COLORING A k-COLORABLE GRAPH

Instance: A graph G that you are promised is k-colorable.

Output: A k' coloring of G where k' is as small as possible.

Note: We will abuse terminology in the table below by calling a lower bound on k' the "Approx Factor." As with the usual approx factors, these are obtained using assumptions such at P ≠ NP or UNIQUE GAMES CONJECTURE. Here is an example: If I say

Assuming P ≠ NP the approx factor for coloring 3-colorable graphs is 4,

that means that

If every 3-colorable graph could be 4-colored in polynomial time then P = NP.

SCHEDULING WITH PRECEDENCE CONSTRAINTS

Instance: An acyclic graph G of jobs. The idea is that if there is an edge (i,j), then job i must be completed before job j. Each job has associated with it a processing time p_i and a weight w_i. The weight indicates how important it is to get this job done earlier.

Output: A linear ordering of the jobs. This ordering will also give a set of completion times. Let the ith job have completion time c_i. Minimize $\sum_{i=1}^{n} c_i w_i$.

Problem	Best App	UGC+	P $\neq$ NP
COLORING 3-COL GRAPH	$N^{0.211}$ [AC06]	$\omega(1)$ [DMR09]	4 [KLS00]
COLORING $2d$-COL GRAPH	$N^{1-\frac{3}{2d+1}}$ [KMS98]	$\omega(1)$ [DMR09]	$2d + 2\lfloor \frac{2d}{3} \rfloor - 1$ [KLS00] $d^{\Omega(\log d)}$ [Kho01]
SCH WITH PREC. CONST.	2 (FL)	2ε [BK09]	NONE

10.6 UGC Implies Integrality Bounds

Usually we aim for theorems like

Assuming P $\neq$ NP. There is no $\varepsilon > 0$ and polynomial-time approximation algorithm for set cover that returns $(1 - \varepsilon)(\ln n)OPT$.

Note that this is a statement about *all* polynomial-time algorithms.

There are times when you have a *particular* polynomial-time approximation algorithm and want to know the limits of how well it can do. In this section we look at algorithms based on relaxation methods and formulate the question of finding limits on how well they can do. We present the following without proof.

1. A relaxed linear programming approach to approximate set cover, and the limit to how good it can do. This example has nothing to do with the UNIQUE GAMES CONJECTURE; however, it is a good example of integrality gaps.
2. A relaxed semi-definite programming approach to approximate MAX CUT, and the limit to how good it can do. This is particularly interesting since the limit to how well this particular algorithm can do is exactly the limit you get by assuming UNIQUE GAMES CONJECTURE. Hence the SEMI-DEFINITE PROGRAMMING algorithm may be optimal.
3. A relaxed semi-definite programming approach to approximate sparsest cut, and the limit to how good it can do. This is fascinating since (a) the unique games conjecture inspired the proof (actually more than inspired); however, the proof stands without assumption, and (b) the proof involved a lot of hard math, particular geometric embeddings.

10.6.1 LINEAR PROGRAMMING and SET COVER

We give an example of how 0-1 programming can be relaxed to linear programming, and then used to approximate set cover.

> LINEAR PROGRAMMING and 0-1 PROGRAMMING
> *Instance:* An integer matrix A and integer vectors $\vec{b}$ and $\vec{c}$.
> *Output:* The max value $\vec{c} \cdot \vec{x}$ can have relative to the constraint $A\vec{x} \leq \vec{b}$. The LINEAR PROGRAMMING problem restricts $x_i \in \mathbb{Q}$. The 0-1 PROGRAMMING problem restricts $x_i \in \{0, 1\}$.
> *Note:* We use LP and ZOP to denote an instance of LINEAR PROGRAMMING and 0-1 PROGRAMMING.

The LINEAR PROGRAMMING problem is in P, whereas the 0-1 PROGRAMMING problem is NP-hard.

One way to deal with NP-hard problems is to do the following:

1. Formulate them as a ZOP.
2. Relax this to an LP problem.
3. Solve this LP.
4. Use the answer to get an approximation for the original 0-1 PROGRAMMING problem (that's the clever part).

We give an example that seems to be folklore; however, Trevisan [Tre11] has a writeup.

Algorithm to approximate SET COVER

1. Input n and $S_1, \ldots, S_m \subseteq \{1, \ldots, n\}$. We denote the instance of set cover X.
2. Formulate the following 0-1 PROGRAMMING, which we denote ZOP(X).
   - The variables are $x_1, \ldots, x_m$. The idea is that $x_i = 1$ means S_i is chosen and $x_i = 0$ means S_i is not chosen. Since this is 0-1 programming, $x_1, \ldots, x_m \in \{0, 1\}$.
   - Constraints. For $j \in \{1, \ldots, n\}$ we need that at least one of the S_i's that has j is chosen. Hence we have the constraint:
   $$\forall j \in \{1, \ldots, n\} : \sum_{j \in S_i} x_i \geq 1.$$
   - Objective function: We want to minimize
   $$x_1 + \cdots + x_m.$$

   Note that the ZOP(X) solves set cover exactly.
3. Relax ZOP(X) to an LP, denoted LP(X). The only difference is that $x_i \in [0, 1]$ instead of $x_i \in \{0, 1\}$.

4. Solve LP(X) to obtain $x_1^*, \ldots, x_n^* \in [0,1]$ (the x_i^* will be rational by a well-known theorem about linear programming).
5. Set x_i to 1 with probability x_i^*.

Trevisan's writeup shows that, with probability ≥ 0.45, the algorithm returns a set cover that is $\leq 2(\ln n + 6)$OPT.

To prove the analysis of the algorithms is roughly tight we would need an instance X of Set Cover in which the algorithm above returned a set cover that is $\geq \ln(n)$OPT.

Definition 10.11. The ***integrality gap*** for the above algorithm is

$$\alpha = \inf_X \frac{\text{LP}(X)}{\text{ZOP}(X)} = \frac{\text{LP}(X)}{\text{OPT}(X)}.$$

(We use inf instead of min since it may be a limit.)

It is easy to see that the algorithm cannot do any better than $\leq \alpha$OPT(X).

Lovász [Lov75] showed that the integrality gap is H_n, the nth harmonic number, which is $\ln n$ in the limit. Hence the LINEAR PROGRAMMING approach cannot do any better than the simple greedy algorithm of Chvatal [Chv79]. This is worth knowing!

Takeaway: If we find the integrality gap α, then it is evidence that the algorithm cannot give an approximation that is any better than α. This shows a limitation on a particular algorithm; however, it is not a hardness result.

10.6.2 SEMI-DEFINITE PROGRAMMING and MAX CUT

SEMI-DEFINITE PROGRAMMING (SDP)
Instance:

- Integers $\{c_{i,j} \mid 1 \leq i,j \leq n\}$.
- Integers $\{a_{i,j,k} \mid 1 \leq i,j \leq n,\ 1 \leq k \leq m\}$.

Output: Vectors $x^1, \ldots, x^n \in \mathbb{R}^n$ such that

- $$\sum_{1 \leq i,j \leq n} c_{i,j}(x^i \cdot x^j)$$

  is minimized.
- Subject to

$$(\forall k)\left[\sum_{1 \leq i,j \leq n} a_{i,j,k}(x^i \cdot x^j) \leq b_k\right].$$

We will consider SEMI-DEFINITE PROGRAMMING to be in polynomial time, though there are some issues about approximation and the reals (that we will not consider).

Goemans and Williamson obtained an approximation algorithm for MAX CUT using an SDP. The rest of this section is about their work.

They first formulate MAX CUT as an optimization problem (this type of optimization problem does not have a nice name like 0-1 PROGRAMMING, however, it is NP-complete). They then formulate an SDP, which is similar to the optimization problem, and use it to get an approximation.

Algorithm to approximate MAX CUT

1. Input Graph $G=(V,E)$. Let $V=\{1,\dots,n\}$.
2. Formulate the following problem, denoted $MC(G)$.
   - The variables are $x_1,\dots,x_n$. The idea is that $x_i=1$ means i is chosen to be in the cut, and $x_i=-1$ means i is not chosen to be in the cut.
   - Constraints. For $1\le i\le n$, $x_i\in\{-1,1\}$.
   - Objective function: We want to maximize:

   $$\frac{1}{2}\sum_{(i,j)\in E}(1-x_ix_j).$$

   Note that the $MC(G)$ solves MAX CUT exactly.
3. The analogous SEMI-DEFINITE PROGRAMMING (called a ***relaxation*** in the literature) is as follows.
   - The variables are ***vectors*** $x_1,\dots,x_n$ in $\mathbb{R}^n$.
   - Constraints: For $1\le i\le n$, $\|x_i\|=1$.
   - Objective: Maximize

   $$\frac{1}{|E|}\sum_{(i,j)\in E}\frac{1-x_ix_j}{2}.$$

   We call this SDP(G).
4. Solve SDP(G) to obtain $\vec{x}_1,\dots,\vec{x}_n$. (The entries might be irrational. There are ways to deal with this; however, we omit this discussion.)
5. Pick a random vector r on the unit sphere.
6. Let $V=V_1\cup V_2$ where $V_1=\{i\mid x_i\cdot r\ge 0\}$. (This is called ***randomized rounding*** in the literature.)

Definition 10.12. The ***integrality gap*** for the above algorithm is

$$\alpha=\sup_G\frac{\text{SDP}(X)}{\text{OPT}(X)}.$$

(We use sup instead of max since it may be a limit.)

Note that this approach to MAX CUT cannot do any better then αOPT(G).

Goemans and Williamson [GW95] showed that α is the α_{MC} we defined in section 10.5.1. Hence their algorithm cannot do better than α_{MX}OPT(G). This constant became more interesting when from UNIQUE GAMES CONJECTURE one obtains α_{MC} for the lower bound on how well *any* algorithm can do.

10.6.3 SEMI-DEFINITE PROGRAMMING and SPARSEST CUT

Leighton and Rao [LR99] used a relaxation of an LP to obtain an $O(\log n)$-approximation for sparsest cut. They also showed that the integrality gap was $\Omega(\log n)$ so the bound was tight. Note again that this is just for their LP. Aumann and Rabani [AR98] and Linial et al. [LLR95] independently used a geometric theorem of Bourgain [Bou85] to obtain another approach to a relaxation of an LP for sparsest cut. This LP is looking for a metric with certain properties. They also obtained matching upper and lower bounds of $\Theta(\log n)$.

Even though no better algorithms came out of using LINEAR PROGRAMMING to look for metrics, this was a breakthrough since it introduced deep geometrical theorems to the field. Goemans [Goe97] and Linial [Lin02] made a conjecture that metrics exist, which would imply an $O(1)$-approximation for sparsest cut (so sparsest cut would be in APX).

Arora et al. [ARV09] used a relaxation of a SEMI-DEFINITE PROGRAMMING to obtain an $O(\sqrt{\log n})$-approximation for sparsest cut. So that is progress! But then Khot and Vishnoi [KV15] showed that the integrality gap for *any* SDP that is based on geometry (and that is probably any SDP) has integrality gap $\omega(1)$. This proof was inspired (actually more than inspired) by the UNIQUE GAMES CONJECTURE; however, the result is absolute and needs no assumption. Realize that this just rules out SDP approaches. While we believe that UNIQUE GAMES CONJECTURE is true and therefore sparsest cut is not in APX, this has not been proven.

10.7 Is UGC True?

We are not going to answer the question "Is UNIQUE GAMES CONJECTURE true?" since neither we, nor anyone else, knows. That's why its a *conjecture*. However, we will give some arguments for it, together with counters to those arguments.

1. There is a SEMI-DEFINITE PROGRAMMING approach to solving the UNIQUE GAME PROBLEM. Khot and Vishnoi have shown, by integrality gap methods, that this approach will not get the UNIQUE GAME PROBLEM in polynomial time. But is there some other algorithmic paradigm that has not been ruled out? There is! Lasserre [Las01a, Las01b] devised a way to iterate SEMI-DEFINITE PROGRAMMING programs to obtain better approximations (also see Rothvob's notes [Rot13b] and the paper of Karlin et al. [KMN11]). Khot et al. [KPS10] have shown results that indicate (though do not prove) that Lasserre's approach with a constant number of rounds cannot solve THE UNIQUE GAMES PROBLEM.

This still leaves open the possibility of a non-constant number of rounds. What makes the Lasserre method so hopeful (or non-hopeful if you think UGC is true) is that it has not been ruled out by integrality gaps or other paradigms.

2. Arora et al. [ABS15] have a subexponential algorithm for the problem. This tends to indicate that UNIQUE GAMES CONJECTURE is false. However, we note that there are some NP-complete problems with subexponential time algorithms, though they are contrived and involve padding the input.
3. Braverman et al. [BKM21] have devised ***the rich 2-to-1 game conjecture*** and showed that it is equivalent to the unique game conjecture. The rich 2-to-1 game conjecture may be easier to prove.
4. (This is what we consider the most compelling reason.) UGC has great explanatory power. Take VERTEX COVER as an example. It seemed like it was very hard to get a $(2-\varepsilon)$OPT approximation for VERTEX COVER. And the community believed that it was impossible. But we couldn't prove it. Ah, but with UNIQUE GAMES CONJECTURE we can! VERTEX COVER is not an isolated example. The table in section 10.5 gives many cases where we get either matching bounds or better bounds. Counterargument: We may one day prove all of these better lower bounds from $\mathrm{P} \neq \mathrm{NP}$. That is possible and has not been ruled out.

10.8 Open Problems

Open Problem 10.13.

1. *Prove or disprove UNIQUE GAMES CONJECTURE.*
2. *For the tables in section 10.5 for which we do not have matching bounds, close the gap. This may be done by either better approximation algorithms or better reductions.*
3. *There may come a time when we think, for lower bounds on approximation, that UNIQUE GAMES CONJECTURE has gone as far as it can go (we may already be there with $\mathrm{P} \neq \mathrm{NP}$). When this happens, find a new conjecture that is reasonable and whose assumption will help close the gaps.*

II

Above NP

11

Counting Problems

11.1 Introduction

Recall that $\varphi \in$ SAT if there exists *at least one* satisfying assignment for φ. For SAT, all we care about is the distinction between 0 and ≥ 1 satisfying assignments. What if we want to know *how many* satisfying assignments there are? This problem is clearly at least as hard as SAT, but is it harder? Probably yes. Valiant [Val79c, Val79a, Val79b] defined the class of functions that capture this notion.

Chapter Summary

1. *We define the complexity class #P, which is the set of functions that count the number of solutions to an NP problem. We also define an appropriate notion of reduction and completeness. We give reasons why problems that are #P-complete are likely much harder than NP-complete problems (see theorem 11.2).*
2. *We define the related notion of* Another Solution Problem: *Given one (or more) satisfying assignment, can you find another? We define an appropriate notion of reduction and completeness. ASP-completeness turns out to be an even stronger property than #P-completeness.*
3. *We show many problems ASP-complete and/or #P-complete.*
4. *We look at the related notion of* Fewest Clues: *How many variables' assignments (say) do you need so that the remaining formula has only one satisfying assignment?*

11.2 #P

Definition 11.1.

1. ***#SAT*** is the following function: On input a Boolean formula φ, output the number of satisfying assignments.

2. More generally, let $A \in \mathsf{NP}$. Hence there exist a polynomial p and a set $B \in \mathsf{P}$ such that
$$A = \{x \mid \exists y : |y| = p(|x|) \wedge (x, y) \in B\}.$$
#A is the function that, on input x, outputs the number of y of length $p(|x|)$ such that $(x, y) \in B$.
3. ***#P*** is the set of all functions of the form $\#A$ where $A \in \mathsf{NP}$.
4. Recall from definition 0.6 that FP is the set of functions that can be computed in polynomial time.

Note: The above definition of $\#A$ is ambiguous as written. There may be many different p, B for a set $A \in \mathsf{NP}$. Here is an example:
$$\text{SAT} = \{\varphi(x_1, \ldots, x_n) \mid \exists x = yz : |y| = |z| = n \wedge \varphi(y) = \text{TRUE} \wedge z \text{ is a square}\}.$$
Using this definition of SAT, $\#\text{SAT}(\varphi(x_1, \ldots, x_n))$ would be
$$(\text{number of satisfying assignments}) \cdot (\text{number of squares of length } n).$$
This is clearly not what we want. We will ignore this issue and assume that the B, p used to define A are natural. For example, #3-COLORING is the number of 3-colorings of a graph, #HAMILTONIAN CYCLE is the number of Hamiltonian cycles in a graph, and so on.

Is #SAT much harder than SAT? Likely yes. Toda [Tod91] showed the following.

Theorem 11.2. *If A is in the polynomial hierarchy (so $A \in \Sigma_i$ or $A \in \Pi_i$), then A reduces to #3SAT. Hence, if #3SAT is in FP, then every set in the polynomial hierarchy is in P. The reduction may use many evaluations of #3SAT.*

Valiant's motivation for studying counting problems was to show that computing the permanent of matrix is likely hard. We elaborate on this in section 11.5.1. Another motivation for studying counting problems comes from puzzle design. We often want to know when there is a unique solution for a puzzle. Hence we want to know when the number of solutions to an NP problem is 1.

11.2.1 Reductions and Completeness

The statement $A \leq_p B$ can be viewed as saying that we can solve the question "$x \in A$?" by making one query to "$y \in B$?" and the answer to the query is the answer to the question. Instead we consider reductions that allow many calls and can use the results any way you want. Formally, this definition would require the definition of an oracle Turing machine. We are not going to define the concept formally. All you need to know is that an oracle Turing machine $M^{\cdots}$ can make queries to whatever is in the superscript (a set or a function). In the calculation of $M^g(x)$, if the machine wants to know $g(y)$, it only has to write down y on a special tape.

Definition 11.3. Let f, g be functions.

1. $f \leq_{p,o} g$ if there is an oracle Turing machine M^g such that $f(x) = M^g(x)$ and the calculation of $M^g(x)$ takes polynomial time.
2. g is ***#P-hard*** if, for all $f \in$ #P, $f \leq_{p,o} g$.
3. g is ***#P-complete*** if $g \in$ #P and g is #P-hard.

We will often want to show that, for some set $A \in$ NP, $\#A$ is #P-complete. We will do this by presenting a certain type of reduction from A to B. We define two such reductions.

Definition 11.4. Let A, B be sets in NP. A ***parsimonious reduction*** from A to B is a reduction f such that the following hold.

1. $x \in A$ if and only if $f(x) \in B$.
2. The number of solutions for $x \in A$ (e.g., the number of satisfying assignments) equals the number of solutions for $f(x) \in B$ (e.g., the number of cliques of size k).

The next terminology is due to E. Demaine when teaching this material [BCC+20], though the idea of consistent expansion of the number of solutions is used in the original #P papers (e.g., [Val79a]).

Definition 11.5. Let A, B be sets in NP. A ***c-monious reduction*** from A to B is a reduction f such that the following hold.

1. $x \in A$ if and only if $f(x) \in B$.
2. The number of solutions for $x \in A$ equals c times the number of solutions for $f(x) \in B$.

Normally we think of c being a constant (e.g., 2 or 4), but it can also be a polynomial-time function of the input x (e.g., n or 2^n).

Exercise 11.6.

1. Show that #SAT is #P-complete.
   Hint: Modify the proof of the Cook–Levin theorem to make the reduction parsimonious.
2. Show that, if there is a parsimonious or c-monious reduction from SAT (or another #P-hard problem) to A, then $\#A$ is #P-hard.
3. Show that, if $\#A$ is #P-complete and $\#A \in$ FP, then #P $=$ FP.

11.2.2 NP-Complete Versus #P-Complete

Many researchers have showed that, for many NP-complete problems A, there is a parsimonious reduction from SAT to A, and hence $\#A$ is #P-complete. We will present some of those proofs, as well as other proofs that problems are #P-complete or #P-hard.

Empirically it seems that, for every natural problem A that is NP-complete, $\#A$ is #P-complete. This is not a theorem, and it probably cannot be a theorem since it is hard to define "natural."

11.2.3 Sometimes $\#A$ is Harder Than A

Are there sets $A \in \mathsf{P}$ for which $\#A$ is #P-complete? Yes. We state four of them.

Theorem 11.7.

1. *Brightwell and Winkler [BW05] showed that counting the number of Eulerian Cycles in a graph is #P-complete. Note that detecting whether a graph has an Euler cycle is in P.*
2. *Jerrum [Jer94] showed that counting the number of labeled trees in a graph is #P-complete. Note that detecting a graph has a tree for a subgraph is trivial; the answer is always yes.*
3. *Valiant [Val79b] showed that counting the number of bipartite matching is #P-complete. We will see this as an easy corollary of* PERMANENT *being #P-complete. Note that finding a matching in a bipartite graph (even a general graph) is in P. Vadhan [Vad01] showed that this problem is still #P-complete when G is restricted to bipartite graphs (1) with maximum degree 4, (2) planar of maximum degree 6, or (3) k-regular for any $k \geq 5$. That paper has many other problems in P whose counting version is #P-complete.*
4. *Valiant [Val79b] showed that counting the number of satisfying assignments in a monotone 2-SAT formula is #P-complete. Note that the problem of determining whether a monotone 2-SAT formula is satisfied is trivial, the answer is always yes. Provan and Ball [PB83] extended this result.*

Exercise 11.8. Find more sets $A \in \mathsf{NP}$ that are also in P such that $\#A$ is in polynomial time.

11.3 ANOTHER SOLUTION PROBLEM (ASP A)

Next we look at the related ANOTHER SOLUTION PROBLEM (ASP) family. The reductions we need for ASP are parsimonious along with one more condition. Thus ASP-completeness is a stronger property than #P-hardness, so we will aim for it when possible.

Ueda and Nagao [UN96] defined ASP A and the appropriate reductions for them; however, we follow the treatment of Yato and Seta [YS03] who defined the more general notion of k-ASP A.

Definition 11.9. Let $A \in \mathsf{NP}$. Hence there exist a polynomial p and a set $B \in \mathsf{P}$ such that

$$A = \{x \mid \exists y : |y| = p(|x|) \wedge (x, y) \in B\}.$$

Then ***ASP A*** is the following problem: Given x and given a y with $|y| = p(|x|)$ and $B(x,y)$, determine whether another solution $y' \neq y$ exists. (ASP stands for another solution problem.) Note that you need not find that other solution, only decide its existence.

More generally, ***k-ASP A*** is the problem: Given k solutions to A, is there another solution? Thus 0-ASP A is the same as A, and 1-ASP A is the same as ASP A.

The ASP concept is useful in puzzle design. Typically, we design a puzzle alongside an intended solution. The (negation of) the ASP problem tells us whether the intended solution is unique.

11.3.1 Sometimes ASP *A* Is Easier Than *A*

There are cases where A is hard but ASP A is easy. ASP 3-COLORING is easily seen to be in P: Given a 3-coloring of a G, just permute the colors to get another 3-coloring of G. On the other hand, if we factor out the $3! = 6$ permutations of the colorings, then 3-COLORING is ASP-complete [Bar04] (in particular, ASP 3-COLORING is NP-complete).

Next we look at a more interesting example. Let CUBIC HAMILTONIAN CYCLE be HAMILTONIAN CYCLE restricted to cubic graphs, which is known to be NP-complete. Then we will see that ASP CUBIC HAMILTONIAN CYCLE $\in$ P.

1. Tutte [Tut46] proved Smith's theorem: Any cubic graph has an even number of Hamiltonian cycles. Hence, if a cubic graph G has a Hamiltonian cycle, then it has another one. The proof was nonconstructive and, hence, did not yield an algorithm to actually find the second Hamiltonian cycle. (Tutte credits Smith with an earlier and different proof of the theorem.)
2. Price [Pri77] and Thomason [Tho78] came up with similar algorithms that, given a graph G and a Hamiltonian cycle H, output another Hamiltonian cycle. They did not analyze the running time.
3. Krawczyk [Kra99] showed that Thomason's algorithm takes exponential time in the worst case. Cameron [Cam01] showed that the same result holds even when restricted to planar cubic graphs.

Because we want to know only *whether* there is another Hamiltonian cycle, and do not need to find it, we have the following theorem.

Theorem 11.10.

1. *ASP 3-COLORING is trivially in P since we can permute the colors. Note that we do not just know there is another solution; we can actually find it.*
2. *Let CUBIC HAMILTONIAN CYCLE be the problem of finding a Hamiltonian cycle in a cubic graph. By the statements above, ASP CUBIC HAMILTONIAN CYCLE $\in$ P trivially because the answer is always yes.*

The question arises: Given a cubic graph and a Hamiltonian cycle in it, how hard is it to *find* another one? This question is in some sense captured by the complexity class PPA, described in section 22.9.1; see also [IP22].

11.3.2 ASP Reductions

Let $A, B \in \mathsf{NP}$. We define a notion of reduction so that, if A reduces to B and ASP $A \notin \mathsf{P}$, then ASP $B \notin \mathsf{P}$.

Definition 11.11. Let $A, B \in \mathsf{NP}$.

1. A is ***ASP reducible to B*** if there is a parsimonious reduction from A to B with the additional property that there is a polynomial-time bijection from solutions of an instance of B to solutions of an instance of A. (This bijection is an algorithmic form of parsimony; in particular, it implies parsimony.)
2. B is ***ASP-complete*** if, for every $A \in \mathsf{NP}$, A is ASP reducible to B. (We do not have a notion of ASP-hardness since these notions only make sense if $B \in \mathsf{NP}$.)

Exercise 11.12. Prove that if A is ASP reducible to B, then the following hold:

- ASP $B \in \mathsf{P}$ implies ASP $A \in \mathsf{P}$.
- ASP A is NP-hard implies ASP B is NP-hard.
- A is ASP-complete and $B \in \mathsf{NP}$ implies B is ASP-complete.

Exercise 11.13. Prove that if A is ASP-complete, then the following hold:

- k-ASP A is NP-complete for any $k \geq 0$.
- A is NP-complete.
- #A is #P-complete.

You can assume that SAT is ASP-complete [YS03].

Exercise 11.14. Fillmat is a logic puzzle published by Nikoli [Nik08, Nik]. (Nikoli is a publisher of puzzles that are culture-independent.) It was shown to be both NP-complete and ASP-complete by Uejima and Suzuki [US15]. Read their paper and either try to prove their theorems or read how they did and put it in your own words.

11.4 ASP-Completeness via Parsimonious Reductions

In this section, we show several problems are ASP-complete using ASP (in particular, parsimonious) reductions. As a consequence, the corresponding counting problems are #P-complete.

11.4.1 3SAT, Clique, and 3SAT-3

Theorem 11.15.

1. *SAT is ASP-complete. This is exercise 11.6, hence we omit the proof.*
2. *There is an ASP reduction from SAT to 3SAT. Hence 3SAT is ASP-complete.*
3. *There is an ASP reduction from 3SAT to 3SAT-3. Hence 3SAT-3 is ASP-complete.*
4. *There is an ASP reduction from 3SAT to* Clique*. Hence* Clique *is ASP-complete.*

Proof.

2) Here is a parsimonious reduction from SAT to 3SAT:

1. Input $\varphi = C_1 \wedge \cdots \wedge C_k$, where each C_i is an OR of literals.
2. Introduce three new variables x, y, z and the clauses

$$(x \vee y \vee \neg z) \wedge (x \vee \neg y \vee z) \wedge (x \vee \neg y \vee \neg z) \wedge$$
$$(\neg x \vee y \vee z) \wedge (\neg x \vee y \vee \neg z) \wedge (\neg x \vee \neg y \vee z) \wedge (\neg x \vee \neg y \vee \neg z).$$

   Hence, in any satisfying assignment, x, y, z are all set to FALSE.
3. For every clause $C_i = (L_1)$ with 1 literal, replace it with $(L_1 \vee x \vee y)$.
4. For every clause $C_i = (L_1 \vee L_2)$ with 2 literals, replace it with $(L_1 \vee L_2 \vee x)$.
5. For every clause $C_i = (L_1 \vee L_2 \vee L_3)$ with 3 literals, make no change.
6. Finally, consider a clause C_i with ≥ 4 literals. We do an example of what to do with a 5-clause $C_i = (L_1 \vee L_2 \vee L_3 \vee L_4 \vee L_5)$. From our example, the general case will be clear.

   Introduce a new variable w local to this clause. Replace C_i with

$$(L_1 \vee L_2 \vee w) \wedge (\neg L_1 \vee L_2 \vee w) \wedge (L_1 \vee \neg L_2 \vee w) \wedge$$
$$(\neg L_1 \vee \neg L_2 \vee \neg w) \wedge (\neg w \vee L_3 \vee L_4 \vee L_5).$$

   If $w =$ FALSE, then we get all solutions where $(L_1, L_2) = ($TRUE, TRUE$)$. If $w =$ TRUE, then we get all solutions where $(L_1, L_2) \in \{($TRUE, FALSE$), ($FALSE, TRUE$), ($FALSE, FALSE$)\}$. Hence the number of satisfying assignments is the same. But we still have a clause with 4 literals.

   Perform the same procedure on the clause with 4 literals.

   We leave it to the reader to formalize the case where there are ≥ 4 literals and then to show that the reduction is parsimonious and ASP.

3) The reduction from 3SAT to 3SAT-3 in theorem 1.11 is parsimonious.

4) The reduction from 3SAT to Clique in theorem 1.32 is parsimonious. □

Exercise 11.16.

1. Give the reduction for theorem 11.15.2 in the case where the formula has $\geq$4 literals.
2. Prove that the reduction in theorem 11.15.2 is parsimonious.
3. Prove that the reduction in theorem 11.15.3 is parsimonious.
4. For the reductions in the last two items, give the additional function needed to make the ASP-reductions.

11.4.2 Planar 3SAT and Variants

Before reading this section, the reader is advised to read chapter 2 because here we discuss the reductions from that chapter with an eye toward either observing that they already are parsimonious or modifying them so they are parsimonious.

Exercise 11.17.

1. Prove that Planar 3SAT is ASP-complete.
   Hint: The reduction in theorem 2.10 is parsimonious, though this needs proof.
2. Prove that Planar Rectilinear 3SAT is ASP-complete.
   Hint: Use the proof of theorem 2.27.
3. In chapter 2, we proved that for several problems X, Planar 3SAT $\leq_p X$. Modify these reductions (if needed) to show that X is ASP-complete and/or #X is #P-complete.

Theorem 11.18. *#Planar Rectilinear 3SAT is #P-complete.*

Proof sketch. The reduction in the proof of theorem 2.27 suffices. □

Exercise 11.19. Prove theorem 11.18 rigorously.

What about Planar 1-in-3SAT? The reduction from 3SAT to Planar 1-in-3SAT shown in figure 2.17 is sadly not parsimonious. There is exactly one case when we can set the variables of the original instance and cannot force the internal variables to a specific value. The figure shows that there are two possible solutions of the Planar 1-in-3SAT instance that correspond to the same solution of the 3SAT instance. If every solution of the 3SAT instance had two corresponding Planar 1-in-3SAT solutions, we would have a c-monious reduction, but this irregularity makes this reduction not parsimonious.

Luckily, there is a different parsimonious reduction from Planar 3SAT to Planar 1-in-3SAT, even if we require Positive and exactly three unique literals in each clause. Hunt et al. [HIMRS98] proved the following.

Theorem 11.20.

1. *There is an ASP reduction from Planar 3SAT to Planar EU3SAT, so Planar EU3SAT is ASP-complete.*

2. *There is an ASP reduction from* PLANAR *3SAT to* PLANAR POSITIVE *1-*IN*-EU3SAT, so* PLANAR POSITIVE *1-*IN*-EU3SAT is ASP-complete.*

11.4.3 PLANAR HAMILTONIAN CYCLE

In the proof of theorem 4.1, we showed that PLANAR 3SAT reduces to HAMILTONIAN CYCLE restricted to planar directed graphs of maximum degree 3. See figure 4.1 for all of the gadgets used. The reduction is not parsimonious. When multiple variables satisfy a clause, any of them can grab the vertices in the clause gadget. Because we can sometimes have 1, 2, or 3 possible solutions that correspond to the same 3SAT solution, we do not have a parsimonious (or c-monious) reduction. We sketch how to modify the reduction to make it parsimonious, while also making the graph undirected, as proved by Seta [Set02].

Theorem 11.21. *HAMILTONIAN CYCLE restricted to planar max-degree-3 graphs is ASP-complete.*

Proof sketch. Figure 11.1 shows gadgets for XOR, crossing XOR, OR, and IMPLIES. Figure 11.2 shows a gadget for 3-OR which we will describe later. Figure 11.3 gives an example of the overall construction. All of these figures are due to Seta [Set02].

We start with the XOR gadget. This gadget admits only one solution and forces that exactly one edge connected by the gadget is used in the cycle.

The OR requires that one or two of the edges are used in the cycle. The implication forces an edge to be used if the other edge is used. The 3-way OR gate implements a 3CNF constraint, that at least one of the three edges is used in the cycle. The difference between this clause gadget compared to the proof of theorem 4.1 is that the 3-way OR gadget has all its edges forced, admitting only one cycle, per solution of the original 3SAT instance. Consequently, this new reduction is parsimonious. □

Recently, the MIT Hardness Group [MIT24a] proved ASP-completeness of HAMILTONIAN CYCLE in grid graphs (a special type of planar graph described in section 4.2.2) of maximum degree 3. This result holds for both directed and undirected graphs. They also proved ASP-completeness of HAMILTONIAN CYCLE when the vertex set is the rectangle $\{1, \ldots, m\} \times \{1, \ldots, n\}$, and either (1) there is a directed edge between every pair of vertices at unit distance or (2) there are undirected edges between a subset of pairs of vertices at unit distance (and the graph still has maximum degree 3).

Notably, these results (at least those for undirected graphs) crucially rely on degree-2 vertices: Theorem 11.10 tells us that ASP CUBIC HAMILTONIAN CYCLE is in P.

11.4.4 SLITHERLINK

Recall that in section 4.3.2 we defined the game of *Slitherlink* and showed that it was NP-complete. That proof used a reduction from HAMILTONIAN CYCLE restricted to grid

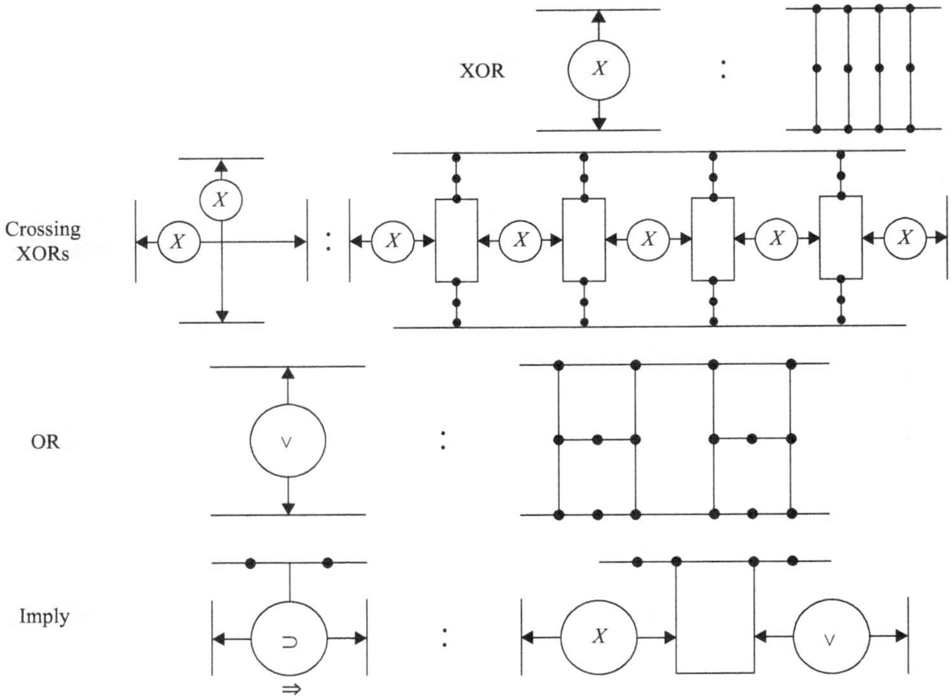

Figure 11.1
Gadgets for XOR, Crossing XOR, OR, IMPLIES.

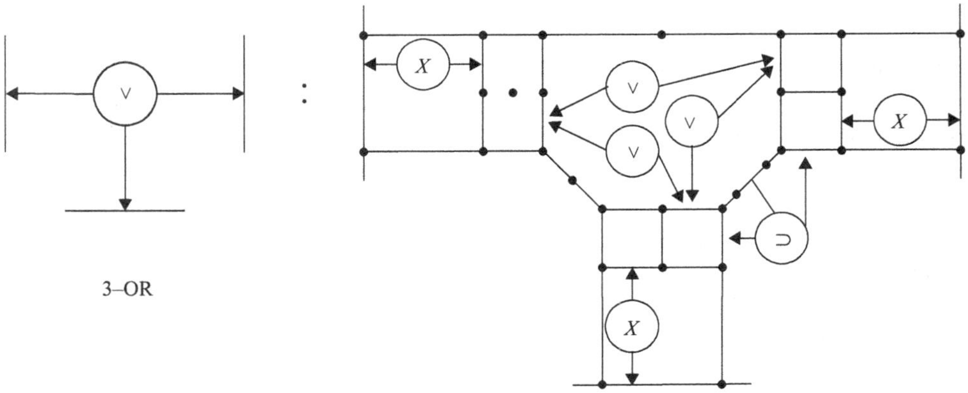

Figure 11.2
Gadget for 3-OR.

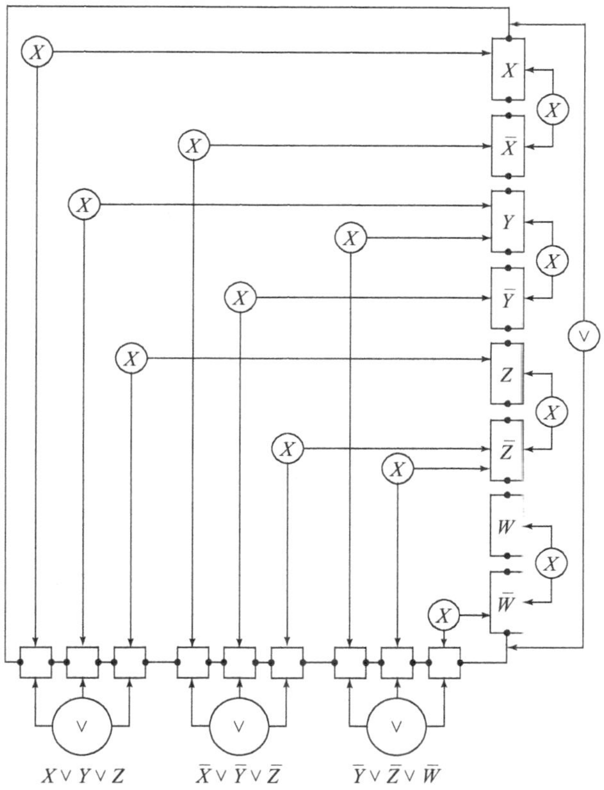

Figure 11.3
ASP reduction from PLANAR 3SAT to HAMILTONIAN CYCLE restricted to planar max-degree-3 graphs.

graphs. The reduction is even parsimonious. Now that we know that this variant of HAMILTONIAN CYCLE is ASP-complete [MIT24a], we immediately obtain that SLITHERLINK is ASP-complete. (Before this, Yato [Yat03] observed that their earlier reduction [Yat00] from planar max-degree-3 graphs is parsimonious, so by theorem 11.21, SLITHERLINK is ASP-complete.)

11.5 #P-Completeness via *c*-monious Reductions

In this section, we show that several functions in #P are #P-complete using *c*-monious reductions. These problems are not ASP-complete (unless P = NP), because their ASP versions are either in P or do not make sense.

11.5.1 PERMANENT

All results in this section are due to Valiant [Val79a].

Definition 11.22. Let $M = (m_{i,j})$ be an $n \times n$ matrix.

1. The ***determinant***, denoted DETERMINANT(M), is $\sum_\pi (-1)^{\text{sign}(\pi)} \prod_{i=1}^n m_{i,\pi(i)}$, where (1) the sum is over all permutations π of the set $\{1, 2, \ldots, n\}$, and (2) the sign of a permutation is a 0/1 value given by the parity of the number of inversions modulo 2. Note that, even though this definition has an exponential (in n) number of terms, computing the determinant is in FP.
2. The ***permanent*** of M, denoted PERMANENT(M), is $\sum_\pi \prod_{i=1}^n m_{i,\pi(i)}$—that is, the unsigned sum over permutations.

It has been known for a long time that computing the determinant is easy. People wondered if computing the permanent was also easy. There were attempts to, in our terminology, show that PERMANENT $\leq_{p,o}$ DETERMINANT, which would show that PERMANENT was easy. But before modern notions of complexity, they could not even state the goal of proving that computing the permanent was hard.

The problem of showing PERMANENT was hard motivated Valiant [Val79a] to define #P. He proved that computing the permanent is #P-complete. His proof is complicated. There are alternative proofs by Ben-Dor and Halevi [BH93] and Aaronson [Aar11]. The proof by Aaronson uses quantum computing. The proof by Ben-Dor and Halevi [BH93] is simpler than Valiant's but is still complicated.

Theorem 11.23.

1. *PERMANENT is #P-complete even when the matrix is restricted to having elements in* $\{-1, 0, 1, 2\}$.
2. *PERMANENT is #P-complete even when the matrix is restricted to having elements in* $\{0, 1\}$.

11.5.2 Bipartite Matchings: Perfect and Maximal

Definition 11.24. Let G be a graph. Refer to figure 11.4.

1. A ***matching*** in G is a set of edges that share no vertices.
2. MATCHING is the set of graphs that have a matching. BIPARTITE MATCHING is the set of bipartite graphs that have a matching. These are silly definitions since all graphs that have a a matching, namely the empty set $\emptyset$. However, we will see that #MATCHING and #BIPARTITE MATCHING are #P-complete.
3. A ***perfect matching*** in G is a matching M such that every vertex in G is the endpoint of some edge in M. In particular, if G is bipartite, then a perfect matching in G is a bijection from the left vertices to the right vertices.

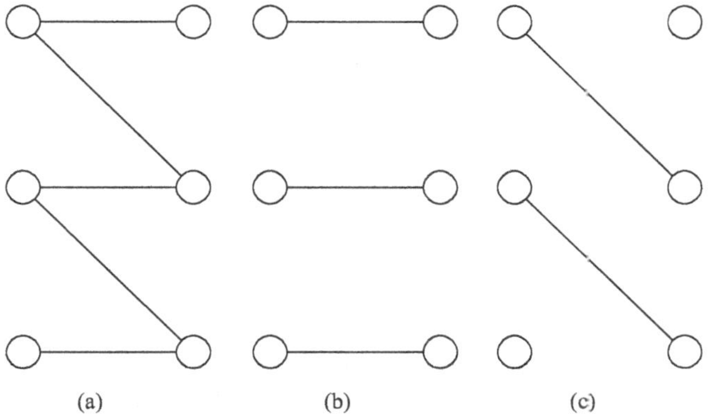

Figure 11.4
(a) Bipartite graph. (b) Perfect matching. (c) Maximal matching.

4. PERFECT MATCHING is the set of graphs that have a perfect matching. BIPARTITE PERFECT MATCHING is the set of bipartite graphs that have a perfect matching. It is known that both PERFECT MATCHING $\in$ P and BIPARTITE PERFECT MATCHING $\in$ P; however, we will see that #PERFECT MATCHING and #BIPARTITE PERFECT MATCHING are #P-complete.
5. A ***maximal matching of G*** is a matching M such that, for any edge e that is not in the matching, $M \cup \{e\}$ is not a matching. Note that a graph with no edges has the empty set $\emptyset$ as a maximal matching.
6. MAXIMAL MATCHING is the set of all graphs that have a maximal matching. BIPARTITE MAXIMAL MATCHING is the set of bipartite graphs that have a maximal matching. These are silly definitions since, by a greedy algorithm, *every* graph has a maximal matching and it can be found quickly; however, we will see that #MAXIMAL MATCHING and #BIPARTITE MAXIMAL MATCHING are #P-complete.

Exercise 11.25. Let M be an $n \times n$ 0-1 matrix, where a_{ij} is the entry of M in row i, column j. Let G be the bipartite graph with both left and right sets of vertices be $\{1, \ldots, n\}$, and there is an edge from i (left) to j (right) if and only if $a_{ij} = 1$. Show that the permanent of M is #BIPARTITE PERFECT MATCHING(G).

Theorem 11.26. *#BIPARTITE PERFECT MATCHING is #P-complete. Hence #PERFECT MATCHING is #P-complete.*

Proof. This follows from exercise 11.25 and theorem 11.23.2. □

The proof of theorem 11.26 uses the premise that the graph is bipartite. None of the proofs of the remaining results in this section need that premise. Even so, we state the results for bipartite graphs.

Theorem 11.27. *#BIPARTITE MATCHING and #BIPARTITE MAXIMAL MATCHING are #P-complete. Hence #MATCHING and #MAXIMAL MATCHING are #P-complete.*

Proof. We show #BIPARTITE PERFECT MATCHING $\le_{p,o}$ #BIPARTITE MAXIMAL MATCHING and #BIPARTITE PERFECT MATCHING $\le_{p,o}$ #BIPARTITE MATCHING at the same time since the reductions are so similar. They are both one-call reductions.

1. Input $G=(X,Y,E)$, a bipartite graph with $|X|=|Y|=n$ (if $|X|\neq|Y|$ then there are no perfect matchings). We want to know #BIPARTITE PERFECT MATCHING(G).
2. Create a bipartite graph by doing the following:
   a. Replace every node v in $X\cup Y$ with n^2 nodes $v_1,\ldots,v_{n^2}$.
   b. If (v,w) is an edge in G, then put an edge between every v_i and w_j.
   c. Call this new bipartite graph G'.
   d. Make the query #BIPARTITE MAXIMAL MATCHING(G'). We now give commentary before saying the next step. Let M be a (maximal) matching of G with i edges. Then it will correspond to $(n^2!)^i$ (maximal) matchings in G'. For $1\le i\le n$ let m_i be the number of maximal matchings in G with i edges. Note that $m_n=$ #BIPARTITE PERFECT MATCHING(G) (in either the matching or maximal matching case). Hence

   $$\#\text{BIPARTITE MAXIMAL MATCHING}(G')=\sum_{i=1}^{n} m_i(n^2!)^i.$$

   Because $m_i\le\binom{n^2}{n}<n^2!$, we can view #BIPARTITE MAXIMAL MATCHING(G') as the base-$n^2!$ number $m_n m_1\cdots m_0$.
   e. Express #BIPARTITE MAXIMAL MATCHING(G') in base $n^2!$ to obtain $m_n\cdots m_0$. Output m_n. □

11.5.3 2SAT

Recall that 2SAT $\in$ P. What about #2SAT? It turns out that it is #P-complete? We do not prove this. We prove that a variant of #2SAT is #P-complete and use this variant to show several other problems #P-complete or #P-hard.

Definition 11.28. THRESHOLD POSITIVE 2SAT is the following problem: Given a Boolean formula in 2CNF form with all literals positive, and a number t (called the ***threshold***), does it have a satisfying assignment with at least t variables set to FALSE?

Theorem 11.29. *#THRESHOLD POSITIVE 2SAT is #P-complete.*

Proof. We describe a parsimonious reduction from PERFECT MATCHING to THRESHOLD POSITIVE 2SAT.

1. Input $G = (V, E)$, a graph on $2k$ vertices. Note that any perfect matching will have k edges.
2. Form a formula as follows:
   a. For every edge e we have a Boolean variable v_e. Our intent is that if e is in the matching then v_e is set to FALSE (you read that right).
   b. For every edges e and e' that are incident we have the clause $(v_e \vee v_{e'})$.
   c. Let the formula be φ.
3. Output (φ, t).

We show a bijection between the perfect matchings of G and the satisfying assignments of φ that have $\geq k$ variables set FALSE.

Given a perfect matching, every variable corresponding to an edge in the matching is set to FALSE, and all of the other variables are set to TRUE. Since the edges in the matching are not incident to each other, there is no clause that has two variables false. Hence this is a satisfying assignment.

We leave it to the reader to show that every satisfying assignment is in the image of this map. □

Exercise 11.30.

1. Show that THRESHOLD POSITIVE 2SAT $\leq_p$ CLIQUE by a parsimonious reduction. (This provides another proof that #CLIQUE is #P-complete.)
2. Show that counting the number of maximal independent sets is #P-hard.
3. Show that counting the number of maximal cliques is #P-hard.

11.6 Further Results

- EDGE MATCHING. Edge-matching puzzles from section 5.5.1 always have a trivial second solution, by rotating a solution by 180°. Bosboom et al. [BCC+20] showed that this is the limit to easy other solutions: 2-ASP EDGE MATCHING is NP-complete, and #EDGE MATCHING is #P-complete, even when restricted to a $1 \times n$ board. We can also forbid global 180° rotation (e.g., by forcing one of the tiles), and then EDGE MATCHING becomes ASP-complete.
- #VERTEX COVER. Counting the number of vertex covers of size $\leq k$ in a bipartite graph. Pravan and Ball [PB83] showed that this problem is #P-complete, along with several other variants.

- #MINIMUM CARDINALITY (s,t)-CUT. Given (G,s,t) where G is a graph and s,t are vertices, find how many minimum-size edge cuts separate s and t. Pravan and Ball [PB83] showed that this problem is #P-complete.
- LINEAR EXTENSION is the problem of, given a partially ordered set (via the elements and the relations), determine whether there is a linear extension of it (a consistent way to order every pair). This problem is trivial because every partial order has a linear extension. But what *counting* the number of extensions? Brightwell and Winkler [BW91] showed that #LINEAR EXTENSION is #P-complete.
- #ANTICHAIN. Given a partial order $(X,\preceq)$, find the number of ***antichains***—that is, sets with all elements pairwise incomparable). This problem was shown #P-complete by Provan and Ball [PB83], along with several other variants.
- Recall that 3-COLORING is the problem of, given a graph G, determine whether it is 3-colorable. Creignou and Hermann [CH99] have shown that #3-COLORING is #P-complete.
- For all of the problems in this exercise, look into the ASP version.

Exercise 11.31. For each of the problems listed above, either try to prove they are #P-complete or look up the papers, read them, understand the reductions, and put it in your own words.

For more #P-complete problems, see the papers of Valiant [Val79c, Val79a, Val79b] and Simon [Sim77].

11.7 FEWEST CLUES PROBLEM

Imagine that you are designing a hard sudoku puzzle. You want to give as few numbers as possible yet still have a unique solution. Demaine et al. [DMS+18] formalized this notion as follows.

Definition 11.32. Let $A=\{x \mid \exists^p y:(x,y)\in B\}$ where $B\in \mathrm{P}$. Let $q(|x|)$ be the size of y. Let $x\in\Sigma^*$. A ***clue for x*** is a string in $(\Sigma\cup\bot)^{q(|x|)}$ such that there is exactly one way to fill in the $\bot$ symbols to get a y such that $(x,y)\in B(x,y)$.

> FEWEST CLUES PROBLEM (FCP(A))
> *Note:* $A=\{x \mid \exists^p y:(x,y)\in B\}$ where $B\in \mathrm{P}$.
> *Instance:* $x\in\Sigma^*$ and $k\in\mathbb{N}$.
> *Question:* Does there exist a clue for x with k non-$\bot$ characters? (If $x\notin A$, then the answer is NO.)
> *Note:* For the function version of the problem, you are given x and want to find the least k such that $(x,k)\in$ FCP(A).

Demaine et al. [DMS+18] showed that FCP versions of PLANAR 1-IN-3SAT, PLANAR 3SAT, and 1-IN-3SAT are Σ_2-complete. They then used these results to get that the FCP version of several games problems are Σ_2-complete: SUDOKU, AKARI, and SHAKASHAKA (these are all Nikoli games [Nik08, Nik]).

11.8 Open Problems

Project 11.33. *For many NP-complete sets A, determine whether A is ASP-complete and/or #A is #P-complete.*

Project 11.34. *For many sets $A \in P$, determine whether #A is #P-complete. Try to formulate (1) criteria on A that makes #A #P-complete, and (2) criteria on A that makes #A $\in$ FP. One must be careful since not every set $A \in P$ has a natural form involving existential quantifiers.*

Here is an ill-posed conjecture.

Conjecture 11.35. *If A is a natural NP-complete problem, then #A is #P-complete.*

The conjecture is ill-posed because *natural* is not well defined. Even so, it seems to be true. So the open question here is to find some rigorous way to state it, and then prove it. To help narrow the conjecture, it would also help to find some examples where it fails. We do not actually know of one, but we suspect there is a contrived NP-complete problem *A* where #*A* is in some conjectured-lower complexity class.

12

Existential Theory of the Reals

12.1 Introduction

Consider the following problem.

> POINT GUARD ART GALLERY PROBLEM
> *Instance:* A polygon and a number $k \in \mathbb{N}$. We think of the polygon as an art gallery that has valuable paintings and, hence, needs to be guarded
> *Question:* Can guards be placed at k points in the polygon such that every point of the museum is visible to some guard?

This is an unusual problem in that (1) it is NP-hard, (2) it is open as to whether it is NP, (3) it does not seem to be in the polynomial hierarchy, and (4) it is in PSPACE but does not seem to be PSPACE-complete. The reason why POINT GUARD ART GALLERY PROBLEM is hard to classify is that it asks if some *reals* with some properties exist, as opposed to some finite object (like a satisfying assignment). In this chapter we discuss many problems that involve asking if a finite set of reals exists.

Chapter Summary

1. *We discuss the problem* EXISTENTIAL THEORY OF THE REALS *(ETR), which will be analogous to CNF SAT. It is a problem that asks if certain reals exist.*
2. *We discuss $\exists\mathbb{R}$, a complexity class that is analogous to NP. The main difference is that a problem $A \in$ NP is defined by a polynomial quantifier over a finite set, whereas $A \in \exists\mathbb{R}$ is defined by a quantifiers over the reals.*
3. *We give many examples of problems that are $\exists\mathbb{R}$-complete. We will discuss why is evidence that they are not in NP, not in the polynomial hierarchy, and not PSPACE-complete.*

The class $\exists\mathbb{R}$ is a vast topic that we will, for reasons of space, only be able to discuss briefly. We will have few proofs, algorithms, or reductions. For more information on these topics, see (1) the paper by M. Schaefer [Sch09], which named the class $\exists\mathbb{R}$ and gave us the modern view of it, (2) the paper by M. Schaefer and Stefankovic [SS17], whose introduction is a good summary of the theory, (3) the expository paper of Matoušek [Mat14], (4) the survey by Cardinal [Car15], (5) the thesis of Bieker [Bie20], and (6) the talk by M. Schaefer [Sch20]. This chapter will contain additional references for particular results.

12.2 Existential Theory of the Reals

Consider the following two questions where the quantifiers range over $\mathbb{R}$.

$$S_1 = \exists x : \exists y : \exists z : x^2 + y^z + z^2 < 0.$$

$$S_2 = \exists x : \exists y : \exists z : x^2 + y^2 + z^2 > 0.$$

Clearly S_1 is false and S_2 is true. We consider the problem where you are given a sentence like S_1 or S_2 and you need to determine whether it is true or false.

> Existential Theory of the Reals (ETR)
> *Instance:* A sentence of the form
>
> $$\exists x_1 : \exists x_2 : \cdots : \exists x_n : C_1 \wedge \cdots \wedge C_k,$$
>
> where each C_i is a polynomial equality or inequality in $x_1, \ldots, x_n$ (it might not use all of them). The polynomials have coefficients in $\mathbb{Z}$.
> *Question:* If the quantifiers range over $\mathbb{R}$, then is the sentence true?

12.3 Complexity of ETR

We state upper and lower bounds on the complexity of ETR. First an upper bound.

It is not obvious that ETR is decidable. Nevertheless, it is, as illustrated by the following sequence of results of increasing strength:

Theorem 12.1.

1. *(Tarski [Tar48]) ETR is decidable. (Tarski had the result in 1930 but did not publish it until 1948.)*
2. *(Collins [Col75]) ETR is in time $2^{2^{O(n)}}$, where n is the number of variables.*
3. *(Canny [Can88]) ETR $\in$ PSPACE, and thus in time $2^{O(n)}$.*

The algorithms in theorem 12.1 will, given a formula, only tell you whether it is TRUE or FALSE. If it is TRUE, then the algorithm does not return the reals $r_1, \ldots, r_n$ that make it TRUE. In fact, this might be impossible since some reals might require an infinite representation. But some reals do not. This leads to our next definition.

Definition 12.2. A ***language for some reals (LFSR)*** is an injection from a countable set to $\mathbb{R}$. We intend the representation of the reals to be natural, though we do not define that rigorously.

Example 12.3.

1. Map $\{0,1\}^* \times \{0,1\}^*$ to $\mathbb{R}$ by (x,y) maps to $x.y$ which is the base 2 number represented by x then a decimal point then y.
2. The countable set A is formed as follows:
   a. Every element of Q is in A.
   b. If $a \in A$ and $q \in \mathbb{Q}$ then $a^q \in A$ with two caveats: (1) if a^q is ambiguous, then add some bits to indicate which a^q it is, and (2) do not include the a^q that are imaginary.
   c. If $a_1, a_2 \in A$, then $a_1 + a_2$, $a_1 \times a_2$, a_1/a_2 (if $a_2 \neq 0$), and $a_1 - a_2$ are all in A.

   The map from A to $\mathbb{R}$ is obvious.

When we discuss having an algorithm that actually outputs reals, we will preface it with saying that there is an LFSR that is being used.

Theorem 12.1.2 is used to prove other problems in PSPACE whose membership there is not at all obvious. We do one example. For more examples, see Bieker [Bie20].

> UNIT-DISTANCE GRAPH
> *Instance:* A graph $G=(V,E)$. Let the vertices be $\{1,\ldots,n\}$.
> *Question:* Does there exist a set of points in the plane $p_1,\ldots,p_n$ such that, for all $1 \le i < j \le n$, $|p_i - p_j| = 1$ if and only if $(i,j) \in E$.

Since the plane has irrational points it does not seem that UNIT-DISTANCE GRAPH is in PSPACE. Nevertheless, it is. The proof that UNIT-DISTANCE GRAPH $\in$ PSPACE is a reduction from UNIT-DISTANCE GRAPH to ETR. We do not know of a proof that does not use ETR. The issue of irrationals does not come up since the algorithm will, on input G, only return YES or NO, and if the answer is YES, it does not return the points in the plane that show G is a unit-distance graph.

Theorem 12.4. *UNIT-DISTANCE GRAPH is in PSPACE.*

Proof. A set of n points in the plane will be represented as a set of n ordered pairs of reals $(x_1,y_1),\ldots,(x_n,y_n)$.

Let $G=(V,E)$ be a graph. G is in UNIT-DISTANCE GRAPH if and only if the following problem is in ETR (the quantifiers are over $\mathbb{R}$)

$$\exists x_1 : \cdots : \exists x_n : \exists y_1 : \cdots : \exists y_n :$$
$$\bigwedge_{(i,j)\in E} [(x_i-x_j)^2+(y_i-y_j)^2=1] \wedge \bigwedge_{(i,j)\notin E} [(x_i-x_j)^2+(y_i-y_j)^2 \neq 1].$$

Since ETR $\in$ PSPACE, so is UNIT-DISTANCE GRAPH. □

Open Problem 12.5.

1. *Does there exist a LFSR and a polynomial p such that the following holds. Let $G=(V,E)$ be a graph. Assume $|V|=n$. Given a graph $G=(V,E)$ that is a UNIT-DISTANCE GRAPH, there is a representation in the plane where the points can be described by elements of the LFSR that are $\le p(n)$ bits long.*
2. *If the answer to part 1 is YES, then find a PSPACE algorithm that will solve the following: Given a graph G that Canny's algorithm has already said is a UNIT-DISTANCE GRAPH, output the points in the plane (using the LFSR) and the edges that realize the graph.*

We give a proof of the next theorem, which is probably folklore from before any written account. We know of the written accounts by Shor [Sho91], J. Buss et al. [BFS99], and Schaefer and Stefankovic [SS17]; however, all of these are of harder results of which the next theorem is a corollary.

Theorem 12.6. *ETR is NP-hard.*

Proof. We show 3SAT $\le_p$ ETR. We use

$$\varphi(x_1,x_2,x_3,x_4)=(x_1 \vee x_2 \vee \neg x_3) \wedge (\neg x_1 \vee x_3 \vee x_4)$$

as a running example throughout the construction.

Here is the reduction.

1. Input $\varphi(x_1,\ldots,x_n)=C_1 \wedge \cdots \wedge C_m$ where each C_i is a $\vee$ of literals. $((x_1 \vee x_2 \vee \neg x_3) \wedge (\neg x_1 \vee x_3 \vee x_4).)$
2. For each Boolean variable x_i we have a real variable r_i and will have as part of the ETR the equation $r_i=0 \vee r_i=1$.
3. For each clause C_j we obtain an inequality E_j as follows. Replace x_i with r_i, and $\neg x_i$ with $(1-r_i)$. E_j is the sum of these expressions is ≥ 1. (In our running example (1) the first clause becomes $r_1+r_2+(1-r_3)\ge 1$, which we simplify to $r_1+r_2-r_3\ge 0$ and (2) the second clause becomes $(1-r_1)+r_3+r_4\ge 1$, which we simplify to $-r_1+r_3+r_4\ge 0$.)

4. The instance of ETR is

$$\exists r_1 : \cdots : r_n : \bigwedge_{i=1}^{n} (r_i = 0 \vee r_i = 1) \wedge \bigwedge_{j=1}^{m} E_j.$$

In our running example we get

$$\exists r_1 : r_2 : r_3 : r_4 : \left(\bigwedge_{i=1}^{4} (r_i = 0 \vee r_i = 1) \right) \wedge (r_1 + r_2 - r_3 \geq 0) \wedge (-r_1 + r_3 + r_4 \geq 0).$$

We leave it to the reader to show that the reduction works. □

We have ETR is NP-hard. Is ETR in NP? This is open but is thought to be unlikely. We contrast CNF SAT to ETR.

1. Let

$$\exists x_1 \in \{\text{TRUE}, \text{FALSE}\} : \cdots : \exists x_n \in \{\text{TRUE}, \text{FALSE}\} : C_1 \wedge \cdots \wedge C_k$$

be an instance of 3SAT (each C_i is a $\vee$ of literals). If Alice wants to convince Bob that this instance is satisfiable, then Alice can give Bob the appropriate $(x_1, \ldots, x_n) \in \{\text{TRUE}, \text{FALSE}\}^n$. Note that $(x_1, \ldots, x_n)$ is a finite object.

2. Let

$$\exists x_1 \in \mathbb{R} : \cdots : \exists x_n \in \mathbb{R} : C_1 \wedge \cdots \wedge C_k$$

be an instance of ETR (each C_i is a polynomial equality or inequality in $x_1, \ldots, x_n$). If Alice wants to convince Bob that this instance is satisfiable, then what can she do? It is possible that there is a language for some reals that works, but that seems unlikely. There may be some short proof that such reals exist without exhibiting them, but that also seems unlikely.

12.4 ETR in the Real World

The ETR problem is hard in theory. But what about in practice? There has been a great deal of research on this, including actual systems built to solve it. They seem to solve special cases. They do not seem to use Tarski's or Canny's algorithms stated in theorem 12.1. The result of Collins [Col75] introduced a method called ***cylindrical algebraic decomposition***, which are used in many of the best performing systems today. Collins's algorithm is the only one that is simultaneously complete and used.

Passmore and Jackson [PJ09] survey decision methods that seem to hit the sweet spot of solving just the kind of problems that need to be solved. More recent techniques can be found in the works of Passmore [Pas11], Jovanovic and de Moura [JdM12], and de Moura and Passmore [dMP13]. One package, the MetiTarski theorem prover, is actually used by NASA in the analysis of flight control software. For more on this, see the package itself [Unk] and the paper by Passmore et al. [PPdM12].

12.5 $\exists\mathbb{R}$ and $\exists\mathbb{R}$-Hardness

Theorems 12.1.2 and 12.6 leave open the question of how to best classify ETR. This question is better thought of as the complexity of the class $\exists\mathbb{R}$. We define $\exists\mathbb{R}$ and then discuss its history and complexity.

Definition 12.7.

1. A decision problem A is ***in*** $\exists\mathbb{R}$ if $A \le_p$ ETR.
2. A decision problem A is **$\exists\mathbb{R}$-*hard*** if ETR $\le_p A$.
3. A decision problem A is **$\exists\mathbb{R}$-*complete*** if $A \in \exists\mathbb{R}$ and A is $\exists\mathbb{R}$-hard.

Note: The prehistory and history of the study of ETR and $\exists\mathbb{R}$ is complicated.

1. L. Blum et al. [BSS88] defined a model of complexity over the reals. One of the complexity classes they defined is now seen as being $\exists\mathbb{R}$. Some of the results on that model (theirs and others) are now seen as saying certain problems are $\exists\mathbb{R}$-complete.
2. Mnëv [Mnë88] proved a problem in geometry was $\exists\mathbb{R}$-complete; however, his paper was phrased in geometric rather than complexity-theoretic terms. Shor [Sho91] proved that same problem NP-hard (a different way) and gave an explanation of Mnëv's proof suitable for complexity theorists. J. Buss et al. [BFS99] defined a class that can now be seen as $\exists\mathbb{R}$. They also showed that it was equivalent to what we call $\exists\mathbb{R}$ but with $k=1$ and C_1 of the form $p(x_1,\ldots,x_n)=0$.
3. Schaefer [Sch09] defined $\exists\mathbb{R}$ and $\exists\mathbb{R}$-complete and argued for the importance of a problem being $\exists\mathbb{R}$-complete. Schaefer and Stefankovic [SS17] showed that $\exists\mathbb{R}$ does not change if you restrict to problems where all of the inequalities are strict. This result led to more problems being proven $\exists\mathbb{R}$-complete.

Erickson, van der Hoog, and Miltzow [EvdHM24] showed that membership in $\exists\mathbb{R}$ is equivalent to the existence of a polynomial-time nondeterministic algorithm (or verification algorithm) in the "real RAM" model of computation, where the computer can do arithmetic on exact real numbers. This is especially useful for proving problems are in $\exists\mathbb{R}$.

By theorems 12.6 and 12.1, $\mathsf{NP} \subseteq \exists\mathbb{R} \subseteq \mathsf{PSPACE}$. It is not known whether $\mathsf{NP} = \exists\mathbb{R}$ or whether $\exists\mathbb{R} = \mathsf{PSPACE}$. We discuss the possible equalities and also the possible relation to PH (the polynomial hierarch).

1. The statement $\exists\mathbb{R} \subseteq \mathsf{NP}$ (and hence $\exists\mathbb{R} = \mathsf{NP}$) does not currently contradict any known theorem. Nevertheless, this is thought to be unlikely. We give three reasons that were

told to us by M. Schaefer. (1) If $\exists\mathbb{R} = \mathsf{NP}$, that probably reflects some deep structural property of existential real quantification that mathematicians have missed so far; that seems unlikely. (2) We noted above that 3SAT has a finite witness, but ETR does not seem to. If $\mathsf{NP} \subseteq \exists\mathbb{R}$, then every element of ETR would have a finite witness; that seems unlikely. We will see an example where the witnesses using finite decimal expansions are quite long when we discuss RECTILINEAR CROSSING NUMBER (this does not preclude some other LFSR working). (3) Nobody has placed any of the many $\exists\mathbb{R}$-complete problems into NP.

2. The statement $\exists\mathbb{R} \subseteq \mathsf{PH}$ does not currently contradict any known theorem. Nevertheless, this is thought to be unlikely, though not with the same confidence as the item above. All three reasons given for why $\exists\mathbb{R} \subseteq \mathsf{NP}$ are also reasons why $\exists\mathbb{R} \subseteq \mathsf{PH}$ is unlikely. In addition, consider the problem SQUARE-ROOT SUM (which we will discuss in section 12.7). It is in $\exists\mathbb{R}$, but not known to be $\exists\mathbb{R}$-hard or even NP-hard. Yet getting this problem into PH has proven to be difficult.
3. The statement $\mathsf{PSPACE} \subseteq \exists\mathbb{R}$ (and hence $\exists\mathbb{R} = \mathsf{PSPACE}$) does not currently contradict any known theorem. Nevertheless, this is thought to be unlikely. If $\exists\mathbb{R} = \mathsf{PSPACE}$, then $\exists\mathbb{R}$ is closed under complement. Hence any existential statement over the reals is equivalent to a universal statement over the reals. This seems unlikely from a logical perspective.
4. The statement $\mathsf{PH} \subseteq \exists\mathbb{R}$ does not currently contradict any known theorem. We do not find this to be so unlikely.

We present an analog to NP-completeness.

1. If A is shown to be NP-complete, then (1) the complexity of A is regarded as settled and (2) A is thought to be not in P. This is the common viewpoint since it is believed that $\mathsf{P} \neq \mathsf{NP}$.
2. If A is shown to be $\exists\mathbb{R}$-complete, then (1) the complexity of A is regarded as settled (2) A is thought to be not in NP, (3) A is thought to be not in PH, and (4) A is thought to be not PSPACE-complete. This is the common viewpoint since it is believed $\exists\mathbb{R}$ is strictly between NP and PSPACE, and not contained in PH. (Note that the community working with $\exists\mathbb{R}$-completeness is much smaller than the community working with NP-completeness, so the notion of a *common viewpoint* is not as compelling.)

Open Problem 12.8.

1. *Prove that $\exists\mathbb{R} \subseteq NP$ has unlikely consequences (e.g., $P \neq NP$).*
2. *Let $i \in \mathbb{N}$ and $i \geq 2$. Prove that $\exists\mathbb{R} \subseteq \Sigma_i$ has unlikely consequences (e.g., $P \neq NP$). (Recall that Σ_i is the ith level of the polynomial hierarchy, see section 0.15.)*
3. *Prove that $PSPACE \subseteq \exists\mathbb{R}$ has unlikely consequences (e.g., $P \neq NP$).*

4. *Show that the above problems are hard, perhaps with oracles. This may be difficult since defining $\exists\mathbb{R}$ with an oracle may be problematic. You may need to use an equivalent definition.*

12.6 $\exists\mathbb{R}$-Complete Problems

In this section we present several problems that are $\exists\mathbb{R}$-complete. These results are interesting for several reasons: (1) these problems are in PSPACE which is not obvious from their description, (2) these problems are likely not in NP, (3) these problems are likely not in PH, (4) these problems are likely not PSPACE-complete, and (5) the complexity of these problems is settled.

12.6.1 $\exists\mathbb{R}$-Complete Problems About Graph Realizability

We present several $\exists\mathbb{R}$-complete problems that ask, given graph G, can G be presented in a certain way?

INTERSECTION GRAPHS OF LINE SEGMENTS (IGLS)
Instance: A graph $G=(V,E)$. Let the vertices be $\{1,\dots,n\}$.
Question: Does there exist a set of line segments in the plane $\ell_1,\dots,\ell_n$ such that, for all $1 \le i < j \le n$, ℓ_i intersects ℓ_j if and only if $(i,j) \in E$.
Note: Kratochvíl and Matoušek [KM94] proved this problem is $\exists\mathbb{R}$-complete, though not in that terminology. See also the expositions of M. Schaefer [Sch09] and Matoušek [Mat14].
Note: Schaefer [Sch21] showed that IGLS remains $\exists\mathbb{R}$-complete when restricted to graphs with maximum degree ≤ 216. The number 216 is an artifact of the proof. It is likely there is a much smaller value will suffice.

INTERSECTION GRAPHS OF CONVEX SETS (IGCS)
Instance: A graph $G=(V,E)$. Let the vertices be $\{1,\dots,n\}$.
Question: Does there exist a set of convex sets in the plane $C_1,\dots,C_n$ such that, for all $1 \le i < j \le n$, C_i intersects C_j if and only if $(i,j) \in E$.
Note: Schaefer [Sch21] showed that IGCS remains $\exists\mathbb{R}$-complete when restricted to graphs with maximum degree ≤ 216. The number 216 is an artifact of the proof. It is likely that there is a much smaller value that will suffice.

Open Problem 12.9. *Let $IGLS_k$ ($IGCS_k$) be IGLS (IGCS) restricted to graphs with max degree $\leq k$. Note the following contrast: (1) $IGLS_2, IGCS_2 \in P$; (2) $IGLS_{216}$, $IGCS_{216}$ are $\exists\mathbb{R}$-complete. Narrow the gap between 2 and 216. It is possible that, for some values of k, $IGLS_k$ ($IGCS_k$) is neither in P nor $\exists\mathbb{R}$-complete. For example, it is possible that $IGLS_3$ is NP-complete.*

UNIT-DISTANCE GRAPH
Instance: A graph $G = (V, E)$. Let the vertices be $\{1, \ldots, n\}$.
Question: Does there exist a set of points in the plane $p_1, \ldots, p_n$ such that, for all $1 \leq i < j \leq n$, $|p_i - p_j| = 1$ if and only if $(i, j) \in E$.
Note: The proof of theorem 12.4 showed that UNIT-DISTANCE GRAPH is in $\exists\mathbb{R}$.
Note: Breu and Kirkpatrick [BK98b] showed that this problem is NP-hard (see the k-sphere graphs problem for what they then conjectured). M. Schaefer [Sch12] showed that this problem is $\exists\mathbb{R}$-complete.

k-SPHERE GRAPHS AND k-DOT GRAPHS. k is a parameter.
Instance: A graph $G = (V, E)$. Let the vertices be $\{1, \ldots, n\}$.
Question: (k-sphere) Does there exist a set of k-dimensional vectors $v_1, \ldots, v_n$ such that, for all $1 \leq i < j \leq n$, $|v_i - v_j| \leq 1$ if and only if $(i, j) \in E$?
Question: (k-dot) Does there exist a set of k-dimensional vectors $v_1, \ldots, v_n$ such that, for all $1 \leq i < j \leq n$, $v_i \cdot v_j \geq 1$ if and only if $(i, j) \in E$?
Note: Looges and Olariu [LO02] showed that, for $k = 1$, the k-SPHERE PROBLEM is in linear time. For $k = 2$ the k-SPHERE PROBLEM is the UNIT-DISTANCE GRAPH problem. After Breu and Kirkpatrick [BK98b] showed that this case is NP-hard, they conjectured that, for $k \geq 3$, the problem is NP-hard. Kang and Müller [KM12] proved that conjecture. They note in the last section of their paper that their proof also shows that, for $k \geq 3$, k-sphere is $\exists\mathbb{R}$-complete.
Note: Fiduccia et al. [FSTZ98] showed that, for $k = 1$, k-DOT is in linear time. He conjectured that, for $k \geq 2$, the problem is NP-hard. Kang and Müller [KM12] proved that conjecture. They note in the last section of their paper that their proof also shows that, for $k \geq 2$, k-dot is $\exists\mathbb{R}$-complete.

RECTILINEAR CROSSING NUMBER
Instance: A graph G and a number c.
Question: Can G be drawn in the plane such that (1) every edge is a straight line (such a drawing is called RECTILINEAR) and (2) there are at most c crossings?
Note: Bienstock [Bie91], showed that RECTILINEAR CROSSING NUMBER is $\exists\mathbb{R}$-complete, though not in that terminology.
Note: Goodman et al. [GPS89] showed the following.

1. There exists d_1 such that for every n, for every graph G on n vertices, there exists a rectilinear drawing of G where the vertices are on rational coordinates and each every coordinate takes $\leq 2^{d_1 n}$ bits.
2. There exists d_2 such that for every n there exists a graph G on n vertices so that for all rectilinear drawings of G that use rational coordinates, the coordinates require at least $2^{d_2 n}$ bits.

These results show that there it might be difficult to actually compute the rectilinear drawing quickly.

Open Problem 12.10. *What is the complexity of* RECTILINEAR CROSSING NUMBER *for fixed c? For c* $= 3$ *the problem is in NP since it is equal to the crossing number. For c* $= 4$ *it is an open question as to whether it is in NP. Is there a fixed c such that the problem is* $\exists\mathbb{R}$*-complete?*

12.6.2 Non-Graph $\exists\mathbb{R}$-Complete Problems

In this section we present several $\exists\mathbb{R}$-complete problems that are not about graphs. This shows that there are $\exists\mathbb{R}$-complete problems of different types.

Theorem 12.11. *(Abrahamsen et al. [AAM21]) POINT GUARD ART GALLERY PROBLEM is* $\exists\mathbb{R}$*-complete.*

Many variations on the art gallery problem are also $\exists\mathbb{R}$-complete; see [AAM21, Sta25].

GENERAL PACKING
Instance: A set of polygons and a square S.
Question: Can these polygons fit into the square S without overlap?
Note: Abrahamsen et al. [AMS20] showed this problem is $\exists\mathbb{R}$-complete.

Theorem 12.12. *(Abrahamsen et al. [AMS20]) GENERAL PACKING is* $\exists\mathbb{R}$*-complete.*

Open Problem 12.13. *In section 5.4.3 we discussed several packing problems:* Rectangle-Rectangle Packing, Square-Rectangle Packing, Square-Square Packing, Triangle-Rectangle Packing, *and* Triangle-Triangle Packing. *They are all NP-hard and in* $\exists\mathbb{R}$. *None of them are known to be in NP. For each one, either prove it is* $\exists\mathbb{R}$*-complete or prove it is in NP.*

Min Convex Cover and min Triangle Cover

Instance: A polygon P represented by an ordered set of rational coordinates in the plane and a number k.

Question: (Min Convex Cover) Do there exists k convex polygons whose union is P?

Question: (Min Triangle Cover) Do there exists k triangles whose union is P?

Note: Abrahamsen [Abr21] showed that both problems are $\exists\mathbb{R}$-complete.

Partial Order Type Realizability

Instance: A number n and a function $f : \{1, \ldots, n\}^3 \rightarrow \{L, C, R\}$ (we will say what this means soon).

Question: Does there exist points $p_1, \ldots, p_n$ in the plane such that, for all $(i, j, k) \in \{1, \ldots, n\}^3$, the following happen:

1. If $f(i, j, k) = L$, then p_k is to the left of the line through p_i and p_j. (L is for left.)
2. If $f(i, j, k) = C$, then p_k is on the line through p_i and p_j. (C is for colinear.)
3. If $f(i, j, k) = R$, then p_k is to the right of the line through p_i and p_j. (R is for right.)

Note: Mnëv [Mnë88] showed this problem is $\exists\mathbb{R}$-complete. Schaefer [Sch21] showed that a restricted version of this problem is $\exists\mathbb{R}$-complete.

In section 22.3.3 we will define ***an r-game*** and ***a Nash equilibrium***. The reader should read that section before proceeding.

There are two obstacles in studying the complexity of Nash equilibrium.

1. Given an r-game, there is always a Nash equilibrium. So the decision problem is trivial. In chapter 22 we study many problems of that type and look at the complexity of *finding a solution* (in this case a Nash equilibrium). In our discussion here, we deal with this issue by asking questions about the Nash equilibrium.

2. If there are two players, then the Nash equilibrium is always rational. However, for three players, the Nash equilibrium can be irrational. That is not an issue for the $\exists\mathbb{R}$ framework.

In chapter 22 we show that finding a 2-player approximate Nash equilibrium is PPAD-complete. Here we consider $r \geq 3$ players.

QUESTIONS ABOUT NASH EQUILIBRIUM
Instance: An r-game where all of the utilities are integers.
Question: Does there exist a Nash equilibrium that has certain properties? We go into detail about which properties in the notes below.
Note: In all of the problems below, $r \geq 3$.

1. If the strategies are restricted to those where no action is played with probability $> \frac{1}{2}$, then is there a Nash equilibrium? Schaefer and Stefankovic [SS17] showed this problem is $\exists\mathbb{R}$-complete.
2. Are there at least two Nash equilibria? Garg et al. [GMVY18] showed this problem is $\exists\mathbb{R}$-complete. They also showed that other questions about Nash equilibrium are $\exists\mathbb{R}$-complete.
3. The input also has $q \in \mathbb{Q}$. Is there a Nash equilibrium where every player has payoff $\geq q$? Bilo and Mavronicolas [BM21] showed this problem is $\exists\mathbb{R}$-complete. They also showed that other questions about Nash equilibrium are $\exists\mathbb{R}$-complete.
4. Berthelson and Hansen [BH22] showed that the results of Garg et al. [GMVY18] hold when restricted to 3-player zero-sum games. They also give a summary of what is known for $r \geq 3$ players.

12.7 Further Results

12.7.1 Possibly Between $\exists\mathbb{R}$ and PSPACE

Schaefer and Stefankovic [SS22] have defined a hierarchy of classes based on $\exists\mathbb{R}$ that is similar to the polynomial hierarchy. They have used this to classify some problems that seem to be strictly between $\exists\mathbb{R}$ and PSPACE.

12.7.2 SQUARE-ROOT SUM: A Hard-to-Classify Problem

We present a problem that is in $\exists\mathbb{R}$, but otherwise its status is unknown and there is no consensus.

SQUARE-ROOT SUM *Instance:* $d_1, \ldots, d_n, k \in \mathbb{N}$. *Question:* Is $\sum_{i=1}^{n} \sqrt{d_i} \geq k$?

SQUARE-ROOT SUM comes up in the study of EUCLIDEAN TSP. In corollary 4.6 we showed that EUCLIDEAN TSP is NP-hard; however, its status in NP is unknown.

Exercise 12.14.

1. Show that, if SQUARE-ROOT SUM is in P, then EUCLIDEAN TSP is in NP.
2. Show that, if SQUARE-ROOT SUM is in NP, then EUCLIDEAN TSP is in NP.

Way back in 1976, Garey et al. [GGJ76] asked whether SQUARE-ROOT SUM is in NP. This question is still open. So perhaps it is $\exists\mathbb{R}$-complete, giving evidence that it is not in NP. This question is also open. It is not even known to be NP-hard.

It is easy to show that SQUARE-ROOT SUM is in $\exists\mathbb{R}$. Since SQUARE-ROOT SUM is in $\exists\mathbb{R}$, SQUARE-ROOT SUM is in PSPACE, though this can also be proved directly. Allender et al. [ABKM06] obtained slightly better upper bounds on SQUARE-ROOT SUM.

Is SQUARE-ROOT SUM one of those problems that is thought to be hard, and by assuming it is hard we can get other problems are hard? Etessami and Yannakakis [EY09] took this approach. With the assumption that SQUARE-ROOT SUM is hard, they proved that some problems involving recursive Markov chains are hard. Perhaps more problems of this type will be found and there will be a notion of SQUARE-ROOT SUM-hard and SQUARE-ROOT SUM-complete.

Open Problem 12.15. *Show that any of the following are true.*

1. *SQUARE-ROOT SUM is in P.*
2. *SQUARE-ROOT SUM is in NP.*
3. *SQUARE-ROOT SUM is $\exists\mathbb{R}$-hard (and hence $\exists\mathbb{R}$-complete).*

13

PSPACE-Hardness

13.1 Introduction

In chapter 4, we discussed two metatheorems, due to Viglietta [Vig14a], that guided us to proofs that some game problems were NP-hard. We expand on that theme.

Chapter Summary

1. *We discuss more metatheorems due to Viglietta (from the same paper) and others that allow us to prove more hardness results.*
2. *We apply these metatheorems to show some games are PSPACE-complete.*

The term *metatheorem* is somewhat vague in a sense because it's hard to state all the assumptions for all games. It will give a general setup for the proof.

13.2 Definitions

Viglietta defines an "avatar" in his proofs, but for our case, we will use the term "player." The player is the character in the game that we can control and move around to complete objectives. One basic assumption we make about the player is that we can choose, at any time, to change the player's direction of movement.

13.3 PSPACE

PSPACE is the set of problems solvable in polynomial space.

Exercise 13.1. Prove the following.

1. NP $\subseteq$ PSPACE.
2. PSPACE $\subseteq$ EXPTIME.
3. PSPACE $=$ NPSPACE.

13.3.1 PSPACE-Complete Problems

To show that a problem is PSPACE-hard, we need a set of problems known to be PSPACE-hard (to reduce from). The classical problem is to simulate a polynomial space algorithm (or simulate a linear space Turing Machine). This problem is not very useful for hardness proofs.

We present a set of simpler problems.

> QUANTIFIED SAT (QSAT), Q2SAT, Q3SAT
> *Instance:* For QBF, a quantified Boolean formula. It may begin with either a ∃ or a ∀. We will take it to be of the form
> $$Q_1x_1 : \cdots : Q_nx_n : \varphi(x_1, \ldots, x_n),$$
> where the Q_i's are quantifiers. For Q2SAT, φ is in 2CNF form. For Q3SAT, φ is in 3CNF form. Other variants can be defined, and we will freely use them.
> *Question:* Is the quantified Boolean formula true?
> *Note:* QSAT is often called QUANTIFIED BOOLEAN FORMULAS (QBF) in the literature. We use QSAT because it is easier to add modifiers in the same way as SAT.

Exercise 13.2.

1. Show that Q2SAT ∈ P.
2. Show that Q3SAT is PSPACE-complete.
   Hint: Given a polynomial-space-bounded Turing machine M and an input x, you need a QSAT φ such that $M(x)$ accepts if and only if φ is true. Construct a sequence of QSAT formulas $\varphi_0(a,b), \varphi_1(a,b), \ldots, \varphi_L(a,b)$ (we leave it to you to figure out L) such that φ_i is satisfiable if and only if there is a way for M to go from configuration a to configuration b in time $\leq 2^i$.

13.3.2 Schaefer-Style Dichotomy Theorem

Recall that T. Schaefer [Sch78] (our theorem 1.23) proved a dichotomy theorem that states exactly which types of SAT problems are in P and which are NP-complete. Schaefer stated a dichotomy theorem for QSAT but did not provide a proof. twenty-three years later, Creignou et al. [CKS01] proved the theorem T. Schaefer stated. For other interesting variants on the issue of dichotomy for QSAT, see the papers of E. Hemaspaandra [Hem04] and Dalmau [Dal99].

We now present the dichotomy theorem for QSAT [CKS01]. Note that QSAT problems are to determine whether

$$Q_1x_1 : \cdots : Q_kx_k : \varphi(x_1, \ldots, x_k)$$

is true. The different QSAT formulas are based on the different forms that φ can have.

Theorem 13.3.

1. *QSAT* $\in$ *P if and only if the unquantified formula* φ *is Horn, dual Horn, 2SAT, or affine.*
2. *QSAT is PSPACE-complete otherwise.*

As with 3SAT, the planar version of Q3SAT is also PSPACE-hard [Lic82]: we can start with the same crossover gadget as seen in the proof of theorem 2.10 that forces some variables to be the same, and then also have it create new variables that must also be quantified. We can do this by adding an $\exists x_i$ for every newly created variable x_i. Notably, PLANAR 1-IN-3QSAT is still hard, and PLANAR NAE 3QSAT is still easy.

For non-Boolean variables, we do not yet have a full dichotomy theorem (like we do for existentially quantified SAT/CSP [Zhu20], as described in section 1.2.9). Zhuk and Martin [ZM22] proved a dichotomy for three-value domains, showing that QSAT is either in P, NP-complete, coNP-complete, or PSPACE-complete. But they also showed that QSAT over four-value domains is sometimes NP $\cap$ coNP-complete.

13.4 Viglietta's Metatheorem 3

(Viglietta's metatheorems 1 and 2 are in sections 4.4.2 and 4.4.3, respectively.)

Convention 13.4. When we say that a game is PSPACE-complete (or any other complexity), we mean that determining who wins is PSPACE-complete. This also applies to one-player games (puzzles) where the question is whether the player can complete the task.

The third metatheorem from Viglietta's paper [Vig14a] states that a game where you have a player who needs to traverse a planar graph from start to finish, and has door and pressure plate objects, is PSPACE-hard. A door can be considered an edge that exists only if a certain condition is met. There are two types of pressure plates—an "open" pressure plate that satisfies the condition of the door and a "closed" pressure plate that causes the door condition to not be satisfied.

(Keep in mind that these pressure plates simply cause the door to open or close but do not require constant pressure to keep them in that state—that is, unlike Portal's door mechanics.)

All of the games we consider are easily seen to be **in PSPACE**. Hence when we show that they are PSPACE-complete, we are actually showing they are PSPACE-complete.

13.4.1 Reduction from Q3SAT

The idea is as follows:

- We start at the position labeled "start" and continue along the path.
- Wherever there is a $\exists$ quantifier, we fix a value for that variable.
- Wherever there is a $\forall$ quantifier, we will check both possible values by:

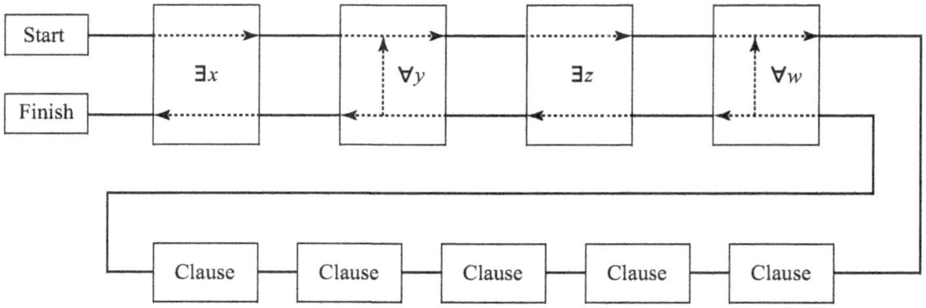

Figure 13.1
Overview of reduction.

1. Traversing toward the clauses (top branch in figure 13.1), we remember that we have seen this variable once and set its value to true.
2. Traversing back from the clauses (bottom branch in the diagram), we check to see if this variable is true: If true, set to false and loop again; if false, continue along the branch.

Remember that we will continue along the bottom only if the quantifier is satisfied; e.g., if one value fails to satisfy the formula for a $\forall$ quantifier, the second loop is no longer necessary, as we already know that the $\forall$ quantifier cannot be satisfied.

Now, we let a door's state determine the value of its *literal*: An open door indicates that its represented literal is selected. Thus, for a variable x, we need two doors for it: x and $\neg x$. Note that, when we say an x door, we actually mean all doors labeled by x such that an x-open switch opens all the x doors, and so on for closing and $\neg x$.

Then, a clause is a hall with three doors possible to the next hall, one for each literal; if any of the three doors is open, the clause is considered to be satisfied.

Finally, we need to come up with our quantifiers:

- existential quantifier: two branching paths, that mutually close each other, and turn the variable on or off
- universal quantifier: snake-like path, with middle gate that "counts" how many times we've gone through

The clause gadget and the quantifier gadgets are both diagrammed in figure 13.2. One key point to note about the quantifier gadgets is that they prevent deadlock by requiring the path to open one door containing pressure plates to close the other doors such that the progression must move forward.

Note that, by construction, a solution will take exponential time: at least $O(2^U)$ time for U universal quantifiers. However, the reduction process should still take polynomial time.

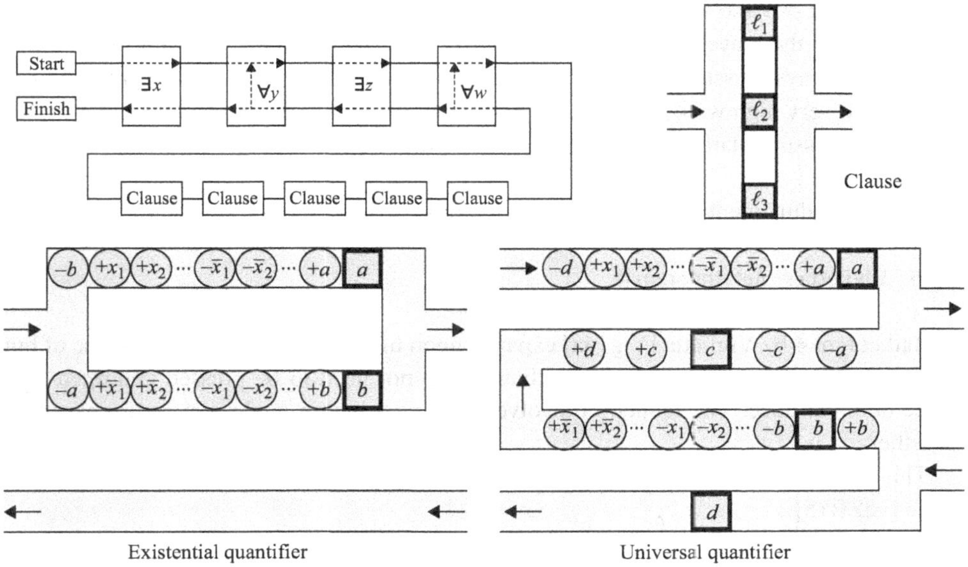

Figure 13.2
Gadgets.

13.4.2 First-Person Shooter Games

Metatheorem 3 allows many first-player shooter games such as *Quake* to be easily proved PSPACE-hard by simply designing the maps and puzzles using pressure plates and doors as described in the reduction above.

13.4.3 Role-Playing Games

Additionally, Metatheorem 3 allows one to prove many role-playing games games (such as EYE OF THE BEHOLDER) to be PSPACE-hard, using the same idea of designing the maps and puzzles correctly.

13.4.4 Script Creation Utility for Maniac Mansion Engine

Many adventure games can also be proven PSPACE-hard in this way. One rather large category of such games—the script creation utility for maniac mansion (SCUMM) engine—can be shown to be PSPACE-hard, allowing all games that use it to be shown to be as well. Some SCUMM engine-based games are *The Secret of Monkey Island* and *Maniac Mansion*. Most Sierra adventure games, such as the Space Quest series, can also be proved hard in this way.

13.4.5 Prince of Persia

In this game, the player is given the ability to jump. So, in order to ensure that pressure plates are always pressured when the player passes over that tile, we put the pressure plate on top of a very high wall, such that the player can only jump that high, so the player must touch the pressure plate (which is important to ensure that the player must make progress in the game).

With this adjustment, one can show that Prince of Persia is PSPACE-hard.

13.5 Viglietta's Metatheorem 4

Metatheorem 4 by Viglietta [Vig14a] expands upon metatheorem 3 to allow the use of buttons instead of pressure plates. These buttons do not need to be pressed and open/close three doors at once. The reduction involves treating a button as three pressure plates put together.

This result has since been improved so that each button opens/closes only two doors at once [vdZB15].

13.5.1 Examples

Some examples of games that fall under this pattern are *Sonic the Hedgehog*, *The Lost Vikings*, and *Tomb Raider*. With this adjustment, they can be shown PSPACE-hard.

13.6 Doors and Crossovers

Another metatheorem, by Aloupis et al. [ADGV15], is a further generalization that relies on doors and crossovers to show PSPACE-hardness.

To do this, we show that we can use doors and crossovers to create an environment that provides the following three types of paths: a traverse path (whereby the player can only pass if the door is opened), an open path (which allows the player to open the door), and a close path (which forces the player to close the door).

This result can be found in the paper of Aloupis et al. [ADGV15], which uses it to analyze various Nintendo games. More recent work of Ani et al. [ABD$^+$21] has removed the need for a crossover.

13.6.1 Legend of Zelda: A Link to the Past

In *Legend of Zelda*, we create a traverse path with just a labeled door: If the labeled door is open, then we can traverse; if not, then we cannot.

One interesting caveat with *Legend of Zelda* is that the game has only toggles: It opens all the connected doors if they're closed and closes them if they were open. This means that the open and close paths are somewhat trickier: The general idea is for the open and close paths

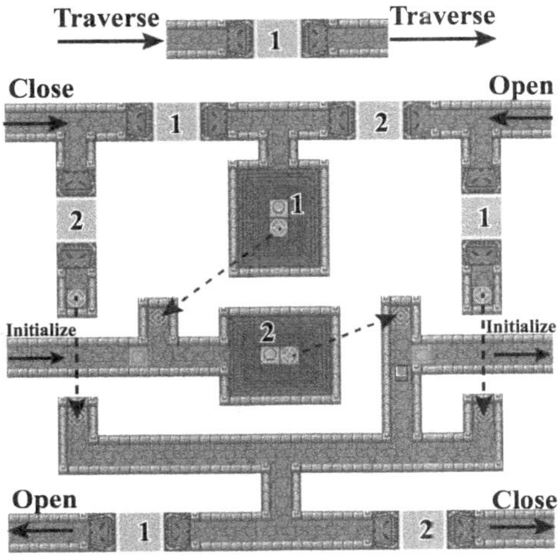

Legend of Zelda: A link to the Past

[Aloupis, Demaine, Guo, Viglietta 2014]

Figure 13.3
Legend of Zelda gadget.

to lead to directed teleporters to the appropriate halls (seen at the bottom of figure 13.3). Then, to toggle the door, we have to traverse from the direct that we want to toggle to and then hit the toggle in a secluded room, for which the only way out is through the directed teleporter right next to it, which takes us back to a hall.

Since we want to initialize all doors to the closed state, we need to create an initializer for this purpose; basically, we just create a chain of gadgets that will toggle all the doors to be closed. At the very end, we have a crystal that can be broken, which will activate the inactive toggles, and deactivate the active toggles. There will be only one crystal, and it serves the purposes of destroying our initializing paths and enabling a one-way gadget. To make sure that it does not appear at the wrong place, we ensure that it appears only between the initializer and final traversal gadgets.

13.6.2 Donkey Kong Country 1, 2, and 3

Donkey Kong Country (*DKC*) is another Nintendo series of games that can be shown to be PSPACE-hard using doors and crossovers. Unfortunately, *DKC* 1, 2, and 3 have mutually exclusive "features," so we will have to address each one individually.

Donkey Kong Country 1 In the *Donkey Kong Country* games, there are bees. If you touch a bee, you die, so the goal is to not die. In *Donkey Kong Country 1*, there is a tire object: If you land on the tire, you bounce back up. For a traversal, we use a barrel shot straight down, so landing on a tire would cause us to be stuck in the game (and thus not win).

Donkey Kong Country 1

[Aloupis, Demaine, Guo, Vigiletta 2014]

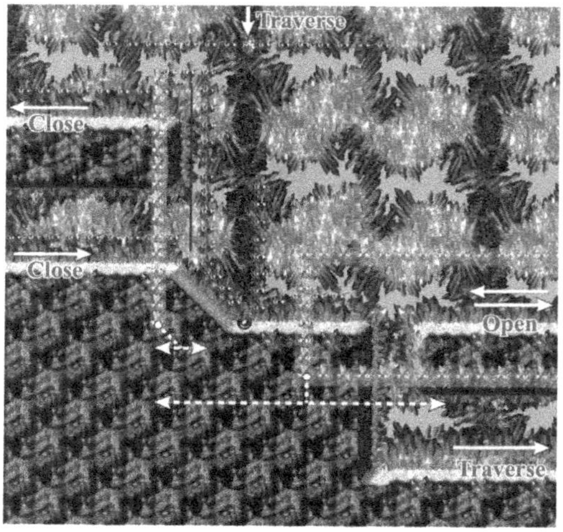

Figure 13.4
Donkey Kong Country 1 gadget for PSPACE-completeness

Furthermore, we will introduce stationary bees as obstacles and moving bees (highlighted in red in figure 13.4—not to be confused with actual red bees in the game), which move in a deterministic pattern, demonstrated by the arrows.

To open the door, we come in from the open path, jump across the ledge, and push the tire just up the hill so that it does not roll back down. Note that the giant box of bees that are labeled "open" move in a left-down-right-up pattern, so that after pushing the tire up the hill, we can still run back down through the traversal path.

To close the door, we come in from the close path, push the tire down the hill with a slight nudge, and scramble up a rope to exit. One caveat here is that, in the real game, we'd have to either make the bees slower or make the climbing faster.

With these adjustments one can show that DONKEY KONG COUNTRY 1 is PSPACE-hard.

DONKEY KONG COUNTRY 2 In DONKEY KONG COUNTRY 2, instead of tires, we have balloons and air currents (the gray steam near the bottom of the map). The balloons float on top of the air currents, and in order to traverse, we have to move it out of the way of the traversal path.

To do this, from the open end, we can jump on the balloon and move it away from the air currents in the middle so that it drops down to point A. We then escape through the entrance of the open path.

To close, we enter from the close path, drop onto the balloon, and move it back into the current before using the barrels to propel ourselves out of the region.

With these adjustments one can show that DONKEY KONG COUNTRY 2 is PSPACE-hard.

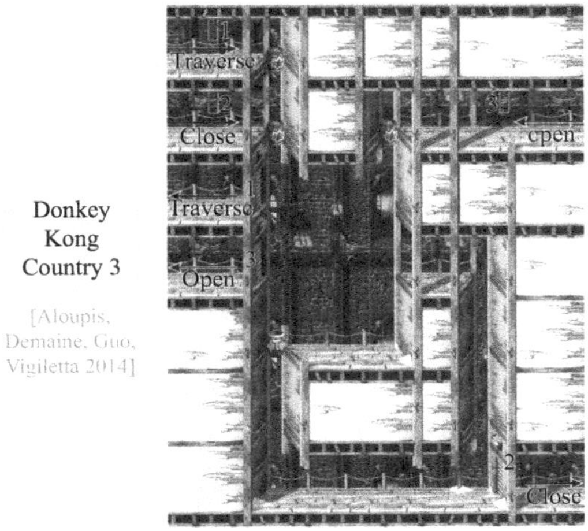

Figure 13.5
Donkey Kong Country 3 gadget for PSPACE-completeness.

Donkey Kong Country 3 In *Donkey Kong Country 3*, instead of tires and balloons, we have tracking barrels, which the player can slide around sideways once the player lands in one, and it always shoots up. The caveat is that the barrel will follow the player if the player tries to jump out of it, so the tracking barrel effective prevents the player from going down.

Thus, the open state is for the barrel in the big gap in figure 13.5 to be on the left: If the player enters to traverse, the player can drop into the barrel, get shot up, and exit through the traversal path.

To close, we enter from the close path, jump into the barrel, and slide it all the way to the right. Since the barrel tracks us wherever we go after we land in it, we are guaranteed that, if we leave through the close path, the barrel must be on the right.

To open, then we just drop into the barrel from the open end, slide to the left, and shoot ourselves out through the open path.

With these adjustments one can show that Donkey Kong Country 3 is PSPACE-hard.

13.6.3 Super Mario Bros.

Recall that in section 1.3.5 we showed that Super Mario Bros. is NP-hard. Actually it is harder than that: Super Mario Bros. is PSPACE-hard.

Using figure 13.6, we see that the traversal path is on the left, the open path is on the bottom, and the close path is the entire right side.

The intuition behind the setup is: We need something that the player cannot pass through, but an obstacle can. We use a rotating firebar to separate the traversal and close paths, and a spiny to block a path. We draw the firebar as length-1 for clarity of what it intends to block,

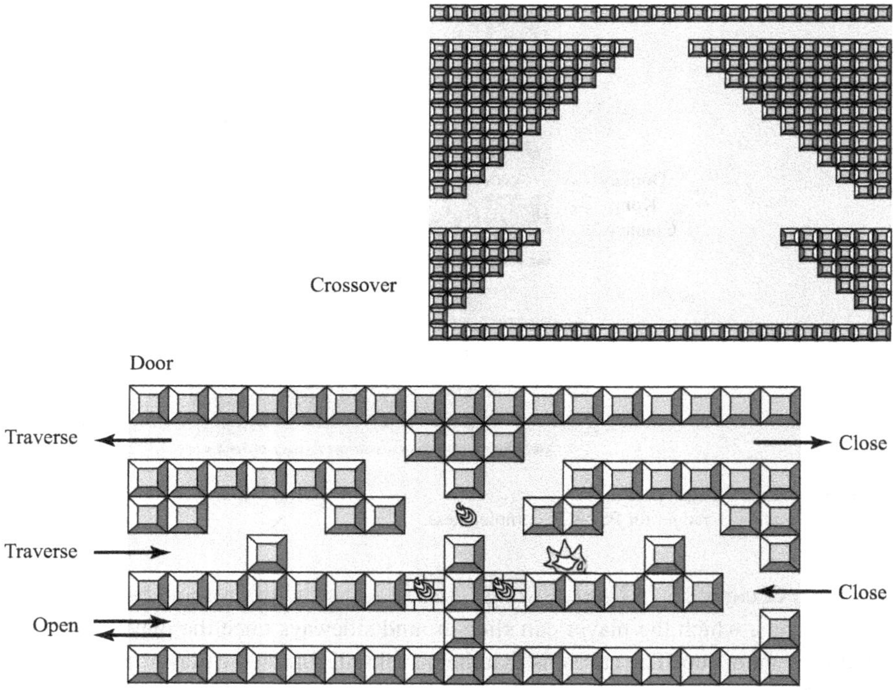

Figure 13.6
Super Mario Bro. gadget for PSPACE-completeness.

though the gadget can still be (carefully) traversed with longer firebars). Note that, in the original game, spinies only fall from the top of the screen, and here we assume they start in certain locations.

To close, we enter from the close side, and if the spiny is on the right, then we go under the spiny, and at the right time, we knock up and over the fireball to the other side, so that we can traverse the close path.

To open, we enter from the open side, knock the spiny over to the right, and leave again.

Then, when we traverse, we can traverse if and only if the spiny isn't on the traverse side.

13.6.4 Lemmings

Lemmings is an old real-time strategy game where the player is in control of a bunch of lemmings, and can imbue any lemming with jobs. For our purposes, we will have two jobs: a builder (to build bridges) and a basher (to cut through dirt, but not steel).

Although many results for *Lemmings* exist, Viglietta showed in another paper [Vig14b] that LEMMINGS is PSPACE-hard using an exponential amount of builders, bashers, and time.

The standard lemming is a ***walker***, and its mechanics are documented in figure 13.7a. In brief, walkers can climb up out of ditches, but only if they're not too tall.

One kind of special lemming that we will use is a ***basher***, who can cut through dirt but not steel. The mechanics are documented in figure 13.7b, and a basher will stop bashing once the steel check finds steel, or a solidarity check finds no more dirt.

What is interesting to note is that disparity between the steel and the solidarity checks.

The last kind of special lemming that we will use is a ***builder***, who can build bridges over gaps (otherwise the lemmings will die from falling between the gaps). The mechanics are documented in figure 13.7c, and again, there are solidarity checks for where to lay the bridge pieces, and whether the Lemming can keep going forward and up, by using a solidarity check near the head.

To apply the doors and crossovers metatheorem, we need to show the primitive paths that we will use: a rising path, a falling path, rise-and-double-back path, fall-and-double-back path, and a crossing path. To prevent lemmings from effectively running away, we place steel pretty much everywhere that isn't marked with a red crossed-box or a gray platform in figure 13.8.

Now, we will need a fork (aka crossover) gadget, to allow for traversals of two different paths. This is achieved by having a gap for lemmings to fall through: If they fall through, they'll go through one path; if they build a bridge instead, they will go through another path (which also requires a basher to clear some obstacles, which are placed to prevent builders from going crazy and building a stairway to heaven).

Finally, our door gadget will be represented in figure 13.9: An open door will have a gap in the middle, while a closed door will have a bridge in the middle that will prevent traversal from the top to the bottom. To open the door, we bash the bridge apart using a basher; to close the door, we layer a bridge on top of the gap using a builder. Since we have a limited (but exponential) number of builder and basher upgrades available to us, we must use each one wisely.

Thus, with our door and crossover gadgets, we have shown that *Lemmings* is PSPACE-hard.

13.7 Bounded SAT Games

QSAT naturally represents winning strategies in two-player games. Does there exist a first move such that, for any possible next move by the opponent, there is a second move such that . . . such that the first player wins? Assuming the total number of moves is polynomially bounded, such questions are naturally represented by quantified formulas with alternating quantifiers:

$$\exists x_1 : \forall x_2 : \exists x_3 : \forall x_4 : \cdots : \varphi(x_1, \ldots, x_n).$$

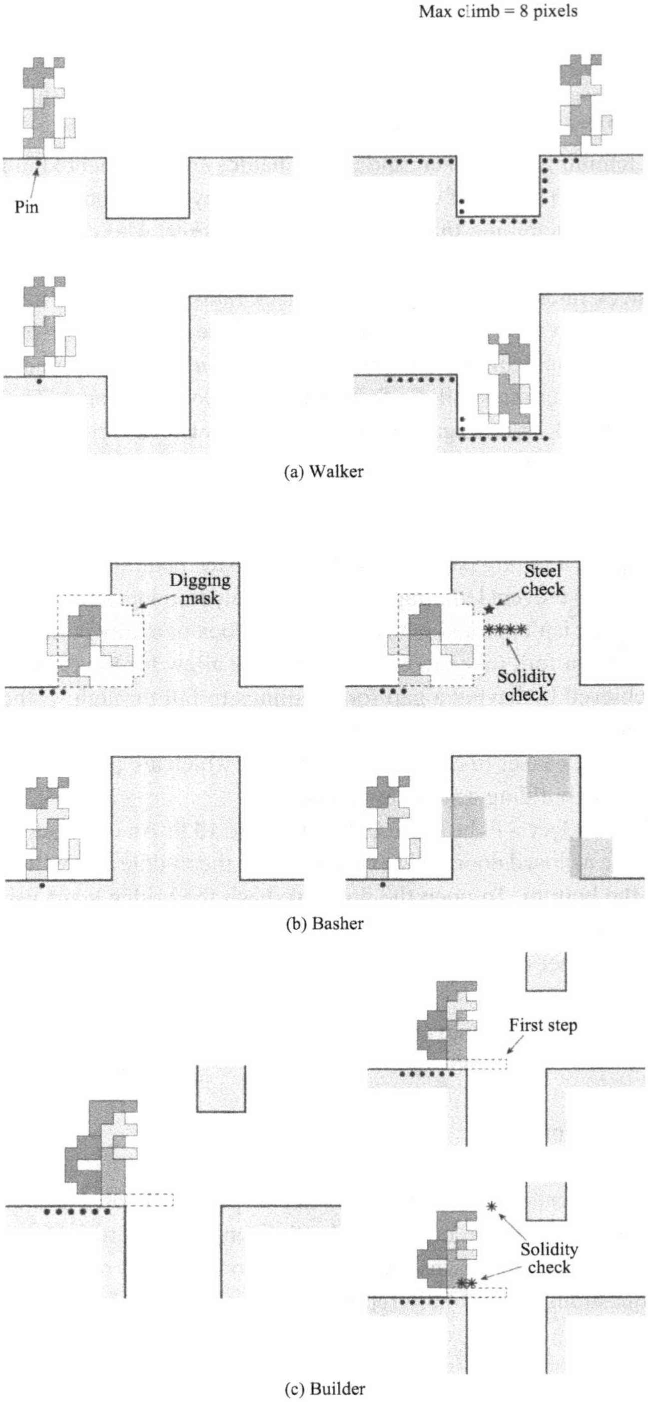

Figure 13.7
Lemmings behaviors.

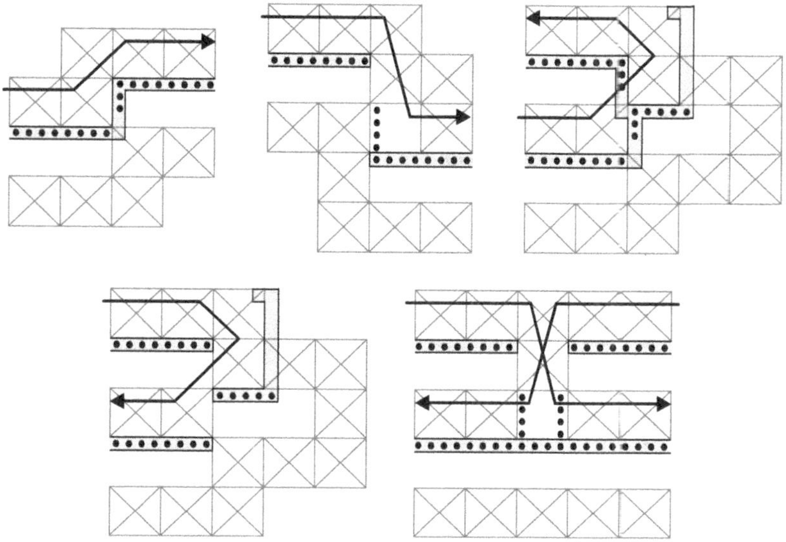

Figure 13.8
Lemmings path gadget

We can formalize this connection with the following game:

Definition 13.5. The ***QSAT game*** is as follows.

1. The board is a Boolean formula $\varphi(x_1, \dots, x_{2n})$.
2. For $i = 0$ to $n - 1$:
   a. Player I sets x_{2i+1} to either TRUE or FALSE.
   b. Player II sets x_{2i+2} to either TRUE or FALSE.
3. If $\varphi(x_1, \dots, x_{2n}) = \text{TRUE}$, then player I wins; else player II wins.

This is an example of a two-player SAT game. It is a *bounded* game because the number of moves is exactly n. Indeed, determining who wins is exactly the QSAT problem (where the quantifiers are alternating, which we can arrange by adding extra quantified variables as needed), so it is PSPACE-complete.

13.8 Stochastic Games

What if we have a two-player game where the opponent player plays *randomly*? Papadimitriou [Pap85] called these "games against nature." A natural decision problem here is whether the deterministic player can win with probability greater than $\frac{1}{2}$. More formally:

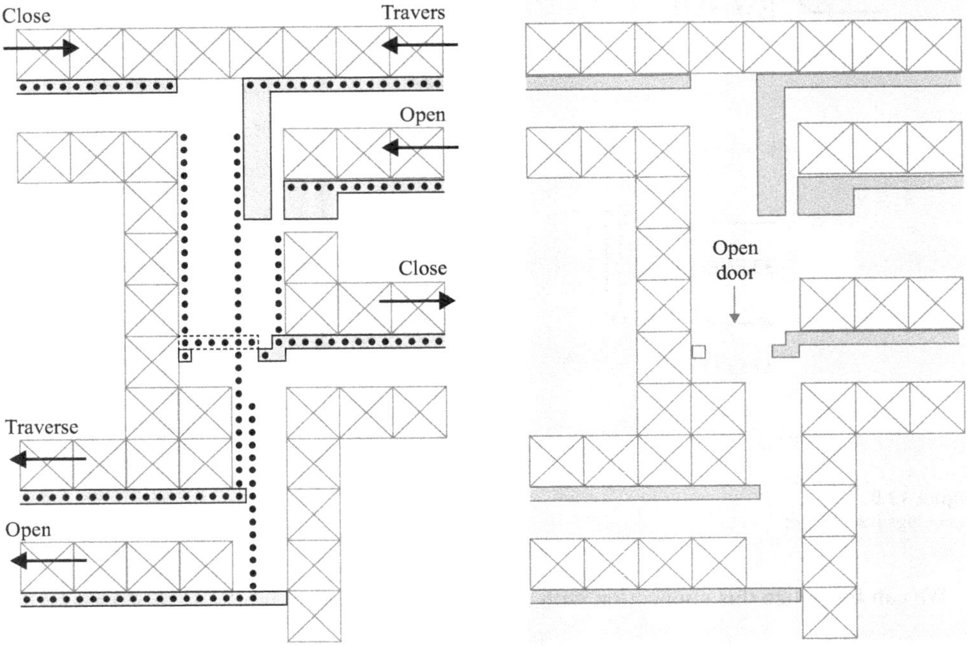

Figure 13.9
Lemmings door gadget for PSPACE-completeness.

> Stochastic SAT
> *Instance:* A quantified Boolean formula of the form
> $$\exists x_1 : \text{Я} x_2 : \exists x_3 : \text{Я} x_4 : \cdots : \Pr(\varphi(x_1, \ldots, x_n)) > \tfrac{1}{2},$$
> where $\text{Я} x_i$ denotes a uniformly randomly chosen x_i.
> *Question:* Is the quantified Boolean formula true?

Papadimitriou [Pap85] proved that Stochastic SAT is PSPACE-complete. Interestingly, there is no difference in complexity between random choice and the for-all choice in standard two-player games.

13.9 Open Problems

1. Define and analyze generalizations of *Nine-Men-Morris*, *Quoridor*, and other games (check to see if they have already been analyzed). Then determine the complexity of these versions. You should also look at variants of these games.
2. (This is a not-well-defined research program.) Most of the results on the hardness of games do not use the game as it is actually played. (See Biderman [Bid20] for a possible exception.) For example, chess and checkers are played on an 8×8 board, not an $n \times n$ board. Develop a framework for the complexity of games that can be used to show that a game, as it is actually played, is hard.

14

Beyond PSPACE but Decidable

14.1 Introduction

In this chapter, we will examine problems whose complexity is beyond PSPACE. Some of them will be EXPTIME-complete. Notably, we know that EXPTIME-hard problems *require exponential time*, so are not in P, without needing any assumptions like $P \neq NP$.

It is also easy to show that $PSPACE \subseteq EXPTIME$, but they are not known to be different. Nevertheless, we (and the theory community) believe they are. Why? Note that if $A \in EXPTIME$, then the algorithm for A can use exponential space. For example, if your algorithm needs to look at *every* subgraph of an n-vertex graph at the same time, then an EXPTIME algorithm can do this, whereas a PSPACE algorithm cannot.

Chapter Summary

1. *We define many games based on SAT.*
2. *We use these SAT games to show that several other games are EXPTIME-complete. We note that these are not games people actually play.*
3. *We look at the complexity of games people really do play. They are mostly EXPTIME-complete.*

The problems we consider will all be games: Given a game position, which player will win (if they both play perfectly).

We will not define "game" formally; however, we will define parameters of games and give examples, which are also summarized in table 14.1.

14.2 Types of Games

We define a wide variety of games with varying numbers of players, information, and number of moves. Table 14.1 summarizes these different types of games and the corresponding complexity classes where the games tend to reside.

Table 14.1
Natural complexity classes for computations with varying numbers of players and length of computation. (Based on [HD09, figure 1.1].)

	0 Players (simulation)	1 Player (puzzle)	2 Players (game)	Team, imperfect info
Unbounded	PSPACE	PSPACE	EXPTIME	UNDECIDABLE
Bounded	P	NP	PSPACE	NEXPTIME

Definition 14.2.

1. A game is ***bounded*** if the number of moves is bounded by a polynomial in the board size. A game is ***unbounded*** otherwise.
2. A game has ***perfect information*** if there is no hidden information. A game has ***imperfect information*** otherwise.
3. A ***0-player game*** is a simulation.

   The ***game of Life***, which we will study in section 15.3, is such a game. It is also unbounded and has perfect information. Bounded 0-player games tend to be P-complete, whereas unbounded 0-player games tend to be PSPACE-complete. P-complete problems are interesting because it is conjectured that such problems cannot be parallelized (see chapter 21 for a short discussion of P-completeness). Life is PSPACE-complete on a finite board (the setting we study most games), but it can also be considered on an infinite board where it becomes undecidable.
4. A ***one-player game*** is a puzzle.

   Jigsaw puzzles, packing puzzles, and all of the other NP-complete puzzles in chapter 5 are one-player games with perfect information. The bounded versions of one-player games with perfect information tend to be NP-complete, whereas the unbounded versions tend to be PSPACE-complete.
5. A ***two-player game*** is just what you think it is.

   Bounded two-player games with perfect information tend to be PSPACE-complete, whereas the unbounded versions tend to be EXPTIME-complete. We saw an example of a bounded two-player SAT game in section 13.7; it is PSPACE-complete. Chess is an example of an unbounded two-player game with perfect information; it is EXPTIME-complete.

6. A ***multiplayer game*** is a game of three or more players.

   Monopoly and *Scrabble* are examples. Without additional constraints (see team games below), multiplayer games generally degenerate into two-player games: one player versus all the opponents.

7. ***Team games*** are when there are two teams and each team can have two or more players.

   Bridge is an example of a team game of imperfect information. It is conjectured to be NEXPTIME-complete. More generally, bounded team games with imperfect information tend to be NEXPTIME-complete, whereas unbounded team games with imperfect information tend to be undecidable. Rengo Kriegspiel Go is a team version of Go where (1) white has two players, (2) black has two players, (3) the player on each team alternate play, (4) the players on a team cannot communicate, and (5) the players cannot see the moves of other players. This is a game that humans play, and is conjectured to be undecidable [Hea06].

14.3 Unbounded SAT Games

Stockmeyer and Chandra [SC79] introduced several SAT games that are unbounded—the number of moves can be exponential in n—and showed they are EXPTIME-complete. They used these EXPTIME-complete problems to prove that other problems are EXPTIME-complete. Because $\mathsf{P} \neq \mathsf{EXPTIME}$ (by a simple diagonalization argument), any EXPTIME-complete problem is *provably (!)* not in P. We do not need to condition on $\mathsf{P} \neq \mathsf{NP}$ or any other assumption.

The games they consider are unbounded in that they could go on forever. This happens because (1) in some games you can change the value of a variable and (2) in some games a player can pass.

All of the games have the following:

1. The game begins with one or two Boolean formulas and a partial assignment (which might not set any variables).
2. The players are RED (who goes first) and BLUE (who goes second).
3. Some of the variables are RED, some of the variables are BLUE, and some of the variables are neither. Only the RED (BLUE) player can set the RED (BLUE) variables.
4. These variables can then be set to TRUE or FALSE. In some cases they can be reset.
5. A player may be allowed to pass.

Stockmeyer and Chandra [SC79] proved the following theorem.

Theorem 14.3. *For the following games, determining which player will win is EXPTIME-complete.*

1. *G1:*
   Board: *A 4CNF formula with variables* RED, BLUE, *and one x that is neither, and a truth assignment for the* RED *and* BLUE *variables, but not x.*
   Move: *A move of* RED (BLUE) *consists of setting all the* RED (BLUE) *variables and also setting x to* TRUE (FALSE).
   Passing: *Not allowed.*
   Win Condition: *If, after a player's move, the formula is* FALSE, *then that player loses.*
2. *G2:*
   Board: *Two 12DNF formulas with all variables* RED *or* BLUE, *and a truth assignment for all the variables. One formula is itself called* RED, *the other* BLUE.
   Move: *A move of* RED (BLUE) *consists of changing* ≤ 1 RED (BLUE) *variable.*
   Passing: *Allowed since you can opt to change 0 variables.*
   Win Condition: *If, after* RED (BLUE) *moves, the* RED (BLUE) *formula is* TRUE, *then* RED (BLUE) *wins.*
3. *G3:*
   Board: *Two 12DNF formulas with all variables* RED *or* BLUE, *and a truth assignment for all the variables. One formula is itself called* RED, *the other* BLUE.
   Move: *A move of* RED (BLUE) *consists of changing one* RED (BLUE) *variable*
   Passing: *Not allowed.*
   Win Condition: *If, after* RED (BLUE) *moves, the* RED (BLUE) *formula is* TRUE, *then* RED (BLUE) *loses.*
4. *G4:*
   Board: *A 13DNF formula with all variables* RED *or* BLUE, *and a truth assignment for all the variables.*
   Move: *A move of* RED (BLUE) *consists of changing* ≤ 1 RED (BLUE) *variable*
   Passing: *Allowed since you can opt to change 0 variables.*
   Win Condition: *If, after* RED (BLUE) *moves, the formula is* TRUE, *then* RED (BLUE) *wins.*
5. *G5*
   Board: *An (unrestricted) formulas with all variables* RED *or* BLUE, *and a truth assignment for all the variables.*
   Move: *A move of* RED (BLUE) *consists of changing* ≤ 1 RED (BLUE) *variable*
   Passing: *Allowed since you can opt to change 0 variables.*
   Win Condition: *If the formula* ever *becomes true, then* RED *wins.*
6. *G6: Identical to G5 except that the formula has to be in CNF form.*

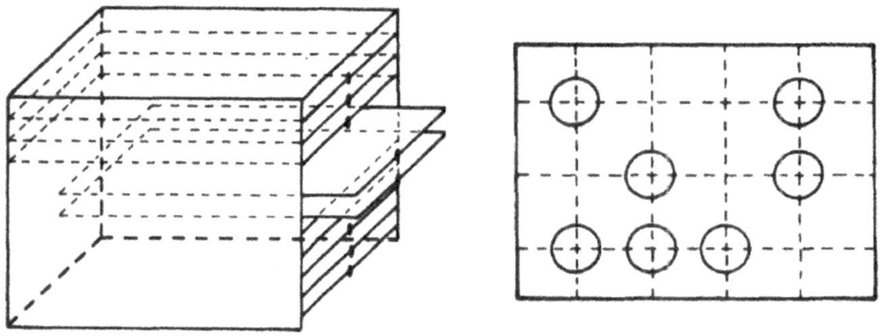

Figure 14.1
Left: A stack of plates. Right: An example of a plate.

14.4 Games People Don't Play

All of the results in this section are due to Stockmeyer and Chandra [SC79]. In each of the games, one player is RED and the other is BLUE. RED goes first. The complexity of a game is the complexity of, given a position in it, determining which player wins.

14.4.1 Peek

Definition 14.4. Peek is the following two-player game:

1. The parameters of the game are n and d. The players are RED and BLUE. RED will go first.
2. The setup is a stack of n plates, each of which has $\le dn$ holes in it. The plates are known to both players. The plates are initially all in a box (see figure 14.1). Later in the game each plates can each be in one of two states: IN or OUT. (Note from the picture that an OUT plate is still partially in the box.)
3. One of the plates cannot be moved. The rest are in two disjoint sets: RED and BLUE.
4. On a players, turn, he can either (1) pass, (2) push one of his OUT plates IN, or (3) push one of his IN plates OUT.
5. The game ends when a hole appears through the entire stack of plates. When this happens, the player who made the last move wins.

Theorem 14.5. *G4 $\le_p$ Peek, so Peek is EXPTIME-complete.*

14.4.2 Ham Game

Definition 14.6. Ham Game is a game where we start with a simple, undirected graph where each edge is colored either RED or BLUE. The players are named RED and BLUE. Player RED

(BLUE) will control the RED (BLUE) edges. Player RED goes first. In addition to a color, each edge also has a state—IN or OUT. Each turn, a player must toggle the state of an edge of his color. Player RED wins if at any point in the game, the edges that are IN form a Hamiltonian cycle. Player BLUE wins if this never happens. The associated problem asks if player RED has a winning strategy.

Theorem 14.7. *G6 $\leq_p$ HAM GAME, hence HAM GAME is EXPTIME-complete.*

14.4.3 BLOCK

BLOCK is another graph game.

Definition 14.8. BLOCK starts with three graphs over the same vertices. Some of these vertices contain tokens that are either RED or BLUE. A vertex can have at most one token. Each turn, a player must slide a token of his color along any path in one of the three graphs, as long as the target vertex and all intermediate vertices on the path do not have any tokens. Each player i has some set of "victory vertices" W_i. If a player can move one of his tokens to a vertex in his set of victory vertices, he wins.

Theorem 14.9. *G3 $\leq_p$ BLOCK, so BLOCK is EXPTIME-complete.*

Proof sketch. Figures 14.2 shows the variable gadgets for both players (white left, black right). Stars are winning vertices, and the dashed, dotted, and solid lines represent edges in each of the graphs. If either player deviates from setting variables, the other can instantly win.

Figure 14.3 shows the clause gadget for the clause $\overline{x_3} \wedge y_5$. Once white activates by moving up, both black and white are forced to move up one at a time—or else the other player wins instantly. If x_3 and $\overline{y_5}$ are blocked, then white cannot move up, in which case black wins and white should not have activated the formula. □

14.5 Games People Play

We look at the complexity of checkers, chess, and Go on an $n \times n$ board. A statement of the form "CHECKERS is EXPTIME-complete" means that the problem of, given a position, which player wins (the one whose turn it is, or the other one) is EXPTIME-complete. CHECKERS mate-in-1 is the problem of determining whether the player who is about to go and make one move that wins the game. Similar for other problems of the form GAME mate-in-1.

We will also look at slight variations on the ruleset that make these games even harder than EXPTIME.

14.5.1 CHECKERS

Questions about checkers run the gamut between P and EXPTIME-complete.

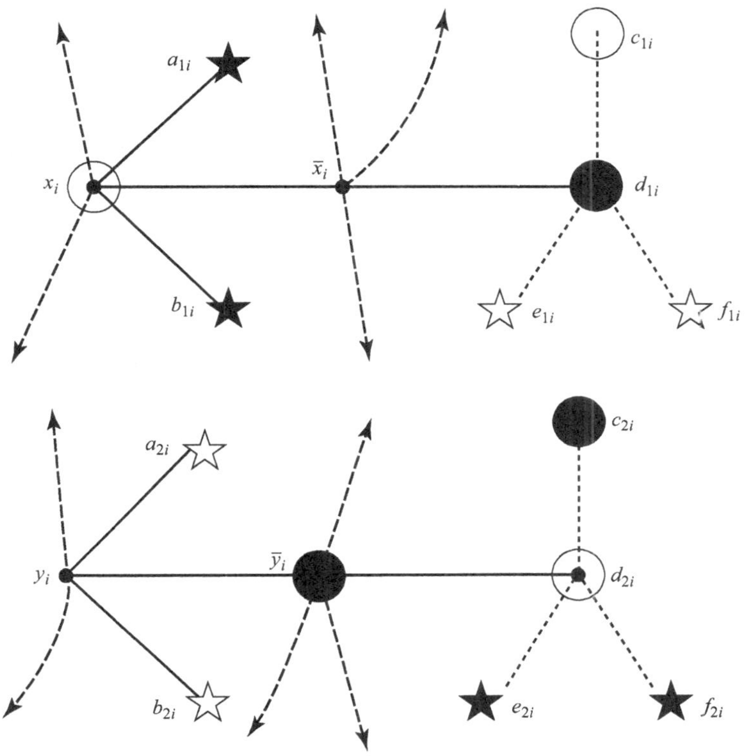

Figure 14.2
Variable gadgets.

Theorem 14.10.

1. *(Fraenkel et al. [FGJ+78]) Checkers mate-in-1 seems to involve a large number of jumps; however, deciding whether a checkers position is a mate-in-1 is in P. The proof reduced this problem to that of finding an Eulerian path on a graph, which is in P.*
2. *(Bosboom et al. [BCD+19]) Deciding whether there is a move that forces the other players to win in one move is NP-complete. (This may be useful if you are playing a child, or a Wookiee, or a Wookiee Child. See https://www.youtube.com/watch?v=rN0T5tyJlo8.)*
3. *(Bosboom et al. [BCD+19]) Checkers where every move has to be a jump (capturing the other player's piece) is PSPACE-complete. The first player who cannot make a jump loses. For the standard beginning position, player I would always lose; however, the problem is hard if you allow any initial position.*
4. *(Robson [Rob84b] G3 $\le_p$) Checkers, hence Checkers is EXPTIME-complete. This is the result you probably care about the most.*

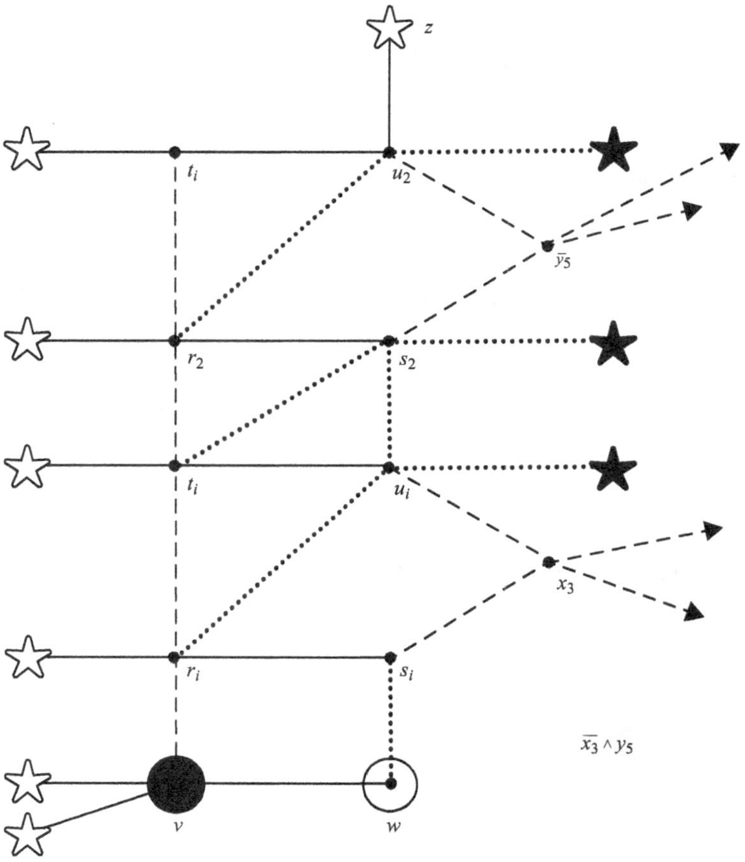

Figure 14.3
Clause gadget for the clause $\overline{x_3} \wedge y_5$.

Proof sketch. We sketch the proof of part 4.

In the proof we exploit the fact that a player must capture a piece when it is possible to. The board is split into two regions: an inner region, where the clauses and variables are represented, and an outer region, which can be triggered by a player to win (figure 14.4).

Initially, players start by moving their own kings between either a true or a false position. If at any point, a player satisfies their own DNF clause represented by pieces in the middle, the other player can activate an attack. A successful attack results in "free moves," where the opponent has a configuration that requires a capture, but the other player does not. After accumulating enough free moves, the player can maneuver his pieces in the outer spiral where the opponent is forced to jump into a position where all of his pieces in the spiral

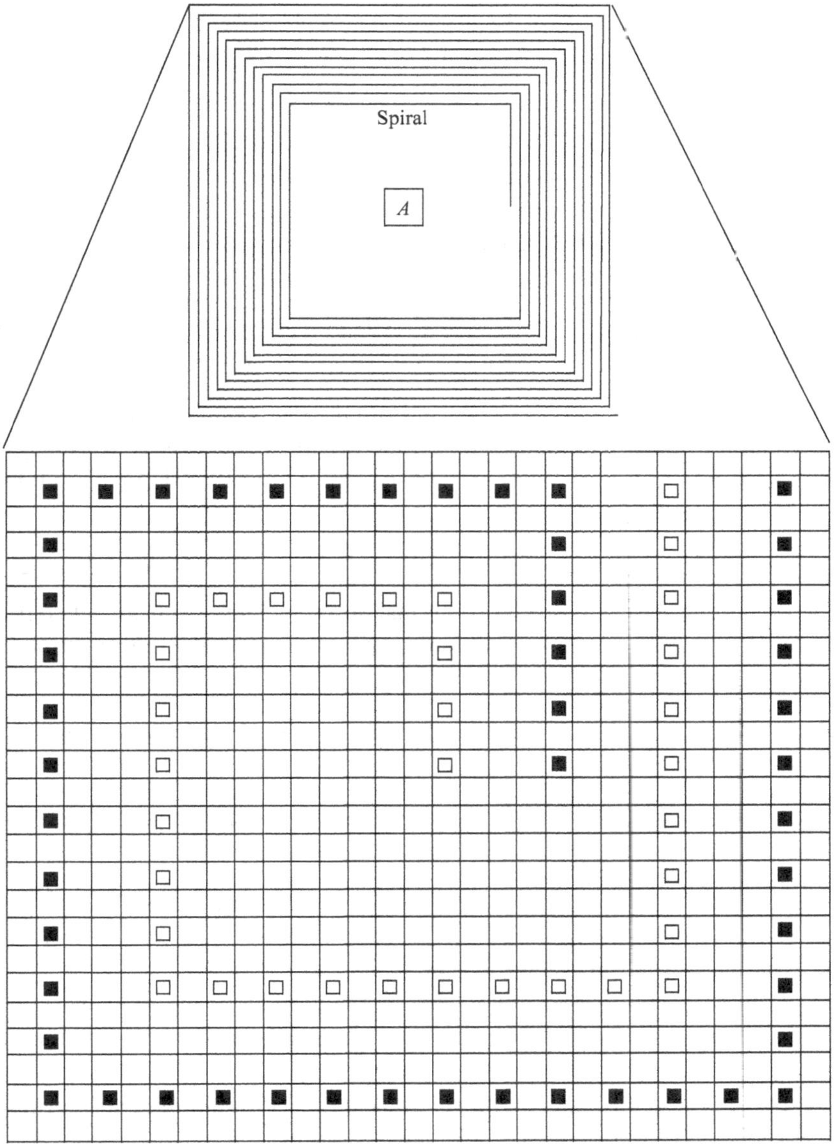

Figure 14.4
The spiral setup around the variables and clauses.

can be taken in one jump in the next turn. This gives enough of a material advantage to guarantee a win. □

14.5.2 Chess

In traditional chess there are some limits on how long the game can go. For example, if the same position is repeated three times, then the game is a draw (it's more complicated than that and depends on which chess federation you are playing in, but we ignore that). In this bounded version, chess is PSPACE-complete [Sto83]. But without this rule, we obtain an unbounded game, and Fraenkel and Lichtenstein [FL81] showed EXPTIME-completeness:

Theorem 14.11. *Let Chess be chess with no rule to limit the length of a game, and only using pawns, bishops, and kings. Then G3 $\leq_p$ Chess. Hence Chess is EXPTIME-complete.*

Another rule in the International Chess Federation is that a game is a draw if it impossible for either player to reach checkmate (even if the opponent cooperates). Deciding whether such a "helpmate" exists is, in fact, PSPACE-complete [BDHW20]. Thus, with this rule, draw-in-0 chess is PSPACE-complete.

14.5.3 Go

In the Japanese ruleset for Go, the ***ko*** rule is that positions cannot immediately repeat. This puts Go in EXPTIME since we only need to compare two states at a time. Robson [Rob83] showed the following:

Theorem 14.12. *G6 $\leq_p$ Go, if Go uses Japanese rules. Hence this version of Go is EXPTIME-complete.*

With the Chinese ***superko*** rule for Go, positions cannot ever repeat. This puts Go in EXPSPACE, but it is not known to be EXPSPACE-complete [Rob84a].

Open Problem 14.13. *Is Go with superko EXPSPACE-complete?*

14.5.4 Hard Variants of Games People Play

The results on checkers, chess, and Go were for the versions of the game where the rules allow the game to go on forever. If we add some additional rules that prohibit this, then the games may get harder.

Robson [Rob84a] studied both the ***no-repeat rule*** and the ***conditional no-repeat rule***. He looked at versions of G1, G2, and G3. We state his results.

Definition 14.14. A problem is in 2EXPTIME if it is in time $O(2^{2^{n^k}})$ for some k.

No-Repeat Rule. This rule makes a player lose if they ever repeat a past game configuration. This condition makes the following games EXPSPACE-complete: G1, G2, G3, Chess, and Checkers. The intuition here is that we have to track and compare against all past game states.

Open Problem 14.15. *Is Go with the no-repeat rule (known as the superko rules) EXPSPACE-complete?*

Conditional No-Repeat Rule. We are not going to define this notion formally; however, we will give an example of a game with this condition.

We introduce two special variables y and z to the game G1. y will be RED and z will be BLUE. A player now loses if they repeat a past game configuration and at most one of y or z have changed since that configuration was played. Adding this rule makes G1 2EXPTIME-complete. Here, the intuition is that we have to track y and z temporarily, in addition to all past game states.

Reif [Rei84] studied both ***private-information games*** and ***blind games***. He looked at versions of G1 and G2. We state his results.

Private-Information Games. In this variation, you can see some but not all of an opponent's state. An example of this game is PEEK with a partial barrier obscuring the state of some of each player's plates to the other player. This condition makes the following games 2EXPTIME-complete: (a) G1 with 5DNF, (b) G2 with DNF, and (c) PEEK.

Blind Games. Here, neither player knows the state of the other player. This condition makes the following games EXPSPACE-complete: (a) G2 with DNF and (b) PEEK.

14.6 Further Results

We give a partial list of problems that are complete in classes that are (likely) above PSPACE. For games listed, the problem is

Given a position in the game, which player wins?

1. Chinese checkers, and other pebble games, were proven EXPTIME-complete by Kasai, Adachi, and Iwata [KAI79]. Variants of pebble games were proven EXPTIME-complete by Kolaitis and Panttaja [KP03].
2. *Shogi*, also known as Japanese chess, is a two-player strategy game. It does share some properties of chess, though the games are not that much alike. It was proven EXPTIME-complete by Adachi et al. [AKI87]. The complexity of variants of *Shogi* was studied by Yato et al. [YSI05].
3. *Quixo* is a complicated variant of tic-tac-toe. It was shown EXPTIME-complete by Mishiba and Takenaga [MT20].
4. *Cops and Robbers* is a game played on a graph where a robber is trying to escape a group of cops trying to encircle them. It was shown EXPTIME-complete by Kinnersley [Kin15].
5. The *Custodian Capture* game is a game where pieces move like rooks and capture by being on either side of a piece. It was shown EXPTIME-complete by Ito et al. [INKT21].

6. Reachability-time games on timed automata were showed EXPTIME-complete by Jurdzínski and Trivedi [JT07].
7. Different versions of *Angry Birds* were shown NP-hard or PSPACE-hard or EXPTIME-hard by Stephenson et al. [SRG20].
8. SKOLEM'S PROBLEM. Given $a_0, \ldots, a_{k-1} \in \mathbb{Z}$ and $u_0, \ldots, u_{k-1} \in \mathbb{Z}$, does the recurrence

$$u_n = a_{k-1}u_{n-1} + \cdots + a_0 u_{n-k}$$

ever produce a 0? (The problem has also been considered with rational coefficients and algebraic coefficients.)

Skolem, Mahler, and Lech independently showed that if a sequence of numbers is generated by a linear recurrence (with real coefficients), then, with a finite number of exceptions, the places where the sequence is 0 forms a regular repeating pattern. Hence the problem we are considering is called SKOLEM'S PROBLEM even though it is not know if Skolem asked it.

Halava et al. [HHHK05] and Ouakine and Worrell [OW12] have nice surveys of what is known. (See those surveys for references). For $k = 1$ and $k = 2$, Skolem's problem is decidable. This is folklore and fairly easy. For $k = 3$ and $k = 4$, Skolem's problem is decidable. These proofs are difficult. For all $k \geq 5$ the problem is open. Halvana et al. claimed to show that for $k = 5$ the problem is decidable; however, Ouakine and Worrell showed that there was a bug in the proof.

Blondel and Portier [BP02] have shown that SKOLEM'S PROBLEM is NP-hard.

15

Undecidable Problems

15.1 Introduction

All of the problems discussed in this book so far (and in later chapters) are ***decidable***. That is, they may take a long time to solve, but *some* algorithm solves them. Are there problems that are undecidable? Yes. The HALTING problem (defined below) is undecidable. More generally, Rice [Ric53] showed that any nontrivial problem about Turing machines is undecidable. For example,

$$\{M \mid \text{Turing Machine } M \text{ halts on all the primes}\}.$$

The problem above is about Turing machines and, hence, may be considered unnatural.

Chapter Summary

1. *We will discuss problems that are undecidable but are perhaps unnatural.*
2. *We will discuss natural problems that are undecidable.*

15.2 Basic Undecidable Problems

To show that a set A is undecidable, we need to already have some basic undecidable problem X and then show $X \leq_r A$ (the r stands for recursive, which means decidable). We present some of these basic undecidable problems.

HALTING
Instance: A Turing Machine M.
Question: Does M halt on 0?
Note: There are many equivalent formulations of HALTING that are all efficiently reducible to each other.

Note: This is the problem that one first proves is undecidable from first principles. Almost all proofs that a problem is undecidable use either HALTING or some other problem known to be undecidable. This is similar to how we view SAT except that we actually *know* that HALTING is undecidable, whereas we need to *assume* SAT is not in P.

POST CORRESPONDENCE PROBLEM (POST CORRESPONDENCE PROBLEM)
Instance: An finite alphabet Σ and two vectors of words over Σ, $\alpha = (\alpha_1, \ldots, \alpha_N)$, $\beta = (\beta_1, \ldots, \beta_N)$.
Question: Does there exist $1 \leq i_1, \ldots, i_k \leq N$ such that $\alpha_{i_1} \cdots \alpha_{i_k} = \beta_{i_1} \cdots \beta_{i_k}$?
Note: k might be much larger than n, so the naïve algorithm of trying all possibilities to see if one works will go on forever if the answer is no.

A ***counter machine*** [Min61, Min67, Kor96] is another model of computation, consisting of a small constant number of ***registers*** that store nonnegative integers, instructions to increment and decrement registers, and the ability to test whether a register is zero and perform different instructions based on that test. We will not define it formally.

COUNTER MACHINES
Instance: A counter machine M.
Question: Does M halt?

Theorem 15.1.

1. *(Post [Pos46]) HALTING $\leq_r$ POST CORRESPONDENCE PROBLEM with an efficient reduction.*
2. *If POST CORRESPONDENCE PROBLEM $\leq_r A$ with an efficient reduction then HALTING $\leq_r A$ with an efficient reduction. This will be important when we claim that a reduction from POST CORRESPONDENCE PROBLEM to a game problem will produce reasonably small, playable, initial settings for the game.*
3. *(Minsky [Min61, Min67], but probably also folklore) HALTING $\leq_r$ COUNTER MACHINES. However, the reduction is slow: To simulate n steps of a Turing machine takes $2^{\Theta(n)}$ steps of a counter machine.*

15.3 Conway's Game of LIFE

In Conway's game of *Life*, we have a grid of cells that are either ***living*** or ***dead***; refer to figure 15.1. A cell's ***neighbors*** are defined to be the eight cells that are horizontally,

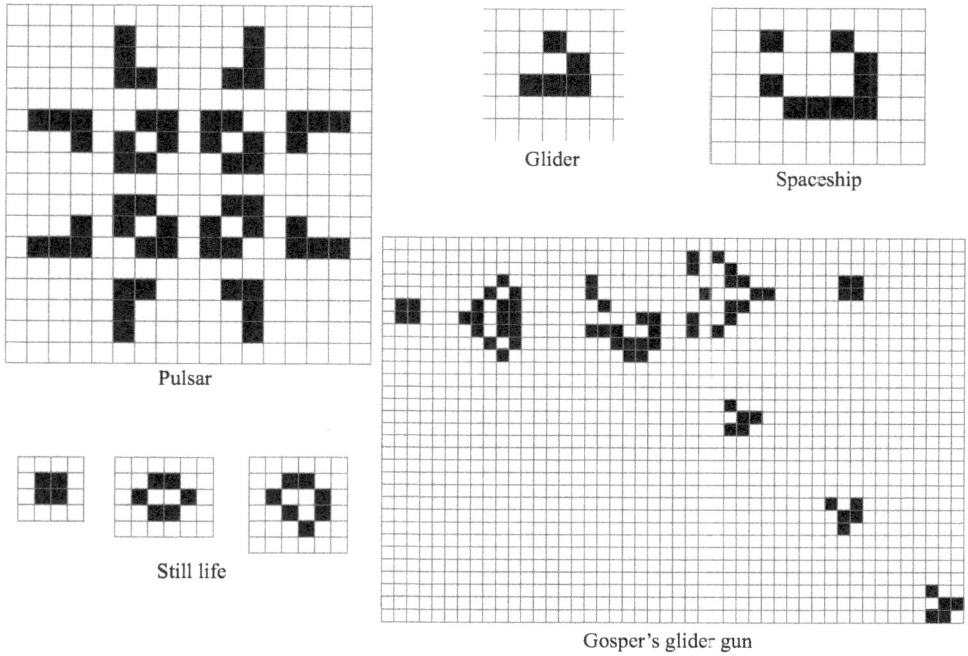

Figure 15.1
A couple of common patterns in *Life*. Black squares are living, and white squares are dead.

vertically, or diagonally adjacent to it. Each iteration of the simulation, the cells update as follows:

1. A living cell stays alive if it has two or three living neighbors. Otherwise it dies.
2. A dead cell becomes living if it has exactly three living neighbors.

> LIFE
> *Instance:* An initial configuration for the game of *Life*. This is a grid where some cells are blank and some have a living cell.
> *Question:* What will happen when the game is run? This is an informal question; however, we will state rigorous theorems.

With these rules, we can end up with periodic patterns, such as pulsars, or static patterns, known as ***still life***. Some of these can be seen in figure 15.1. For more examples of how various patterns behave, there are many interactive simulations online.

There are a couple of different decision problems we can ask. The first, whether it is an example of still life is easy; we just compute the next step. We can also ask whether the configuration is periodic and whether all the cells will die out, which are harder. These are unbounded simulations, and with polynomial space, they are PSPACE-complete, whereas with infinite space but a finite starting area for live starting area, they are undecidable.

1. Given a configuration, is it a still life? This is trivial: Just do one step of the game and see if anything changed.
2. Given a configuration and a box that it is inside of, so the game does not affect anything outside of that box, will it become periodic? Will all of the cells die? Rendell [Ren00b] showed that both of these are PSPACE-complete.
3. Given a configuration, will it become periodic? Will all of the cells die? Rendell [Ren00b, Ren11] showed that both of these questions are undecidable. The statement "The Game of Life is Undecidable" refers to this result.

These results are generally proven by showing that Conway's game of *Life* can simulate Turing machines. This works for both bounded games (e.g., inside a box) for PSPACE-hardness, and unbounded games for undecidability.

The game of *Life* is undecidable. Perhaps a lesson for us all.

15.4 RECURSED Video Game

Recursed is a 2D puzzle platform game involving chests, pink flames, green glows, crystals, ledges, jars, rings, and other objects. You start in a room. If you open a chest in that room, you do not get any object or treasure. Instead you are *in another room!* Jars also lead to rooms, but in a different way. Suffice to say, the game is complicated. We hasten to point out that *this is a fun game that people actually play*.

Recursed is a one-player game where the goal is for the player to get the crystal. RECURSED is the decision problem where you are given an initial setup for the game *Recursed* and need to determine whether the player can win.

Demaine, Kopinsky, and Lynch [DKL20] showed the following.

Theorem 15.2. *RECURSED is undecidable.*

Some notes about the result and the proof:

1. Theorem 15.2 was proven by showing POST CORRESPONDENCE PROBLEM $\leq_r$ RECURSED and using theorem 15.1. Since the basic problem used is POST CORRESPONDENCE PROBLEM, and the reduction is efficient, the instances of RECURSED that are produced are fairly small.
2. Most complexity-of-games results such as those about CHESS and GO rely on (1) the *board* getting bigger and bigger, and (2) such large constants that, even for short inputs,

the resulting games are not playable. The result for recurse is novel in that (1) the board stays the same size at 15×20, and (2) the games for short inputs are playable. A caveat: The number of recursions from the chests gets bigger and bigger.

15.5 Other Undecidable Games

1. The puzzle video game *Braid* is playable and fun. Hamilton [Ham14] showed that determining whether the player can win a given *finite* level of *Braid* is undecidable. Hamilton's reduction is from COUNTER MACHINES, by simulating a counter by a stack of enemies at the same location. As a result, the resulting simulation of Turing machines is slow.
2. The puzzle video game *Baba Is You* is undecidable [AH24], even when restricted to 8×17 levels. Again the reduction is from COUNTER MACHINES, representing counters by arbitrarily large stacks of items at the same location.
3. The Nintendo video game series *New Super Mario Bros.* and *Super Mario Maker* are all undecidable [MIT24e], even when restricted to constant-size levels. Again the reduction is from COUNTER MACHINES, representing counters by arbitrarily large stacks of enemies at the same location.
4. *Magic: The Gathering* [UA93] (henceforth "*Magic*") is a popular card game. Note that *Magic* is two-player, as opposed to *Recursed* or *Braid*, which are one-player. We denote the problem of determining whether a particular player can win *Magic* by MAGIC. Churchill et al. [Chu12] showed that MAGIC is undecidable by using COUNTER MACHINES. As a result, the resulting simulations of Turing machines were slow. Later Churchill et al. [CBH21] presented a reduction using Turing machines that was efficient. Yin and Churchill [YC24] developed even more practical Turing simulations. Biderman [Bid20] showed that the problem of mate-in-n for *Magic* is not arithmetic, which means that it is beyond the arithmetic hierarchy (much harder than decidable—see section 0.19).

15.6 Diophantine Equations: HILBERT10

In 1900 David Hilbert proposed 23 problems for mathematicians to work on. We state Hilbert's tenth problem in today's terminology: Is the following problem decidable?

> HILBERT10
> *Instance:* A polynomial $p \in \mathbb{Z}[x_1, \ldots, x_n]$.
> *Question:* Does there exist $a_1, \ldots, a_n \in \mathbb{Z}$ such that $p(a_1, \ldots, a_n) = 0$?

Note: We denote the problem where the degree $\leq d$ and the number of variables is $\leq n$ by HILBERT10(d, n).

Note: The question of decidability is equivalent to the case where $a_1, \ldots, a_n \in \mathbb{N}$. If you compare the results we state to those in the literature, they might not be the same since the literature often states results for the $\mathbb{N}$ version.

Hilbert had hoped this problem would lead to number theory of interest. It did lead to some, but the combined efforts of Davis, Putnam, and Robinson [DPR61] and Matijasevic [Mat70] (see also a survey article by Davis [Dav73] and a book by Matijasevic [Mat93]) showed that the problem is undecidable. Gasarch [Gas21] has a survey about what happens for particular degrees d and number of variables n.

15.7 Mortal Matrices

We present another undecidable problem that does not refer to Turing machines.

MORTAL MATRICES

Instance: A set of m $n \times n$ matrices $M_1, \ldots, M_m$ over $\mathbb{Z}$.

Question: Do there exist $i_1, \ldots, i_N$ such that $M_{i_1} \times \cdots \times M_{i_N}$ is the all zero matrix? Note that we are allowed to use M_i many times.

Theorem 15.3.

1. *(Cassaigne et al. [CHHN14])* MORTAL MATRICES *is undecidable for (1) six* 3×3 *matrices, (2) four* 5×5 *matrices, (3) three* 9×9 *matrices, (4) two* 15×15 *matrices.*
2. *(Bournez and Branicky [BB02])* MORTAL MATRICES *is decidable for two* 2×2 *matrices.*
3. *(Bell et al. [BHP12])* MORTAL MATRICES *for* 2×2 *matrices is NP-hard.*

Note that the question of decidability for two 3×3 matrices is open.

15.8 Further Results

Poonen [Poo14] has collected many other undecidable problems that do not involve Turing machines. We list a few of those that do not involve too much background knowledge.

1. WORD PROBLEM FOR GROUPS. Given a group, by being given a finite set of generators and relations, and given two words in the group w_1, w_2, determine whether $w_1 = w_2$.

Novikov [Nov55] and Boone [Boo58] independently showed that there is a group G so that the word problem for that group is undecidable.

2. Group isomorphism problem. Given two groups, by being given a finite set of generators and relations, determine whether they are isomorphic. Adian [Adi55] and Rabin [Rab58] independently showed that it is undecidable to determine whether a given group is the trivial group.
3. Wang tiling. Wang [Wan60] posed the following problem: Given square 1×1 tiles where the edges are colored, is there a tiling of the plane such that whenever two squares touch, the edge they share has the same color. The tiles cannot be rotated. Berger [Ber66] showed that the problem is undecidable. This result implies that tiling the plane with a given set of polygons is undecidable [Gol70], using a reduction similar to the jigsaw reduction in section 5.5.3. This result has been improved to require just three polygons [DL25].
4. Context-free grammar equivalence. Given two context-free languages L_1 and L_2, do they generate the same language? This was shown undecidable by Greibach [Gre68]. See also Hopcroft [Hop69].
5. Airport travel. de Marcken [dM21] formulated a problem about planning a trip that is undecidable.

16

Constraint Logic

16.1 Introduction

Constraint logic is a technique that has been used to prove problems NP-hard, PSPACE-hard, EXPTIME-hard, and Undecidable. This chapter will give just a taste of this vast area. The interested reader should see the book by Hearn and Demaine [HD09] and/or the PhD thesis of Hearn [Hea06].

In general, constraint logic aims to define one model of computation for which natural problems are complete for the natural complexity class in a variety of scenarios:

1. The *number of players* influencing the computation can vary, from ***0 players*** (a standard computation or a simulation like LIFE) to ***1 player*** (a nondeterministic computation or a puzzle with choices) to ***2 players*** (a perfect-information game where players alternate choices) to ***team games*** (with three or more players divided into two teams, and imperfect information between players).
2. The *length* of the computation can vary from ***bounded***, where the number of operations/moves is at most polynomial in the size of the computation, and ***unbounded***, where there is no a priori bound so the computation may go on for exponentially long.

Table 14.1 summarizes the natural complexity classes for each combination of these parameters. Constraint logic, interpreted appropriately for each setting (as described below), gives a complete problem for each.

16.2 Constraint Logic Graphs

In each case, constraint logic defines computation in terms of a directed weighted graph representing the state of a machine:

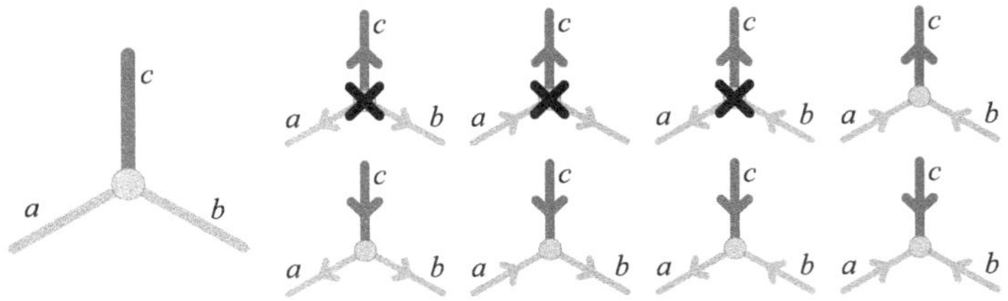

(a) AND vertex: the top (blue) edge can be directed out if and only if the left *and* right (red) edges are directed in. We put an *X* on those that are not valid.

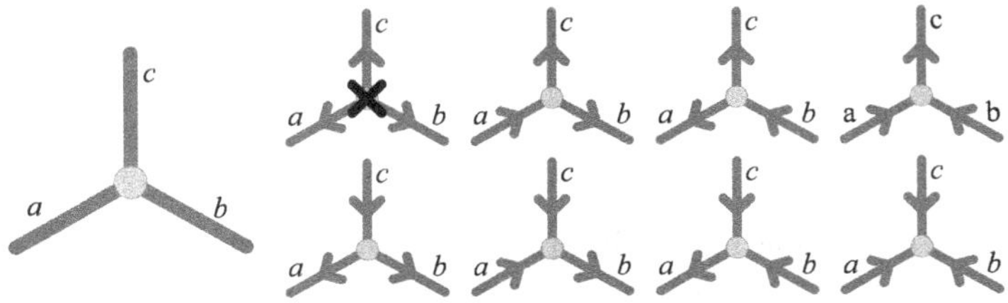

(b) OR vertex: the top (blue) edge can be directed out if and only if the left *or* right (blue) edge is directed in. We put an *X* on those that are not valid.

Figure 16.1
The two main types of constraint logic vertices, and all eight possible configurations, with illegal configurations labeled with an ×. (Based on [HD09, figure 2.1].)

Definition 16.1.

1. A ***constraint graph*** is an undirected weighted graph G where every weight is either 1 ("red," drawn thinner) or 2 ("blue," drawn thicker). (The constraint graph is the "machine.")
2. A ***configuration*** is a constraint graph together with an orientation (direction) for each edge. (A configuration represents a state of the machine.)
3. A ***valid configuration*** satisfies the ***constraints*** that every vertex has incoming weight at least 2.

The key is the constraints that define "valid configuration." How do these constraints represent computation? Figure 16.1 gives two examples of vertices that represent (in a certain sense) AND and OR constraints. The top edge is constrained to direct out only when we have AND (both) or OR (at least one) of the bottom edges directed in. Intuitively, this makes it possible to express Boolean logic, and reduce from various versions of SAT.

16.3 Constraint Graph Satisfaction

We start with the simplest form of constraint logic, corresponding to one-player bounded computation. This problem is just a problem in graph theory. It is "bounded" in the sense that each edge orientation can be chosen only once.

> Constraint Graph Satisfaction (CGS)
> *Instance:* A constraint graph G.
> *Question:* Is there a valid configuration for G?
> *Note:* CGS can be considered a one-player game: the player is given the constraint graph and tries to find a way to orient the edges to obtain a valid configuration.

Hearn and Demaine [HD09] proved the following:

Theorem 16.2.

1. *3SAT* $\leq_p$ *CGS, hence CGS is NP-complete.*
2. *CGS restricted to planar graphs is NP-complete.*
3. *CGS restricted to planar graphs where every vertex is an* AND *or an* OR *according to figure 16.1 is NP-complete.*
4. *CGS restricted to grid graphs is NP-complete.*

CGS has not been applied to prove natural problems are NP-complete. Nonetheless, theorem 16.2 is a good starting point for the study of constraint logic.

16.4 Nondeterministic Constraint Logic

Next we consider the problem of, given two valid configurations, whether you can move from one to the other. This is essentially a one-player unbounded game. It is "unbounded" because each edge can flip (exponentially) many times.

> Nondeterministic Constraint Logic (NCL)
> *Instance:* A constraint graph G and two valid configurations C_1 and C_2 of G.
> *Question:* Is there a sequence of valid configurations that (1) begins with C_1, (2) ends with C_2, and (3) every adjacent pair differ in that one edge's orientation was flipped?
> *Note:* The term "nondeterministic" is used because an NSPACE algorithm can solve the problem by guessing the next valid configuration. Because PSPACE = NSPACE, the problem is in PSPACE.

Hearn and Demaine [HD09] proved the following:

Theorem 16.3.

1. *NCL is PSPACE-complete.*
2. *NCL restricted to planar graphs is PSPACE-complete.*
3. *NCL restricted to planar graphs where every vertex is an* AND *or an* OR *according to figure 16.1 is PSPACE-complete.*
4. *NCL restricted to grid graphs is NP-complete.*

The following two puzzles were proved hard by a reduction from NCL.[1]

> SLIDING BLOCKS
> The sliding block game is a puzzle game (both physical and digital) played on a square grid containing rectangular blocks. The player can slide each block horizontally and vertically so long as it does not collide with other blocks.
> *Instance:* An initial position for the sliding-block game, and a special block and position for that block.
> *Question:* Can the player win by moving the special block to the specified position?

> RUSH HOUR
> *Rush Hour* is a puzzle game (both physical and digital) played on a square grid containing horizontal $1 \times k$ and vertical $k \times 1$ cars (for varying k), and the player can slide each car along its long axis so long as it does not collide with other cars.
> *Instance:* An initial position for the game *Rush Hour* and a special car and position for that car.
> *Question:* Can the player win by moving the special car to a specified position?

Theorem 16.4.

1. *(Hearn and Demaine [HD09])* SLIDING BLOCKS *is PSPACE-complete. The case where all of the rectangles are* 1×2 *is still PSPACE-complete.*
2. *(Flake and Baum [FB02], Hearn and Demaine [HD09])* RUSH HOUR *is PSPACE-complete, even when restricted to cars of length* 2 *and* 3.

1. Flake and Baum's RUSH HOUR hardness proof [FB02] predates and provided inspiration for NCL.

3. *(Tromp and Cilibrasi [TC05]) Rush Hour is PSPACE-complete, even when restricted to cars of length* 2.
4. *(Brunner et al. [BCD+21]) Rush Hour with cars of length* 1 *(which now must be explicitly specified as either horizontal or vertical) and fixed blocks that cannot move at all is PSPACE-complete.*

Proof sketch. (2) Figure 16.2 gives a sketch of the proof of theorem 16.4.2 for 1×2 and 1×3 blocks. By theorem 16.3.3, we just need to build an AND vertex and an OR vertex of a constraint graph, and show how to connect them into an arbitrary planar graph. Figure 16.2 gives two Sliding Blocks gadgets that implement AND and OR vertices, and shows some sample move sequences for how to reverse edge directions by sliding blocks around. Here a block being retracted inside the gadget represents an NCL edge directed *away* from the vertex, and a block extending outside the gadget represents an NCL edge directed *into* the vertex (the opposite of what might seem natural). The naturally limited capacity of each gadget to store blocks ends up implementing the "incoming weight of at least 2" constraint. These gadgets can then be put together into a grid, following the grid graph of theorem 16.3.4. For grid graphs we need a "straight" and "turn" gadget that simply propagate an edge in a desired direction; these gadgets can be constructed from an OR gadget by closing off one side. When the gadgets are against the outer boundary, we also need to extend the corner blocks to fill the empty corners in figure 16.2; otherwise, they will be filled by the adjacent gadget. See [HD09] for details, and for the more complicated 1×2 construction.

□

Van der Zanden [vdZ15] strengthened these theorems to show NCL remains PSPACE-complete when the graph has ***bounded bandwidth***—that is, the vertices can be placed on the integer line so that every edge has length at most c for some constant c. For example, this graph class is smaller than graphs of bounded pathwidth or graphs of bounded treewidth. This result often lets us prove PSPACE-completeness of games and puzzles when restricted to a narrow $O(1) \times n$ board.

Theorem 16.5.

1. *NCL restricted to planar bounded-bandwidth graphs where every vertex is an* AND *or an* OR *according to figure 16.1 is PSPACE-complete.*
2. *Rush Hour and Sliding Blocks restricted to* $O(1) \times n$ *boards are PSPACE-complete.*

Next we describe some "reconfiguration" problems that were proved hard by a reduction from NCL. In general, the ***reconfiguration*** version of an NP problem is, given two certificates, to determine whether you can change one into the other by a sequence of "local" moves. The natural notion of locality depends on the problem.

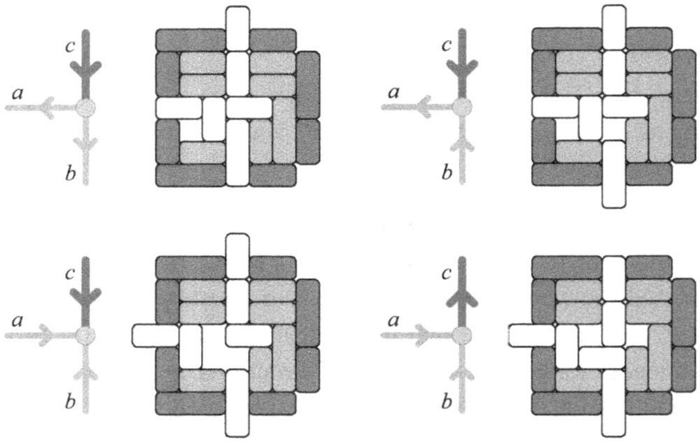

(a) AND gadget: The top middle block can be retracted into the gadget if and only if the left middle block *and* the bottom middle block can be extended out of the gadget (i.e., retracted into the neighboring gadget).

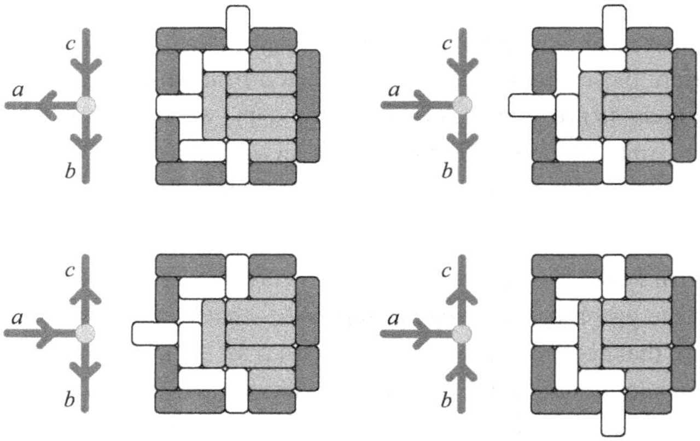

(b) OR gadget: The top middle block can be retracted into the gadget if and only if the left middle block OR the bottom middle block can be extended out of the gadget (i.e., retracted into the neighboring gadget).

Figure 16.2
The reduction from NCL to SLIDING BLOCKS consists of just two gadgets (plus an argument about how they fit together, omitted here). Dark blocks are never useful to slide. (Based on [HD09, figure 1.2].)

Reconfiguration SAT
Instance: A Boolean formula $\varphi(x_1, \ldots, x_n)$ and two satisfying assignments $\vec{x}, \vec{y}$ for φ.
Question: Is there a sequence of satisfying assignments for φ that (1) begins with $\vec{x}$, (2) ends with $\vec{y}$, and (3) every two consecutive satisfying assignments differ in only one variable?

Reconfiguration NAE 3SAT
Instance: NAE 3SAT formula $\varphi(x_1, \ldots, x_n)$ and two satisfying assignments $\vec{x}, \vec{y}$ for φ.
Question: Is there a sequence of satisfying assignments for φ that (1) begins with $\vec{x}$, (2) ends with $\vec{y}$, and (3) every two consecutive satisfying assignments differ in only one variable?
Note: Here a "satisfying assignment" for φ means that, in each clause, not all literals have the same assignment.

Theorem 16.6.

1. *(Gopalan et al. [GKMP09]) Reconfiguration SAT is PSPACE-complete.*
2. *(Gopalan et al. [GKMP09]) Reconfiguration NAE 3SAT is PSPACE-complete.*
3. *(Cardinal et al. [CDE+20]) Reconfiguration NAE 3SAT restricted to planar formulas is PSPACE-complete.*

Moret [Mor88] showed that Planar NAE 3SAT is in P, so it is surprising that Reconfiguration Planar NAE 3SATis PSPACE-complete.

Holzer and Jakobi [HJ12] show the following maze puzzles hard using a reduction from NCL. (The reader needs to look at their paper for formal definitions of the mazes.)

Theorem 16.7.

1. *Determining whether a a Rolling Block Maze can be solved is PSPACE-complete.*
2. *Determining whether an Alice Maze can be solved is PSPACE-complete.*

For more games and puzzles proved PSPACE-complete using NCL, see Hearn [Hea06], Hearn and Demaine [HD09], Hearn [Hea09], and Holzer and Jakobi [HJ12].

16.5 Deterministic Constraint Logic

Next we look at a 0-player version of constraint logic, which corresponds to a more typical computation. To make it zero player, we add rules to constraint logic to ensure that all moves are deterministic, rather than chosen by a player. Aside from orientation of edges, we also track whether an edge is "active" or "inactive." An edge is ***active*** if it was just flipped in the last round. Otherwise, it is ***inactive***. We call a *vertex* ***active*** if its active incoming edges have a total weight of at least 2.

> DETERMINISTIC CONSTRAINT LOGIC (DCL)
> *Instance:* A constraint graph G and an edge e.
> *Question:* Consider a sequence of rounds, where in each round the following things happen:
>
> - Inactive edges pointing to active vertices get reversed.
> - Active edges pointing to inactive vertices get reversed.
> - The edges that have been reversed are the new active edges.
>
> Does edge e ever get reversed?

Demaine et al. [DHHL22] proved.[2]

Theorem 16.8.

1. *DCL is PSPACE-complete.*
2. *DCL restricted to planar graphs is PSPACE-complete.*

Demaine et al. [DHHL22] use theorem 16.8 (or rather, a framework used to prove theorem 16.8) to prove PSPACE-completeness for predicting the behavior of several reversible deterministic systems. We give one example.

> BILLIARDS
> *Instance:* A set of billiard balls on a table. The table also has fixed blocks that, when a ball hits it, reflect off at the same angle. We are also given, for each ball, the direction the ball will initially roll and the ball's initial velocity. Finally, we are given a ball and a place on the table.
> *Question:* Will the specified ball ever get to the specified place?

2. This result was claimed in Hearn and Demaine [HD09]; however, the proof had a very subtle flaw. The flaw is discussed, and a correct proof is given by Demaine et al. [DHHL22].

Theorem 16.9.

1. *(Fredkin and Toffoli [FT82])* BILLIARDS *is PSPACE-complete.*
2. *(Demaine et al. [DHHL22]) DCL $\leq_p$* BILLIARDS*, hence* BILLIARDS *is PSPACE-complete. This proof (1) is simpler than that of Fredkin and Toffoli and (2) has only two balls moving at any one time and they are close together.*

16.6 Two-Player Constraint Logic

We have looked at zero-player and one-player Constraint Logic. Next we turn to two-player constraint logic (2CL). We define two games corresponding to bounded and unbounded length:

Definition 16.10.

1. A ***two-player constraint graph*** is a constraint graph where each edge is labeled "white" or "black," in addition to being labeled "red" (weight 1) or "blue" (weight 2). The two colorings are independent of each other.
2. The ***bounded/unbounded two-player constraint logic game*** is as follows:
   a. Initially the two players, called "white" and "black," are given a two-player constraint graph and a valid configuration. Each player is also given a target edge.
   b. The players alternate making moves, with white going first. A move consists of flipping the orientation of an edge of the same color as the player, such that the resulting configuration is still valid.
   c. In the *bounded* game, each edge can be reversed at most once.
   d. The first player to reverse their target edge wins.

> BOUNDED/UNBOUNDED TWO-PLAYER CONSTRAINT LOGIC (BOUNDED/UNBOUNDED 2CL)
> *Instance:* Two-player constraint graph, valid orientation, and two target edges (so a starting board for the game).
> *Question:* Which player wins the bounded/unbounded 2-player constraint logic game?

Hearn and Demaine [HD09] proved the following:

Theorem 16.11.

1. *Bounded 2CL is PSPACE-complete.*
2. *Bounded 2CL restricted to planar graphs is PSPACE-complete.*
3. *Unbounded 2CL is EXPTIME-complete.*
4. *Unbounded 2CL restricted to planar graphs is EXPTIME-complete.*

We describe three games that have reductions from Bounded 2CL:

Definition 16.12.

1. ***Amazons*** is played by two players—black and white—with an equal number of black and white queens on a chess board. Each turn, a player must make a "queen move" with one of his queens, and then shoot an arrow onto any square reachable by a "queen move" from the new position of the queen. Arrows do not kill or even injure; however, they are used to restrict how the queens (of either side) can move. Here, a queen move is any nonzero move in a straight line diagonally, horizontally, or vertically. Queens may not move through or shoot through squares occupied by arrows or other queens. The player who is able to move last wins. We denote the problem of determining who wins by Amazons.
2. ***Konane*** is played with two colors of stones—black and white—on a board. Each turn, a player can perform with a single piece, one or more jumps over opponent pieces, as long as they all lie in a straight line. The jumped pieces are removed. The last player to move wins. We denote the problem of determining who wins by Konane.
3. ***Cross Purposes*** is played with black and white stones on the intersection of a Go board. A black stone represents towers of two blocks, and a move consists of "pushing" a black stone over, resulting in two white stones either above, below, left, or right of the stone. The two players are named "Vertical" and "Horizontal," with Vertical moving first. Vertical may only tip over a stone up or down, and Horizontal may only tip over the stone left or right, so long as the two spaces are unoccupied before hand. The last player to be able to move wins. We denote the problem of determining who wins by Cross Purposes.

It is easy to see that Amazons, Konane, and Cross Purposes are in PSPACE. Hearn [Hea09] proved the following theorem by a reduction from 2CL.

Theorem 16.13. *Amazons, Konane, and Cross Purposes are PSPACE-complete.*

Figures 16.3 and 16.4 are the gadgets used to show Konane is PSPACE-complete.

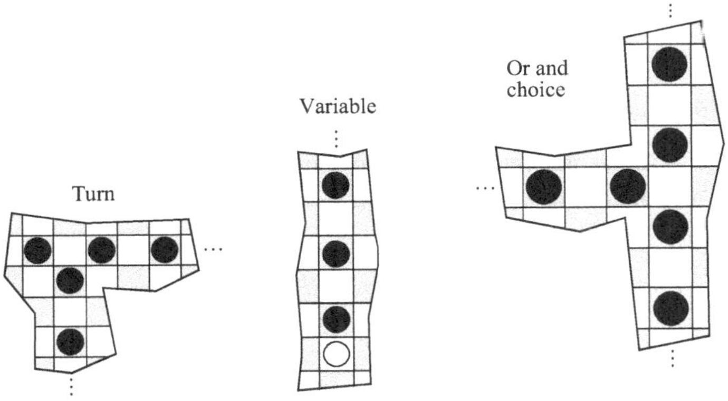

Figure 16.3
Gadgets to prove *Konane* PSPACE-complete.

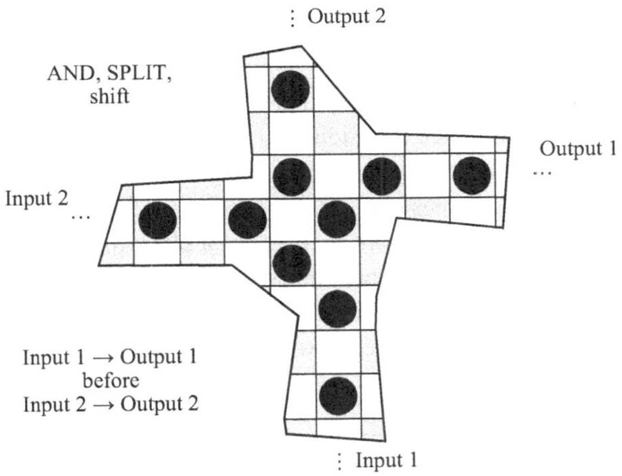

Figure 16.4
More gadgets to prove KONANE PSPACE-complete.

Bilò et al. [BGL+18] show the following using a reduction from 2CL (the reader needs to look at their paper for formal definitions of the game).

Theorem 16.14. *Peg Duotaire (a two-player peg jumping game) is PSPACE-complete.*

Unbounded 2CL has not yet been used to show natural problems EXPTIME-complete; however, it is part of a framework for lower bounds developed by Demaine et al. [DHL20] and Ani et al. [ADHL22].

Project 16.15. *Prove Go, Chess, or Checkers EXPTIME-complete using Unbounded 2CL.*

16.7 Team Private Constraint Logic

Definition 16.16. The ***bounded/unbounded team private constraint logic game*** is a version of the constraint logic game played with three players—one black player and two white players—on a ***team private constraint graph***. Like a two-player constraint graph, every edge is marked as flippable only by the white team or only by the black team. A new feature is that the state of the edges in the game is semi-private information: Each edge is marked as visible to (only) the white players, one of the white players, or the black player.

The same moves allowed in 2CL games apply, including in the bounded case a limit of one reversal per edge. Players face the additional restriction that all moves that they make must be legal given visible information. Passing is also allowed, which makes it difficult to gain any information about changes to the state of the board based on the possible moves of other players. This is a necessary condition, because a black player may move an invisible edge, which may not change the available moves for a white player, thus giving them some knowledge about the black player's moves. By allowing passing, the white player does not know whether the state of an invisible edge was changed, or if the black player decided to pass on a turn.

> Bounded/Unbounded Team Private Constraint Logic (BTPCL/UTPCL)
> *Instance:* A team private constraint graph.
> *Question:* Which player wins the bounded/unbounded team private constraint logic game?

Hearn and Demaine [HD09] proved the following:

Theorem 16.17. *BTPCL $\in$ NEXPTIME.*

Proof. If there exists a winning strategy for the white player, the strategy is deterministic and is a function of the visible state. There are at most n different visible edges, so the number of possible states is exponential in the number of visible edges. Thus the white players may guess a strategy at the beginning of the game, which takes exponentially many bits, and the white players play deterministically with that strategy to verify that it is a solution. The game may run for many rounds, but the information used to describe the moves is exponential, hence determining who wins is in NEXPTIME. □

Theorem 16.18. *BTPCL is NEXPTIME-complete.*

Theorem 16.19. *UTPCL is RE-complete and in particular Undecidable.*

The last result provides a game with a finite board (the constraint graph) where the question of who wins is Undecidable. This result is surprising because Turing machines or word-RAM algorithms can run for an arbitrary amount of time and have an arbitrary amount of space to store information and their state, whereas the game has finite board space and cannot represent the state of an arbitrarily large computation. Effectively, the state of the simulated machine is in the heads of the players as they consider what move would be optimal.

III

Below NP

17

3SUM Conjecture: A Method for Obtaining Quadratic Lower Bounds

17.1 Introduction

Part I of this book is about showing that problems are probably not solvable in polynomial time. But even within polynomial time, there are distinctions. We will show that certain problems are probably not in subquadratic time.

Chapter Summary

1. *We discuss the problem 3SUM and the conjecture that it cannot be solved in subquadratic time.*
2. *We define an appropriate notion of reduction.*
3. *We use the assumption that 3SUM cannot be solved in subquadratic time, and the reductions in the last item, to show that many problems cannot be solved in subquadratic time.*
4. *We look at the consequences of assuming that another problem, ORTHOGONAL VECTORS, cannot be done in subquadratic time.*

17.2 3SUM Problem

Gajentaan and Overmars [GO12] used the following problem to show that other problems are probably not in subquadratic time.

> 3SUM
> *Instance:* n integers.
> *Question:* Do three of the integers sum to zero?
> *Note:* We consider any arithmetic operation to be unit cost.

The following theorem gives better and better algorithms for 3SUM.

Theorem 17.1.

1. *3SUM can be solved in $O(n^3)$ time.*
2. *3SUM can be solved in $O(n^2 \log n)$ time.*
3. *3SUM can be solved in randomized $O(n^2)$ time.*
4. *3SUM can be solved in deterministic $O(n^2)$ time.*

Proof. Let A be the original input of n integers.

1) The trivial algorithms suffice to get $O(n^3)$ time: Check all $O(n^3)$ 3-sets of A to see if any of them sum to zero.

2) First compute all the pairwise sums of A and sort them into an array B. This takes $O(n^2 \log n)$ steps. Then, for each element of A, use binary search to see if its negation is in B. This takes $O(n \log n)$ steps. If you find a negation in B, then the answer is YES; otherwise NO.

3) Let p be a prime close to n^2. We will have a hash table of size p. The number z will go into cell $z \pmod p$ of the hash table.

For all $1 \le i < j \le n$, put $-(x_i + x_j)$ (along with (x_i, x_j)) into the hash table. This takes $O(n^2)$ steps. Then, for each element of $x \in A$, hash it into the table. See if there is at least one pair already there. If there is, then with high probability there are $O(1)$ pairs there. See if any of the sums in that entry of the hash table, sum with x to 0. If so, then output YES and halt. If for no $x \in A$ you get a YES, then output NO. For each x, with high probability every x will be involved with $O(1)$ checks, so the expected run time is $O(n^2)$.

4) First sort A. This takes $O(n \log n)$ steps. Place a pointer at both the front and the end of A. Then, for each $x \in A$, do the following: If the sum of the integers at the two pointers and x is smaller than zero, then move the first array's pointer forward; if the sum is larger than zero, then move the second array's point backward; otherwise, we have the three integers sum to zero, and we output YES and are done. If the two pointers crossover, then move onto the next integer in A. This algorithm clearly takes $O(n^2)$ time. □

Exercise 17.2. Code up all four algorithms in theorem 17.1. Run them on data and see which ones do well when.

Is there an algorithm for 3SUM that runs in time better than $O(n^2)$? This depends on your definition of "better." The following are known:

1. If the integers are in $[-u, u]$, then 3SUM can be solved in $O(n + u \log n)$ time. We leave this as an exercise.
2. Baran et al. [BDP08] showed the following: Assume the word-RAM model, which can manipulate $\log n$-bit words in constant time. Then there is a randomized algorithm for

3SUM that takes time

$$O\left(\frac{n^2(\log\log n)^2}{(\log n)^2}\right).$$

3. Gronlund and Pettie [GP18] have shown that there is a randomized algorithm for 3SUM that takes time

$$O\left(\frac{n^2\log\log n}{\log n}\right)$$

   and a deterministic algorithm that takes time

$$O\left(\frac{n^2(\log\log n)^{2/3}}{(\log n)^{2/3}}\right).$$

4. Chan [Cha20] has shown there is a deterministic algorithm for 3SUM that runs in time

$$O\left(\frac{n^2(\log\log n)^{O(1)}}{\log^2(n)}\right).$$

5. Gronlund and Pettie [GP18] have also shown that *there exists* a decision tree algorithm that had depth (so time) $O(n^{1.5}\sqrt{\log n})$ for 3SUM. Their proof does not show how to actually construct the decision tree in subquadratic time.

Exercise 17.3. Show that if the integers are in $[-u, u]$, then 3SUM can be solved in $O(n + u\log n)$ time.

While the algorithms above are impressive and very clever, none are that much better than $O(n^2)$. We need a definition for "much better than $O(n^2)$."

Definition 17.4. An algorithm is ***subquadratic*** if there exists $\varepsilon > 0$ such that it runs in time $O(n^{2-\varepsilon})$.

Despite enormous effort, nobody has obtained a subquadratic algorithm for 3SUM. In the next section we turn this around: We state a conjecture that 3SUM cannot be done in subquadratic time. From that conjecture we obtain quadratic lower bounds on other problems.

17.3 3SUM CONJECTURE

Gajentaan and Overmars [GO12] (essentially) defined 3SUM-hardness and showed that many problems are 3SUM-hard. Many of the problems are in computational geometry.

Conjecture 17.5. *3SUM CONJECTURE: There is no subquadratic algorithm for 3SUM.*

We will later see ways in which this conjecture is similar to the conjecture that SAT is not in P and ways in which it is different.

We define a notion of reduction between problems.

Definition 17.6. Let A and B be sets or functions (they will almost always be sets).

1. $A \leq_{sq} B$ means that if there is a subquadratic algorithm for B, then it can be used to obtain a subquadratic algorithm for A. The sq stands for ***subquadratic***.
   a. (The usual way to do this.) On input x produce in $O(n\text{polylog}n)$ time (usually just linear) a y such that $x \in A$ if and only if $y \in B$. Note that $\leq_{sq}$ is transitive.
   b. (This is sometimes needed.) On input x produce in $O(n\text{polylog}n)$ time a set $y_1, \ldots, y_k$ (k is independent of n) such that $x \in A$ can be determined from the answers to $y_1 \in B$?, $\ldots$, $y_k \in B$? in $O(n\text{polylog}n)$ time. Note that $\leq_{sq}$ is still transitive.
   c. We leave it to the reader to modify the above definitions for when A and B are functions.
2. $A \equiv_{sq} B$ if $A \leq_{sq} B$ and $B \leq_{sq} A$.

Definition 17.7. Let A be a problem.

1. A is 3SUM-hard if 3SUM $\leq_{sq} A$.
2. A is 3SUM-complete if A is 3SUM-hard and $A \leq_{sq}$ 3SUM.

Because of conjecture 17.5, we think that if A is 3SUM-hard, then there is no subquadratic algorithm for it. In brief:

1. When you read "A is 3SUM-complete," you should think: A is in quadratic time but not in subquadratic time.
2. When you read "A is 3SUM-hard," you should think: A is not in subquadratic time.

By the 3SUM CONJECTURE we think that, if A is 3SUM-hard, then there is no subquadratic algorithm for A.

The definition of NP-hard is used to show that problems are not in P, contingent on the conjecture that P $\neq$ NP. The definition of 3SUM-hard is used to show that problems do not have subquadratic algorithms, contingent on the conjecture that 3SUM does not. We list out similarities and differences between the two theories.

1. Both use reductions and build up a large set of problems that are thought to be hard. The number of NP-hard problems is far larger than the number of 3SUM-hard problems.
2. Contrast the following:
   a. A problem B is NP-hard if for all $A \in$ NP, $A \leq_p B$. SAT is a natural NP-hard problem. As a consequence, one can show C is NP-hard by showing SAT $\leq_p C$.
   b. A problem B is 3SUM-hard if 3SUM $\leq_{sq} B$. Note that we *do not* have a result for 3SUM that is analogous to the Cook–Levin theorem. We suspect 3SUM requires quadratic time and use it as such.

If we did not know the Cook–Levin theorem, but really thought SAT was hard, we could still have a large set of problems that are NP-complete and think they were hard. That is the position we are in with 3SUM-hard.

Exercise 17.8. Let 2SUM be the problem of: Given an array A of integers, does there exist $x, y \in A$ such that $x + y = 0$?

1. Show that 2SUM is in $O(n \log n)$ time.
2. Show that 2SUM is in $O(n)$ time.
3. F3SUM is the following problem: Given a set of integers A, determine whether there exist $x, y, z \in A$ such that $x + y + z = 0$, and if there is, then output an (x, y, z) that works. Show that F3SUM $\leq_{sq}$ 3SUM.

17.4 Three Variants of 3SUM

Recall that if the integers are in $[-u, u]$, then 3SUM can be solved in time $O(n + u \log n)$. For $u = O(n^{2-\varepsilon})$ this yields a subquadratic algorithm for 3SUM. What if u is bigger? The hashing technique used in theorem 17.1.3 to obtain a randomized $O(n^2)$ algorithm for 3SUM can be used to prove the following.

Theorem 17.9. *The 3SUM problem restricted to* $[-n^3, n^3]$ *is 3SUM-complete.*

Proof sketch. Clearly the restricted problem reduces to 3SUM.

The proof that 3SUM reduces to the restricted problem uses a reduction that hashes the set of n elements to $[-n^3, n^3]$. □

Theorem 17.9 is interesting since it rules out the possibility that 3SUM is only hard when it involves large numbers.

Pătraşcu [Păt10] considered the following problem and proved it was 3SUM-hard.

CONVOLUTION 3SUM
Instance: A set of n integers $a_1, \ldots, a_n$.
Question: Is there $i \neq j$ such that $a_{i+j} = a_i + a_j$?
Note: Proving CONVOLUTION 3SUM is in $O(n^2)$ is much easier than showing 3SUM is in $O(n^2)$.

Theorem 17.10. *CONVOLUTION 3SUM is 3SUM-complete.*

Proof sketch. CONVOLUTION 3SUM is easily in $O(n^2)$, so CONVOLUTION 3SUM $\leq_{sq}$ 3SUM trivially.

The proof that 3SUM $\leq_{sq}$ CONVOLUTION 3SUM uses a family of hash functions. □

Gajentaan and Overmars [GO12] defined the following problem and showed it was 3SUM-hard. It is used in many proofs that problems are 3SUM-complete.

> 3SUM′
> *Instance:* Three sets A, B, C of n integers.
> *Question:* Is there $a \in A$, $b \in B$, and $c \in C$ such that $a+b=c$?

Theorem 17.11. *3SUM′ is 3SUM-complete.*

Proof. 3SUM $\leq_{sq}$ 3SUM′.

This is easy to see by a reduction from 3SUM with the original instance being S; just put $A=S$, $B=S$, $C=-S$.

3SUM′ $\leq_{sq}$ 3SUM.

Given A, B, C we set S to be all the elements of $A + large$, $B + 2 \times large$, and -$C - 3 \times large$, where *large* just means a large number that is added to each element of the corresponding sets. □

Note that the reduction 3SUM′ $\leq_{sq}$ 3SUM used large numbers.

17.5 3SUM-Hard Problems in Computational Geometry

All of the results in this section, are by Gajentaan and Overmars [GO12]. For each problem in this section try to prove that it is in $O(n^2)$ time, so the 3SUM-hardness will mean that the complexity is $\Theta(n^2)$ (assuming the 3SUM CONJECTURE). In some cases we will give a reference. The results of this section are summarized in figure 17.1.

17.5.1 GEOMBASE and GEOMBASE′

> GEOMBASE
> *Instance:* n points in $\mathbb{Z}^2$ with y-coordinate in $\{0, 1, 2\}$.
> *Question:* Does there exist a nonhorizontal line hitting three points of this set? (This is illustrated in figure 17.1.)

Theorem 17.12. *GEOMBASE is 3SUM-complete.*

Proof. 3SUM′ $\leq_{sq}$ GEOMBASE :

Given A, B, C, an input to 3SUM′, we consider the following input to GEOMBASE:

$$\{(a,0) \mid a \in A\} \cup \{(b,2) \mid b \in B\} \cup \{(c/2,1) \mid c \in C\}.$$

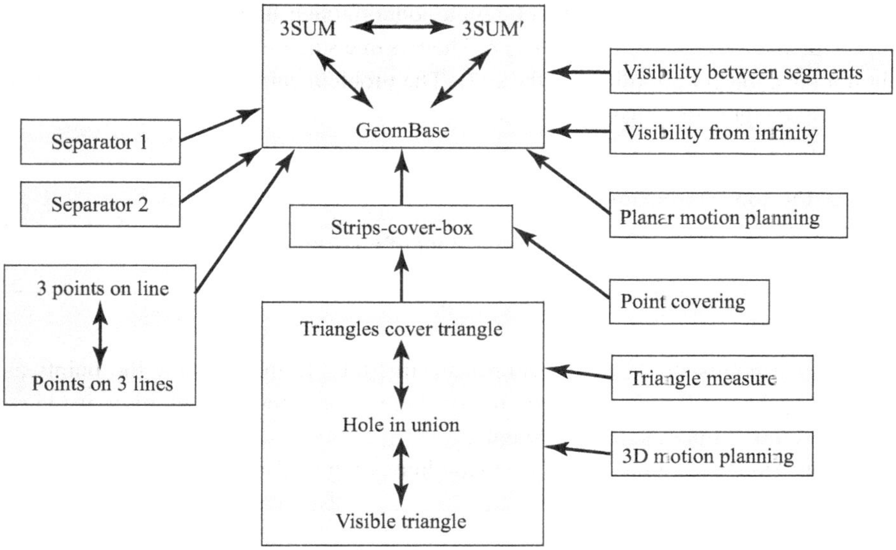

Figure 17.1
The relationships of problems to 3SUM.

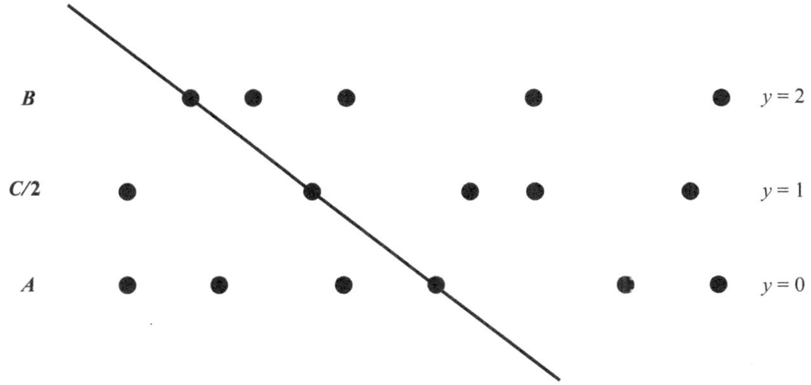

Figure 17.2
The GEOMBASE problem.

It is easy to show that three points lie on a nonhorizontal line if and only if there exist $a \in A$, $b \in B$, and $c \in C$, such that $a+b=c$. There is one small issue. The problem we create might not have integer coordinates (the $c/2$). The problem can easily be scaled to create a version with integer coordinates.

GEOMBASE $\leq_{sq}$ 3SUM′:
Reverse the above reduction. □

We will sometimes need the following variant of GEOMBASE.

> GEOMBASE′
> *Instance:* n points in $\mathbb{Z}^2$ with y-coordinate in $\{0,1,2\}$, and ε. View the points as holes in the $y=0,1,2$ lines and enlarge them to be ε-long. In addition, we view the $y=0,1,2$ lines as finite segments.
> *Question:* Does there exist a nonhorizontal line going through three of the holes?

Theorem 17.13. *GEOMBASE $\leq_{sq}$ GEOMBASE′*

Proof sketch. This proof is illustrated in figure 17.3. □

17.5.2 COLLINEAR

> COLLINEAR
> *Instance:* n points in $\mathbb{Z} \times \mathbb{Z}$.
> *Question:* Are three of the points collinear?

Theorem 17.14. *3SUM $\leq_{sq}$ COLLINEAR, hence COLLINEAR is 3SUM-hard.*

Proof. Given an instance of 3SUM A, we map it to the instance of COLLINEAR $\{(x, x^3) \mid x \in A\}$, as shown in figure 17.4.

We show that three points on the curve will lie on a line if and only if there are three integers summing to zero in the original set.

Indeed, notice that

$$\frac{b^3-a^3}{b-a} = \frac{c^3-a^3}{c-a} \iff b^2+ba+a^2 = c^2+ca+a^2$$
$$\iff (b-c)(b+c+a)=0 \iff b+c+a=0,$$

where in the last equality we assume that the three numbers are distinct. There is a slight issue here. We are assuming that the input to 3SUM are distinct integers. We leave it to the reader to show that this version of 3SUM is 3SUM-complete. □

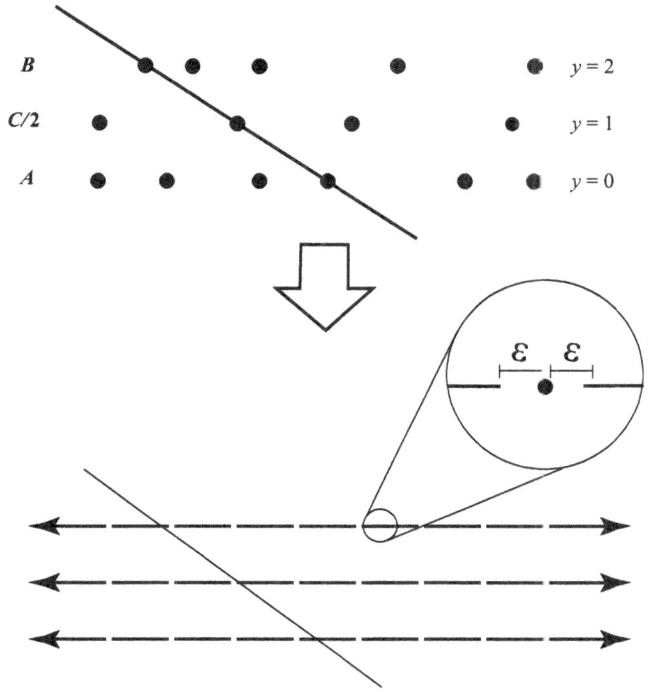

Figure 17.3
Reduction from GEOMBASE to GEOMBASE′.

17.5.3 CONCURRENT

Concurrency is the geometric dual of collinearity.

> CONCURRENT
> *Instance:* n lines.
> *Question:* Is there a point on three of the lines? (Such lines are called ***concurrent***.)

Theorem 17.15. *COLLINEAR $\equiv_{sq}$ CONCURRENT and hence CONCURRENT is 3SUM-hard.*

Proof. COLLINEAR $\leq_{sq}$ CONCURRENT

Given a set of points X, we map it to the set of lines $\{ax + by + 1 = 0 \mid (a, b) \in X\}$. (This is called ***projective plane duality***.) Then, this preserves point/line incidence; if three points were collinear, the three corresponding lines are incident, and vice versa. Therefore, we have a subquadratic reduction.

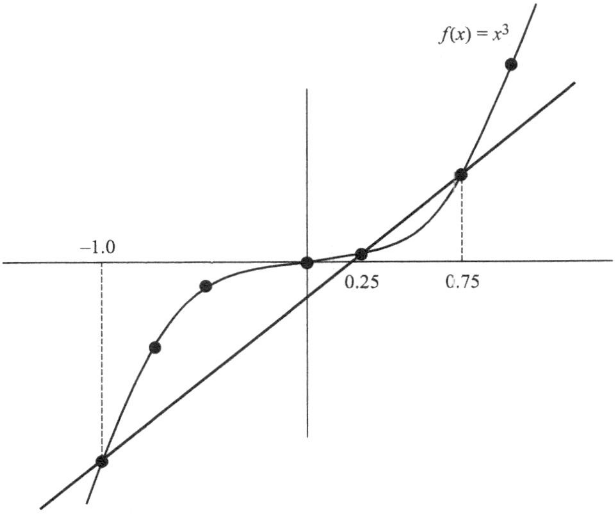

Figure 17.4
Reduction from 3SAT to COLLINEAR.

CONCURRENT $\leq_{sq}$ COLLINEAR
Use the reverse of the COLLINEAR $\leq_{sq}$ CONCURRENT reduction. □

Note: All the d dimensional versions of the problems mentioned so far, are $d+1$-sum hard.

17.5.4 SEPARATOR

> SEPARATOR
> *Instance:* n line segments in the plane.
> *Question:* Is there a line that separates the n line segments into two nonempty groups? (This line is not allowed to intersect any of the segments.)

Exercise 17.16. GEOMBASE′ $\leq_{sq}$ SEPARATOR, hence SEPARATOR is 3SUM-hard. (**Hint:** See figure 17.5.)

17.5.5 Covering Problems

Definition 17.17. A ***strip*** is just a fixed region in between two parallel lines.

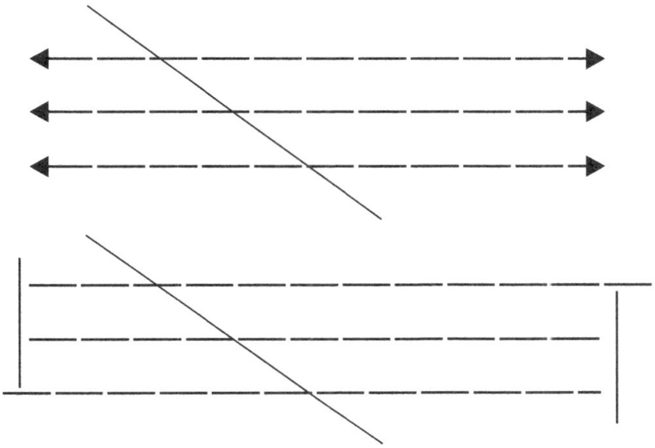

Figure 17.5
Reduction from GEOMBASE′ to SEPARATOR.

> STRIPS COVER BOX(STRIPS)
> *Instance:* A set of n strips and an axis-aligned rectangle R,
> *Question:* Does a union of the strips cover R?

Theorem 17.18. *GEOMBASE′ $\leq_{sq}$ STRIPS, hence STRIPS is 3SUM-hard.*

Proof sketch. We are given an instance of GEOMBASE′. Rotate it 90 degrees. Look at the line segments between the holes and the six half-infinite lines. For each line segment, do the following (called ***point-line-duality***). For each point (m, b) on the segment, associate the line $y = mx + b$. The union of those lines is a strip. This gives us our strips. (See figure 17.6.)

Look at a nonvertical line L as a candidate for a line that goes through three holes in the rotated instance of our original problem. Let p be its dual. L succeeds in going through three holes if and only if p lies in none of the strips. We are not done yet: We need to define the rectangle for STRIPS.

Now take the six half-infinite lines. The point-line duality gives six half-planes $H_1, \ldots, H_6$. Let P be the intersection of $\overline{H_1}, \ldots, \overline{H_6}$. We will now define the rectangle and six more strips.

One can show that P is bounded. Let R be some rectangle that contains it. Also, we turn each H_i into a strip using a line outside R such that the intersection of the strip with R is the same as the intersection of the half-plane with R. □

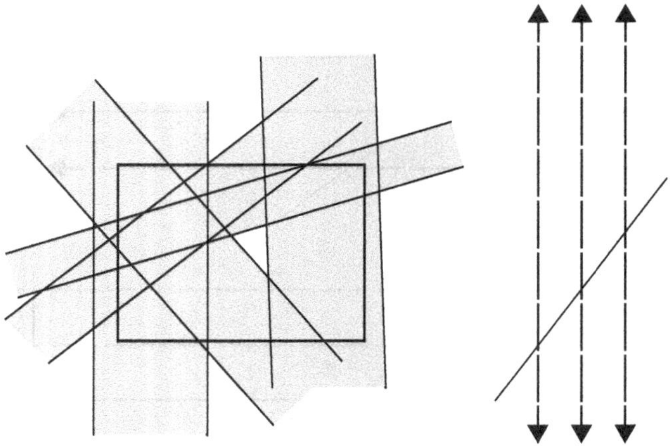

Figure 17.6
Reduction from GEOMBASE′ to STRIPS.

We can also do the same type of question with triangles.

> TRIANGLE COVER TRIANGLE (TRICOVTRI)
> *Instance:* A set of n triangles and a target triangle.
> *Question:* Does the set of triangles cover the target triangle?

Theorem 17.19. *STRIPS $\leq_{sq}$ TRICOVTRI, hence TRICOVTRI is 3SUM-hard.*

Exercise 17.20. Proof theorem 17.19.

> HOLE IN UNION PROBLEM (HOLEINU)
> *Instance:* A set of n triangles in the plane. They can overlap.
> *Question:* Does the set have a hole? That is, is there a closed region within the union that is not covered by any of the triangles?

Theorem 17.21. *TRICOVTRI $\leq_{sq}$ HOLEINU, hence HOLEINU is 3SUM-hard.*

Proof. We are given an instance of TRICOVTRI that is a triangle t and a set of triangles S. Note that every $t' \in S$ can have some parts of it inside t and some parts outside of t. For every $t' \in S$ chop off the parts that are outside of t to obtain t''. If t'' is not a triangle (which is likely), then cut it into a set of $O(1)$ triangles. Take the union over $t' \in S$ of all of these triangles. Call this set U. This is our instance of HOLEINU. Clearly the original set of triangles covers t if and only if there is no hole in the union of the triangles in U. □

We have yet another question regarding triangles that is similar to covering problems.

TRIANGLE MEASURE PROBLEM (TRIMEAS)
Instance: A set of n triangles in the plane.
Output: The measure (that is, the area) of their union.

Theorem 17.22. *TRICOVTRI $\leq_{sq}$ TRIMEAS, so TRIMEAS is 3SUM-hard.*

Proof. We are given an instance of TRICOVTRI, which is a triangle t and a set of triangle S. We form the set U as in the proof of theorem 17.21. The set U is our instance of TRIMEAS. Find its area. Also find (this is easy) the area of t. The union of the triangles in S covers t if and only if the area of the union over U equals the area of t. □

POINT-COVERING PROBLEM (POINT COVER)
Instance: n half-planes and a number k.
Question: Is there a k-way intersection—that is, is there a point in the plane covered by at least k of the half-planes?

If $k \leq \frac{n}{2}$, we claim that the answer is always YES. First rotate the plane so that none of the half-planes are bounded by a vertical line. Now every half-plane's bounding line intersects the y-axis, and the half-plane includes either the interval of the y-axis above that intersection or the interval below the intersection. Equivalently, the half-plane includes either the point $(0, +\infty)$ or the point $(0, -\infty)$. Now divide the half-planes into two classes accordingly. The larger of the two classes has size at least $\frac{n}{2}$, and all of those half-planes meet at a point sufficiently far along the y-axis.

But for larger values of k, the problem becomes harder:

Theorem 17.23. *STRIPS $\leq_{sq}$ POINT COVER, hence POINT COVER is 3SUM-hard.*

Proof. Assume there are n strips. Let S be a strip. Let (H_1, H_2) be the two half-planes that are the complement of the strip. If $x \notin S$, then $(x \in H_1) \oplus (x \in H_2)$. Hence, if x is not in *any* strip, then it must be in n half-planes. One might think we are done; however, the point might not be in the box.

Add to the set of half-planes the four half-planes whose intersection is the box.

There exists x in the box that is not covered by any strip if and only if there exists an x in the intersection of $n+4$ of the half-planes. □

17.5.6 Visibility Between Segments

> Visibility Between Segments (VisBetSeg)
> *Instance:* A set of n horizontal line segments in the plane and two particular segments s_1 and s_2.
> *Question:* Are there points on s_1 and s_2 that can see each other? (This is illustrated in figure 17.7.)

Theorem 17.24. *GeomBase′ $\leq_{sq}$ VisBetSeg, hence VisBetSeg is 3SUM-hard.*

Exercise 17.25. Prove theorem 17.24.

We consider a three-dimensional version of the previous problem with triangles.

> Visible Triangle (VisTri)
> *Instance:* A set of n horizontal triangles in $\mathbb{Z}^3$ (3-dimensions), a special triangle T, and a given point in $\mathbb{Z}^3$.
> *Question:* Can we see triangle T from that given point? (We assume that the n horizontal triangles are not transparent.)

McKenna [McK87] showed that VisTri can be solved in $O(n^2)$ time using hidden surface removal.

Theorem 17.26. *TriCovTri $\equiv_{sq}$ VisTri, hence VisTri is 3SUM-hard.*

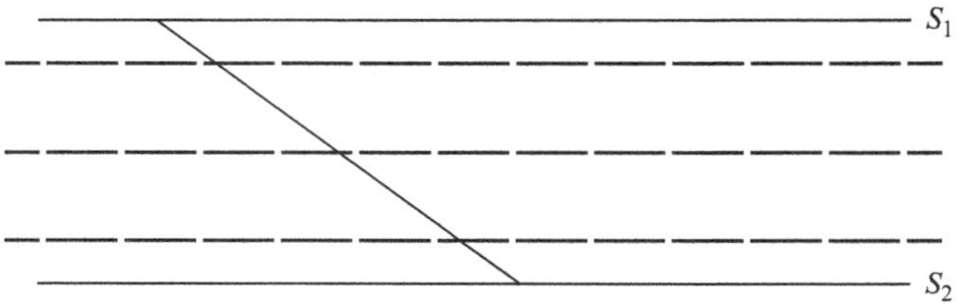

Figure 17.7
Reduction from GeomBase′ to VisBetSeg.

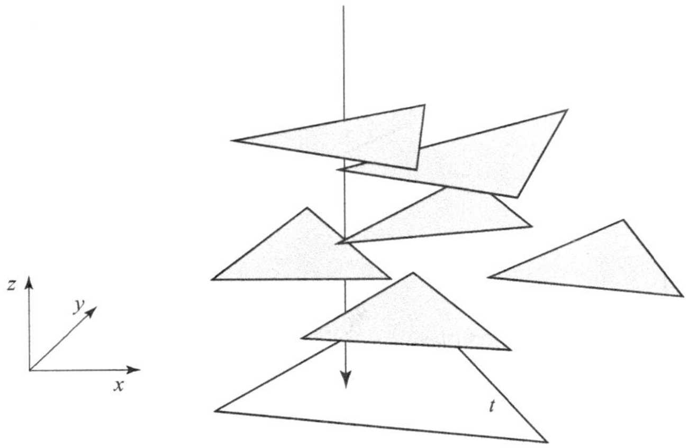

Figure 17.8
Reduction from VISTRI to TRICOVTRI.

Proof. We show VISTRI $\leq_{sq}$ TRICOVTRI. We can assume T has z-coordinate zero, and then make all the other triangles have different heights above T. Then, we can let the point be the point of infinity, and we have that T is visible from infinity if and only if the triangles do not cover T. This is depicted in figure 17.8

We also have a reverse reduction from this problem to TriCovTri. □

17.5.7 PLANAR MOTION PLANNING

We also consider a group of motion planning problems.

> PLANAR MOTION PLANNING (PLMOTPLAN)
> *Instance:* n line segments, some horizontal and some vertical (we think of the line segments as obstacles) and two points called ***the source*** and ***the goal***.
> *Question:* Can we move a robot (represented in the form of a line segment), allowing translations and rotations, from the source to the goal without colliding into any obstacles?

Vegter [Veg90] has shown that PLMOTPLAN is in $O(n^2)$ time.

Theorem 17.27. *GEOMBASE $\leq_{sq}$ PLMOTPLAN, hence PLMOTPLAN is 3SUM-hard.*

Proof. The reduction is evident from figure 17.9. □

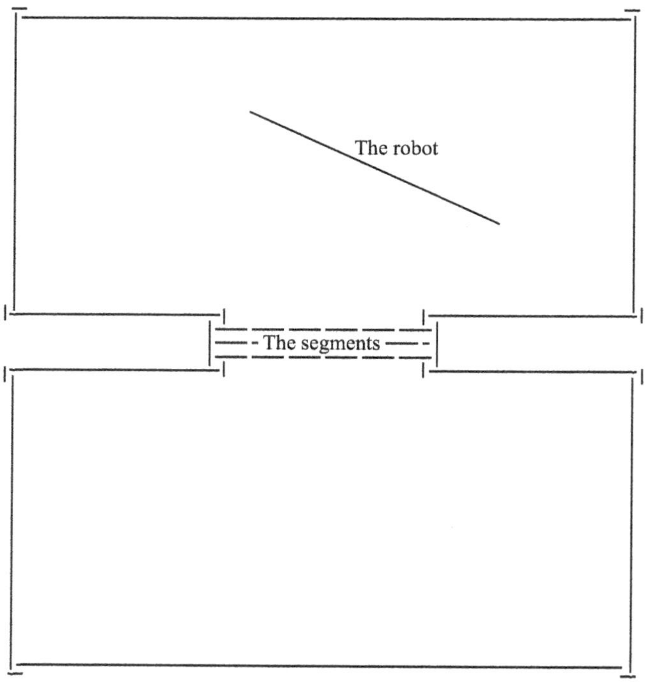

Figure 17.9
Reduction from GEOMBASE to PLMOTPLAN.

We can then extend the previous problem into 3*D*-space, to get 3-dimensional Motion Planning.

> 3-DIMENSIONAL MOTION PLANNING (3DMOTPLAN)
> *Instance:* A set of n horizontal non-intersecting triangle obstacles in $\mathbb{Z}^3$ and a robot represented as a vertical line segment.
> *Question:* Can the robot move through the obstacles without collision, using translations only?

There is an algorithm to solve this problem in $O(n^2 \log n)$ time. This result seems to be folklore. Gajentaan and Overmars [GO12] say that it can be done using a paper by Ke and O'Rourke [KO87].

Theorem 17.28. *TRICOVTRI* $\leq_{sq}$ *3DMOTPLAN, hence 3DMOTPLAN is 3SUM-hard.*

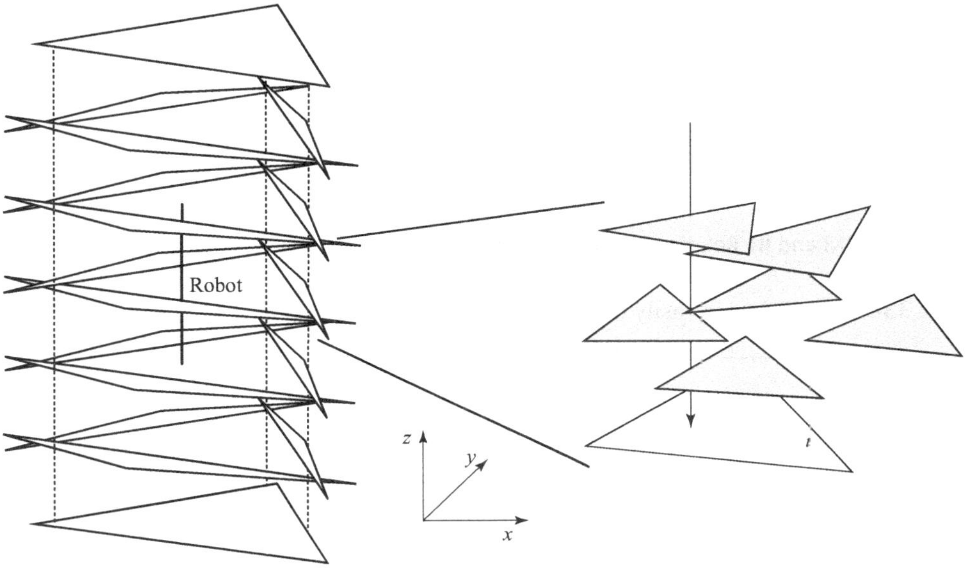

Figure 17.10
Reduction from TRICOVTRI to 3DMOTPLAN.

Proof sketch. First we create a cage to prevent the robot from leaving the original triangle T in the original problem instance. Then, we see that we can go from a given source from the top of the cage to the bottom of the cage, if and only if there is a point not covered by a triangle, and we are done. This is depicted in figure 17.10. □

17.6 Other Lower Bounds from 3SUM-Hardness

Pătraşcu [Păt10] showed that some graph problems are 3SUM-hard:

Theorem 17.29. *Consider the following problem: Given a weighted graph and a number x, we want to know whether some triangle has weight x (weight is the sum of the edges).*

1. *There is an $O(|E|^{3/2})$ algorithm for this problem for dense graphs (this is obvious).*
2. *If there is an $O(|E|^{3/2-\varepsilon})$ algorithm for this problem, then there is an $O(n^{2-\delta})$ algorithm for 3SUM.*

Theorem 17.30. *Consider the following problem: Given an (unweighted) graph, we want to know whether there are $|E|$ triangles.*

1. *There is an $|E|^{3/2}$ algorithm for this problem for dense graphs (this is obvious).*
2. *If there is an $O(|E|^{4/3-\varepsilon})$ algorithm for this problem then there is an $O(n^{2-\delta})$ algorithm for 3SUM.*

Open Problem 17.31. *Narrow the gap between the upper and lower bound in theorem 17.30.*

17.7 *d*SUM and Its Relation to Other Problems

The 3SUM problem can easily be generalized.

*d*SUM *Instance:* A set of n integers. *Question:* Do some d of the integers sum to 0?

Exercise 17.32. Show that there is a randomized $O(n^{\lceil d/2 \rceil})$ algorithm for *d*SUM.

Pătraşcu and R. Williams [PW10] showed that improving the algorithm in exercise 17.32 is equivalent to other problems being improved:

Theorem 17.33. *Let $d < n^{0.99}$. If dSUM with numbers of $O(d \log n)$ bits can be solved in $n^{o(d)}$ time, then 3SAT can be solved in $2^{o(n)}$ time (thus violating the ETH).*

We really want to say

Theorem 17.33 is evidence that dSUM is not in time $O(n^{o(d)})$.

However, Pătraşcu and R. Williams are more ambivalent (page 1066) (comments in square brackets are ours for clarification).

> We have expended significant effort attempting to either find an improved algorithm [for *d*SUM and other problems they have results about], or give interesting evidence against its possibility. In this paper, we present several hypothesis which appear plausible [*d*SUM is not in time $n^{o(d)}$ is one of them], given the current state of knowledge. We prove that if any of the hypothesis are true then CNF SAT has an improved algorithm. One can either interpret our reductions as new attacks on the complexity of CNF SAT, or lower bounds (ruling out all hypothesis) conditional on the hardness of CNF SAT.

Borassi et al. [BCH16] have shown that several problems cannot be done in subquadratic time using hypothesis SETH.

17.8 Lower Bounds on Data Structures via the 3SUM CONJECTURE

Imagine that you want a data structure for (1) a graph with n vertices and m edges or (2) a collection of m sets within a universe of n elements. There are three issues:

- How long will it take to set up the data structure? This is called ***preprocessing***.
- How much space will the data structure need?
- How long will it take to update the data structure? There are many update operations you might allow. For graphs vertex operations (add or delete) or edge operations (add or delete). For sets adding an element, deleting an element, adding a set, deleting a set, merging sets, maintaining the min or max (if the elements are numbers). There are other operations as well.
- How long will it take to answer a query? There are many queries you might be interested in. For graphs one might want to query *is-there-an-edge, reachability, degree,* and others. For sets, *membership* is the key one, though there are others.
- Assume that you want to make L queries where L is large. It may be that some queries take a long time; however, while doing the computation to answer the query, you modify the data structure a lot so that later queries are much faster. We do not want to look at the worst case. We want to say that L queries took $\alpha(n)L$ time. The function $\alpha(n)$ is the ***amortized query time***.
- One can also look at ***amortized update time***.

In the context of data structures, *fast* is polylog (or even $O(1)$) and *slow* is n^δ. Often there is a trade-off.

1. Pătraşcu [Păt10] was the first person to use the 3SUM CONJECTURE to get lower bounds on dynamic data structures. We state one of his results:

   Dynamic reachability The problem is to find a data structure for directed graphs that allows (1) updates: insertion and deletion of edges (but not vertices), and (2) queries: Given vertices u, v determines if there is a directed path from u to v. There exists $\delta > 0$ such that, for any data structure for this problem, either updates or queries take $\Omega(n^\delta)$. All of the later papers build on this paper.

2. Abboud and V. Vassilevska Williams [AW14] considered many problems where the 3SUM CONJECTURE (or other assumptions) were used to get lower bounds on data structures. We state two of the problems they considered that have the same lower bound assuming the 3SUM CONJECTURE.

   ***s*-*t*-reachability** The problem is to find a data structure for a directed graph and two distinguished vertices s, t that allow (1) updates: insertion and deletion of edges (but not vertices), and (2) queries: Is there a directed path from s to t?

Bipartite perfect matching The problem is to find a data structure for undirected bipartite graphs that allows (1) updates: insertion and deletion of edges (but not vertices), and (2) queries: Is there a perfect matching?

For both problems, the following holds: For all α there is no data structure that does updates in $O(m^{\alpha})$ and queries in $O(m^{2/3-\alpha})$.

3. Kopelowitz et al. [KPP16] prove several lower bounds. We state one of them.

The static set disjointness problem The Universe U has n elements: (1) store subsets of U statically so there are no updates; (2) queries: Given two sets, are they disjoint?

They show that if query time is $O(1)$, then preprocessing must take $\Omega(n^{2-o(1)})$,

17.9 Orthogonal Vectors Conjecture

In this chapter we have used the 3SUM Conjecture as a hardness assumption to obtain quadratic lower bounds. We now look at another hardness assumption to obtain quadratic lower bounds.

> Orthogonal Vectors
> *Instance:* n vectors in $\{0,1\}^d$ where $d = O(\log n)$.
> *Question:* Do two of the vectors have an inner product that is 0 mod 2?

A naïve algorithm for Orthogonal Vectors solves it in time $O(n^2 d)$ by trying all possible pairs. Abboud et al. [AWY15] obtained the following slight improvement.

Theorem 17.34. *There is a randomized algorithm of Orthogonal Vectors that runs in time* $O(n^{2-\Omega\left(\frac{1}{\log(d/n)}\right)})$.

There are no known algorithms for the problem that run in time $n^{2-\varepsilon}$ for some $\varepsilon > 0$. This leads to the following conjecture, due to R. Williams [Wil05]. See also Abboud et al. [AVW14], Backurs et al. [BI15], and Abboud et al. [ABV15]

Conjecture 17.35. ***Orthogonal Vectors Conjecture (OVC)****: For all* $\varepsilon > 0$*, there is a* $c \geq 1$ *such that Orthogonal Vectors cannot be solved in time* $n^{2-\varepsilon}$ *on instances with* $d = \lceil c \log(n) \rceil$.

We now have several conjectures with rather concrete bounds in them: ETH, SETH, 3SUM Conjecture, and now OVC. Clearly SETH $\Longrightarrow$ ETH. Are any other implications known? Yes. As we mentioned in section 6.7.1, R. Williams [Wil05] showed the following.

Theorem 17.36. *SETH $\Longrightarrow$ OVC.*

The lack of any subquadratic algorithm for ORTHOGONAL VECTORS, and theorem 17.36, are evidence for OVC. In addition, OVC holds in several restricted computational models. Kane and R. Williams [KW19] show:

1. ORTHOGONAL VECTORS has branching complexity $\tilde{\Theta}(n \cdot \min(n, 2^d))$ for all sufficiently large n, d. ($\tilde{\Theta}(f(n))$ means that we ignore polylog factors.)
2. ORTHOGONAL VECTORS has Boolean formula complexity $\tilde{\Theta}(n \cdot \min(n, 2^d))$ over all complete bases of $O(1)$ fan-in.
3. ORTHOGONAL VECTORS requires $\tilde{\Theta}(n \cdot \min(n, 2^d))$ wires, in formulas composed of gates computing arbitrary symmetric functions of unbounded fan-in.

What does OVC imply and vice-versa?

Theorem 17.37. *Assume OVC. Then the following problems do not have subquadratic algorithms.*

1. *(Backurs and Indyk [BI15]) Edit distance: Given two strings x, y, how many times do you need to delete or insert or replace a letter from either so that, at the end, the resulting strings are the same. (There are many variants depending on what operations are allowed.)*
2. *(Bringmann [Bri14]) Fréchet distance: This is a measure of similarity between two curves that takes into account the location and ordering of the points on the curve. The formal definition is rather long, so we omit it. The problem is, given two curves, find the Fréchet distance between them.*
3. *(Backurs and Indyk [BI16]) Regular expression matching: Given a regular expression p and a string t, does p generate some substring of t.*

While the above results show that many problems are not subquadratic assuming ORTHOGONAL VECTORS is not subquadratic, this does not necessarily imply that they are *equivalent* to ORTHOGONAL VECTORS. Chen and R. Williams [CW19] study equivalences between ORTHOGONAL VECTORS and different problems. They showed the following.

Theorem 17.38. *Each of the following problems is subquadratic-equivalent to ORTHOGONAL VECTORS.*

1. *Min-IP: Given n blue vectors in $\{0, 1\}^d$, and n red vectors in $\{0, 1\}^d$, find the red-blue pair of vectors with minimum inner product.*
2. *Max-IP: Given n blue vectors in $\{0, 1\}^d$, and n red vectors in $\{0, 1\}^d$, find the red-blue pair of vectors with maximum inner product.*

3. *Equals-IP: Given n blue vectors in* $\{0, 1\}^d$*, and n red vectors in* $\{0, 1\}^d$*, and an integer k, find a red-blue pair of vectors with inner product k, or report that none exists.*
4. *Red-Blue-Closest Pair: Let* $p \in [1, 2]$ *and* $d = n^{o(1)}$*. Use the p-norm for distance (see section 6.7.1 for a discussion of the p-norm). Approximating the closest red-blue pair among n red points and n blue points in* $\mathbb{R}^d$*.*

17.10 Further Results

Barequet and Har-Peled [BH01] proved that the following problems (and more) are 3SUM-hard. They are from computational geometry. Some care must be taken to define these problems rigorously since the inputs are real numbers. We ignore such issues.

1. Given two simple polygons P and Q, determine whether P can be translated to fit inside Q.
2. Given two simple polygons P and Q, determine whether P can be translated and rotated to fit inside Q.
3. Given two simple polygons P and Q, determine whether P can be rotated around a given point to fit into Q.
4. Given P, a finite sets of reals, and a set S of intervals of real numbers, determine whether there is a $u \in \mathbb{R}$ so that $P + u \subseteq S$.

Aronov and Har-Peled [AH08] study the problem of finding the "deepest" point in an arrangement of disks, where the ***depth of a point*** denotes the number of disks that contain it. They show that this problem is 3SUM-hard.

Abboud et al. [AVW14] consider the local alignment problem: Given two input strings and a scoring function on pairs of letters, one is asked to find the substrings of the two input strings that are most similar under the scoring function. They show that if there exists $\varepsilon > 0$ and an $O(n^{2-\varepsilon})$ algorithm for this problem, then there exists a $\delta > 0$ such that *a*ll of the following are true:

1. 3SUM can be done in $O(n^{2-\delta})$ time (so the 3SUM CONJECTURE is false).
2. CNF SAT can be done in $O((2-\delta)^n)$ time (so SETH is false).
3. The following problem can be done on $O(4-\delta)^n)$ time (which people who have looked at it do not think it can be): the max weight k-clique problem, which is, given a weighted graph, find a k-clique of max weight or say there is no k-clique.

18

APSP CONJECTURE: A Method for Obtaining Cubic Lower Bounds

18.1 Introduction

In chapter 17 we used the assumption that 3SUM cannot be solved in subquadratic time to prove that many other problems can not be solved in subquadratic time. In this chapter we use the assumption that the ALL-PAIRS SHORTEST PATHS Problem (APSP) cannot be solved in subcubic time to prove that there are many other problems that can not be solved in subcubic time.

Chapter Summary

1. *We discuss the problem APSP and conjecture that it cannot be solved in subcubic time.*
2. *We define an appropriate notion of reduction.*
3. *We use the assumption that APSP cannot be solved in subcubic time, and the reductions in the last item, to show that some problems cannot be solved in subcubic time.*

18.2 APSP: ALL-PAIRS SHORTEST PATHS

Recall that big-O notation is used when you want to ignore constants. We need a notation for when we want to ignore log factors.

Notation 18.1. $f = \tilde{O}(g)$ means that there exist $n_o, c \in \mathbb{N}$ such that, for all $n \geq n_o, f(n) \leq (\log n)^c g(n)$.

Notation 18.2.

1. We denote a weighted directed graph by $G = (V, E, w)$, where w is the weight function. The weights will always be integers. They will almost always be nonnegative.
2. We use n for the number of vertices and m for the number of edges.

3. Let $G = (V, E, w)$ be a weighted directed or undirected graph with weights in $\mathbb{N}$. Let $x, y \in V$. Then $\text{dist}_G(x, y)$ is the length of the shortest path between x and y. We will sometimes use $\text{dist}(x, y)$ when G is clear.

Note: This chapter will deal with weighted directed graphs. Most of what we prove is true for weighted undirected graphs. In project 18.16 we will invite the reader to redo the entire chapter with variants of weighted directed graphs.

> ALL-PAIRS SHORTEST PATHS (APSP)
> *Instance:* A weighted directed graph $G = (V, E, w)$ with weights in $\mathbb{N}$.
> *Output:* For all pairs of vertices x, y, $\text{dist}_G(x, y)$.
> *Note:* We will consider any arithmetic operation to have unit cost.

This problem is a very important and central problem in graph theory.

Floyd [Flo62] and Warshall [War62] (Roy [Roy59] had some of the ideas) independently proved the first part of the next theorem. (The second part is folklore.) This was an early use of the technique now known as dynamic programming.

Theorem 18.3.

1. *APSP can be solved in $O(n^3)$ time.*
2. *Let m be the number of edges. APSP can be solved in $O(nm + n^2 \log n)$ time. (If the graph has n^α edges, where $\alpha \geq 1$, then this algorithm is $O(n^{1+\alpha})$. For $\alpha = 2$ (the worst case), this is $O(n^3)$; however, for $\alpha < 2$, this algorithm is subcubic.)*

Proof. 1) The algorithm is as follows:

```
Data: A graph G = (V, E, w) given as an adjacency matrix w(i,j)
Result: The matrix dist.
∀i,j ∈ V: dist(i,j) ← w(i,j) ;
for k=1 to n do
    for i=1 to n do
        for j =1 to n do
            if dist(i,j) > dist(i,k) + dist(k,j) then
                dist(i,j) ← dist(i,k) + dist(k,j) ;
            end
        end
    end
end
```

Algorithm 1: The Floyd-Warshall algorithm.

2) Another way to compute the all pair shortest path is by invoking Dijkstra's algorithm for each vertex. Dijkstra's algorithm finds the shortest path from a given vertex v to all other vertices in the graph. Hence, invoking it n times by choosing a different starting vertex v each time will result in calculating the all pair shortest path. A single invocation of Dijkstra's algorithm takes $O(m + n\log n)$. Hence, n iterations of this algorithm takes a time of $O(nm + n^2 \log n)$. □

Is there an algorithm for APSP that runs in time better than $O(n^3)$? This depends on your definition of "better." The following are known:

1. Fredman [Fre76] gave a $O\left(\frac{n^3\sqrt[3]{\log\log n}}{\sqrt[3]{\log n}}\right)$ deterministic algorithm.
2. R. Williams [Wil18] gave a randomized algorithm that ran in time $\frac{n^3}{2^{\Omega(\sqrt{\log n})}}$. There were many results between Fredman (1976) and R. Williams (2018).
3. Chan and R. Williams [CW21] found an algorithm for APSP in deterministic time $\frac{n^3}{2^{\Omega(\sqrt{\log n})}}$ time, matching the randomized algorithm listed above.
4. Zwick [Zwi98] showed that APSP can be approximated well: For all $\varepsilon > 0$ there exists a $(1+\varepsilon)$-approximation to the APSP problem running in time $\tilde{O}(\frac{n^\omega}{\varepsilon})$, where ω is the exponent in the running time of the fastest known matrix multiplication problem (currently $\omega < 2.3719$).

While the algorithms above are impressive and very clever, the first three are not that much better than $O(n^3)$, and the fourth is an approximation. We need a definition for "much better than $O(n^3)$."

Definition 18.4. An algorithm is ***subcubic*** if there exists $\varepsilon > 0$ such that it runs in time $O(n^{3-\varepsilon})$.

Despite enormous effort, nobody has obtained a subcubic algorithm for APSP. In the next section we turn this around: We state a conjecture that APSP cannot be done in subcubic time. From that conjecture we obtain cubic lower bounds on other problems.

18.3 APSP CONJECTURE

The following conjecture was first made explicit in a paper by Abboud and V. Vassilevska Williams [AW14]; however, it was used implicitly before then. The first time might have been in 2004 by Roditty and Zwick [RZ04]. Another notable earlier use is a 2010 paper by R. Williams and V. Vassilevska Williams [WW10].

Conjecture 18.5. *APSP CONJECTURE: There is no subcubic algorithm for APSP.*

We define a notion of reduction between problems.

Definition 18.6. Let A and B be sets or functions (they will almost always be sets).

1. $A \leq_{sc} B$ means that if there is a subcubic algorithm for B, then it can be used to obtain a subcubic algorithm for A. The *sc* stands for ***sub-cubic***.
    a. (The usual way to do this.) On input x produce in $O(n\text{polylog}n)$ (usually just linear) a y such that $x \in A$ if and only if $y \in B$. Note that $\leq_{sc}$ is transitive.
    b. (This sometimes is needed.) On input x produce in $O(n\text{polylog}n)$ $y_1, \ldots, y_k$ (k is a constant) such that $x \in A$ can be determined from the answers to $y_1 \in B?, \ldots, y_k \in B?$ in $O(n\text{polylog}n)$ time. Note that $\leq_{sc}$ is still transitive.
    c. We leave it to the reader to modify the above definitions for when A and B are functions.
2. $A \equiv_{sc} B$ if $A \leq_{sc} B$ and $B \leq_{sc} A$. We often use the terminology ***A and B are subcubic equivalent***.

Definition 18.7. Let A be a problem.

1. A is ***APSP-hard*** if APSP $\leq_{sc} A$.
2. A is ***APSP-complete*** if A is APSP-hard and $A \leq_{sc}$ APSP.

Because of conjecture 18.5 we think that if A is APSP-hard, then there is no subcubic algorithm for A. In brief:

1. When you read "A is APSP-complete," you should think: A is in cubic time but not in subcubic time.
2. When you read "A is APSP-hard," you should think: A is in not in subcubic time.

The definition of NP-hard is used to show that problems are not in P, contingent on the conjecture that P $\neq$ NP. The definition of APSP-hard is used to show that problems do not have subcubic algorithms, contingent on the conjecture that APSP does not. We list out similarities and differences between the two theories.

1. Both use reductions and build up a large set of problems that are thought to be hard. The number of NP-hard problems is far larger than the number of APSP-hard problems.
2. Contrast the following:
    a. A problem B is NP-hard if for all $A \in$ NP, $A \leq_p B$. SAT is a natural NP-hard problem. As a consequence, one can show C is NP-hard by showing SAT $\leq_p C$.
    b. A problem B is APSP-hard if APSP $\leq_{sc} B$. Note that we *do not* have a result for APSP that is analogous to the Cook–Levin theorem. We suspect that APSP requires cubic time and use it as such.

If we did not know the Cook–Levin theorem but really thought SAT was hard, we could still have a large set of problems that are NP-complete and think they were hard. That is the position we are in with APSP-hard.

18.4 Problems of Interest: Centrality Measures

We define several measures on graphs that are called ***centrality measures*** (the reason for the name will be clear once we define the measures). These measures appear in a variety of applications such as social networks, biological networks, transportation and allocation problems, and others. Hence, calculating these measures efficiently has a lot of real life implications. All of these measures have trivial $O(n^3)$ algorithms (they all begin by first doing APSP). We will later show reductions between them and some other problems. The goal is to get them to be subcubic equivalent to APSP; however, alas, that is an open problem.

We now define three centrality measures.

RADIUS and CENTER
Instance: A weighted directed graph $G=(V,E,w)$.
Output: (RADIUS) $\min_v \max_w d(v,w)$. (This quantify is called ***the radius of G***.)
Output: (CENTER) A vertex v such that $\max_w d(v,w)$ is minimized. (This vertex is called ***the center of G***.)

DIAMETER
Instance: A weighted directed graph $G=(V,E,w)$.
Output: $\max_{u,v} \text{dist}(u,v)$. (This number is called ***the diameter of G***.)
Note: The diameter is the maximum distance between two vertices.

MEDIAN
Instance: A weighted directed graph $G=(V,E,w)$.
Output: $\min_v \sum_u \text{dist}(v,u)$? (This number is called ***median of the G***.)

18.5 Other Measures

We now define another measure called ***betweenness centrality***. This measure determines how useful a vertex is to shortest paths in the graph.

Definition 18.8. Let $G = (V, E, w)$ be a weighted directed graph and $s, t, x \in V$.

1. BETWEENNESS CENTRALITY$_{s,t}(x)$ is the fraction of shortest paths between s and t that contain x.
2. BETWEENNESS CENTRALITY(x) is $\sum_{s,t \in V}$ BETWEENNESS CENTRALITY$_{s,t}(x)$.
3. POSITIVE BETWEENNESS CENTRALITY$(x) = \begin{cases} 1 & \text{if BETWEENNESS CENTRALITY}(x) > 0 \\ 0 & \text{otherwise} \end{cases}$

   If the graph G is not understood, we sometimes write POSITIVE BETWEENNESS CENTRALITY(G,x). In simpler terms: POSITIVE BETWEENNESS CENTRALITY$(x) = 1$ if and only if x is on some shortest path.

> POSITIVE BETWEENNESS CENTRALITY
> *Instance:* A weighted directed graph $G = (V, E, w)$ and an $x \in V$.
> *Question:* Does POSITIVE BETWEENNESS CENTRALITY$(G, x) = 1$? In other words, is x on some shortest path?

The last problem we present is not a measure; however, it will be useful.

> NEGATIVE TRIANGLE
> *Instance:* A weighted directed graph $G = (V, E, w)$ with weights in $\{-M, \ldots, M\}$.
> *Question:* Is there a triangle with weight-sum negative?

18.6 Subcubic Equivalence

It is clear that all the problems in sections 18.4 and 18.5 are $\leq_{sc}$APSP and hence in time $O(n^3)$. Do any of them have a subcubic algorithm? The following theorem, due to the combined efforts of Abboud et al. [AGV15] and R. Williams and V. Vassilevska Williams [VW18], shows that, assuming the APSP CONJECTURE, no. We omit the proof.

Theorem 18.9. *RADIUS, MEDIAN, BETWEENNESS CENTRALITY, and NEGATIVE TRIANGLE are APSP-complete.*

Note: See Boroujeni et al. [BDE+19b] for evidence that DIAMETER is subcubic-complete. The status of DIAMETER is still open.

We will prove some other subcubic reductions. From theorem 18.9 and what we prove in the next few sections, we will the diagram of reductions in figure 18.1. The dotted lines are easy reductions, whereas the solid lines are harder reductions.

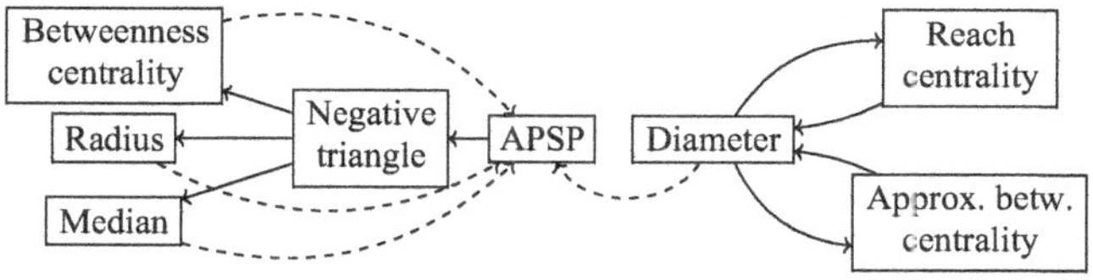

Figure 18.1
Subcubic equivalence between various problems.

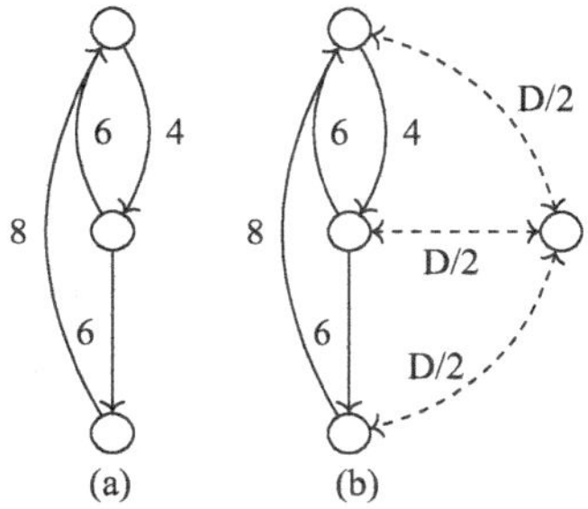

Figure 18.2
(a) Original graph G. (b) Transformed graph G'.

18.7 DIAMETER and POSITIVE BETWEENNESS CENTRALITY Are Subcubic Equivalent

Abboud et al. [AGV15] proved both theorems in this section.

Theorem 18.10. *DIAMETER* $\leq_{sc}$ *POSITIVE BETWEENNESS CENTRALITY.*

Proof. Figure 18.2 is an example of the reduction. Here is the reduction.

1. Input a weighted directed graph $G = (V, E, w)$. Without loss of generality, let us assume that all distances, and hence the diameter, are even. (Multiply all distance by 2 initially, and divide the finally obtained diameter by 2.) Let M be the maximum weight.
2. (This is not part of the algorithm; this is a definition we will use later.) For $1 \leq D \leq M$ with D even, let $G_D = (V_D, E_D)$ be the following graph:

   $V_D = V \cup \{x\}$—that is, add a new vertex labeled x.

   $E_D = E \cup \{(x, u, D/2) \mid u \in V\}$.

   The largest possible value of D such that POSITIVE BETWEENNESS CENTRALITY $(G_D, x) = 1$ is the required diameter of the original graph G. Note that POSITIVE BETWEENNESS CENTRALITY$(G_2, x) \geq \cdots \geq$ POSITIVE BETWEENNESS CENTRALITY (G_M, x).

3. Perform a binary search on D to find the largest D such that POSITIVE BETWEENNESS CENTRALITY$(G_D, x) = 1$. Output that D. In the example of figure 18.2 we get $D = 10$. □

Theorem 18.11. *POSITIVE BETWEENNESS CENTRALITY* $\leq_{sc}$ *DIAMETER.*

Proof. We present the reductions.

1. Input a weighted directed graph $G = (V, E, w)$ and $x \in V$. Let M be the max weight and let $\tilde{D} = 3M|V|$, which is much bigger than any shortest path in G. For all pairs of vertices that do not have an edge between them, we put an edge of weight $\tilde{D}$. This guarantees that those edges will not be used in any shortest path.
2. By running Dijkstra's algorithm on (G, x), find, for all vertices v, $d_G(v, x)$ and $d_G(x, v)$. Note that this takes $O(|E| + |V|\log(|V|))$, which is subcubic in $|V|$.
3. We create a weighted graph $G' = (V', E', w')$ as follows:

   $V' = \{v_a \mid v \in V - \{x\}\} \cup V \cup \{v_b \mid v \in V - \{x\}\}$ (so there are three copies of every $v \in V - \{x\}$ and one copy of x).

   We think of the v_a vertices being on the left, then the V vertices being in the middle, then the v_b vertices being on the right.

   $E' = E_a \cup E \cup E_b$, where E_a, E, E_b are the following sets of weighted edges:

   $E_a = \{(v_a, v) \mid v \in V\}$ with $w'(v_a, v) = \tilde{D} - d_G(v, x)$.

   E is the original set of edges from G. So the middle graph is the original graph.

   $E_b = \{(v, v_b) \mid v \in V$ with $w'(v, v_b) = \tilde{D} - d_G(x, v)$.

   See figure 18.3 for an example.
4. Let $D = $ DIAMETER(G').
5. (This is not part of the algorithm. This is commentary.) The longest path in G' has to go from some s_a to some t_b. We can assume the path is of the form

   $$s_a \rightarrow s \Rightarrow t \rightarrow t_b,$$

   where $s_a \rightarrow s$ is and edge, $s \Rightarrow t$ is a path in G, and $t \rightarrow t_b$ is an edge. Hence, in G', the distance of the path from s_a to t_b is

   $$(\tilde{D} - \text{dist}_G(s, x)) + \text{dist}_G(s, t) + (\tilde{D} - \text{dist}_G(x, t))$$
   $$= 2\tilde{D} + \text{dist}_G(s, t) - \text{dist}_G(s, x) - \text{dist}_{G'}(x, t).$$

   Since $\text{dist}_G(s, t) - \text{dist}_G(s, x) - \text{dist}_G(x, t) \leq 0$,

   $$\forall s, t \in V : 2\tilde{D} + \text{dist}_G(s, t) - \text{dist}_G(s, x) - \text{dist}_G(x, t) \leq \text{dist}_{G'}(s_a, t_b) \leq 2\tilde{D}.$$

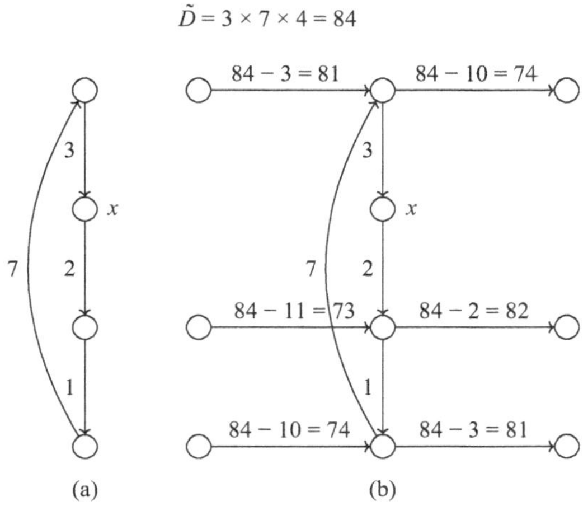

Figure 18.3
Reduction from POSITIVE BETWEENNESS CENTRALITY to DIAMETER.

Note the following:

a. If POSITIVE BETWEENNESS CENTRALITY$(x) = 1$, then there exist $s, t \in V$ such that $\text{dist}_G(s,t) = \text{dist}_G(s,x) + \text{dist}_G(x,t)$, hence $\text{dist}_{G'}(s_a, t_b) = 2\tilde{D}$.

b. If POSITIVE BETWEENNESS CENTRALITY$(x) = 0$, then for all $s, t \in V$, $\text{dist}_{G'}(s,t) < \text{dist}_G(s,x) + \text{dist}_G(x,t)$, hence $\text{dist}_{G'}(s_a, t_b) < 2\tilde{D}$.

6. If $D = 2\tilde{D}$, then output YES; else output NO. □

18.8 NEGATIVE TRIANGLE

R. Williams and V. Vassilevska Williams [VW18] showed the following.

Theorem 18.12. *NEGATIVE TRIANGLE* $\leq_{sc}$ *RADIUS.*

Proof. Here is the reduction. The graph we construct will be undirected. Since formally RADIUS is a set of directed graphs, one can take each edge $\{x, y\}$ and make two directed edges (x, y) and (y, x).

1. Input a weighted directed graph $G = (V, E, w)$ with weights in $\{-M, \dots, M\}$. Let $Q = 3M$.
2. We create a graph $G' = (V', E')$ as follows (See figure 18.4).

   x is a new vertex. V_a, V_b, V_c, V_d are each copies of V.

   If $v \in V$, then v_a (v_b, v_c, v_d) is the analog of v in

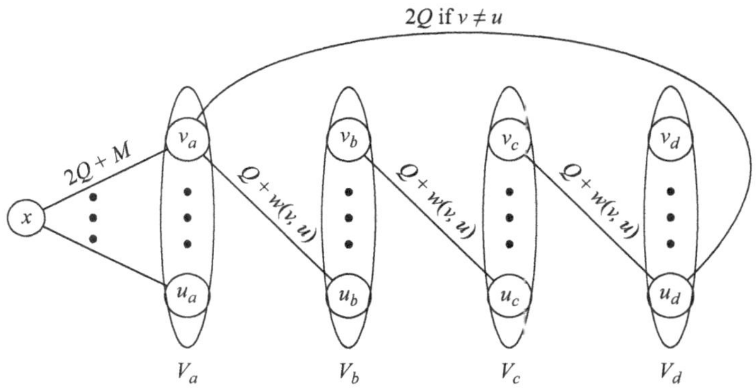

Figure 18.4
Reduction from NEGATIVE TRIANGLE to RADIUS.

V_a (V_b, V_c, V_d).

$V' = \{x\} \cup V_a \cup V_b \cup V_c \cup V_d$.

There is an edge of weight $2Q + M$ from x to each vertex of V_a.

For all $(u, v) \in E$, we put edges of weight $Q + w(u, v)$ between (1) u_a, v_b, (2) u_b, v_c, (3) u_c, v_d.

For all $u, v \in V$ with $u \neq v$ we put edges of weight $2Q$ between u_a and v_d. Note that we do not require that $(u, v) \in E$.

There are no edges within V_a or V_b or V_c or V_d.

3. If the radius of G' is $< 9M$, then output YES (there is a negative triangle). Else output no. We prove this works below.

Claim 18.13. RADIUS$(G') < 9M$ if and only if G has a negative triangle.

Proof of Claim: We will make and prove four observations that will essentially prove this claim. Let $R = \text{RADIUS}(G')$.

- **Observation 1**: Any vertex of the form v_b, v_c, or v_d is more than $9M$ away from x. Hence, if $R < 9M$, then the center of the graph is contained in V_a.

  We look at v_b, v_c, and v_d.

  If $v_b \in V_b$, then there exists u_a such that

$$\begin{aligned} \text{dist}_{G'}(v_b, x) &= \text{dist}_{G'}(v_b, u_a) + \text{dist}_{G'}(u_a, x) \\ &= (Q + w(v, u) + (2Q + M) \\ &= 10M + w(v, u) \\ &\geq 9M. \end{aligned}$$

  A similar arguments work for v_c and v_d.

- **Observation 2**: Let $v_a \in V_a$ and let $z \in V' - \{v_b, v_c, v_d\}$. Then $\text{dist}_{G'}(v_a, z) \leq 8M$.

  There are cases:

  $z = x$. Then $\text{dist}_{G'}(v_a, x) = 2Q + M = 7M < 8M$.

  $z = u_b$. Then $\text{dist}_{G'}(v_a, u_b) = Q + w(v, u) \leq 3M + M = 4M < 8M$.

  $z = u_c$. Then there exists $w \in V$ such that

$$\begin{aligned} distGp(v_a, u_c) &= \text{dist}_{G'}(v_a, u_b) + \text{dist}_{G'}(u_b, u_c) \\ &= (Q + w(v, w) + (Q + w(w, u) = 6M + w(v, w) + w(w, u) \\ &\leq 8M \end{aligned}$$

  $z = u_d$. Then the shortest distance to x uses the edge from v_a to u_d, which has distance $2Q = 6M$.

  $z = u_a$. Then $\text{dist}_{G'}(v_a, u_a)$ is determined by going from v_a to u_b and then from u_b to u_a, so

$$\text{dist}_{G'}(v_a, u_a) \leq Q + w(v, u) + Q + w(u, v) \leq 3M + M + 3M + M = 8M.$$

- **Observation 3**: If vertex v in graph G was present in a negative triangle, then $\text{dist}_{G'}(v_a, v_b) < 3Q = 9M$.

  Let the triangle by v, w, x. So $w(v, w) + w(w, x) + w(x, v) < 0$. Then $\text{dist}_{G'}(v_a, v_b)$ can be bounded by the route that goes from v_a to w_b, then w_b to x_c, then x_c to v_b, so

$$\begin{aligned} \text{dist}_{G'}(v_a, v_b) &\leq (Q + w(v, u)) + (Q + w(u, x)) + (Q + w(x, v)) \\ &= 3Q + w(v, u) + w(u, x) + w(x, v) < 9M. \end{aligned}$$

- **Observation 4**: Finally, if a vertex v is not in a negative triangle in the original graph, then $\text{dist}_{G'}(v_a, v_b) \geq \min\{3Q, 4Q - 2M\} = 9M$.

  We leave this to the reader.

The claim follows easily from the above four observations, and the theorem follows easily from the claim. □

Exercise 18.14. Give a subcubic reduction from negative-triangle to median.

18.9 Connection to SETH

The strong exponential time hypothesis (SETH) states that there is no $\delta < 1$ such that SAT can be solved in time $O(2^{\delta n})$. We mentioned this (in a different form) in hypothesis 6.7. Note that SETH implies $\mathsf{P} \neq \mathsf{NP}$.

Roditty and V. Vassilevska Williams [RV13] obtained an upper bound (that is, an algorithm) for computing the approximate value of the diameter and a lower bound assuming SETH. We state both.

Theorem 18.15.

1. *There is an expected run time* $\tilde{O}(m\sqrt{n})$ *algorithm for* 1.5*-approximation for* DIAMETER. *Note that if* $m = \Theta(n^{1.5})$ *(which could be called quasi-sparse but never is) this is an* $\tilde{O}(n^2)$ *algorithm.*
2. *Assume SETH. There is no* ε *such that there is an* $O(m^{2-\varepsilon})$ *time, 1.5-approximation algorithm, for the diameter of a graph.*

Project 18.16. *This chapter dealt with directed weighted graphs. There are three other options: undirected weighted graphs, directed unweighted graphs, and undirected unweighted. For each of those options try to redo this entire chapter. Which theorems are true with similar (or even the same) proofs?*

18.10 Further Results

18.10.1 More APSP-Complete and APSP-Hard Problems

1. The MATRIX PRODUCT VERIFICATION PROBLEM is as follows: Given matrices A, B, C verify that $AB = C$ where the product is over the (min, +)-semiring. R. Williams and V. Vassilevska Williams [VW18] showed this problem is APSP-complete.
2. The REPLACEMENT PATHS PROBLEM is as follows: Given weighted directed graph G, vertices s, t, and a shortest (s, t)-path P compute the length of the shortest (s, t)-path that does not use any edge from P. R. Williams and V. Vassilevska Williams [VW18] showed this problem is APSP-complete. They also show that other problems related to shortest paths are APSP-complete.
3. CODIAMETER: Given a graph G, the goal of CoDiameter is to report a vertex that does not participate in an edge of length equal to the diameter of G. Boroujeni et al. [BDE$^+$19b] showed that this problem is APSP-complete by a reduction from APSP. Boroujeni also defines CORADIUS, CORADIUS, CONEGATIVETRIANGLE, COMEDIAN, and shows them all APSP-complete.
4. The TREE EDIT PROBLEM is (informally) as follows: Given two trees, what is the least number of changes needed to get one from the other? Bringmann et al. [BGMW20] show that this problem is APSP-hard.
5. The METRICITY PROBLEM is as follows: Given an $n \times n$ nonnegative matrix A, determine whether it defines a metric space on $[n]$—that is, if A is symmetric, has zeros on the diagonal, and entries satisfy the triangle inequality. R. Williams and V. Vassilevska Williams [VW18] showed this problem is APSP-hard.

18.10.2 Assuming Hardness of Unweighted APSP

The next two problems have as their hypothesis that the unweighted APSP problem is hard. Both results are by Lincoln et al. [LPV20].

1. The ALL EDGES MONOCHROMATIC TRIANGLE PROBLEM is as follows. Given an n-node graph G with edges labeled a color from 1 to n^2, decide, for each edge, if it belongs to a monochromatic triangle (a triangle whose three edges have the same color). If this problem has a $T(n)$ time algorithm, then the unweighted APSP has an $O(T(n)\log n)$ time algorithm.
2. The MIN-MAX PRODUCT PROBLEM is as follows. Given two matrices A, B compute the matrix C where $C_{i,j} = \min_k \max(A_{ik}, B_{kj})$. If this problem has a $T(n)$ algorithm, then the unweighted APSP problem has an $O(T(n)\log n)$ time algorithm.

18.10.3 Unusual Hardness Assumptions, Proofs, or Results

1. (Abboud et al. [AGI+19]) Let ω be (as usual) the exponent for matrix multiplication. Let 4-CLIQUE be the problem of, given a 4-partite graph, does it have a 4-clique. Eisenbrand and Grandoni [EG04] have an $O(n^{\omega+1})$ algorithm for 4-clique. The 4-CLIQUE CONJECTURE is that this algorithm is optimal. The ALL-PAIRS MIN CUT problem asks: Given a graph G compute, for every pair of vertices s, t, a min s-t cut (a partition of the vertices where s and t are in different parts and the number of edges between the parts is minimal). The ALL-PAIRS MIN CUT-k problem asks for the min-cut if it is $\leq k$, and if not then just report that it is $\geq k$. Assuming the 4-CLIQUE CONJECTURE, ALL-PAIRS MIN CUT-k has a super-cubic lower bound of $n^{\omega+1-o(1)}k^2$.
2. BOOLEAN MATRIX MULTIPLICATION (BMM): (R. Williams and V. Vassilevska Williams [VW18]) A ***combinatinatoral algorithm for BMM*** is an algorithm that takes advantage of the viewpoint that a Boolean matrix represents a bipartite graph. This definition is not rigorous. Indeed, it is hard to make a rigorous definition; however Das et al. [DKS18] have tried. No known combinatorial algorithm for BMM is subcubic. If BMM has a subcubic combinatorial algorithm, then so does the triangle detection problem in graphs. All known algorithms for triangle detection take cubic time, hence using the hardness of triangle detection as an assumption is reasonable.
3. (Boroujeni et al. [BDE+19b]) CoAPSP VERIFICATION: Given a graph G and a matrix D, either find a pair (i,j) such that $D_{i,j}$ is equal to the distance between vertices i and j in G, or determine that there is no such pair. There is a subcubic reduction from DIAMETER to CoAPSP VERIFICATION Recall that DIAMETER is thought to require cubic time; however, we do not know whether DIAMETER is APSP-hard.
4. $\{-1, 0, 1\}$-APSP: Given a weighted directed graph with edge weights in $\{-1, 0, 1\}$, compute the APSP. Despite the complication of having negative edge weights, this problem has a subcubic ($O(n^{2.52})$) algorithm given by Zwick [Zwi02]. This problem seemed to require cubic time but did not. Consider this a cautionary note.

18.10.4 Using the OR of APSP, SETH, and 3SUM Conjectures

In chapter 6 we used ETH and SETH as assumptions. In chapter 17 we used the 3SUM CONJECTURE as an assumption. In this section we used the APSP CONJECTURE as an assumption. We think the AND of these assumptions is true. Abboud et al. [AVY18] assumed the OR of the assumptions is true. They looked at four problems involving data structures for graphs or directed graphs. All four of them had some update operations and one type of query. In all four, the lower bound was the same: You must have either amortized query time $n^{1-o(1)}$, or amortized updated time $n^{1-o(1)}$, or preprocessing time $n^{3-o(1)}$. We present the four problems.

1. Type of graph: directed. Update: edges insertions or deletions. Query: the number of strongly connected components.
2. Type of graph: directed with one node s specified. Update: edges insertions or deletions. Query: the number of nodes reachable from s.
3. Type of graph: undirected with one nodes s specified. Update: vertex insertions or deletions. Query: the number of nodes adjacent to s.
4. Type of graph: directed with weights in $\{1, \ldots, n\}$ and two nodes s, t specified. Update: edge insertions or deletions. Query: the max flow from s to t.

18.11 Open Problems

Some of the open problems that are closely related to these topics are as follows:

Open Problem 18.17.

- *Are DIAMETER and APSP subcubic equivalent?*
- *Is there a theorem for approximating RADIUS and MEDIAN similar to the one given by Roditty and V. Vassilevska Williams [RV13] for DIAMETER.*
- *Lastly, the big open problem: Is there a subcubic time algorithm for the APSP problem?*

19

Lower Bounds for Online Algorithms

19.1 Introduction

For most of the problems in this book, there is an input and an output. More precisely, the program gets an input, a finite string representing a formula or graph or something else, the program reads the string, and outputs YES or NO or something else. And then the computation is done.

But there are problems that are not in this format. For example, when requests for memory come in, the program has to decide how to satisfy the request, but then more requests come in. This is an example of an online problem.

Online algorithms are a class of algorithms that, unlike the classical algorithms, do not have access to the entire input at once. Instead the input is revealed gradually to the algorithm and the algorithm must take action while the input is revealed.

Chapter Summary

1. *We define online algorithms and a measure of how well they do: the competitive ration. This is (roughly speaking) the ratio between how well the algorithm does in the worst case and how well an algorithm can do that knows the future. The exact definition is different for a max problem and a min problem.*
2. *We give examples of online algorithms for bin packing and what their competitive ratio is.*
3. *We study algorithms and lower bounds (on the competitive ratio) for the following problems: variants of the secretaries problem, caching, online matching, and online set cover.*
4. *We discuss Yao's lemma, which is used for lower bounds on the competitive ratio for randomized online algorithms.*

For online algorithms, information plays a significant role. Hence many of the hardness results for online algorithms focuses solely on the information observed by the algorithm, and not the time complexity of the algorithm.

19.2 Online Algorithms

Online algorithms differ from the classical algorithms that we studied so far with respect to the timing of observing the input and presenting the output. In all of the problems that we have studied so far, (1) the algorithm is given the entire input all at once and (2) the answer is a static string (e.g., YES or NO or an assignment that satisfies the max number of clauses). However, in an ***online problem*** the input is given in pieces gradually. For example, the input is a sequence of requests to book flight tickets, a sequence of requests for a ride share, and so on. Moreover the answer is a sequence of responses given each time you see one piece of information. Examples of the output would be ticket prices as soon as a booking request is made, or assignment of cars as soon as riders request them.

In analyzing online algorithms, often, we do not care about the running time of the algorithm. The main source of the hardness here is the lack of information, not time complexity assumptions such as $\mathsf{P} \neq \mathsf{NP}$. Since the algorithm does not know the rest of the input when making a decision, it may not be able to make the optimum decisions.

In online problems, we want to find the best solution we can. In order to measure the quality of an online algorithm, we compare the outcome of the online algorithm with the best solution given the whole input (aka, best offline solution). This is the notion of ***competitive ratio***, which is defined as follows.

Definition 19.1. Let $\text{OPT}(\sigma)$ be the cost of an optimal solution given the whole input σ, aka, an optimal offline solution. Let $\textsc{Algorithm}(\sigma)$ denote the cost of our online algorithm on input σ. We say our online algorithm is α-competitive if:

1. (for minimization problems): For all σ, $\textsc{Algorithm}(\sigma) \leq \alpha \text{OPT}(\sigma)$. Note that $\alpha \geq 1$ and the smaller it is, the better the algorithm is.
2. (for maximization problems): For all σ, $\textsc{Algorithm}(\sigma) \geq \alpha \text{OPT}(\sigma)$. Note that $\alpha \leq 1$ and the larger it is, the better the algorithm is.

For simplicity in this chapter, when we say "algorithm" we mean an online algorithm.

19.3 Warm-Up: Bin Packing

In this section, as a warm up, we start by providing a simple example of an online problem and a simple lower bound for it.

> **BIN PACKING**
> *Instance:* b (the bin size) and then a sequence of natural numbers. We call the natural numbers ***items***.
> *Output:* As soon as an item arrives you must either put it in an existing bin or start a new bin and put it into the new bin. The sum of the items in a bin must be $\le b$. The goal is to use as few bins as possible while packing all the items.

We give two online algorithms for the above example. In both cases $b = 10$.

First Fit If there is a bin that the item will fit into, then put the item into the least indexed such bin. If there is no such bin, then start a new bin and put the item there.

On input 3, 3, 3, 3, 3, 3, 4, 4, 4 the result is as follows.

1. Bin 1: 3,3,3.
2. Bin 2: 3,3,3.
3. Bin 3: 4,4.
4. Bin 4: 4.

This is not an optimal solution. An optimal solution would be (3,3,4), (3,3,4), (3,3,4), which uses only three bins.

Next we consider another online algorithm for this problem.

First Fit with a 1-Rule If there is a bin that the item will fit in, and if it is put there the amount in the item is <9, then put the item in that bin. (We are trying to avoid bins that have just one unit of space empty since that might be hard to use.) If there is no such bin, then start a new bin and put the item there.

On input 3, 3, 3, 3, 3, 3, 4, 4, 4 the result of the above algorithm is optimal:

1. Bin 1: 3,3,4.
2. Bin 2: 3,3,4.
3. Bin 3: 3,3,4.

However, there are other inputs where this algorithm is not optimal. For example, on input 3,3,3,2,8 the result is as follows.

1. Bin 1: 3,3,2.
2. Bin 2: 3.
3. Bin 3: 8.

This is not an optimal solution. An optimal solution would be (3, 3, 3), (2, 8), which uses only two bins.

Next we prove that it is indeed impossible to provide *any* online algorithm for this problem that provides optimal solutions for all inputs. Specifically, we show that there is no online algorithm with competitive ratio better than 1.5.

Theorem 19.2. *There is no online algorithm for* Bin Packing *with a competitive ratio better than* 1.5.

Proof. Assume there is a fixed online algorithm. We will create an input that is bad for that algorithm. We will take $b = 10$.

Consider the following two sequences of inputs: $I_1 : (4, 4, 1, 1)$ and $I_2 : (4, 4, 6, 6)$. Note that the first two items of I_1 and I_2 are the same; hence, the online algorithm makes the same decisions for the first two items of I_1 and I_2. There are two cases:

1. **Case 1:** The first two items are assigned to different bins. Now feed in items of size 1 and 1, so the input is I_1. Note that the algorithm has already used two bins. The input I_1 can be done with one bin. Hence, in this case the competitive ratio is at least 2.
2. **Case 2:** The first two items are assigned to the same bin. Now feed in items of size 6 and 6, so the input is I_2. The first 6 has to go into a second bin, and the second 6 has to go into a third bin. Note that the algorithm has already used three bins. The input I_2 can be done with two bins. Hence, in this case the competitive ratio is at least $\frac{3}{2} = 1.5$.

□

Exercise 19.3.

1. Show that first fit algorithm always has competitive ratio ≤ 2.
2. Johnson et al. [JDU+74] show that the competitive ratio of first fit bin packing is $\frac{17}{10}$. Either try to obtain their results (or a weaker version) or go read the paper.

19.4 Prophet Inequality, Secretary, and Prophet Secretary

Consider the following problem. We have a sequence of numbers arriving one by one in an online fashion. Upon arrival of each item we need to decide whether we accept that number and stop or reject that number and continue. The goal is to maximize the number that we accept. It is not hard to see that it is not possible to have any online algorithm with a nontrivial competitive ratio for this problem. Say there are two numbers and the first number that we observe is 1. If we reject this number, the next number might be 0, and the competitive ratio would be 0. If we accept the first number, the next number might be a very large number C and the competitive ratio would be $\frac{1}{C}$. Hence this problem is not that interesting.

We give two classic problems that capture the problem in a more interesting way, and then a third (newish) problem that combines the two problems. Esfandiari et al. [EHLM17] give the history and context for the first two, and is the origin of the third one.

PROPHET INEQUALITY

Instance: Distributions $D_1, \ldots, D_n$ and a sequence $x_1, \ldots, x_n$ of reals such that x_i is drawn using D_i.

Output: As soon as a number arrives you either take it and stop the process or let it go. The goal is to maximize the expected value of the selected number.

Note: The problem is called PROPHET INEQUALITY since the point of it is to compare a prophet, who sees the future, with an onlooker, who does not.

SECRETARY PROBLEM

Instance: A sequence $x_1, \ldots, x_n$ of distinct reals. They are in a random order. So the probability of (say) the numbers being in increasing order is $\frac{1}{n!}$.

Output: As soon as a number arrives, you either take it and stop the process or let it go. The goal is to maximize the probability of getting the largest number.

Note: The problem is called THE SECRETARY PROBLEM since it comes from the following scenario: You want to interview n candidates for a job (as a secretary). You interview them one by one. As soon as a candidate is interviewed, you either hire them and stop the process or let them go. The goal is to maximize the probability of hiring the best candidate.

After you see one, you can compare them to the ones you have already seen and must right away hire them or not. Once you turn someone down, you cannot later hire them. If you turn down the first $n-1$, then you must hire the nth. The goal is to maximize the probability of getting the best one.

Exercise 19.4. Give a strategy for THE SECRETARY PROBLEM that maximizes the probability of hiring the best candidate. What is that probability?

PROPHET SECRETARY

Instance: Distributions $D_1, \ldots, D_n$ and a sequence $x_1, \ldots, x_n$ of reals such that x_i is drawn from $D_{\sigma(i)}$, where σ is some unknown permutation chosen uniformly at random.

Output: As soon as a number arrives, you either take it and stop the process or let it go. The goal is to maximize the expected value of the selected number.

We need to define competitive ratio carefully.

Definition 19.5. Let ALGORITHM be an algorithm for either PROPHET INEQUALITY or PROPHET SECRETARY. The competitive ratio of ALGORITHM is

$$\frac{E[\text{ALGORITHM}]}{E[\max\{x_1,\ldots,x_n\}]}.$$

where ALGORITHM is a random variable that indicates the outcome of the online algorithm and the expectation is taken over the randomness of the input.

We give two simple hardness results: one for PROPHET INEQUALITY and one for PROPHET SECRETARY.

Krengel and Sucheston [KS46] proved the following (using a different notation).

Theorem 19.6. *There is no algorithm for PROPHET INEQUALITY with competitive ratio better than* 0.5.

Proof. We prove this by contradiction. Assume that there exists an algorithm for PROPHET INEQUALITY with competitive ratio $0.5+\varepsilon$. We will use the following two distributions:

D_1: Always return 1.
D_2: Return $\frac{1}{\varepsilon}$ with probability ε, and 0 otherwise.

The algorithm can only be one of the following:

- See the first input is 1. Take it. Then the $E[\text{ALGORITHM}]=1$.
- See the first input is 1. Ignore it. Then the second input is taken. Then $E[\text{ALGORITHM}]=\varepsilon\times\frac{1}{\varepsilon}=1$.

Hence $E[\text{ALGORITHM}]=1$.
On the other hand, we have

$$E[\max\{x_1,x_2\}]=\frac{1}{\varepsilon}\times\varepsilon+1\times(1-\varepsilon)=2-\varepsilon.$$

Hence the competitive ratio is at most $\frac{1}{2-\varepsilon}<0.5+\varepsilon$, which is a contradiction.

□

Esfandiari et al. [EHLM17] showed the following.

Theorem 19.7. *There is no algorithm for PROPHET SECRETARY with a competitive ratio better than* 0.75.

Exercise 19.8. Proof theorem 19.7.

19.5 Caching Problem

In this section we consider the caching problem. The main memory (RAM) and the cache are decomposed into pieces called pages. There are n pages in RAM and k pages in cache. When a page from the RAM is requested, if it exists in the cache, it will be read from there. Otherwise it is read from the RAM and it is put into the cache. However, when we want to put a page in the cache, another page from the cache has to be kicked out and put back into the RAM. Roughly speaking we want the pages in the cache to be ones that are going to be requested either soon or a lot.

For simplicity, we assume that the first k pages have already been requested and are in the cache. Next we formally define the problem.

CACHE

Instance: A sequence of requests for pages from RAM. The cache has size k, and initially there are k pages in the cache.

Output: Every time a request is made, we first check whether the page is already in the cache. If so, then the cost is 0 and we do not need to do anything (though we may keep track of the fact that the request was made). If the page is not in the cache, then we bring it into the cache and put some page that was in the cache back into the main memory. In this case, the cost is 1. The goal is to minimize the total cost, or equivalently minimize the number of times that a requested page is not in cache.

Definition 19.9. A ***fault*** is when a request is made that is not in the cache.

There are several basic algorithms that one may consider for the caching problem such as:

1. **FIFO:** first in, first out. Remove from cache the element that has been there the longest.
2. **LIFO:** last in first out. Remove from cache the element that has been there the shortest.
3. **LRU:** least recently used. Remove from cache the element that has been requested the longest ago.
4. **LFFO:** least frequency first out. Remove from cache the element that has had the least requests.

We show that the LRU algorithm is k-competitive.

Theorem 19.10. *LRU is k-competitive.*

Proof. Let $S = \sigma_1, \ldots, \sigma_N$ be a sequence of requests. We need to show that, for every k faults that LRU makes, the optimal algorithm would make at least one fault.

We assume that the first k different page requests are put into the cache by both the optimal algorithm and LRU. We only consider the page requests after the first k.

We partition S (except for the first k distinct requests) as $\Sigma_1, \ldots, \Sigma_L$ where for all $i \geq 1$, in Σ_i LRU makes exactly k faults. We show that, for all $i \geq 1$, each Σ_i can be mapped to a fault of the optimal algorithm.

Let $1 \leq i \leq L$. Let σ be the last request made in Σ_{i-1}. There are three cases.

1. **Case 1:** The k page requests in Σ_i are distinct from each other and from σ. Hence σ together with the requests in Σ_i have $k+1$ distinct page requests. Hence the optimal algorithm will have a fault.
2. **Case 2:** There is some page τ that LRU faults on twice in Σ_i. Say $\sigma_i = \tau$ and $\sigma_j = \tau$. Then when the σ_i request is made τ is brought into cache; however, by the time the σ_j request is made, τ has been evicted. When it was evicted, it was the least recently requested page. Hence between σ_i and τ being evicted, there must have been $k+1$ distinct requests. Hence the optimal algorithm will have a fault.
3. **Case 3:** All of the faults in Σ_i are distinct from each other, but one of them is σ. Hence σ must have been evicted. From here the reasoning is as in case 2.

□

Now that we have an online k-competitive algorithm, the question arises: Is there a better one? Sadly no.

Theorem 19.11. *There is no deterministic algorithm with competitive ratio better than k for the caching problem.*

Proof. Let ALGORITHM be a deterministic online algorithm for the cache problem. We use an adversarial argument. That is, we feed the algorithm a sequence of page requests that will force it to have competitive ratio $\sim k$. We denote what we feed it $\sigma_1, \ldots, \sigma_N$. We will think of N as large, much larger than k. We will assume k divides N.

1. Initially request $k+1$ distinct pages. This will cause a fault. Let τ_1 be the evicted page.
2. Ask for τ_1. Let τ_2 be the evicted page.
3. Ask for τ_2. Let τ_3 be the evicted page (it could be that $\tau_1 = \tau_3$).
4. Ask for τ_3. Let τ_4 be the evicted page.
5. Do this N times.

If there are N requests, then there will be N faults (after the first k, which we do not count).

The optimal would have been to evict the page that will be requested furthest in the future. Realize that this means that, after a fault, there will not be another fault for at least k requests. Hence OPT causes $\le \frac{N}{k}$ requests.

Since ALGORITHM causes $\sim N - k$ faults and OPT causes $\frac{N}{k}$ faults, the competitive ratio is bounded by:

$$\frac{N-k}{N/k} \sim k.$$

□

The above technique to prove lower bounds on deterministic algorithms is typical: For every deterministic algorithm construct a sequence of requests that cause many faults for that algorithm (which cannot see the future) but fairly few faults for the optimal algorithm (which can see the future).

But what about randomized algorithms? Fiat et al. [FKL+91] showed the following:

Definition 19.12. For all k, H_k is $\sum_{i=1}^{k} \frac{1}{i}$ which is $\ln k + \Theta(1)$.

Theorem 19.13. *Let k be the size of the cache.*

1. *There is a randomized paging algorithm, namely the marking algorithm, with competitive ratio* $2H_k = 2\ln k + O(1)$.
2. *If ALGORITHM is any randomized paging algorithm, then the competitive ratio is* $\geq H_k = \ln k - O(1)$.

We present the ***marking algorithm*** but omit the proof that it is optimal.

1. Initially there are k pages in cache. They are not marked.
2. Whenever a request is made, whether or not it is in cache, it is marked.
3. If a request is made for a page not in cache, then a page is chosen uniformly at random from the unmarked pages to be evicted.
4. When all k pages in cache are marked, all marks except the most recent one are removed.

19.6 Yao's Lemma

Proving a lower bound on the expected runtime of randomized algorithms seems hard! An adversary argument won't work since that just gives one input that the algorithm is bad at.

Yao's lemma [Yao77] allows us to obtain lower bounds on the expected runtime of a randomized algorithm by looking at lower bounds on deterministic algorithms.

Assume you have some problem (e.g., CACHE). There is an associated cost that we are trying to minimize. Picture the following two scenarios:

1. Let $\mathcal{A}$ be a set of deterministic algorithms. Let q be a distribution over a set of inputs.

   Fix an $a \in \mathcal{A}$. Use q to pick the input x. We denote the expected cost $E[c(a,x)]$. We pick the a that minimizes this. Hence we have the quantity $\min_{a\in\mathcal{A}} E[c(a,x)]$.
2. Let $\mathcal{X}$ be a set of inputs. Let p be a distribution over a set of algorithms.

   Fix an $x \in \mathcal{X}$. Use p to pick the algorithm a. We denote the expected cost $E[c(a,x)]$. We pick the x that maximizes this. Hence we have the quantity $\max_{x\in X} E[c(a,x)]$.

The following is Yao's lemma:

Lemma 19.14. $\max_{x\in X} E[c(a,x)] \geq \min_{a\in\mathcal{A}} E[c(a,x)]$.

To get the intuition, we recap what each side of the equation is.

- $\max_{x\in X} E[c(a,x)]$. We are picking the input x to make the expected cost (picking the algorithm via distribution p) as high as possible.
- $\min_{a\in\mathcal{A}} E[c(a,x)]$. We are picking the algorithm a to make the expected cost (picking the input via distribution q) as low as possible.

19.7 Online Matching

> ONLINE MATCHING
>
> *Instance:* A bipartite graph $((U, V), E)$ is going to be the final input. Initially we have the set V in advance, which are called ***offline vertices***. The vertices in U arrive one by one, which are called ***online vertices***. At the time we get u_i, we get all of its neighbors as well.
>
> *Output:* When we receive an online vertex u_i, we need to match u_i to one of its neighbors (which will be an offline vertex) that has not already been matched, if there is one. This decision is irrevocable. The goal is to maximize the number of matches.

The ***greedy algorithm*** for this problem will, given a new u_i, match it to the least indexed, still available, $v \in V$ such that $(u_i, v) \in E$. (This algorithm does not look particularly greedy. We discuss a truly greedy algorithm for the weighted case in the exercises.)

Exercise 19.15. Let $G = ((U, V), E)$ be a bipartite graph. We will assume here and throughout this paper that $|U| = |V|$.

1. Show that the greedy algorithm for online matching always returns a maximal matching. Note that this is maximal, meaning that no edge can be added.
2. Show that for any bipartite graph, a maximal matching is $\geq \frac{1}{2}|V|$.

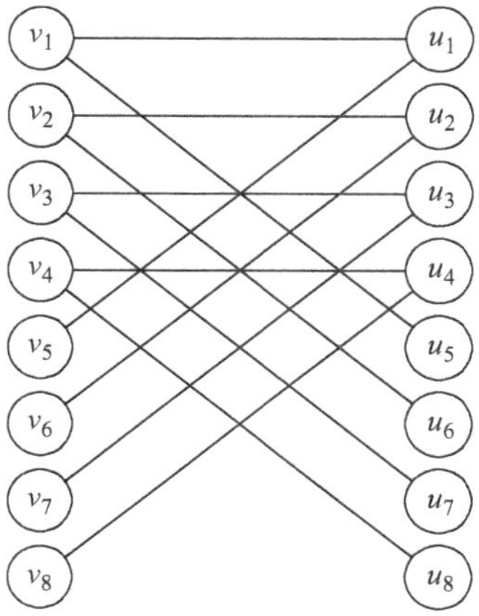

Figure 19.1
Greedy online matching has competitive ratio at most $\frac{1}{2}$.

3. For all n, give an example of a graph where $|U| = |V| = n$ and an arrival order for U where (1) the graph has a matching of size n, but (2) the greedy algorithm produces a matching of size $\frac{n}{2}$.

   Hint: Use the bipartite graph where (a) for $1 \le i \le \frac{n}{2}$ there are edges (u_i, v_i) and $(u_i, v_{i+(n/2)})$, and (b) for $\frac{n}{2} + 1 \le i \le n$ there is an edge $(u_i, v_{i-(n/2)})$. See figure 19.1.
4. Show that the greedy algorithm has competitive ratio $\frac{1}{2}$. (This follows from the parts 2 and 3).
5. Show that *any* deterministic algorithm will have competitive ratio $\le \frac{1}{2}$.
   Hint: Use an adversary argument.

We now consider a promising randomized algorithm, just to show that it does not do well after all.

Theorem 19.16. *Consider the randomized algorithm that picks a match for u_i at random from the vertices that are available.*

1. *If you run this algorithm on the graph from exercise 19.15.3, the expected competitive ratio is $\frac{3}{4}$.*
2. *Let $G = (V, U, E)$ be the following bipartite graph (see figure 19.2). $V = \{v_1, \ldots, v_n\}$. $U = \{u_1, \ldots, u_n\}$. For all i, there is an edge (u_i, v_i). For all $1 \le i \le \frac{n}{2}$ for all $\frac{n}{2} + 1 \le j \le n$, there is an edge (u_i, v_j). If you run the algorithm on this graph, then the expected competitive ratio is $\frac{1}{2}$.*

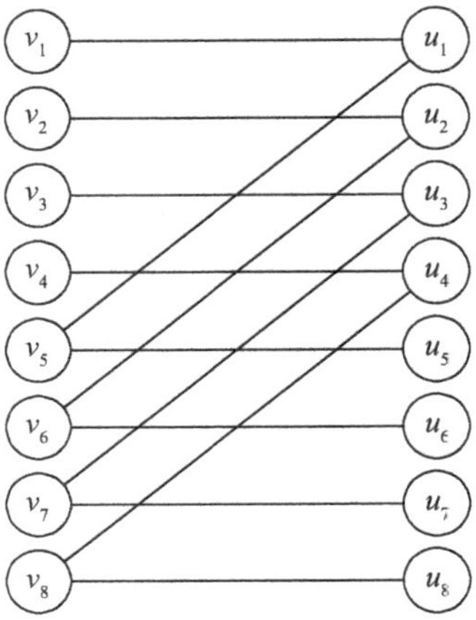

Figure 19.2
Random online matching has competitive ratio at most $\frac{1}{2}$.

Proof. 1) For $1 \leq i \leq \frac{n}{2}$ when u_i arrives, the probability that it will match to $v_{i+(n/2)}$ is $\frac{1}{2}$. Hence the expected number of $v_{n/2}, \ldots, v_n$ that are available for $u_{n/2}, \ldots, u_n$ is $\frac{n}{4}$. So the expected number of matches is $\frac{n}{2} + \frac{n}{4} = \frac{3n}{4}$. The optimal is n. Hence the competitive ratio is $\frac{3}{4}$.

2) The vertices $u_1, u_2, \ldots, u_n$ arrive in that order. For $1 \leq i \leq \frac{n}{2}$, if u_i gets matched to v_i, that's good (in terms of maximizing matches) since it does not take a vertex that is the only neighbor of some u_i with $i \geq \frac{n}{2} + 1$. Hence we call such a vertex ***good***.

$$\Pr[u_1 \text{ is good}] = \frac{1}{n/2+1}$$

$$Pr[u_2 \text{ is good}] = Pr[u_1 \text{ is good}]\frac{1}{n/2+1} + Pr[u_1 \text{ is bad}]\frac{1}{n/2} \leq \frac{1}{n/2}.$$

We leave it as an exercise to show that

$$Pr[u_i \text{ is good}] = 1/(n/2 - i + 2).$$

Thus we have:

$$E(\text{ALGORITHM}) \leq \sum \frac{1}{n/2 - i + 2} + n/2 = O(\log n) + n/2.$$

Note that there is a matching of size n. Hence the competitive ratio is

$$\frac{O(\log n) + (n/2)}{n} \sim \frac{1}{2}.$$

□

To recap: Any deterministic greedy algorithm, and the simple randomized algorithm, have competitive ratio 2. Is there a randomized algorithm with competitive ratio <2? Yes! Karp et al. [KVV90] (see also Birnbaum and Mathieu [BM08] and Plotkin [Plo13]) showed the following.

Theorem 19.17. *Let e be Euler's number, roughly 2.7185.*

1. *There is a randomized algorithm for online matching with competitive ratio* $\frac{e-1}{e} \sim 0.632$.
2. *All randomized algorithms for online matching have competitive ratio* $\leq \frac{e-1}{e} \sim 0.632$.

19.8 Online Set Cover

In this section we consider the online set cover problem. We first recall the standard offline set cover problem.

> SET COVER
> *Instance:* $n \in \mathbb{N}$ and E a collection of subsets of $\{1, \ldots, n\}$. We will let $m = |E|$.
> *Question:* What is the size of the smallest collection of sets from E that contain (cover) all of the elements of F.

Next we define the online set cover problem (OLS).

> OLS
> *Instance:* $n \in \mathbb{N}$ that is known ahead of time. E, a collection of subsets of $\{1, \ldots, n\}$. E is known ahead of time. We will let $m = |E|$. Elements of $\{1, \ldots, n\}$, together with a list of which sets in E cover them, arrive one by one.
> *Output:* When we receive an elements $x \in F$ and a set of subsets $E_1, \ldots, E_k$ that each cover x, pick one. This decision is irrevocable. The goal is to minimize the number of sets in E that are used to cover all of the elements received. (Note that we may well not receive all of the elements.)

This online problem is very different from both CACHE and ONLINE MATCHING. For those two problems, the offline version was in P. For SET COVER the offline version is NP-hard.

We can obtain an easy lower bound on OLS by assuming P $\neq$ NP.

Theorem 19.18. *Assume P $\neq$ NP. The competitive ratio for any deterministic polynomial time algorithm for OLS is* $\geq (1 - o(1)) \ln n$ *when* $m \sim n$.

Exercise 19.19. Proof theorem 19.18.
(**Hint:** Use the lower bound on approximate set cover.)

We state the known lower bounds on the competitive ratio for OLS. Note that there are both results with and without assumptions.

Theorem 19.20.

1. *(Alon et al. [AAA+09]) There exists a deterministic algorithm for OLS with competitive ratio* $O(\log n \log m)$.
2. *(Alon et al. [AAA+09]) Any deterministic algorithm for OLS has competitive ratio* $\Omega(\frac{\log n \log m}{\log\log m + \log\log n})$. *Note that this does not use any assumption such as* $P \neq NP$. *(The bound needs the condition* $\log n \leq m \leq e^{n^{1/2-\delta}}$*.)*
3. *(Korman [Kor04]) Assume either* $P \neq NP$ *or* $NP \not\subseteq BPP$. *There exists a constant d such that there is no polynomial time randomized algorithm for OLS with competitive ratio* $\leq d \lg n \lg m$.

19.9 Further Results

1. In section 19.7 we looked at online matching for bipartite graphs where the *vertices* arrive. We found that (a) deterministic algorithms always have competitive ratio $\leq \frac{1}{2}$, (b) there is a randomized algorithm with competitive ratio $\frac{e-1}{e}$, and (c) $\frac{e-1}{e}$ is the best one can do.

   What about general graphs? What if edges arrive? Gamlash et al. [GKM+19] showed that (a) for vertex arrivals in general graphs, there is a randomized algorithm with competitive ratio $\frac{1}{2} + \Omega(1)$, and (b) for edge arrivals, randomization does not help.
2. ROLE MATCHMAKING is a problem where players of different skill levels arrive and must be assigned to a team as soon as they arrive. The goal is to have the teams be balanced so that no team dominates. This can get very complicated since different skills is not one-dimensional. For example, in soccer, a team may need a good goalkeeper more than a great midfielder. This problem has immediate applications to many popular online video games such as *League of Legends* and *Dota 2*. Alman and McKay [AM17] view this as a dynamic data structures problem. The show (a) assuming the 3SUM CONJECTURE conjecture, any data structure for this problem requires $n^{1-o(1)}$ time per insertion or $n^{2-o(1)}$ time per query, and (2) there is an approximation algorithm that takes $O(\log n)$ per operation.

20

Lower Bounds on Streaming Algorithms

20.1 Introduction

The following problem is easy in the standard model of computation: Given a graph G and a number k, does G have a connected component of size $\geq k$? But what if the following hold?

1. The graph is given to you by a stream of edges from G in a random order (n is known).
2. n is large, so we want to avoid storing the entire graph. Our main complexity measure will be how much space you need.

We wil later see that this problem requires $\Omega(n)$ space.

This is an example of a ***streaming problem***.

Chapter Summary

1. *We will define several types of streaming models and give examples of streaming algorithms for these models.*
2. *We will introduce* ***communication complexity****, which is the main tool used for lower bounds on space for streaming problems.*
3. *We will obtain lower bounds on space for several streaming problems.*

20.2 Streaming Algorithms

Informally, streaming algorithms take a large set of data and process it without seeing the entire stream, and with limited space. The study of this field was initiated by a seminar paper of Alon et al. [AMS99].

Streaming algorithms are used to process very long data streams. Online social networks, such as Facebook and X (formerly known as Twitter), produce such streams.

Definition 20.1. ***Streaming algorithms*** are algorithms for processing data streams in which the input is presented as a sequence of items (often numbers or edges of a graph) and can be examined in only a few passes (typically just one) by using relatively little space (much less than the input size), and also limited processing time per item. We often produce approximate answers based on a summary or "sketch" of the data stream in memory. A common technique for creating a "sketch" is sampling at random.

Example 20.2. This problem is called ***Missing 1***. The input will be all of the numbers in $\{1, \ldots, n\}$ appearing once, except there is one number that will not appear. The output will be that one number. Space is $O(\log n)$, and there is only one pass through the data. Because of the space limitation, you cannot keep all or even most of the stream in memory.

Here is the algorithm: As the numbers come in, keep a running sum s of them. This sum will take only $O(\log n)$ space. Once you have the sum s, compute $\frac{n(n+1)}{2} - s$. That's your number!

Exercise 20.3.

1. Show that MISSING 1 requires $\Omega(\log n)$ space.
2. Define MISSING k in the obvious way. Get upper and lower bounds on how much space is needed to solve MISSING k.
3. For MISSING k (including $k = 1$), can you use less space if you allow more passes through the stream?

The data stream of the input can be a sequence of items such as integers or edges of a graph. If they are integers, then there are two common models: (1) in the ***turnstile model*** we allow the integers to be negative, and (2) in the ***cash register model*** the inputs must be positive.

In many cases when the streaming data are items, a vector $\vec{a} = (a_1, \ldots, a_n)$ is initialized to the zero vector $\vec{o}$ and the input is a stream of updates in the form of $< i, c >$, in which a_i is incremented by an integer c—that is, $a_i = a_i + c$. Note that c can be negative here. Below are four examples of streaming problems for items based on the vector $\vec{a}$.

1. Evaluate the kth frequency moment: $F_k(\vec{a}) = \sum_{i=1}^{n} a_i^k$.
2. Find heavy hitters, which is to find all the elements i with frequency $a_i > T$.
3. Count the number of distinct elements.
4. Calculate entropy: $E(\vec{a}) = \sum_{i=1}^{n} \frac{a_i}{m} \log \frac{a_i}{m}$, where $m = \sum_{i=1}^{n} a_i$.

Note here that in all streaming algorithms, we want to minimize space, and then update time, even through multiple passes. The accuracy of the algorithm is often defined as an ***(ε, δ)-approximation***. It means the algorithm achieves an error of less than ε with probability $1 - \delta$.

We will need the following notation.

Notation 20.4. $f = \tilde{O}(g)$ if there exists c, n_o, k such that for all $n \geq n_0, f(n) \leq c(\log n)^k g(n)$.

20.3 Streaming for Graph Algorithms

There are two main versions of graph streaming:

1. Insertion only. Edges are added to the graph over time.
2. Dynamic. Edges are added to the graph but can also be deleted. If (i,j) appears twice, then the first time means to add it and the second time means to delete it. We will not be discussing this model.

Here we present four models of graph streaming algorithms:

1. ***Streaming***: In this model the input is a stream of edges. The algorithm knows how many vertices there are but does not get to store them (in contrast to ***semi-streaming***). Typically the number of vertices is n and we want to prove an $\Omega(n)$ lower bound. Unless we specify otherwise, assume that a theorem is about the streaming model.
2. ***Semi-streaming***: In this model the input is a stream of edges. Hence the vertices are already in storage. Since these algorithms are using $\tilde{O}(n)$ space for the edges, the main concern is that these algorithms take $\ll \tilde{O}(n^2)$ space. There are variants where an edge can be removed.
3. ***Parameterized streaming***: This model is a combination of fixed-parameter tractable algorithms and streaming algorithms. There is a parameterized problem (e.g., VERTEX COVER with parameter k). You want a streaming algorithm for it where the space is a function of the parameter (e.g., $\tilde{O}(k)$) or involves the parameter in some way (e.g., $k\text{polylog } n$).
4. ***Sliding window***: This model is a more generalized one, in which the function of interest is computing over a fixed size window of the stream according to the time and the update. Note that when new items are added to the window, items from the end of the window are deleted.

In fact, there are lots of interesting lower bounds for massive graph problems, especially in the semi-streaming model where edges E are streaming in as inputs (often in adversarial order, but sometimes in a random order) with the bounded storage space, which is $\tilde{O}(n) = O(n \cdot \text{polylog } n)$, where $n = |V|$.

20.4 Semi-Streaming Algorithm for MAXIMUM MATCHING

> MAXIMUM MATCHING
> *Instance:* $G = (V, E)$, a graph.
> *Output:* The largest set of disjoint edges in G. (Any set of disjoint edges is called ***a matching***.)

We will look at semi-streaming algorithms for maximum matching with regard to (a) number of passes, (b) space used, and (c) approximation factors.

We look at an easy streaming algorithm to establish a baseline.

Theorem 20.5. *There is a semi-streaming algorithm for* MAXIMUM MATCHING *where (a) as soon as an edge comes in you need to decide whether or not to put it in the matching, and you cannot change your mind; (b) there is only one pass through the data; (c) the space used is $\tilde{O}(n)$; and (d) the algorithm has approximation factor $\frac{1}{2}$—that is, $Alg(\text{worst case}) \geq 1/2 \cdot Opt$.*

Proof.

1. The vertices are stored. Each vertex requires $O(\log n)$ bits to store, so the total storage is $\tilde{O}(n)$. Initially all vertices are unmarked and the set M is empty.
2. Whenever a new edge $e=(x, y)$ comes in, if both x and y are unmarked, add e to M and mark both end vertices.

Since each selected edge ruins at most two edges in the optimal solution, the approximation factor of this algorithm is 2—that is, $Alg(\text{worst case}) = 1/2 \cdot Opt$. □

Better results are known. Farhadi [FHM+20] showed the following:

1. There is a semi-streaming algorithm for maximum matching in bipartite graphs that uses (a) 1 pass, (2) $\tilde{O}(n)$ space, and (3) has approximation factor 0.6.
2. There is a semi-streaming algorithm for maximum matching in general graphs that uses (a) 1 pass, (2) $\tilde{O}(n)$ space, and (3) has approximation factor 0.545.

Before going into some hardness proofs, let's get insight into communication complexity, which is used as the most common technique for computing lower bounds in streaming algorithms.

20.5 Communication Complexity

In this section we present just the communication complexity we need for our lower bounds. An excellent reference that includes proofs of all the theorems in this section is the book by Kushilevitz and Nisan [KN97].

Assume Alice has $x \in \{0, 1\}^n$ and Bob has $y \in \{0, 1\}^n$. They want to know whether $x=y$. Alice *could* just send Bob the entire string x, and then Bob could send back 0 if $x \neq y$ and 1 if $x=y$. The entire protocol takes $n+1$ bits. Can they accomplish this with less bits? What if we allow randomization and a small probability of error? This is the beginning of *communication complexity*.

Exercise 20.6.

1. Show that any deterministic protocol for determining whether $x = y$ requires $\geq n$ bits.
2. Show that there is a randomized protocol for determining whether $x = y$ that uses private coins, $O(\log n)$ bits, and has probability of error $\leq \frac{1}{n}$.
   Hint: View the sequence $a_{n-1} \cdots a_0$ as the polynomial $a_{n-1}x^{n-1} + \cdots + a_0$ over mod p for a suitably chosen prime p.

Definition 20.7. Let $f : X \times Y \to Z$ be a function. The two-party communication model consists of two players, Alice and Bob. Alice is given an input $x \in X$ and Bob is given an input $y \in Y$. Their goal is to compute $f(x,y)$. Neither knows the other player's input. We will be concerned with how many bits they need to communicate in order to compute f.

1. A ***protocol for f*** is just what you think it is: an algorithm for Alice and Bob to compute f. Formally it is a decision tree where what a player sends is based on what they have already seen and their own input.
2. The ***best protocol*** for f is the one with the smallest worst case for number of bits needed.
3. The ***communication complexity*** of f is the worst case of the best protocol.
4. There are many types of protocols: (1) deterministic, (2) randomized with a small probability of error, and (3) ***one-way protocol***, which means that only one player sends information and the other one receives information, where both roles, sender and receiver, need to be specified and only the receiver needs to be able to compute f.

INDEX, INDEX SAME, and DISJOINT (short for disjointness) are three well-known problems in the area of communication complexity.

INDEX
Instance: Alice has a string $x \in \{0, 1\}^n$ and Bob has a natural number $i \in [n]$.
Outcome: Bob wants to know x_i. We use the one-way model so Alice sends Bob a string, and just from that, Bob needs to find x_i.

INDEX SAME
Instance: Alice has a string $x \in \{0, 1\}^n$ and Bob has a natural number $i \in [n-1]$.
Outcome: Bob wants to know whether $x_i = x_{i+1}$. We use the one-way model, so Alice sends Bob a string and just from that Bob needs to determine whether $x_i = x_{i+1}$.

> DISJOINT
> *Instance:* Alice has a string $x \in \{0,1\}^n$, and Bob has a string $y \in \{0,1\}^n$.
> *Outcome:* They both want to know whether the sets that x and y represented (as bit vectors) are disjoint. The communication is two-way and they can have as many rounds as they want.

Definition 20.8. A ***randomized protocol*** is one that uses public coins and has probability of correctness $\geq \frac{2}{3}$.

Theorem 20.9.

1. *(Kremer et al. [KNR99]) INDEX requires $\Omega(n)$ bits (in the one-way communication model). This lower bound also holds for both deterministic and randomized protocols.*
2. *(Follows from part 1) INDEX SAME requires $\Omega(n)$ bits (in the one-way communication model). This lower bound also holds for both deterministic and randomized protocols.*
3. *(Kalyanasundaram and Schintger [KS92], but see Razborov [Raz92] for a different approach) DISJOINT requires $\Omega(n)$ bits (even though Alice and Bob may use many rounds of communication). This lower bound holds for both deterministic and randomized protocols. The lower bound still holds when restricted to (x, y), where $\sum x_i = \sum y_i = \lfloor n/4 \rfloor$.*

Exercise 20.10.

1. Prove that INDEX requires $\geq n$ bits in the one-way deterministic model.
2. Prove that if INDEX SAME can be done with $o(n)$ bits, then INDEX can be done with $o(n)$ bits.
   Hint: Given (x, i), an instance of INDEX, create $(x', 2i - 1)$ an instance of INDEX SAME so that $x_i = 1$ if and only if $x'_i = x'_{i+1}$. Do this by replacing $x_i = 0$ with 01 and $x_i = 1$ with 11.
3. Prove that INDEX SAME cannot be done with $o(n)$ bits. (This is just the contrapositive of part 2. We list it here so we can refer to it.)

20.6 Lower Bounds on Graph Streaming Problems

In this section we reduce communication problems to several streaming graph problems. Since the communication problems have unconditional lower bounds (see theorem 20.9), we obtain unconditional lower bounds on the space needed for the streaming graph problems.

We only use the lower bounds on the deterministic communication complexity. However, in all cases, there is a lower bound on the randomized communication complexity. Hence we really obtain lower bounds on the space needed for randomized streaming algorithms for graph problems.

All of the results in this chapter are due to Tiseanu [Tis13].

20.6.1 Lower Bound Using Index: Max Connected Component

> Max Connected Component(k), $k \geq 3$
> *Instance:* A forest $G = (V, E)$.
> *Question:* Is there a connected component of size $\geq k$? Note that k is not part of the input.

Theorem 20.11. *Let $k \geq 3$. Any single-pass streaming graph algorithm that solves the Max Connected Component(k) problem on a forest needs $\Omega(n)$ space.*

Proof. Proved by reduction from Index to Max Connected Component(k).

Assume there is an $o(n)$ space streaming algorithm ALG for Max Connected Component(k). Note that it works for any arrival order of edges. We use ALG in the following protocol for Index. The protocol will take $o(n)$ space, which contradicts theorem 20.9.1. Hence we will have that no such ALG exists.

1. Alice has $x_1 \cdots x_n \in \{0, 1\}^n$ and Bob has $i \in [n]$. Our goal is that Alice gives Bob a string of length $o(n)$ and then Bob knows if $x_i = 1$.
2. Alice and Bob construct different parts of a graph. The vertices are $V_l \cup V_r \cup V_d$, where

$$V_l = \{l_1, l_2, \ldots, l_n\}$$
$$V_r = \{r_1, r_2, \ldots, r_n\}$$
$$V_d = \{d_1, d_2, \ldots, d_{k-2}\}$$

   a. Alice constructs the graph on vertices $V_l \cup V_r$ by letting

$$E_A = \{(l_j, r_j) \mid x_j = 1\}.$$

   b. Bob constructs the graph on vertices $\{r_i\} \cup V_d$ by letting

$$E_B = \{(r_i, d_1)\} \cup \{(d_i, d_{i+1}) \mid 1 \leq i \leq k-3\}.$$

   (See figure 20.1 for an example when $x = 1011$ and $k = 4$.)

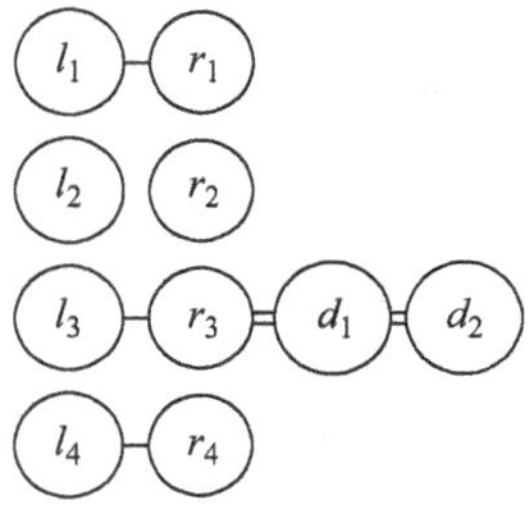

Figure 20.1
Instance of MAX CONNECTED COMPONENT(k) based on INDEX(1011, 3).

3. Alice runs E_A through the streaming algorithm ALG. Since it is an $o(n)$ space algorithm, when she is done, there are $o(n)$ bits in memory. She tells Bob these bits. Note that this is just $o(n)$ bits.
4. Bob initializes the memory to those bits and then runs the streaming algorithm on E_B.
5. (Comment, not part of the algorithm.) If $x_i = 1$, then the path l_i, r_i, $d_1, \ldots, d_{k-3}$ is a connected component of size k. If $x_i = 0$, then the longest connected component is of size $k - 1$.
6. If when Bob finishes the streaming algorithm the answer is YES, then he knows that $x_i = 1$; if the answer is NO, then he knows that $x_i = 0$. The total one-way communication is $o(n)$. □

20.6.2 Lower Bound Using INDEX SAME: IS TREE

IS TREE
Instance: A graph G
Question: Is G a tree?

Theorem 20.12. *Any single-pass streaming algorithm solving Is TREE needs $\Omega(n)$ space.*

Proof. Essentially we use a similar proof strategy to the proof of theorem 20.11: prove by reduction from INDEX SAME to Is TREE. Assume there is an $o(n)$ space streaming algorithm ALG for Is TREE. We use ALG in the following protocol for INDEX SAME. The protocol will take $o(n)$ space, which contradicts theorem 20.9.2. Hence we will have that no such ALG exists.

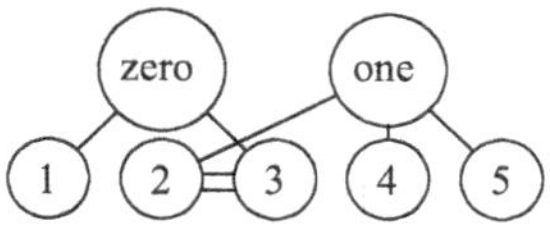

– are Alice's edge. = are Bob's edge.

Figure 20.2
Instance of IsTREE(G) based on INDEX SAME(01011, 2).

1. Alice has $x_1 \cdots x_n \in \{0, 1\}^n$ and Bob has $i \in [n-1]$. Our goal is that Alice gives Bob a string of length $o(n)$ and then Bob knows if $x_i = x_{i+1}$ or not.
2. Alice and Bob construct different parts of a graph. The vertices are

$$V = \{zero, one, 1, 2, \ldots, n\}.$$

(The above list is not a typo. We really do use *zero* and *one* for vertices.)

a. Alice constructs the graph on vertices V

$$E_A = \{(zero, i) \mid x_i = 0\} \cup \{(one, i) \mid x_i = 1\}.$$

b. Bob constructs the graph with just one edge $E_b = \{(i, i+1\})$. (See figure 20.2 for an example when $x = 01011$ and $k = 2$.)

3. Alice runs E_A through the streaming algorithm ALG. Since it is an $o(n)$ space algorithm, when she is done there are? $o(n)$ bits in memory. She tells Bob these bits. Note that this is just $o(n)$ bits.
4. Bob initializes the memory to those bits and then runs the streaming algorithm on E_B.
5. (Comment, not part of the algorithm.) If $x_i = x_{i+1} = 0$, then $(zero, x_i)$, (x_i, x_{i+1}), and $(x_{i+1}, zero)$ are all edges so the graph is not a tree. Similar for $x_i = x_{i+1} = 1$. If $x_i \neq x_{i+1}$, then there are no cycles (exercise), so G is a tree.
6. If when Bob finishes the streaming algorithm the answer is YES (G is a tree), then he knows that $x_i \neq x_{i+1}$; if NO, then he knows that $x_i = x_{i+1}$. The total one-way communication is $o(n)$. □

20.6.3 Lower Bound Using INDEX: PERFECT MATCHING

> PERFECT MATCHING
> *Instance:* A graph G.
> *Question:* Does G have a perfect matching? (A disjoint set of edges such that every vertex is in one of the edges.)

Theorem 20.13. *Any single-pass streaming algorithm for* PERFECT MATCHING *needs* $\Omega(m) = \Omega(n^2)$ *space.*

Proof. We prove this by reduction from INDEX to PERFECT MATCHING.

We assume that the string x in INDEX is an $n \times n$ array and Bob wants to compute x_{ij}. Then an instance of INDEX can be written as (x, i, j). Then we need space $\Omega(n^2)$ to solve this INDEX(x, i, j) problem.

Assume there is an $o(n^2)$ space streaming algorithm ALG for PERFECT MATCHING. We use ALG in the following protocol for INDEX. The protocol will take $o(n)$ space, which contradicts theorem 20.9.1. Hence we will have that no such ALG exists.

1. Alice has the doubly indexed string x (so all of the x_{ij} as $1 \leq i, j \leq n$). Bob has $(i, j) \in [n] \times [n]$. Our goal is that Alice gives Bob a string of length $o(n)$ and then Bob knows if $x_{ij} = 1$.
2. Alice and Bob construct different parts of a graph. The vertices are

$$V = \{l_1, l_2, \ldots, l_{n-1}\} \cup \{r_1, r_2, \ldots, r_{n-1}\} \cup \{1, 2, \ldots, n\} \cup \{1', 2', \ldots, n'\}.$$

   They will be arranged as seen in figure 20.3.

   a. Alice constructs the graph on vertices V

$$E_A = \{\{i, j'\} \mid x_{i,j} = 1\}.$$

   b. Bob constructs the graph on vertices V

$$E_B = \{\{l_k, k\} | 1 \leq k < i\} \cup \{\{l_k, k+1\} | i \leq k \leq n-1\} \cup \\ \{\{r_k, k\} | 1 \leq k < j\} \cup \{\{r_k, k+1\} | j \leq k \leq n-1\}.$$

   (See figure 20.3 for an example where

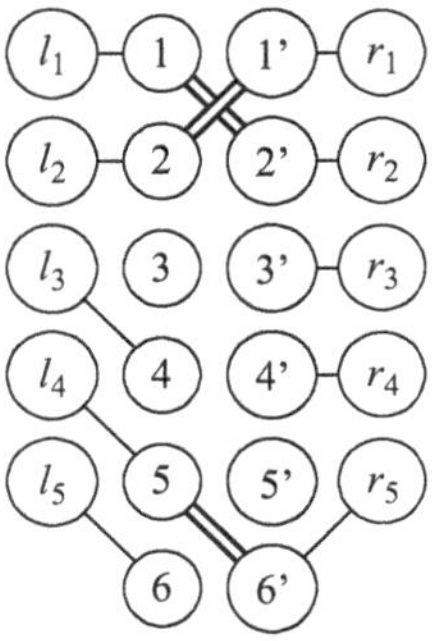

Figure 20.3
Instance of PERFECT MATCHING(G) based on INDEX(x, 3, 5).

$$x = \begin{matrix} 0 & 1 & 0 & 0 & 0 & 0 \\ 1 & 0 & 0 & 0 & 0 & 0 \\ 0 & 0 & 0 & 0 & 0 & 0 \\ 0 & 0 & 0 & 0 & 0 & 0 \\ 0 & 0 & 0 & 0 & 0 & 1 \\ 0 & 0 & 0 & 0 & 0 & 0 \end{matrix}$$

and Bob's has (3, 5).)

3. Alice runs E_A through the streaming algorithm ALG. Since it is an $o(n^2)$ space algorithm, when she is done, there is $o(n^2)$ bits in memory. She tells Bob these bits. Note that this is just $o(n^2)$ bits.
4. Bob initializes the memory to those bits and then runs ALG on E_B.
5. (Comment, not part of the algorithm.) We leave the argument that

$$x_{ij} = 1 \textit{ if and only if } G \textit{ has a perfect matching}$$

to the reader.
Hint: All connected components of the graph are of length 2 or 4 except for the nodes i and j'. There status depends on x_{ij}.
6. If when Bob finishes the streaming algorithm the answer is YES (G has a matching), then he knows that $x_{ij} = 1$; if NO, then he knows that $x_{ij} = 0$. The total one-way communication is $o(n^2)$. □

20.6.4 Lower Bounds Using INDEX: SHORTEST-PATH

> SHORTEST PATH
> *Instance:* An unweighted graph G and two vertices v, w.
> *Question:* What is the length of the shortest path from v to w?

We show that not only does any one-pass streaming algorithm for this problem require $\Omega(n^2)$ space, even approximating the problem requires $\Omega(n^2)$ space.

Theorem 20.14. *Any single-pass streaming algorithm that approximates* SHORTEST PATH *with factor better than $\frac{5}{3}$ needs $\Omega(n^2)$ space. (So the algorithm produces a number that is $<\frac{5}{3} \times$ the length of the shortest path.)*

Proof. The proof here is similar to the same with the proof of the PERFECT MATCHING problem. We prove this by reduction from INDEX to SHORTEST-PATH.

We assume that the string x in INDEX is an $n \times n$ array and Bob wants to compute x_{ij}. Then an instance of INDEX can be written as (x, i, j). Then we need space $\Omega(n^2)$ to solve this INDEX(x, i, j) problem.

Assume there is an $o(n^2)$ space streaming algorithm ALG for a better than $\frac{5}{3}$ approximation to shortest-path (note that it is strictly better than $\frac{5}{3}$ optimal).

We use ALG in the following protocol for INDEX. The protocol will take $o(n)$ space, which contradicts theorem 20.9. Hence we will have that no such ALG exists.

1. Alice has x, which is an $n \times n$ matrix of 0's and 1's. Bob has $(i,j) \in [n] \times [n]$. Our goal is that Alice gives Bob a string of length $o(n)$ and then Bob knows if $x_{ij} = 1$.
2. Alice and Bob construct different parts of a graph G. v and w will be vertices of G and (G, v, w) will be in the input to the shortest-path problem. The vertices are

$$V = \{i, j, v, w\} \cup \{1, \ldots, n\} \cup \{1', \ldots, n'\}.$$

They will be arranged as seen in figure 20.4.

a. Alice constructs the graph on vertices V

$$E_A = \{\{i, j'\} \mid x_{i,j} = 1\}.$$

b. Bob constructs the graph on vertices V

$$E_B = \{(v, i), (j, w)\}.$$

(See figure 20.4 for an example where

$$x = \begin{matrix} 0 & 0 & 0 & 0 & 0 & 0 \\ 0 & 0 & 1 & 0 & 0 & 0 \\ 0 & 0 & 1 & 0 & 0 & 1 \\ 0 & 0 & 0 & 0 & 0 & 0 \\ 0 & 0 & 0 & 1 & 0 & 1 \\ 0 & 0 & 0 & 0 & 0 & 0 \end{matrix}$$

and Bob's has $(2, 5)$.)

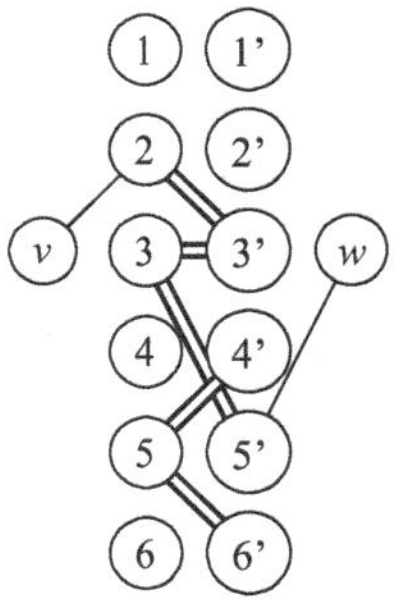

= are Alice's edges. – are Bob's edge.

Figure 20.4
Instance of SHORTEST-PATH(v, w) based on INDEX$(x, 3, 4)$.

3. Alice runs E_A through the streaming algorithm ALG. Since it is an $o(n^2)$ space algorithm, when she is done, there is $o(n^2)$ bits in memory. She tells Bob these bits. Note that this is just $o(n^2)$ bits.
4. Bob initializes the memory to those bits and then runs the streaming algorithm on E_B.
5. (Comment, not part of the algorithm.)

   If $x_{ij} = 1$ then there is a path from v to w of length 3: (v, i), (i, j), (j, w). Hence a $<\frac{5}{3}$-approximation algorithm will return a number $< 3 \times \frac{5}{3} = 5$.

   If $x_{ij} = 0$, then the shortest path from v to w is at least 5: The path begins with (v, i). The next edge must be of the form (i, j'). Since $j' \neq j$, j' is not adjacent to w. So the next edge must be of the form (j', k). From node k, the shortest distance possible to w is 2. Hence the shortest path is ≥ 5. Therefore a $<\frac{5}{3}$-approximation algorithm will return a number ≥ 5.
6. When Bob finishes the streaming algorithm the answer is α. If $\alpha < 5$, then Bob knows $x_{ij} = 1$. If $\alpha \geq 5$, then Bob knows $x_{ij} = 0$. The total one-way communication is $o(n^2)$. □

For the next exercise we will consider the following streaming problem.

> TREE DIAMETER
> *Instance:* A number $k \geq 3$ and a tree T. Note that the number is given first and can be easily stored, while the tree will be streamed.
> *Question:* Is the diameter of T at least k?

Exercise 20.15. Find $k_0, k_1 \in \mathbb{N}$ such that the following hold.

1. If $k \leq k_0$, then TREE DIAMETER has a single-pass $O(\log n)$ space algorithm.
2. If $k \geq k_1$, then any single-pass streaming algorithm for TREE DIAMETER requires at least $\Omega(n)$ space.

Try to make k_0 and k_1 close together.

20.7 Further Results

20.7.1 Graph Problems

For all of the problems listed, the model is streaming with edge arrivals and n is the number of vertices.

1. For MAXIMAL MATCHING, Esfandiari et al. [EHL+18] gives an algorithm that, with high probability, approximates the size of a maximum matching within a constant factor using $\tilde{O}(n^{2/3})$ space.

2. For Weighted Maximal Matching, Crouch and Stubbs [CS14] gives a $(4+\varepsilon)$ approximation algorithm that applies in semistreaming, sliding window, and MapReduce models. Chen et al. [CHK+21] looked at the problem where you want to have a matching of size k. They showed a lower bound in the one-pass streaming model of $\Omega(k^2 W \log(W))$, where W is the number of distinct weights.
3. Parameterized Vertex Cover with parameter k was studied by Chitnis et al. [CCHM15]. They proved a tight lower bound on the space of $\Omega(k^2)$ for randomized streaming algorithm.
4. Minimum Spanning Tree Estimation: Given a weighted undirected graph and ε, find a spanning tree that has weight $\leq(1+\varepsilon)$OPT where OPT is the weight of the minimal spanning tree. Assadi and N [AN21] proved that any algorithm for this problem that use $n^{o(1)}$ space requires $\Omega(1/\varepsilon)$ passes. The result still holds if the weights are constant. Assadi and N [AN21], Assadi [Ass22], and Aassadi and Raz [AR20] have studied the streaming complexity of other graph problems as well.
5. The Gap Cycle Counting Problem: Let k be small. A graph G is streamed, which is either a disjoint union of $\frac{n}{k}$ k-cycles or a disjoint union of $\frac{n}{2k}$ $2k$-cycles. Determine which is the case. Assadi [Ass22] showed that any p-pass streaming algorithm requires $n^{1-1/k^{\Omega(1/p)}}$ space.
6. Assadi and Raz [AR20] show that two-pass graph streaming algorithm for the s-t reachability problem for directed graphs requires space $n^{2-o(1)}$.
7. Goel et al. [GKK12] consider the maximum matching problem. They show that any one-pass algorithm cannot achieve better than 2/3 approximation. There have been improvements to the bound since this work, and most recently, [Kap21] showed that no algorithm can do better than a $\frac{1}{1+\ln 2}$ approximation.

20.7.2 Non-Graph Problems

1. The Longest Increasing Subsequence: Given an ordered sequence of numbers $\vec{x}=(x_1,\ldots,x_n)$, find an increasing subsequence that is of maximal length. This is a streaming problem if, as the x_i's arrive, you decide if they will be in the increasing subsequence or not. Saks and Seshadhri [SS13] showed that, for all $\delta>0$, a deterministic, single-pass streaming algorithm for additively approximating this problem to within an additive δn requires $O(\log^2 n/\delta)$ space. They also considered the Longest Common Subsequence problem (given $\vec{x}$ and $\vec{y}$, find a maximal sequence that is a subsequence of both strings) and gave an analogous result for that one as well.
2. Maximum Coverage: Given n, k, and a set of m sets $S_i \subseteq \{1,\ldots,n\}$, find the k subsets that maximize the size of their union. There is a straightforward greedy $(1-e^{-1})$-approximation algorithm that runs in polynomial time. McGregor and Tu [MV19] give two single-pass streaming algorithms and one multi-pass streaming algorithm for

approximations to this problem. For the multi-pass case, they also have some lower bounds.

3. BASIC COUNTING: given $0 < \varepsilon < 1$ and then stream of bits, maintain a count of the number of 1's in the last N elements within a factor of $(1+\varepsilon)$. Datar et al. [DGIM02] showed that this problem requires $\Omega(\varepsilon^{-1}\log^2 N)$ space for any randomized algorithms.
4. THE APPROXIMATE NULL VECTOR PROBLEM: Given $x_1, \ldots x_{d-1}$ vectors in $\mathbb{R}^d$, output a vector that is approximately orthogonal to all of them. Dagan et al. [DKS19] show that any one-pass streaming algorithm for this problem requires $\Omega(d^2)$ space.

20.7.3 Frequency Moments

FREQUENCY MOMENTS
Instance: A data stream of numbers from $[m]$: $y_1, y_2, \ldots, y_n$.
Output: There are several.
For $k \in [m]$ the frequency of $k \in [m]$ is $x_k = |\{j \mid y_j = k\}|$.
The ***frequency vector*** is the vector $x = (x_1, x_2, \ldots, x_m)$.
Let $p \in \mathbb{N} \cup \{\infty\}$. The ***pth frequency moment*** of the input stream is defined as follows:

$$F_p = \begin{cases} \sum_{i=1}^{m} x_i^p & \text{if } p \in \mathbb{N} \\ \max_i x_i & \text{if } p = \infty \end{cases}$$

Note that F_0 is the number of distinct elements of the input. F_1 is the number of elements (with repetition).

Indyk and Woodruff [IW05] proved (among other things) the following.

Theorem 20.16. *Let $p > 2$.*

1. *There exists a randomized streaming algorithm that $1+\varepsilon$-approximates F_p in space $O(m^{1-2/p})$.*
2. *Any randomized streaming algorithm that $1+\varepsilon$-approximates F_p requires $\Omega(m^{1-2/p})$ space.*

21

Parallel Algorithms: The MPC and AMPC Models

21.1 Introduction

Throughout this book we have dealt with *sequential* computations. In this chapter we look at *parallel* computations.

Chapter Summary

1. *We give an overview of various models of parallel computation and how they relate to each other.*
2. *We define the* ***massively parallel computation (MPC) model*** *and give examples of algorithms in this model.*
3. *We discuss unconditional lower bounds on how well certain problems can be done in parallel on the MPC model. These lower bounds are unconditional in that they do not depend on any hardness assumption.*
4. *We discuss conditional lower bounds on how well certain problems can be done in parallel on the MPC model. These depend on a hardness assumption that the 1VS2-CYCLE problem is hard to do on the MPC model.*
5. *We define the* ***adaptive massively parallel computation (AMPC) model****. The 1VS2-CYCLE problem is easy on the AMPC model, and hence, it cannot be used as a hardness assumption.*
6. *We discuss some unconditional lower bounds for problem on the AMPC model.*

21.2 An Overview of Models of Parallelism

There are many models of parallelism. We briefly discuss several models before we explore two recent models: MPC and AMPC.

The PRAM (parallel random access machine) model allows the machines to have access to a shared memory. This feature (or bug) leads to the need for subcategories of PRAM depending on if concurrent reads or concurrent writes are allowed. There is a vast literature on both algorithms and lower bounds for PRAMs. For algorithms we recommend the survey by Karp and Ramachandran [KR88] and the papers it references. For lower bounds we recommend the papers of Cook and Dwork [CDR86] and Li and Yesha [LY89] and the papers they reference.

The model NC (Nick's class after Nick Pippenger) works as follows. A decision problem A is in NC if it can be solved by a PRAM with a poly number of machines in polylog time. The details of which type of PRAM do not matter as they are all poly-time equivalent. There are two versions of NC.

1. ***Non-Uniform:*** There is a polynomial $p(n)$ and a polylog function $d(n)$ such that, for all n, there exists a PRAM with $\leq s(n)$ machines that runs in $\leq d(n)$ time. Note that we do not say how we actually produce the PRAM, just that it exists.
2. ***Uniform:*** There is a polynomial $p(n)$, a polylog function $d(n)$, and a polynomial time algorithm A such that A on input 1^n produces a PRAM with $\leq s(n)$ machines and $\leq d(n)$ time. Virtually all known PRAM algorithms are uniform.

If the parallelism is uniform, then NC $\subseteq$ P. There are some problems in P that are thought to be inherently sequential. There is a notion of P-completeness and an analog of the Cook–Levin theorem. The basic P-complete problem is the following: Given a poly time Turing machine M and an input x, what is $M(x)$. Other (more natural) P-complete problems are (a) the circuit value problem: Given a circuit and an n-bit input, is the output 1? and (b) linear programming: Maximize a linear function of n variables relative to a set of linear constraints, where the all the coefficients are integers and the desired output is a vector of rationals. For more on NC and P-completeness, see the book by Greenlaw et al. [GHR95].

A more recent, and more realistic, model of parallel computing is the binary-forking model. This was originally defined informally in the textbook by Cormen et al. [CLRS03] in the chapter on parallel algorithms. A formal model was later developed by Blelloch et al. [BFGS20]. Goodrich et al. [GJS21] did follow-up work on both (1) low-depth algorithms and (2) lower bounds (especially of interest with respect to algorithmic lower bounds).

We study two more recent models: ***MPC*** and ***AMPC***. We will define it, give examples of problems it solves well, and then discuss lower bounds on it. We will note some relations to the PRAM when they come up. When we refer to PRAMs, we will mean those that allow concurrent reads and writes. We omit details about how they resolve contentions.

We state some open problems about comparing the old models to the new models. They will make more sense later when you have read about the new models.

Open Problem 21.1.

1. *How do the different types of PRAM compare to the MPC and AMPC models?*
2. *How does NC compare to MPC or AMPC?*
3. *Can any P-complete problems be solved in polylog rounds by an MPC or AMPC?*

21.3 The Massively Parallel Computation Model

Beame et al. [BKS17] invented the following model.

Definition 21.2. The ***MPC model*** consists of the following:

- The input data size N. (For a graph $G=(V,E)$, this is $\max\{|V|,|E|\}$, which is usually $|E|$.)
- The number of machines M available for computation. The machines will communicate with each other in rounds (we elaborate on this later).
- The memory size, s words, each machine can hold.

A reasonable assumption here is that a single word stores a constant number of bits. Thus, in practice, each machine is capable of storing $\Theta(s)$ number of bits. Moreover, in the literature, it is most often assumed that

$$M \cdot s = \Theta(N). \tag{21.1}$$

This is the most interesting scenario since the number of available resources for the algorithm ($M \cdot s$) is asymptotically equal to the size of the input data.

Input and output work as follows.

1. The input data is split across the M machines *arbitrarily*. If the input is a graph, then initially each edge is given to some machine. Note that a machine may well have many edges.
2. There will be some machines designated as ***output machines***. At the end of the computation, the answer will be stored at the output machines. We give an example. If the problem is connected components, then we want to assign to every vertex v a number n_v such that two vertices are in the same component if and only if they are assigned the same number. We want to store all the pairs (v, n_v). Since the number of nodes may be much larger than the number of processors, the output machines will each store many (though that is bounded by the space s) (v, n_v) pairs.

Computation is performed in synchronous rounds. In each round, every machine performs some computation on the data that resides locally, then sends/receives messages to any other machine. Since the local memory of each machine is bounded by s words, we assume that a single machine can send and receive at most s words per round; however, each of these words can be addressed to or received from different machines. The three main parameters that are investigated in this model are:

- The number of machines. We often omit this since it will be roughly the ratio of the size of the input and the local memory.
- The number of communication rounds it takes for an algorithm to solve a problem, often called *running time*. The local computations are ignored in the analysis of the running time of MPC algorithms because communication is the bottleneck. In practice, these local computations frequently run in linear or near-linear time; however, there is no bound on their length formally.
- The size of local memory s. Problems are easier to solve with larger s. In the extreme when $s \gg N$, one can just put the entire input on a single machine and solve it locally. Typically, s is polynomially smaller than N—for example, $s = N^{\varepsilon}$ for some constant $\varepsilon < 1$.

If one of the machines had most of the input, then, since we allow machines unlimited power, the problem could likely be done rather quickly. This is not what we want to model. We want to model problems where the input is so large that no machine can have most of it. Hence we have the following definition.

Definition 21.3. An MPC algorithm is ***sublinear*** if each machine gets space $s \ll N$. So the space is much less than the input size. For graph problems (on dense graphs) it will mean that there exists $\delta < 1$ such that $s \leq n^{1+\delta}$. We will often use the term ***sublinear*** rather than specify the space precisely.

Note the following:

Fact 21.4.

1. The MPC algorithm is nonuniform. For every N, we have a different MPC algorithm for input length N. These algorithms will be very similar.
2. An MPC-computation can be viewed as a directed graph in layers. The first layer has the input machines. The second layer has the machines that the first layer communicates with. Put a directed edge between a machine in the first layer and a machine that it sends to. Keep doing this for layers $2, 3, \ldots, R+1$. In section 21.6 we use this viewpoint. The layers are a mental construct—in reality one machine is on many, perhaps all, of the layers.
3. Any PRAM algorithm working in time t can be simulated by an MPC algorithm in $O(t)$ rounds and with sublinear memory per machine. See Goodrich et al. [GSZ11] for a simulation of one of the *PRAM* models.

21.4 Some Algorithms in the MPC Model

21.4.1 Connected Components

> CONNECTED COMPONENTS (CONN COMP)
> *Instance:* An undirected graph $G = (V, E)$.
> *Question:* Which vertices are in the same connected component? A solution is a labeling of vertices $\ell(v)$ such that $\ell(u) = \ell(v)$ if and only if vertices u and v are in the same connected component.

The running time of an algorithm for CONN COMP will depend on the number of vertices n, the number of edges m, and the diameter of the graph D. We saw this concept in section 18.4; however, we include it here for your convenience.

Definition 21.5. Let G be a graph. The ***diameter D of G*** is the length of the longest shortest path between two vertices. Formally:

$$D = \max_{u,v \in V} \text{The length of the shortest path from u to v.}$$

Behnezhad et al. [BDE+19a] proved the following theorem.

Theorem 21.6. *For all ε the following holds. There is a randomized sublinear MPC algorithm for CONN COMP such that, on a graph $G = (V, E)$ ($|V| = n$, $|E| = m$, Diameter D):*

1. *The total space used is $O(m)$. Since the algorithm is sublinear, each machine gets $n^{1+\delta}$ bits. Hence there are $\frac{m}{n^{1+\delta}}$ machines.*
2. *The number of rounds is $O(\log D + \log\log_{m/n}(n))$.*
3. *The algorithm succeeds with high probability.*
4. *The algorithm does not need to know D.*

We will not provide a proof of theorem 21.6. Instead, we will give a short overview of a slightly weaker result due to Andoni et al. [ASS+18].

Theorem 21.7. *There is a randomized sublinear MPC algorithm for CONN COMP such that, on a graph, $G = (V, E)$ ($|V| = n$, $|E| = m$, Diameter D):*

1. *The total space used is $O(m)$. Since the algorithm is sublinear, each machine gets $n^{1+\delta}$ bits. Hence there are $\frac{m}{n^{1+\delta}}$ machines.*
2. *The algorithm takes $O(\log D \cdot \log\log_{m/n}(n))$ rounds.*
3. *The algorithm succeeds with high probability.*
4. *The algorithm does not need to know D.*

Behnezhad et al. [BDE$^+$19a] write the following about the ideas that lead to theorem 21.7:

Graph exponentiation.

Consider a simple algorithm that connects every vertex to vertices within its 2-hop (i.e., vertices of distance 2) by adding edges. It is not hard to see that the distance between any two vertices shrinks by a factor of 2. By repeating this procedure, each connected component becomes a clique within $O(\log D)$ steps. The problem with this approach, however, is that the required memory of a single machine can be up to $\Omega(n^2)$, which, for sparse graphs, is much larger than $O(m)$.

Solution to CONN COMP Built off the Graph Exponentiation Technique.

Suppose that every vertex in the graph has degree at least $d \gg \log n$. Select each vertex as a leader independently with probability $\Theta(\frac{\log n}{d})$. Then contract every non-leader vertex to a leader in its 1-hop (which w.h.p. exists). This shrinks the number of vertices from n to $O(n/d)$. As a result, the amount of space available per remaining vertex increases to $\Omega(\frac{m}{n/d}) = \Omega(\frac{nd}{n/d}) = d^2$. At this point, a variant of the aforementioned graph exponentiation technique can be used to increase vertex degrees to d^2 (but not more), which implies that another application of leader contraction decreases the number of vertices by a factor of $\Omega(d^2)$. Since the available space per remaining vertex increases doubly exponentially, $O(\log \log n)$ phases of leader contraction suffice to increase it to n per remaining vertex. Moreover, each phase requires $\Omega(\log D)$ iterations of graph exponentiation, thus the overall round complexity is $O(\log D \cdot \log \log n)$.

21.4.2 MAXIMAL INDEPENDENT SET

> MAXIMAL INDEPENDENT SET
> *Instance:* A graph $G = (V, E)$.
> *Output:* An independent set I such that no independent set is a proper superset of I.

Note the following:

- MAXIMAL INDEPENDENT SET is very different from the MAXIMUM IND SET. MAXIMAL INDEPENDENT SET is in P by a simple greedy algorithm. MAXIMUM IND SET is NP-complete.
- The greedy algorithm for MAXIMAL INDEPENDENT SET is inherently sequential. Hence it is not obvious that there is a fast parallel algorithm for MAXIMAL INDEPENDENT SET. However, there is.

Luby [Lub86] gave a framework for MAXIMAL INDEPENDENT SET algorithms for the PRAM.

Luby's algorithm for MAXIMAL INDEPENDENT SET:

1. Fix a random permutation $\pi : [n] \to [n]$ of vertices.
2. A vertex v adds itself to I if, for all neighbors u of v, $\pi(v) < \pi(u)$.
3. Remove selected vertices and their neighbors.
4. Repeat until reaching an empty graph.

The following theorem comes from a simple analysis of the above method. The details can be found in Luby's paper.

Theorem 21.8. *There exists an algorithm in the PRAM model that, given a graph G on n vertices and m edges, will, with high probability, solve MAXIMAL INDEPENDENT SET in $O(\log n)$ depth and $O(n)$ machines. The algorithm will always solve the problem, though it may (quite rarely) take more time or more work than specified.*

Ghaffari and Uitto [GU19] showed that the local nature of the MAXIMAL INDEPENDENT SET problem can be efficiently exploited in the MPC model. Namely they proved the following.

Theorem 21.9. *Let $\varepsilon \in (0, 1)$. There is a randomized MPC algorithm for MAXIMAL INDEPENDENT SETwith the following properties:*

1. *Each machine uses $O(n^{\varepsilon})$ memory.*
2. *The number of rounds is $O(\sqrt{\log n} \cdot \log \log n)$ (the constant on the O depends on ε).*
3. *The probability that an MAXIMAL INDEPENDENT SET is found is $\geq 1 - \frac{1}{10n}$*

Theorem 21.9 is the best-known result for general graphs. Ghaffari, Grunau, and Jin [GGJ20] showed that the MAXIMAL INDEPENDENT SET problem, restricted to some specific class of graphs (i.e. trees or graphs with bounded arboricity), has a randomized sublinear MPC algorithm working in $O(\log \log n)$ rounds. Ghaffari, Kuhn, and Uitto [GKU19] showed a (conditional) lower bound on the MAXIMAL INDEPENDENT SET problem in the MPC model of $\Omega(\log \log n)$ rounds. This is the best known lower bound on MAXIMAL INDEPENDENT SET in the MPC model.

21.4.3 FAST FOURIER TRANSFORM and PATTERN MATCHING

Hajiaghayi et al. [HSSS21] obtained a constant-round MPC algorithm for fast fourier transform and used that to get a constant-round MPC algorithm for pattern matching (with wildcards). Everything in this section is from that paper.

FAST FOURIER TRANSFORM (FFT)
Instance: Complex numbers $x_0, \ldots, x_{n-1}$. (These will have rational real and complex parts, so they are finite length.)
Output: The n complex numbers $X_0, \ldots, X_{n-1}$, where

$$X_k = \sum_{n=0}^{N-1} x_n e^{-i2\pi kn/N} \qquad k = \{0, \ldots, N-1\}.$$

(We can either use an approximation or keep the roots of unity in symbolic form so that the length is manageable.)

Theorem 21.10. *Let $\varepsilon > 0$. There is a deterministic MPC algorithm for FFT with local memory $O(n^{\varepsilon})$, total memory $O(n \text{polylog} n)$, and $O(\frac{1}{\varepsilon})$ rounds.*

PATTERN MATCHING AND VARIANTS
Instance: A text T and a pattern P. They are both over an alphabet Σ. The text is usually much longer than the pattern. We describe three types of patterns in the OUTPUT part.
Output:

- $P \in \Sigma^*$. Return all occurrences of P in T.
- $P \in \{\Sigma \cup ?\}^*$. The pattern occurs if it matches all of the characters in Σ. We do not care what happens at the ? spot. For example, acb?a occurs in **acbaa**bbbbb**acbba** as shown. Return all matches of this type.
- $P \in \{\Sigma \cup +\}^*$. The pattern occurs if it matches all of the characters in Σ. At the + spots a single character can appear many times. For example, acb+a occurs in

  acbaaaaaaaa bbbbb **acbbbbbbbbbbbbbbbbbbba**

  as shown. Return all matches of this type.
- $P \in \{\Sigma \cup *\}^*$. The pattern occurs if it matches all of the characters in Σ. At the * spots any string can appear. For example, acb*a occurs in

  acbabbcaabba bbbbb **acabaaaccabbaaadaacadaa**

  as shown. Return all matches of this type.

Theorem 21.11. *Let $0 < \varepsilon < 1$.*

1. *There exists an MPC algorithm for string matching with pattern $P \in \Sigma^*$ with $M = O(n^{\varepsilon})$, $s = O(n^{1-\varepsilon})$, and $r = O(1)$.*
2. *There exists an MPC algorithm for string matching with $P \in \{\Sigma \cup ?\}^*$ with $M = O(n^{\varepsilon} polylog n)$, $s = O(n^{1-\varepsilon} polylog n)$, and $r = O(1)$.*

It is not known if pattern matching with $P \in \{\Sigma \cup *\}^*$ has a polylog round MPC algorithm with sublinear M and s.

21.5 Unconditional MPC Lower Bounds

In this section, we are going to sketch an approach to find unconditional lower bounds for problems on the MPC model. This approach was introduced by Roughgarden et al. [RVW18]. They do the following.

1. Define the s-Shuffle model.
2. Give a relation between the MPC model and the s-Shuffle model.
3. Show that an s-Shuffle computation can be represented by polynomials.
4. Obtain lower bounds on the s-Shuffle model for some problems by looking at polynomials.
5. Use these lower bounds and the relation between the MPC model and the s-Shuffle model to get lower bounds on the MPC model.

The lower bounds are not tight; however, they are important because they are unconditional. We will skip right to a theorem that relates MPC to polynomials, and gets lower bounds using those polynomials.

21.5.1 Polynomials

Definition 21.12.

1. We will be using polynomials on many variables. We will only care about what they output when the variables are replaced by 0's and 1's. Hence the polynomials will not have any exponents. Moreover, when we say $p \in \mathbb{Z}[x_1, \ldots, x_m]$, we will implicitly mean that there are no exponents.
2. The ***degree*** of a polynomial is the number of variables in the largest monomial.
3. Let $f : \{0, 1\}^m \to \{0, 1\}$. Let $p \in \mathbb{Z}[x_1, \ldots, x_m]$. ***p represents f p represents f*** if, for all $\vec{b} \in \{0, 1\}^m, f(\vec{b}) = p(\vec{b})$.

4. If f is a graph property on graphs with n vertices, then the variables are $x_{i,j}$ (with $1 \leq i < j \leq n$) and represent edges.

Theorem 21.13. *For every $f:\{0,1\}^m \rightarrow \{0,1\}$, there exists $p \in \mathbb{Z}[x_1,\ldots,x_m]$ of degree m such that f represents g.*

Proof. For every $\vec{b} \in \{0,1\}^m$, create a polynomial that is 1 if and only if the input is $\vec{b}$. For example, if $n=4$ and the bit sequence is 0110, then we associate the polynomial

$$\mathsf{Poly}(\vec{b}) = (1-x_1)x_2x_3(1-x_4).$$

The polynomial p is

$$p(x_1,\ldots,x_m) = \sum_{\vec{b}:f(\vec{b})=1} \mathsf{Poly}(\vec{b}).$$

□

Note: We will connect the smallest degree of a polynomial that represents f to the complexity of f.

Exercise 21.14. Let AND_n be the function $x_1 \wedge \cdots \wedge x_n$. Let OR_n be the function $x_1 \vee \cdots \vee x_n$.

1. Give a polynomial of degree n that represents AND_n. Same for OR_n.
2. Show that any polynomial that represents AND_n has degree $\geq n$. Same for OR_n.
3. Show that any s-space MPC for AND_n needs at least $\lceil \log_s n \rceil$ rounds. Same for OR_n.

Now, we are going to show that the output of an s-space MPC computation with n input bits, and k output bits, running in R rounds, can be produced by k polynomials with degree of at most s^R such that each polynomial produces one of the s outputs.

Theorem 21.15. *Let $f:\{0,1\}^n \rightarrow \{0,1\}^k$.*

1. *If there is an r-round s-space MPC for f, then there are k polynomials $\{p_i(x_1,\ldots,x_n)\}_{i=1}^k$ of degree at most s^r such that, for all $x \in \{0,1\}^n$, for $1 \leq i \leq k$, $p_i(x) = f(x)_i$.*
2. *If f cannot be represented by a polynomial with degree less than d, then any s-space MPC that computes f has least $\lceil \log_s d \rceil$ rounds. (This follows from part 1).*

21.5.2 Lower Bounds in the MPC Model for Monotone Graph Properties

We will use two known results to obtain a lower bound on s-space MPCs for monotone graph properties.

Definition 21.16.

1. A ***monotone graph property*** is a property of graphs such that if more edges are added, then the property still holds. *Examples*: connectivity, non-planarity.
2. A ***decision tree for a graph property*** is a decision tree for the property where the queries are "$(i,j) \in E$?".
3. In this section (and only this section), a polynomial is over $\mathbb{Z}[x_1, \ldots, x_m]$.

Theorem 21.17. *Let P be a monotone graph property.*

1. *Rosenberg [Ros73] proved that any decision tree for P requires $\Omega(n^2)$ depth. (The constant for the Ω was increased by Rivest and Vuillemin [RV78] and Kahn et al. [KSS84]. See also C. Miller [Mil13].)*
2. *Buhrman and de Wolf [BdW02] proved that the decision tree complexity of a polynomial in $\mathbb{Z}[x_1, \ldots, x_n]$ of degree d is at most $O(d^4)$.*
3. *The degree of a polynomial representing a monotone graph property is at least $\sqrt{n}$. (This follows from parts 1 and 2.)*
4. *Any s-space MPC for P requires $\Omega(\log_s n)$ rounds. (This follows from part 3 and theorem 21.15.)*

With more effort, the following has also been proved.

Theorem 21.18. *For the problem of graph connectivity, the following hold.*

1. *Every MPC algorithm with space s needs at least $\lceil \log_s \binom{n}{2} \rceil$ rounds.*
2. *If $s = N^\varepsilon$ (a realistic choice), then the number of rounds needed is $\Omega(1)$.*

21.6 Conditional MPC Lower Bounds

The unconditional lower bound in theorem 21.18.3 was that if $s = N^\varepsilon$ (which is a realistic value of s), then the number of rounds is $\Omega(1)$. This is a weak lower bound. Hence we now turn to conditional lower bounds.

The framework for these lower bounds was introduced by Ghaffari et al. [GKU19]. Everything in this section is from that paper unless otherwise noted. We will do the following:

1. We will define the 1vs2-CYCLE problem and formally state the conjectured lower bound for it in the MPC model.
2. Assuming the conjectured lower bound for the 1vs2-CYCLE, and additional assumptions, we will state lower bounds for other problems in the MPC model.

1vs2-CYCLE
Instance: An undirected graph $G = (V, E)$. We are promised that it is either one cycle or the union of two cycles.
Question: Is the graph one cycle or the union of two cycles?

The following conjecture is widely believed:

Conjecture 21.19. **The 1vs2-CYCLE Conjecture** *Any MPC algorithm for the 1vs2-CYCLE problem using machines with $s = n^{\varepsilon}$ requires $\Omega(f(\varepsilon)\log n)$ rounds for some f. (Note that since the input is one cycle or two cycles, $m = O(n)$, so sublinear now means $s = n^{\varepsilon}$.)*

The lower bound is conjectured to hold even for the simpler promise problem of distinguishing whether an input graph is a cycle of length n or two cycles of length $n/2$.

Theorem 21.20. *Assume the 1vs2-CYCLE conjecture. Any sublinear MPC for undirected connectivity requires $\Omega(\log n)$ rounds.*

Proof. It is clear that if we have one cycle, the graph is connected; otherwise, the graph is unconnected. Thus, if we find an algorithm for undirected connectivity in $o(\log n)$ rounds, then we will have an algorithm in $o(\log n)$ rounds for 1vs2-CYCLE, which contradicts the 1vs2-CYCLE conjecture. □

We can try to find conditional lower bounds for other problems by making a reduction from 1vs2-CYCLE or other problems that have a known conditional lower bound. But, alas, we need to introduce a restriction on the type of algorithms we can get lower bounds on.

We informally discuss ***component-stable MPC algorithms***. Imagine an MPC algorithm for maximal matching on bipartite graphs. Imagine that the graph G has connected components G_1 and G_2. Let x_1 be a vertex of G_1 and x_2 be a vertex of G_2. Imagine running the algorithm; x_1 is matched with y_1, and x_2 is matched with y_2. What if we then replace G with a $G_1 \cup H$ for some $H \neq G_2$? Run the algorithm on $G_1 \cup H$. You would expect that x_1 would still be matched to y_1. ***But this is not guaranteed!*** Intuitively, ***component-stable MPC algorithm*** for matching would be one where, in the above example (and more complex examples with more components), x_1 still gets matched with y_1. More generally, a ***component-stable MPC algorithm*** for a graph problem is one where the output of each node x (e.g., what a node is matched with) depends only on the connected component that x is in, and not on other connected components. For a formal (though complicated) definition of ***component-stable MPC algorithms***, see Czumaj et al. [CDP21]. We will not give a definition.

Virtually all MPC algorithms on graphs are component-stable. Hence it makes sense to get lower bounds on such algorithms. Also, there are techniques (beyond the scope of this book) to get lower bounds on such algorithms. We present three problems and and then three conditional lower bounds on component-stable algorithms for them.

MAXIMAL MATCHING *Instance:* An undirected graph $G=(V,E)$ *Output:* Return an independent set I such that no independent set is a proper superset of I. *Output:* For every $v \in V$, there is an output machine. At the end of the computation, the machine will have either a 1 (for $v \in I$) or a 0 (for $v \notin I$).

SINKLESS ORIENTATION *Instance:* An undirected graph $G=(V,E)$ *Output:* Return an orientation of the graph (a direction for every edge) so that the graph has no vertices of out-degree 0. (Such vertices are called ***sinks***.) *Output:* For every $\{u,v\} \in E$, there is an output machine. At the end of the computation, the machine will have either (u,v) or (v,u) to indicate the orientation.

$(\Delta+1)$-GRAPH COLORING *Instance:* An undirected graph $G=(V,E)$ and its max degree Δ. *Output:* Return a proper $(\Delta+1)$-coloring of G (such always exists). Colors are $\{1,\ldots,\Delta+1\}$. *Output:* For every $v \in V$, there is an output machine. At the end of the computation, the machine will have the color of v.

The only known MPC algorithms for maximal matching, sinkless orientation, Δ-graph coloring, and many other graph problems are component-stable. Hence it is of interest to get lower bounds on component-stable MPC algorithms for these and other problems.

Theorem 21.21. *Assume the 1VS2-CYCLE conjecture. Assume that all MPCs discussed in this theorem are sublinear and use a component-stable algorithm.*

1. *Maximal matching requires $\Omega(\log\log n)$ rounds. (Same holds for a constant approximation for maximal matching. This lower bound holds even when restricted to trees.)*
2. *Sinkless orientation requires $\Omega(\log\log\log n)$ rounds.*
3. *Let c be a constant. c-coloring a cycle requires $\Omega(\log\log^* n)$ rounds.*

4. *Any constant approximation for* Vertex Cover *requires* $\Omega(\log\log n)$ *rounds.*
5. *(Using additional assumptions)* $(\Delta+1)$*-coloring a graph requires* $\Omega(\sqrt{\log\log n})$ *rounds.*

21.7 Adaptive Massively Parallel Computation Model

The adaptive massively parallel computation (AMPC) model was introduced by Behnezhad et al. [BDE+21]. It is essentially the MPC model but with shared memory. We quote their motivation:

Our model is inspired by the previous empirical studies of distributed graph algorithms [BBD+17, KLM+14], using MapReduce and a distributed hash table service [CDG+08].

We proceed informally.

In the AMPC model the machines have access to a shared memory. This model assumes that all messages sent in a single round are written to a distributed data storage, which all machines can read from within the next round. More specifically, the computation consists of rounds. In the ith round, each machine can read data from a random access memory D_{i-1} and write to D_i (both D_{i-1} and D_i are common for all machines). Within a round, each machine can make up to S reads (henceforth called "queries") and S writes and can perform arbitrary computation. (Note that this S is different from the s parameter for the MPC model.)

Clearly, every MPC algorithm can be easily simulated in the AMPC model. However, the opposite direction is not usually the case. Furthermore, due to known simulations of PRAM algorithms by MPC, the AMPC model can also simulate existing PRAM algorithms. The key property of the AMPC model is that the queries a machine makes in each round may depend on the results of the previous queries it made in that same round, which is why the model is called *adaptive*. For example, if g is a function from X to X and for each $x \in X$, D_{i-1} stores a key-value pair $(x, g(x))$, then in round i a machine can compute $g^k(y)$ in a single round, provided that $k = O(S)$.

21.7.1 AMPC Power: 1vs2cycle Revisited

In this section, we discuss the computation power of the AMPC model. For several fundamental problems, there are AMPC algorithms solving them with significantly lower round complexities than the best-known MPC algorithms. Behnezhad et al. [BDE+19a] includes a number of such algorithms for some of the most fundamental graph problems, such as graph connectivity and maximal independent set. We focus on the 1vs2-Cycle problem since this illustrates that the AMPC is likely more powerful than the MPC model.

Recall that in the MPC model, it is conjectured that 1vs2-Cycle requires a logarithmic number of rounds. However, the following theorem (from [BDE+19a] with help from [BDE+21]) shows that this conjecture does not hold in the adaptive model.

Theorem 21.22. *There is an AMPC algorithm solving the* 1VS2-CYCLE *problem in* $O(1/\varepsilon)$ *rounds w.h.p. using* $O(n^{\varepsilon})$ *space per machine and* $O(n)$ *total space.*

Proof. At first, we discuss the overview of their algorithm. In each round of the algorithm, we sample each vertex with probability $n^{-\varepsilon/2}$. Then, we contract the original graph to the samples by replacing the paths between sampled vertices with single edges. To do so, we traverse the cycle in both directions for each sampled vertex until we hit another sampled vertex, which can be done in a single round using the adaptivity of the model. In each round, with high probability, the number of vertices shrinks by a factor of $n^{\varepsilon/2}$. Therefore, after $O(1/\varepsilon)$ rounds, the number of remaining vertices and edges is reduced to $O(n^{\varepsilon})$. Hence, the graph fits in the memory of a single machine, and we can solve the remaining problem in a single round.

We present the algorithm. We first need a procedure:

Shrink($G = (V, E), \varepsilon, t$)
Begin Algorithm
  For $i = 0, \ldots, t$
    $V' \leftarrow$ a subset of $V(G)$ s.t. each vertex is included independently with probability $n^{-\varepsilon/2}$
    $E' \leftarrow \emptyset$
    For $v \in V'$
      $l_v \leftarrow$ first sampled vertex that we reach traversing the graph starting with $(v, nei_G^1(v))$
      $r_v \leftarrow$ first sampled vertex that we reach traversing the graph starting with $(v, nei_G^2(v))$
      $E' \leftarrow E' \cup \{(v, l_v), (v, r_v)\}$
    $G \leftarrow G'(V', E')$
  EndFor
Return G
End Algorithm

And now the algorithm.

1VS2-CYCLE($G(V, E)$)
Begin Algorithm
  $G' \leftarrow$ **Shrink**$(G, \varepsilon, O(1/\varepsilon))$
  Solve the 1v2-Cycle problem on G' using a single machine
  (Comment: Note that $|V(G')| \in O(n^{\varepsilon})$ w.h.p.)
End Algorithm

Each iteration of the outer loop in Shrink is a single AMPC round, and the correctness of this algorithm is a result of combining the following lemmas from Behnezhad et al. [BDE⁺21].

Lemma 21.23. Let G be a graph consisting of cycles, and let N be the initial number of vertices in G. Consider a cycle with size $k = \Omega(N^{\varepsilon})$ in some iteration of the loop of Shrink

$(G, \varepsilon, O(1/\varepsilon))$. The size of this cycle shrinks by at least a factor of $N^{\varepsilon/2}$ after this iteration w.h.p.

Lemma 21.24. Let G be an N-vertex graph consisting of cycles and let $G' = \text{Shrink}(G, \varepsilon, O(1/\varepsilon))$. Then G' can be obtained from G by contracting edges, and the length of each cycle in G' is $O(n^{\varepsilon})$.

Lemma 21.25. In each round, the total communication of each machine is $O(N^{\varepsilon})$ w.h.p., where N is the initial number of vertices in G. □

21.7.2 Unconditional AMPC Lower Bounds

Recall that in section 21.5 we presented the polynomial method for lower bounds on MPC algorithms. Charikar et al. [CMT20] modified the polynomial method to get lower bounds on AMPC algorithms. Everything in this section is from that paper.

They introduced the following extension of degree.

Notation 21.26.

1. In the definitions below, $\Delta \subseteq \{0, 1\}^n$. For example, Δ could be the set of two graphs: one the cycle on n vertices and the other the disjoint union of two cycles of length $n/2$. We say things like:

   Let g be a partial Boolean function that maps Δ *to* $\{0, 1\}$

   to emphasize that g can be partial.
2. We use the same convention as in section 21.5 whereby one can input a graph on n vertices into a polynomial on $\binom{n}{2}$ variables by viewing the graph as a sequence of $\binom{n}{2}$ bits, where $x_{i,j}$ is 1 if (i,j) is an edge and 0 otherwise.

Definition 21.27. Let $g : \Delta \to \{0, 1\}$. Then

$$deg_{\text{partial}}(g) = \min\{deg(p) \mid p(x) = g(x) \text{ for all } x \in \Delta\}.$$

The next theorem reduces lower-bounding the deterministic AMPC round complexity of g to lower-bounding $deg_{\text{partial}}(g)$:

Theorem 21.28. *Let g be a partial Boolean function that maps* Δ *to* $\{0, 1\}$. *Let M be a deterministic AMPC algorithm that computes g. Let S be the number of queries and writes allowed per round. The number of rounds is at least*

$$\frac{1}{2} \log_S deg_{\text{partial}}(g).$$

In particular, if g is a total Boolean function, then any such algorithm requires $\frac{1}{2} \log_S deg(g)$ *rounds.*

We omit the proof.

Theorem 21.28 provides lower bounds for deterministic algorithms. What about randomized algorithms? To obtain lower bounds on randomized algorithms, we need the following definition.

Definition 21.29. Let $g : \Delta \to \{0, 1\}$.

1. g is approximately represented by a polynomial p if

$$\forall x \in \Delta : |p(x) - g(x)| \leq \frac{1}{3}.$$

2. $\widetilde{deg}_{\text{partial}}$ is

$$\widetilde{deg}_{\text{partial}}(g) = \min\{deg(p) : p \text{ approximately represents } g\}.$$

The following theorem reduces lower-bounding the randomized AMPC round complexity of g to lower-bounding $\widetilde{deg}_{\text{partial}}(g)$:

Theorem 21.30. *Let g be a partial Boolean function that maps Δ to $\{0, 1\}$. Let M be a randomized AMPC algorithm that computes g with probability of error $\leq \frac{1}{3}$. Let S be the number of queries and writes allowed per round. The number of rounds is at least*

$$\frac{1}{2} \log_S \widetilde{deg}_{\text{partial}}(g).$$

In particular, if g is a total Boolean function, then any such algorithm requires $\frac{1}{2} \log_S \widetilde{deg}(g)$ rounds.

We omit the proof.

> PARITY
> *Instance:* $x \in \{0, 1\}^n$. Let $x = x_1 \cdots x_n$.
> *Question:* What is $\sum_{i=1^n} x_i \pmod 2$?
> *Note:* We denote PARITY on n bits by PARITY$_n$.

Paturi [Pat92] proved the following:

Theorem 21.31. *If p is a polynomial that approximates PARITY$_n$, then* $\deg(p) \geq n$.

By combining theorems 21.31, 21.28, and 21.30, one obtains the following:

Theorem 21.32.

1. *If M is a deterministic AMPC algorithm for PARITY$_n$, then the number of rounds is* $\geq \frac{1}{2} \log_S(n)$.

2. *If M is a randomized AMPC algorithm with error $\leq \frac{1}{3}$ for* Parity$_n$ *then the number of rounds is* $\geq \frac{1}{2}\log_S(n)$.
3. *In both of the items above, if* $S = n^{\varepsilon}$, *the number of rounds is* $\Omega(\frac{1}{\varepsilon})$

Now that we have *one* problem with provable unconditional lower bounds, we can use a reduction get others!

Theorem 21.33.

1. *If M is a deterministic AMPC algorithm for 1vs2-*Cycle, *then the number of rounds is* $\geq \frac{1}{2}\log_S(n/2)$.
2. *If M is a randomized AMPC algorithm with error $\leq \frac{1}{3}$ for 1vs2-*Cycle, *then the number of rounds is* $\geq \frac{1}{2}\log_S(n/2)$.
3. *In both of the items above, if* $S = n^{\varepsilon}$, *the number of rounds is* $\Omega(\frac{1}{\varepsilon})$.

Proof. We prove the theorem using a reduction from Parity$_N$ to 1vs2-Cycle. (We use N since we will use n for the number of vertices in a graph.)

Let $x = \{x_1, \ldots, x_N\} \in \{0,1\}^N$ be an instance of Parity$_N$. We construct the graph $G(x)$ as follows:

1. For any x_i we add vertices v_i^1 and v_i^2.
2. For $i = 1, \ldots, N$
   a. if $x_i = 0$, add edges $(v_i^1, v_{(i \bmod N)+1}^1)$ and $(v_i^2, v_{(i \bmod N)+1}^2)$; and
   b. if $x_i = 1$ add edges $(v_i^1, v_{(i \bmod N)+1}^2)$ and $(v_i^2, v_{(i \bmod N)+1}^1)$.

Figure 21.1 shows the graph G (the right two vertices are the left two vertices) obtained if the input is 01001. Note that Parity$(01001) = 0$ and G is the union of two cycles of length n. We leave it to the reader to show the following:

- If Parity$(x) = 0$, then $G(x)$ is the union of 2 cycles of length n.
- If Parity$(x) = 1$, then $G(x)$ is the 1 cycle of length $2n$.

We have a reduction where a string of length N maps to a graph on $2N$ vertices. Hence, by theorem 21.32, any AMPC algorithm (of the two types we are talking about) for 1vs2-Cycle on $2N$ vertices requires $\geq \frac{1}{2}\log_S(N)$ rounds. Hence any AMPC algorithm (of the two types we are talking about) for 1vs2-Cycle on n vertices requires $\geq \frac{1}{2}\log_S(n/2)$ rounds. □

Theorem 21.34.

1. *If M is a deterministic AMPC algorithm for 1vsk-*Cycle, *then the number of rounds is* $\geq \frac{1}{2}\log_S(n/k^2)$.

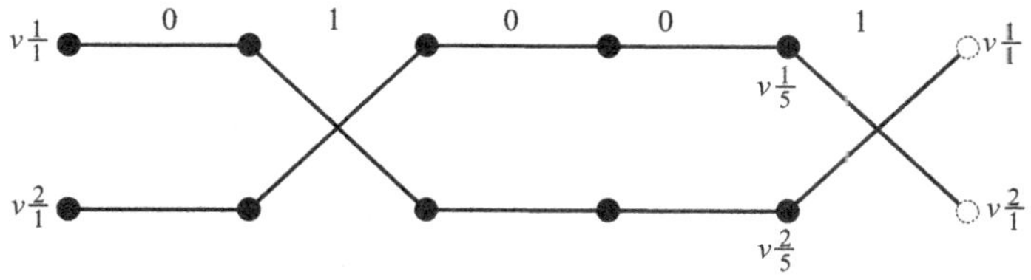

Figure 21.1
Reduction from a parity instance $x = 01001$ to a 1vs2-CYCLE instance with 10 vertices.

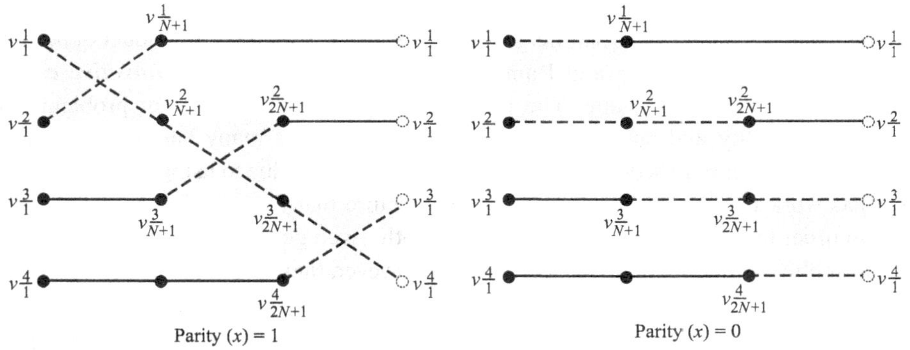

Figure 21.2
Reduction from a parity instance to a 1vsK-CYCLE instance with $k = 4$.

2. *If M is a randomized AMPC algorithm with error $\leq \frac{1}{3}$ for 1vsK-CYCLE, then the number of rounds is $\geq \frac{1}{2}\log_S(n/k^2)$.*
3. *In both of the items above, if $k = n^{\delta}$ and $S = n^{\varepsilon}$, the number of rounds is $\Omega(\frac{1-2\delta}{\varepsilon})$.*

Proof sketch. We give a sketch of the proof since it is similar to the proof of theorem 21.33.

We prove the theorem using a reduction from PARITY$_N$ to 1vsK-CYCLE. (We use N since we will use n for the number of vertices in a graph.)

Let $x = \{x_1, \ldots, x_N\} \in \{0, 1\}^N$ be an instance of PARITY$_N$. we want to construct a graph $G(x)$ such that:

- If PARITY$(x) = 0$, then $G(x)$ is the union of k cycles of length n.
- If PARITY$(x) = 1$, then $G(x)$ is the cycle of length kn.

We leave it to the reader to use figure 21.2 to guide their construction and then to finish the proof. □

21.8 Future Directions

Despite the recent interest in the MPC and AMPC models, there are many fundamental open problems. For example, the MPC-complexity and AMPC-complexity of both 2SAT and directed s-t connectivity remain unknown. These problems are related in the sense that solving 2SAT is usually done by reducing it to directed s-t connectivity. For an up-to-date survey of what is known about lower bounds on the MPC model, see the paper of Nanongkai and Scquizzato [NS22].

It is also a challenge to find more problems where AMPC algorithms are provably better than MPC algorithms. s-t connectivity may be such a problem.

An interesting parameter for parallel algorithms is ***work***, which is the product of number-of-processors and time. Karp and Ramachandran [KR88] define the ***transitive closure bottleneck*** to illustrate the issue. The bottleneck is that many important problems like directed reachability and single-source-shortest-path (and its many variants) are in NC (polynomial number of processors, logarithmic time) but only due to using matrix squaring, which has work $O(n^3)$. This leads to a later conjecture that these problems *require* work $\Omega(n^3)$ in order to be solved in polylogarithmic depth. Even getting poly number of processors and sublinear time seemed like a challenge; however, there are some results where this is, perhaps surprisingly, achieved. The problems we consider are: ***Single Source Graph Reachability (SSGR)*** and ***Single Source Shortest Path (SSSP)***. In all cases, the input is a directed graph and the algorithm is randomized.

1. Fineman [Fin18] showed that SSGR can be done in $\tilde{O}(m)$ work and $\tilde{O}(n^{2/3})$ time.
2. Jambulapati et al. [JLS19] showed that SSGR can be done in $\tilde{O}(m)$ work and $\tilde{O}(n^{1/2+o(1)})$ time.
3. Rozhon [RHM+22] showed that SSSP can be done in $\tilde{O}(m)$ work and $\tilde{O}(m)$ time.

(1) Jambulapati et al. [JLS19] obtained an algorithm for single-source-shortest path in $O(m)$ work and $O(n^{1/2+o(1)})$ time, and (2) Fineman [Fin18] obtained an algorithm for reachability with expected work $\tilde{O}(m)$ and time $\tilde{O}(n^{1/3})$.

22

Nash Equilibria and Polynomial Parity Arguments for Directed Graphs

22.1 Introduction

There are problems that are in P for an odd reason. We give an example. Every directed acyclic graph has a sink. Hence the problem, "Given a directed acyclic graph, does it have a sink?" is easy: Always say YES. But this does not help you *find* that sink. If the input is a graph, then finding the sink is in polynomial time. But what if the input is a circuit that represents a graph on 2^n vertices?

How do we study the complexity of a problem where it is known that there is a solution but not known how to find it? That is the subject of this chapter.

Chapter Summary

1. *We discuss several problems where the answer is "Yes, there is a solution," but it seems hard to find them. One of them is NE, which was the motivation for this study.*
2. *We define a complexity class PPAD for these problems and an appropriate notion of reduction for the class.*
3. *We discuss the proof that NE is PPAD-complete. Along the way, we prove other problems also PPAD-complete.*
4. *We discuss other similar classes and problems that are hard for them.*

22.2 End of Line Problem

Definition 22.1. Let $G = (V, E)$ be a directed graph. A vertex $v \in V$ is ***unbalanced*** if $\text{indeg}(v) \neq \text{outdeg}(v)$.

Theorem 22.2. *Let G be a directed graph.*

1. $\sum_{v\in V} indeg(v) = \sum_{v\in V} outdeg(v)$.

2. *If there exists an unbalanced vertex, then there exists another unbalanced vertex. (This follows from part 1.)*

Consider the following problem.

> UNBALANCED
> *Instance:* A directed graph G and an unbalanced vertex v.
> *Question:* Is there another unbalanced vertex?

This problem is in P for the odd reason that, by theorem 22.2, there is *always* another unbalanced vertex. What if we actually want to *find* that other unbalanced vertex? We certainly can find it in polynomial time by looking at every vertex. But note that the algorithm does not use the proof of theorem 22.2.

What if we were given a short representation (size poly in n) of a large directed graph (size 2^n)? Then you cannot just look at every vertex. We will also restrict the graph to have both in-degree and out-degree ≤ 1. We now define this formally.

> END OF LINE (EOL)
> *Instance:* Two circuits P (for previous) and N (for next) on $\{0, 1\}^n$ such that for all $x \in \{0, 1\}^n$ the following holds.
>
> - $P(x)$ returns either NO or an element of $\{0, 1\}^n$.
> - $N(x)$ returns either NO or an element of $\{0, 1\}^n$.
>
> We interpret the circuits as describing a directed graph by the following:
>
> - If $P(x) = y$, then there is an edge from x to y.
> - If $N(x) = y$, then there is an edge from y to x.
>
> *Output:* If exactly one of $P(0^n)$, $N(0^n)$ is not NO, then find another unbalanced vertex. (Such an unbalanced vertex exists by theorem 22.2.)
> *Note:* This is quite different from *finding* the unbalanced vertex that has a path to 0^n. Papadimitriou [Pap94, theorem 2] showed that problem is PSPACE-hard. Subtle changes in a problem definition may alter things tremendously.

We do an example. Say $n = 3$. Then there is a circuit on three inputs that represents a graph on eight vertices. Figure 22.1 shows that graph. Note that both vertex-111 and vertex-010 are answers. Also note that this gives a very compact way to represent a graph since a circuit on n inputs represents a graph on 2^n vertices.

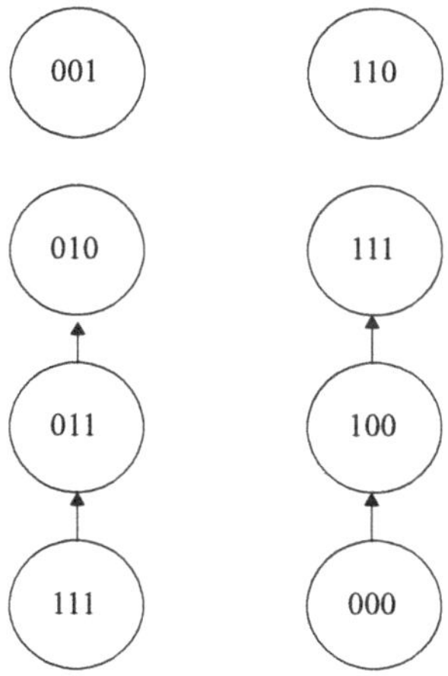

Figure 22.1
The graph produced by the EOL input.

Theorem 22.2 tells us that there is another unbalanced vertex, but, since the proof is nonconstructive, it does not help us find that vertex. So it seems that EOL is hard. Is it NP-hard? Unlikely:

Theorem 22.3. *If SAT reduces to EOL using a polynomial number of queries, then NP = coNP.*

Proof. Assume there is such a reduction. Then $\varphi \notin$ SAT if and only if that reduction returns NO.

If the reduction says NO, then look at that computation. It will have instructions and queries. The key is that the queries to EOL all have answers that can be verified (you can verify that a vertex v is unbalanced by looking at $P(v)$ and $N(v)$). So the reduction can be encoded in a string that has the answers to the queries and the verification of those answers. This string is of length polynomial in $|\varphi|$.

Hence, if $\varphi \notin$ SAT, then there is a short string that verifies the reductions answer of NO. This puts $\overline{\text{SAT}}$ into NP, so SAT is in coNP. □

In this chapter we will assume that EOL is hard and use that to show that other problems are hard. The motivation for this will be to show that finding Nash equilibrium for two-player games is hard. The proof that Nash equilibrium exists is, like the proof that an unbalanced vertex exists, nonconstructive.

We will analyze existence theorems by Nash, Brouwer, and Sperner that arise from game theory, topology, and combinatorics, respectively. These three theorems, as dissimilar as they seem, can be related to one another and rely on very basic combinatorial principles.

In this chapter, we will define the setting for each of the theorems and give an overview of their connection.

The motivation to study these problems is that, unlike (say) 3SAT, we already know that a solution exists. How efficiently can we find a solution? If the search space is polynomial, this is easy: Simply search exhaustively. However, some of these problems, like EOL, have an exponential size search space.

22.3 Game Theory

Problems in game theory have the following.

1. A set of players (often two).
2. For each player a choice of strategies.
3. For each player choices of strategies, a payoff for each player.

The key question is to figure out the best strategy. It may be randomized—for example, flip a coin and, based on the coin flip, choose a strategy. (You could flip many coins and they need not be fair.)

We give several examples of problems in game theory.

22.3.1 Prisoners' Dilemma

Two prisoners are on trial for a crime, and each face the choice of confessing to the crime or remaining silent. If they both remain silent, the authorities will not be able to charge them for this particular crime, and they will both face two years in prison for minor offenses. If one of them confesses, his term will be reduced to one year, but he will have to bear witness against the other, who will be sentenced to five years. If they both confess, they will both get a small break and be sentenced to four years in prison (rather than five). We summarize the four outcomes and the utility with the matrix in table 22.1.

The only ***stable solution*** in this game is when both confess. In each of the other three outcomes, a prisoner can switch from being silent to confessing in order to improve his own payoff. The social optimum in this case is when both remain silent; however, this outcome is not stable. In this game, there is a unique optimal selfish strategy for each player, independent of what other players do. A pair of strategies where neither player has an incentive to change, independent of the other player's strategy, will be called a ***Nash equilibrium*** (NE). In this case confess-confess is a NE.

Table 22.1
The prisoner's dilemma payoff matrix.

	Confess	Silent
Confess	(4, 4)	(1, 5)
Silent	(5, 1)	(2, 2)

Table 22.2
The penalty shot payoff matrix.

	Kick left	Kick right
Dive left	$(1, -1)$	$(-1, 1)$
Dive right	$(-1, 1)$	$(1, -1)$

One way to specify a game in algorithmic game theory is to explicitly list all possible strategies and utilities of all players. Expressing the game in this form is called the ***standard form*** or ***matrix form***. This form is convenient to represent two-player games with a few strategies, as demonstrated by the prisoner's dilemma game.

22.3.2 The Penalty Shot Game

Consider the penalty shot game: Player 1 needs to decide where to shoot a penalty (left or right) and player 2 needs to decide where to dive (left or right). The payoff of this game is easy to describe: If player 1 scores, he gets 1 point and player 2 gets -1 points. If player 1 misses, player 2 gets 1 point and player 1 gets -1 points. This game is summarized in table 22.2.

Imagine that player 1 initially decides to shoot left. If player 2 knows this, then he will dive left. If player 1 knows that player 2 will dive left, he will shoot right. This paragraph could go on forever.

Is there a pair of strategies so that neither player has a reason to deviate from his strategy? If we insist that the strategies are deterministic, then no: Both players will want to deviate. However, there is a pair of *randomized* strategies: Both players flip a fair coin to determine what to do.

22.3.3 Formal Game Theory

Definition 22.6. A ***finite game*** consists of the following elements:

1. A set P of r players.
2. For each $p \in P$ a set S_p of n pure strategies for p. A pure strategy means a deterministic strategy. In the two-player case, these will be the rows for player 1 and columns for player 2.
3. A utility or payoff function that assigns a real value to player p for every possible strategy set. Formally, for every $p \in P$, we have a function

$$u_p : \times_{q \in P} S_q \to \mathbb{R}.$$

We use u for utility. (In the two-player case, this is the matrix of pairs as seen in both the prisoner's dilemma (table 22.1) and the penalty shot game (table 22.2).

Definition 22.7. Let P be a game and p be a player with the set S_p of pure strategies. A ***mixed strategy for player p*** is a distribution over S_p. Formally if the strategies are $s_1, \ldots, s_n$, then a mixed strategy is a set of reals $r_1, \ldots, r_n$ such that $\forall i : 0 \le r_i \le 1$ and $\sum_{i=1}^{n} r_i = 1$.

For the prisoner's dilemma and the penalty shot game, we looked at pairs of strategies where neither player has an incentive to change their mind. We define this formally.

Definition 22.8. A ***Nash equilibrium (NE) for a game*** is a collection of mixed strategies $s_1, s_2, \ldots, s_n$ such that for every player $p \in P$, for every mixed strategy s'_p for p,

$$E(u_p(s_1, s_2, \ldots, s_p, \ldots, s_n)) \ge E(u_p(s_1, s_2, \ldots, s'_p, \ldots, s_n)).$$

Informally, this definition says that a tuple of strategy (one for each player) is a NE if no player has incentive (i.e., cannot be better off) to change his strategy based on the strategies of the other players, in terms of expected utility.

Nash proved the following:

Theorem 22.9. *Every game has a NE.*

The proof is (you guessed it) nonconstructive. Hence this fits the theme (and, in fact, was the original motivation) of this material: We know that a solution exists, but it seems hard to find it.

Example 22.10.

1. In the prisoner's dilemma, the confess-confess pair is a NE of pure strategies.
2. In the penalty shot game, the scenario where both players flip a fair coin to determine their move is a NE of mixed strategies

We are interested in the complexity of finding the NE.

> r-NASH
> *Instance:* An r-player game where all of the utilities are integers.
> *Output:* Return an approximation to a NE. The discussion below will clarify why we settle for an approximation. We omit details of how the approximation works; however, you would need to have the error tolerance ε as part of the input. (We use the term NASH to mean r-NASH for some r.)

Is it possible that a NE uses irrational probabilities? Is it possible that a NE uses rationals but the numerator or denominator are exponential in the length of the problem? Either of these would make asking for an exact NE a hopeless request. We state facts known about NE.

1. In Nash's original paper [Nas50] he gave an example of a three-player game where all of the NE used irrational numbers.
2. Lemke and Howson [EJ64] showed that every two-player game with integers utilities has a rational NE.
3. Cottle and Dantzig [CD68] showed that the rational NE for a two-player game has numerators and denominators that are of length polynomial in the length of the original problem.
4. In 1928, von Neumann [Neu28] showed that in two-player zero-sum games (one where if player i has utility x, then player $1-i$ has utility $-x$), there is always a NE. His proof was by the minimax theorem, which can be interpreted as a polynomial-time algorithm. In fact, this technique is a special case of strong linear programming duality. In our notation we say that 2-NASH restricted to zero-sum games is in P.

What is the complexity of 2-NASH? It seems to be hard to compute. However, by the next exercise, it is unlikely to be NP-hard.

Exercise 22.11. If 2-NASH is NP-hard, then NP = coNP.
Hint: This is similar to the proof of theorem 22.3.

22.4 Brouwer's Fixed Point Theorem

Brouwer's famous fixed point theorem is as follows.

Theorem 22.12. *Let D be a subset of Euclidean space. Let f be a function from D to D. Assume f, D satisfy the following three properties:*

1. *D is a convex set.*
2. *D is compact.*
3. *f is continuous.*

Then there exists $x \in D$ such that $f(x) = x$.

We note that Brouwer's theorem is tight in that if any of the conditions are not met, then the theorem is not true. We also note that the proof of Brouwer's theorem is nonconstructive in that the proof will not help you find the fixed point quickly.

It is worth noting that Nash used Brouwer's theorem to show his result for general games. Roughly, the proof involves a function $f : [0, 1]^n \rightarrow [0, 1]^n$ that indicates the motivation a player has to deviate from his current strategy. The NE corresponds to the fixed point of the mapping.

The computational problem that arises from Brouwer's fixed point theorem is to, given f, D satisfying the conditions of theorem 22.12, find the fixed point. There is a major problem with this problem. f is continuous! D is a subset of the reals! This entire book has been about discrete problems! Not to worry; there is a discrete version of this problem, suitable for study. It is due to Papadimitriou [Pap94, Page 511]. We present it here:

BROUWER$_{p,d}$ (p is a polynomial with coefficients in $\mathbb{Z}$ and $d \in \mathbb{N}$. Both p and d are parameters.)

Instance: A Turing machine M and $n \in \mathbb{N}$. (n is in unary.) Let

$$D = \left\{ \left(\frac{a_1}{n}, \ldots, \frac{a_d}{n} \right) : a_1, \ldots, a_d \in \{0, \ldots, n\} \right\}.$$

1. The inputs to the TM will be elements of $D \subseteq [0,1]^d$.
2. Note that there are only n^d, inputs which is a polynomial number of inputs.

Promise: For all $\vec{x} \in D$,

1. $M(\vec{x})$ runs in time $\le p(n)$;
2. $M(\vec{x}) \in \mathbb{Q}^d$;
3. $|M(\vec{x})| \le \frac{1}{n^2}$; and
4. $M(\vec{x}) + \vec{x} \in [0,1]^d$.

Since there are only n^d inputs and M runs in $p(n)$ time, the promise can be checked in polynomial time. Do this. If the promise does not hold, then output BAD INPUT.

Output: (Assume the promise holds) Let f be the function from D to D defined by $f(\vec{x}) = \vec{x} + M(\vec{x})$. The function f can be extended to a unique piecewise linear map $\hat{f}$ that maps $[1,n]^d$ to $[1,n]^d$. The function $\hat{f}$ satisfies the conditions of Brouwer's fixed point theorem, so there exists an $\vec{x} \in [1,n]^d$ such that $\hat{f}(\vec{x}) = \vec{x}$. Find that point.

Note: There is an issue that $\vec{x}$ might have coordinates that are irrational or the ratio of large naturals. Hence, formally, the input must also contain an error bound. We omit these details.

22.5 Sperner's Lemma

Sperner's lemma is about colorings of n-dimensional objects; however, we will look at the special case where the dimension is 2.

We define some terms and then state Sperner's lemma.

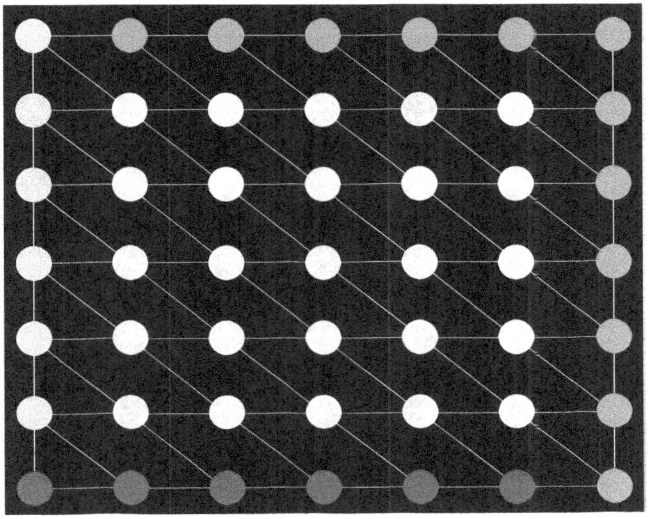

Figure 22.2
Valid edge coloring for the Sperner edge coloring.

Definition 22.13. Let *COL* be a 3-coloring of the lattice points of an $n \times n$ grid.

1. *COL* is ***valid*** if all vertices on the bottom row are RED, all vertices in the leftmost column are YELLOW, and all other boundary vertices are BLUE. (There is no condition on the non-boundary vertices.) Figure 22.2 gives an example of a valid coloring. (If you are reading the black and white version of this book, then the left side is supposed to be all yellow except for the bottom vertex, which is red; the bottom is all red except for the rightmost vertex, which is purple; the right side is all purple; and the top is all purple except for the leftmost vertex, which is yellow. all of the other vertices have their color unspecified.)
2. For all squares in the grid, draw the line from the upper left to the bottom right. Hence we now have many triangles. A ***trichromatic triangle*** is a triangle where all of the vertices have different colors.

Theorem 22.14. *For all valid 3-coloring of the lattice points of an $n \times n$ grid, there is a trichromatic triangle. In fact, there will be an odd number of them.*

Proof. We give two proofs.

Proof One

First, we will add an artificial trichromatic triangle by adding a blue vertex next to the bottom left corner of the grid, where the yellow and red boundaries meet (see figure 22.3). We now define a directed walk on the triangles of the grid graph inductively. We start in

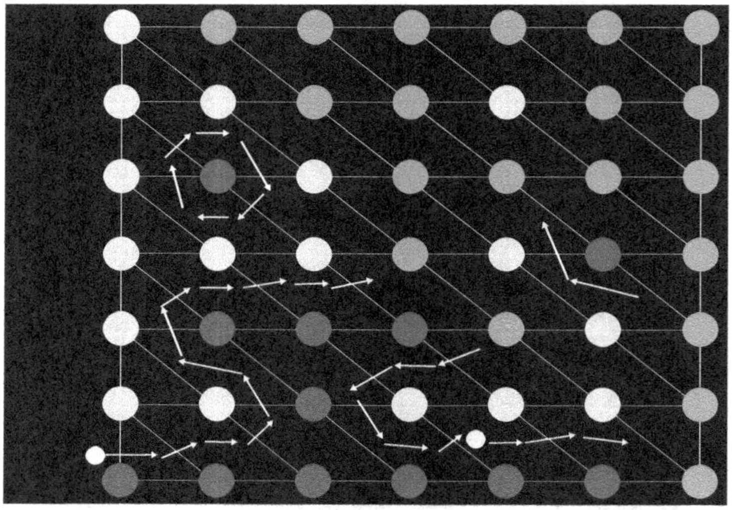

Figure 22.3
A valid walk in a valid coloring of a triangulation of the lattice for Sperner's lemma.

the artificial triangle and leave it though the yellow-red edge with yellow to the left. If we arrive in a trichromatic triangle, we are done. If not, then the other vertex is red or yellow. Hence there will be a way to leave by going over a yellow-red edge with yellow on the left. Keep doing this: Leave a triangle through the yellow-red edge with yellow on the right and either (a) you are in a trichromatic triangle so you are done, or (b) the other vertex is yellow or red, so repeat. We claim that this procedure will find another trichromatic triangle.

Note that this walk can not exit the grid graph with legal boundary coloring: The only red-yellow edge on the boundary is the one on the bottom left corner, and in order to cross it, we would have to have the yellow vertex on our right. This is not allowed. Moreover our walk will not produce a cycle. For the sake of contradiction, suppose that it did. Consider a triangle where the loop closes. This triangle must have had a red-yellow edge that we crossed the first time with yellow to the left. However, on our way back in, any edge we cross will either have red to the left or yellow to the right. Neither of these options is admissible. Therefore, there are no cycles and we never leave the grid graph.

This, together with the fact that the number of triangles is finite, implies that at some point we must encounter a trichromatic triangle, since that is when our walk stops. Therefore, at least one such triangle must exist.

What about any other trichromatic triangles? Well, we can perform the same procedure with the other internal trichromatic triangles. Start another walk from one such triangle, and, by the same argument above, we will end at another.

Therefore, the total number of trichromatic triangles in our modified graph is an even number that is at least 2. However, one of these triangles was artificially introduced. Therefore, the number of trichromatic triangles inside the grid graph must be an odd number that is at least 1.

End of Proof One

Proof Two

This proof is more basic. Consider a directed graph where each vertex represents a triangle. There is an edge (u, v) if triangles u, v are adjacent and the edge that they share has yellow on the left.

We claim that every vertex must have in-degree ≤ 1 and out-degree ≤ 1. This can be done by a case analysis of the possibilities. It is important to note that if a triangle has exactly one red-yellow edge (hence it's trichromatic), then its vertex will have either exactly in-degree 1 and out-degree 0 or out-degree 1 and in-degree 0.

But what can we say about a directed graph where every vertex has in-degree and out-degree at most 1? Well, there can only be three types of (weakly) connected components: isolated vertices, cycles, or directed paths. These paths correspond to the paths we discovered in the walk above and imply that the trichromatic triangles come in pairs. In a more elementary way, the underlying principle is that if a directed graph has an unbalanced (i.e., in-degree is not the same as out-degree) vertex, then there must be another one.

In our example, the unbalanced vertices correspond exactly to trichromatic triangles. Since in our construction we introduce one such vertex, we are guaranteed that our graph will contain another one and a total even number of them. This property is also known as the parity argument for directed graphs, and plays a key role in the definition of the PPAD class. □

Sperner's lemma and Brouwer's theorem have a similar flavor. Indeed, we can derive Brouwer's theorem from Sperner's lemma.

Theorem 22.15. *Sperner's lemma implies Brouwer's theorem.*

Proof sketch. We can consider the problem of finding approximate fixed points for a function f from the unit square to itself. Given ε we want to find x such that $|f(x) - x| < \varepsilon$. We will create a grid on the unit square where the lines are δ apart (later we will set δ very small). We color each point p of the grid as follows:

- Yellow if the vector from p to $f(p)$ points right, though it can be at an angle. Note that all of the grid points on the left will be yellow, which fits the premise of Sperner's lemma. (We also color yellow if it points straight up or straight down.)
- Red if the vector from p to $f(p)$ points up, though it can be at an angle. Note that all of the grid points on the bottom will be red, which fits the premise of Sperner's lemma. (We also color red if it points left or right.)

- Blue in all other cases. Note that all of the grid points on the top or the right will be blue, which fits Sperner's lemma.

We can then use a compactness argument and let $\delta \to 0$ to prove the existence of an approximate fixed point (which will be inside the trichromatic triangle). This proof, however, might not preserve the parity or even the number of trichromatic triangles. □

We now define the problem SPERNER for 2-dimensions, called 2-SPERNER.

> 2-SPERNER
> *Instance:* A circuit that has input $\{0,1\}^n \times \{0,1\}^n$ and output $\{0,1,2\}$ (the three colors). (We do not test if the coloring is valid.)
> *Output:* Either return a trichromatic triangle or return that the coloring is not valid. (It is okay to return a trichromatic coloring even if the coloring is not valid. It is likely that an algorithm will proceed, find an alleged trichromatic coloring, and then test it is trichromatic, return it, if not then return that the coloring is not valid.)

We omit the definition of the general n-dimensional SPERNER problem since it is somewhat technical. It can be found in Papadimitriou [Pap94] (pp. 507–510). We use the term "SPERNER" to refer to the n-dimensional problem for all n.

22.6 Total Search Problems in NP

In the next definition, we state two types of problems we have discussed in this book and two we have not, to show the contrast.

Definition 22.16.

1. ***Decision problems***: Let A be a set. The problem is, given x, determine whether x is in A. Example: SAT. Note that this is NP-hard (actually NP-complete).
2. ***Functions***: Let f be a function. The problem is, given x, find $f(x)$. Example: Given φ, output the least lexicographic satisfying assignment if there is one, and 0 if there is not one. Note that this is NP-hard.
3. ***Search problems*** (this is new): Let R be a relation. The problem is, given x, find *some* y such that $R(x,y)$, or output 0 if there isn't any. Example: Given φ find some y such that $\varphi(y) = \text{TRUE}$, and output 0 if there is no such y. Note that this is NP-hard.
4. ***Total search problems*** (this is new): Let R be a relation *where we are promised that, for all x, there is a short y with R(x,y)*. The problem is, given x, find *some* short y such that

$R(x,y)$. Examples: EOL, 2-NASH, BROUWER, and SPERNER. (We do not include 3-NASH or r-NASH since, as noted earlier, these may have irrational NE. We will later look at approximating them.)

If any of EOL, 2-NASH, BROUWER, or SPERNER are NP-hard, then NP = coNP (we proved this for EOL in theorem 22.3, and the proofs for the others are similar). Hence it is unlikely that any of them are NP-hard. We need another way to show they are hard.

In this chapter we will study problems where (1) you know there is a solution, (2) finding it seems hard, but (3) finding it does not seem to be NP-hard.

Exercise 22.17. Show that if the search problem that finds *some* satisfying assignment is in polynomial time, then the function that finds the lexicographic least satisfying assignment is in polynomial time.

We will now define an analog of NP for search problems and total search problems.

To get an intuition for FNP, let us first define FP (this is different from the FP we defined in section 0.3). We follow the definition by Rich [Ric08] (from the section "The Problem Classes FP and FNP").

Definition 22.18. A relation R is in FP if the following occur.

1. $R(x,y)$ can be computed in time polynomial in $|x|+|y|$.
2. The function that, given x, determines if there is a y such that $R(x,y)$ holds, can be computed in time polynomial in $|x|$.
3. The function that, given x such that there is y with $R(x,y)$, finds such a y can be computed in time polynomial in $|x|$.

Note that if there is such a y, it will be of length bounded by a polynomial in $|x|$.

Definition 22.19. If $R \in$ FP, then the ***set associated with R*** is

$$\{x \mid \exists y : R(x,y)\}.$$

For FP we often think of the set first and the relation later. For example, 2-coloring corresponds to the following relation in FP:

$$\{(G,\rho) \mid \rho \text{ is a 2-coloring of } G\}.$$

FNP will be the NP-analog of FP. We will not demand that the y (if it exists) can be determined. We will demand that the y (if it exists) will be short since that will no longer follow from the definition.

Definition 22.20. A relation R is in FNP if the following occur.

1. $R(x,y)$ can be computed in polynomial time.
2. There is a polynomial p such that, for all x, if there is a y such that $R(x,y)$, then there is also such a y with $|y| \leq p(|x|)$.

Note that we are not claiming that y can easily be found.

Definition 22.21. If $R \in$ FNP, then the ***set associated with R*** is

$$\{x \mid \exists y : R(x,y)\}.$$

For FNP we often think of the set first and the relation later. For example, 3-coloring corresponds to the following relation in FNP.

$$\{(G, \rho) \mid \rho \text{ is a 3-coloring of } G\}.$$

We want to capture the fact that the problems we care about are total.

Definition 22.22. A relation R is in TFNP if $R \in$ FNP, and the following additional property holds: For every x, there is a y such that $R(x,y)$. Note again that we are not claiming that such a y can be found easily.

Note: Unlike FP and FNP, it would be silly to associate with $R \in$ TFNP a set since that set would always be Σ^*.

22.7 Reductions and PPAD

We define reductions for FNP.

Definition 22.23. Let $R, R' \in$ FNP, and let L, L' be the associated sets. ***R is polynomial-time reducible to R′*** if the following occur.

1. There exists a polynomial time computable f such that $x \in L$ if and only if $f(x) \in L'$.
2. There exists a polynomial time computable g such that if $R'(f(x),y)$ holds, then $R(x, g(y))$ holds. (So if you have a witness for $f(x)$, you can recover one for x.)

Exercise 22.24. Show that, if R reduces to R' and $R' \in$ FP, then $R \in$ FP.

We could define a notion of FNP-hard based on this reduction. Alas, it is unlikely that EOL, NASH, SPERNER or BROUWER are FNP-hard. We showed in theorem 22.3 that if EOL is what we now call FNP-hard, then NP = coNP. The same holds for the other problems.

These four problems all *seem* hard. Let's turn this around! Let's *assume* that one of them is hard and define a complexity class based on that assumption.

Papadimitriou [Pap94] defined the following class.

Definition 22.25. Let $R \in FNP$.

1. R is in PPAD (polynomial parity arguments on directed graphs) if R reduces to EOL.
2. R is PPAD-hard if EOL reduces to R.
3. R is PPAD-complete if R is both in PPAD and PPAD-hard.

Theorem 22.26. *SPERNER, NASH, and BROUWER are all in PPAD.*

Proof sketch. The proof of Sperner's lemma uses theorem 22.2 in the case of graphs with in-degree and out-degree ≤ 1. This proof can be modified to obtain a reduction from SPERNER to EOL.

Theorem 22.15 showed (a sketch of) how to prove Brouwer's fixed point theorem from Sperner's lemma. That proof can be modified to obtain a reduction from BROUWER to SPERNER.

The existence of a NE can be proven from Brouwer's fixed point theorem. That proof can be modified to obtain a reduction from NASH to BROUWER. □

It turns out that all three of these problems are PPAD-complete.

We discuss one more problem that is PPAD-complete.

Definition 22.27. Let C be a cake. Let $P_1, \ldots, P_n$ be n people. They each have a utility function that maps areas of the cake to values. The entire cake maps to 1, and a single point maps to 0. If A and B are disjoint parts of the cake, then, for any utility function U, $U(A \cup B) = U(A) + U(B)$.

1. An ***allocation*** of C is a partition $C = C_1 \cup \cdots \cup C_n$ of C where, for all $1 \leq i \leq n$, P_i gets piece C_i.
2. An allocation is ***proportional*** if every person, using their own utility function, gets $\geq \frac{1}{n}$.
3. An allocation is ***envy-free*** if every person, using their own utility function, thinks that nobody has a strictly larger piece than they have.

We state theorems about *existence* of an envy-free allocation using only $n-1$ cuts and *finding* such an allocation.

Theorem 22.28.

1. *(Stromquist [Str80]) For any n utility functions, there exists an envy-free allocation that only uses $n-1$ cuts. The cuts could be at irrational points. He also discusses finding an approximation to an envy-free division.*
2. *(Deng et al. [DQS12]) To discuss the complexity of finding an envy free division, we need a notion of how a valuation can be input. Let the valuation be a polynomial time computable function (see the paper for details). The problem of, given n evaluation functions, find an envy-free allocation that only uses $n-1$ cuts, is PPAD-complete.*

22.8 2-Nash Is PPAD-Complete

Daskalakis et al. [DGP09] proved that 3-Nash is PPAD-complete (this was actually in Daskalakis's PhD thesis). At the time, it was thought that perhaps 2-Nash is in P since (1) 2-Nash always has a rational NE and (2) the zero-sum case is in P. Hence it was a surprise when Chen and Deng [CD06] proved that 2-Nash is PPAD-complete. Later Daskalakis et al. [DGP09] found a way to obtain 2-Nash PPAD-complete from their techniques, and that is in their paper.

We will show part of the proof that 2-Nash is PPAD-complete. We mostly follow the approach of Daskalakis et al. [DGP09]. Hence all the theorems we present are due to them, unless otherwise noted.

The full reduction requires a sequence of reductions:

1. Reduce a PPAD problem to a PPAD-type problem in $[0,1]^3$.
2. Reduce the PPAD-type problem to the 3D-Sperner problem.
3. Reduce Arithmetic Circuit SAT to Poly Matrix Nash. We will define this formally later. It involves a *restricted* Nash problem, though for *many* players. So, initially, it looks incomparable to 2-Nash.
4. Reduce the 3D-Sperner problem to Arithmetic Circuit SAT. We will define this formally later. It involves arithmetic circuits and asks if there is a way to set the variables vertices so that the circuit is consistent.
5. Reduce Poly Matrix Nash to 2-Nash.

We will focus on the reduction from Arithmetic Circuit SAT to Poly Matrix Nash.

22.8.1 PPAD-Completeness of 2-Nash (High Level)

We first find cycles and paths and place them in a cube $[0,1]^3$ without intersections. Then we represent this as a Sperner problem. Next we convert the problem to an Arithmetic Circuit. Finally, that Arithmetic Circuit is converted into a Nash problem.

22.8.2 Arithmetic Circuit SAT

An arithmetic circuit is a circuit that has arithmetic operations at the gates. Normally there would be *input vertices*; however, here we instead have *variable vertices*. The problem will be to find a way to set the variable vertices so that the circuit is consistent.

Definition 22.29. An ***arithmetic circuit*** is a circuit with the following types of gates.

- ***Variable vertices*** $x_1, x_2, \ldots, x_n$. These have in-degree 1 and out-degree 0 or 1 or 2. (Yes, you read that right—the in-degree is 1, not 0. These are not input vertices.)
- Gates of six types ($:=, +, -, a, xa, >$), which we describe in the next definition. They are pictured, together with their in-degree and out-degree, in figure 22.4. For the $>$ gate

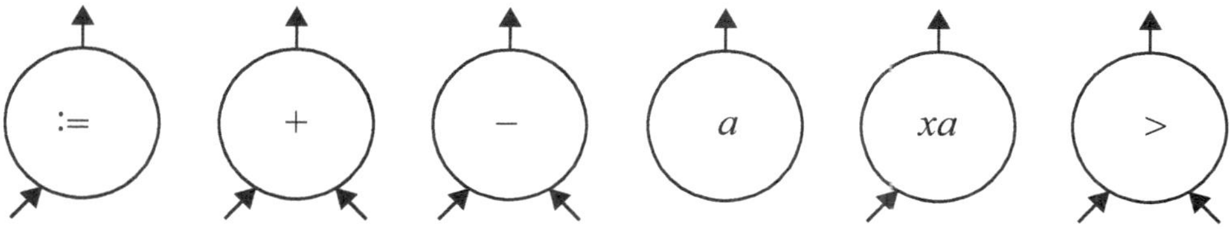

Figure 22.4
The operation vertices of ARITHMETIC CIRCUIT SAT.

and the − gate, the order of inputs matters, which is why the inputs are labeled 1 and 2. For the other gates, the order does not matter.

- Directed edges connecting variables to gates and vice versa.
- Variable vertices have in-degree 1; gates have in-degree 0, 1, or 2 inputs depending on type; gates and vertices have arbitrary fan-out.

Definition 22.30. We use $y \leftarrow f(x_1, \ldots, x_n)$ to mean that if $x_1, \ldots, x_n$ are the inputs to the gate, then y is the output. The gates we define are pictured in figure 22.4.

- Assignment: $y \leftarrow x_1$. In-degree 1.
- Addition: $y \leftarrow \min\{1, x_1 + x_2\}$. In-degree 2.
- Subtraction: $y \leftarrow \max\{0, x_1 - x_2\}$. In-degree 2.
- Equal a constant: $y \leftarrow \max\{0, \min\{1, a\}\}$. In-degree 0.
- Multiply by a constant: $y \leftarrow \max\{0, \min\{1, ax\}\}$. In-degree 1.
- Greater than:

$$y \leftarrow \begin{cases} 0, & \text{if } x_1 < x_2 \\ 1, & \text{if } x_2 < x_1 \\ \text{any value}, & \text{if } x_2 = x_1 \end{cases} \tag{22.1}$$

In-degree 2.

> ARITHMETIC CIRCUIT SAT
> *Instance:* An Arithmetic circuit with variable vertices $x_1, \ldots, x_n$.
> *Output:* An assignment of rationals to the variable vertices so that all of the gate operations are satisfied.
> *Note:* In figure 22.5, we show an example of ARITHMETIC CIRCUIT SAT. (There is a detail here that we are skipping: The rationals cannot have too large a numerator or denominator.)

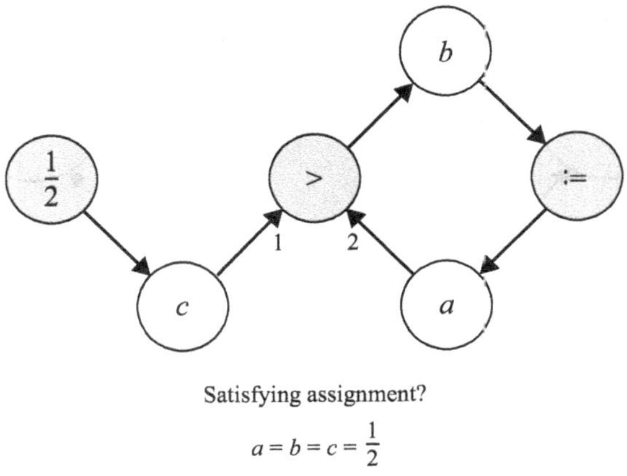

Figure 22.5
The operation vertices of ARITHMETIC CIRCUIT SAT.

We derive values of a, b, c that make the arithmetic circuit consistent.

Node c is assigned to value $1/2$ and, thus, must have this value. Node b has an assignment into vertex a, so $a = b$. If $c > a$, then the $>$ gate outputs a 1, so $b = 1$, so $a = 1$, but that cannot happen since $c = \frac{1}{2}$. If $c < a$, then the $>$ gate outputs a 0, so $b = 0$, so $a = 0$, but that cannot happen since $c = \frac{1}{2}$. Hence we must have $c = a$. Then $a = b = c = \frac{1}{2}$.

In the example, the solution existed and was unique. It is quite possible that there is no solution, one solution, or many solutions.

22.8.3 Graphical Games

Rather than reducing directly to 2-NASH, it will be useful to reduce to a more general kind of game. Kearns et al. [KLS01] defined graphical games.

Definition 22.31. A *graphical game* has the following.

1. A directed graph.
2. Each player's payoff depends only on his own strategy and the strategy of his in-neighbors.

Janovkaya [Jan68] defined a special case of graphical games called polymatrix games.

Definition 22.32. A ***polymatrix game*** is a graphical game with edgewise separable utility functions. For player v,

$$u_v(x_1, \ldots, x_n) = \sum_{(w,v)\in E} u_{w,v}(x_w, x_v) = \sum_{(w,v)\in E} x_v^T A^{(v,w)} x_w.$$

Here, $A^{(v,w)}$ are matrices, x_v is the mixed strategy of v, and x_w is the mixed strategy of w. Now, our strategy for reducing from ARITHMETIC CIRCUIT SAT to NASH will be via POLY MATRIX NASH, so our next goal is to reduce ARITHMETIC CIRCUIT SAT to POLY MATRIX NASH

We define the problem of finding a Nash equilibrium for a polymatrix game.

> POLY MATRIX NASH
> *Instance:* A polymatrix game.
> *Output:* An approximation to a Nash equilibrium.

22.8.4 Reduction: ARITHMETIC CIRCUIT SAT to POLY MATRIX NASH

The key to this reduction is the invention of *game gadgets*, small polymatrix games that model arithmetic at their Nash equilibrium. As an example, consider the following gadget for addition.

Addition Gadget. Suppose each player has two strategies; call these $\{0, 1\}$. Then, a mixed strategy is a number in $[0, 1]$—that is, the probability of playing 1. We construct a gadget with players w, x, y, z and edges (x, w), (y, w), (z, w), (w, z). Player w has expected return

$$\Pr[x:1] + \Pr[y:1] \text{ for playing } 0$$

$$\Pr[z:1] \text{ for playing } 1.$$

It is easy to construct this payoff matrix: $u_w(0) = x + y$ and $u_w(1) = z$.

Player z is paid for "playing the opposite" of w: $u_z(0) = 0.5$ and $u_z(1) = 1 - w$.

Lemma 22.33. In any Nash equilibrium of a game containing the above gadget,

$$\Pr[z:1] = \min\{\Pr[x:1] + \Pr[y:1], 1\}.$$

Proof. Suppose $\Pr[z:1] < \min(\Pr[x:1] + \Pr[y:1], 1)$. Then w will play 0 with probability 1, but then z should have played 1 with probability 1, a contradiction.

Conversely, suppose $\Pr[z:1] > \Pr[x:1] + \Pr[y:1]$. Then w will play 1 with probability 1, but then z should have played 0 with probability 1, again a contradiction.

Thus, $\Pr[z:1] = \Pr[x:1] + \Pr[y:1]$. □

More gadgets. We need such a gadget for all the possible gates in ARITHMETIC CIRCUIT SAT. If z is the output of the gadget and x, y are the inputs, we need (conflating x with $\Pr[x:1]$)

- copy: $z = x$
- addition: $z = \min(1, x + y)$
- subtraction: $z = \max(0, x - y)$

- set equal to constant: $z = a$
- multiply by a constant: $z = ax$
- comparison: $z = 1$ if $x > y$, $z = 0$ if $x < y$, unconstrained otherwise

Another example: Comparison gadget. Players x, y, z. $u_z(0) = y$, $u_z(1) = x$. Then $x > y \Rightarrow \Pr[z:1] = 1$ and similarly for $x < y$.

Now that if we have all these gadgets, we can sketch a proof of the reduction we seek (in this section).

Theorem 22.34. *ARITHMETIC CIRCUIT SAT reduces to POLY MATRIX NASH. Hence POLY MATRIX NASH is PPAD-complete.*

Proof sketch. We omit the construction of the remaining gadgets, but they are not much more complicated than the addition gadget. From here, given an arbitrary instance of ARITHMETIC CIRCUIT SAT, we can create a polymatrix game by composing game gadgets corresponding to each of the gates. At any Nash equilibrium of the resulting polymatrix game, the gate conditions are satisfied, which completes the reduction.

By theorem 22.26, NASH is in PPAD. Since POLY MATRIX NASH is a subcase of NASH, it is also in PPAD. □

22.8.5 Reduction: POLY MATRIX NASH to 2-NASH

Theorem 22.35. *POLY MATRIX NASH reduces to 2-NASH.*

Proof sketch. Since we can construct the game gadgets we used to reduce to POLY MATRIX NASH using only bipartite game gadgets (input and output players are on one side, and auxiliary vertices are on the other side), without any loss of generality we can also assume the polymatrix game is bipartite. This implies that the graph is 2-colorable, say by two colors "red" and "blue." Now, we can think of this as a two-player game between the "red lawyer" and the "blue lawyer," where each lawyer represents all the vertices of their color.

Each lawyer's set of pure strategies is the union of the pure strategy sets of her clients; importantly, this is not the same as having a strategy set equal to the product of the strategy sets of the clients, as this would cause an exponential blowup in the problem size (and thus invalidate the reduction). One way of picturing this is that the red lawyer selects one of her clients to represent, and then selects a strategy for that client.

The payoff of $(u:i)$, $(v:j)$ (the red lawyer plays strategy i of client u, and the blue lawyer plays strategy j of client v) is $A_{i,j}^{(u,v)}$ for u and $A_{j,i}^{(v,u)}$ for v. Informally, the payoff of the lawyers is just the payoffs of the respective clients had they played the chosen strategies. Also, note that the payoff is 0 for either lawyer if their client choice does not have an edge incoming from the other lawyer's client choice (and 0 for both lawyers if they choose non-neighboring clients).

Wishful thinking. If (x, y) is a Nash equilibrium of the lawyer game, then the marginal distributions that x assigns to the strategies of the red vertices and the marginals that y assigns to the blue vertices comprise a Nash equilibrium.

This does not work because lawyers might gravitate only to the "lucrative" clients—we need to enforce that lawyers will (at least approximately) represent each client equally. This is true not only because a lawyer might choose to *never* represent a client (and hence the marginals are undefined) but also because if the red lawyer's clients are not equally represented, the optimal marginals for the blue lawyer are distorted correspondingly.

Enforcing equal representation. Lawyers play a "high-stakes" game on the side. Without loss of generality, assume that each lawyer represents n clients (can create "dummy" clients to equalize number of clients). Label each lawyer's clients $1, \ldots, n$ arbitrarily. Payoffs of the high-stakes game: Suppose the red lawyer plays any strategy of client j and blue lawyer plays any strategy of client k, then if $j \neq k$ then both players get 0 If $j = k$, then red lawyer gets $+M$ while blue lawyer gets $-M$.

Claim 22.36. In any Nash equilibrium of the high-stakes game, each lawyer assigns probability (close to) $1/n$ to the set of pure strategies of each of his clients.

We omit the proof.

Now, the game we need for the reduction is simply the sum of the original lawyer game and the high-stakes game. Choose $M >> 2n^2 u_{max}$, where u_{max} is the maximum absolute value of payoffs in the original game. We can show that the total probability mass is distributed almost evenly among the different vertices.

Lemma 22.37. If (x, y) is an equilibrium of the lawyer game, for all u, v:

$$x_u = \frac{1}{n}\left(1 \pm \frac{2u_{max}n^2}{M}\right)$$

$$y_v = \frac{1}{n}\left(1 \pm \frac{2u_{max}n^2}{M}\right).$$

However, within a particular vertex u, only the original low-stakes game matters, since the different strategies of u are all identical from the perspective of the high-stakes game.

Lemma 22.38. The payoff difference for the red lawyer from strategies $(u:i)$ and $(u:j)$ is

$$\sum_v \sum_\ell (A_{i,\ell}^{(u,v)} - A_{j,\ell}^{(u,v)}) y_{v:\ell}.$$

Corollary. If $x_{u:i} > 0$, then for all j:

$$\sum_v \sum_\ell (A_{i,\ell}^{(u,v)} - A_{j,\ell}^{(u,v)}) y_{v:\ell} \geq 0.$$

Define $\hat{x}_u(i) := \frac{x_{u:i}}{x_u}$ and $\hat{y}_v(j) := \frac{y_{v:j}}{y_v}$ (these are the marginals given by lawyers to different vertices). □

Observation: If we had $x_u = 1/n$ for all u and $y_v = 1/n$ for all v, then $\{\{\hat{x}_u\}_u, \{\hat{y}_v\}_v\}$ would be a Nash equilibrium. Since we have $\pm\frac{2u_{max}n^2}{M}$ deviation, we get approximate Nash equilibrium instead. Fortunately, ARITHMETIC CIRCUIT SAT is still PPAD-hard with ε approximation, so we still get PPAD-hardness for NASH from this reduction.

NASH really asks for an approximation to the NE. This means that if (x, y) is the NE (where x and y are vectors of probabilities that add to 1), then the algorithm produces (x', y'), where x' is close to x and y' is close to y. We briefly discuss a different kind of approximation.

Definition 22.39. An *ε-Nash equilibrium (henceforth just ε-equilibrium)* is a pair of mixed strategies (x, y) such that the following holds.

1. If the row player deviates from x, and the column player still uses y, then the row player benefits by at most ε.
2. If the column player deviates from y, and the row player still uses x, then the column player benefits by at most ε.
3. For each player, the payoff at (x, y) is at most ε less than the optimal.

There are essentially matching upper and lower bounds for the time needed to find an ε-equilibrium:

Theorem 22.40.

1. *Lipton et al. [LMM03] showed that, for all $\varepsilon > 0$, there is an algorithm that finds an ε-equilibrium that runs in time $O(n^{\varepsilon^{-2}\log n})$.*
2. *Braverman et al. [BKW15] showed that, assuming ETH, there exists ε^* such that any algorithm that finds an ε^*-equilibrium and requires time $O(n^{\log n})$.*

Remark: Consider a two-player game with payoff matrices R and C. Zero-sum games correspond to having $R + C = 0$ (the zero matrix), but we can also consider games where the rank of $R + C$ is r, for some constant r. Adsul et at. [AGMS11, AGM$^+$21] showed that rank 1 games have a polynomial-time algorithm for finding a NE (just like zero-sum games), but a result of Mehta in [Meh14] shows that, in rank 3 games, it is already PPAD-hard to find a NE. The case or rank 2 seems to be open.

22.9 Other Arguments of Existence and Resulting Complexity Classes

The purely combinatorial theorem with a nonconstructive proof at the core of the definition of PPAD is:

If a directed graph has an unbalanced vertex, then it has another unbalanced vertex.

This statement led to problems (EOL, NASH, BROUWER, SPERNER) that are in P, but finding the witness seems hard. We then defined PPAD to pin down that finding the witness is probably hard.

In this section we discuss other purely combinatorial theorems with nonconstructive proofs. In all cases (1) the theorems can be interpreted as problems that are in P, (2) finding a witness to verify a relation seems hard, and (3) a complexity class is inspired.

22.9.1 There Are an Even Number of Vertices of Odd Degree

The oldest theorem in graph theory, due to Euler, is the following (we refashion it for our purposes).

Theorem 22.41. *If a graph has a vertex of odd degree, then it must have another.*

The proof of theorem 22.41 is nonconstructive and inspires the following problem and complexity class.

ODD DEG
Instance: A circuit C on $\{0, 1\}^n$ that, on input x, outputs a set of elements of $\{0, 1\}^n$. We interpret this as a graph on $\{0, 1\}^n$ where the circuit C, when run on x, outputs the neighbors of x. We also have that 0^n has an odd number of neighbors.
Output: A vertex $v \in \{0, 1\}^n - \{0^n\}$ that has an odd degree.

Papadimitriou [Pap94] defined the following classes to make this statement rigorous. (Our definition differs from Papadimitriou's, but the two definitions are equivalent.)

Definition 22.42. Let $R \in FNP$.

1. R is in PPA (polynomial parity argument) if R reduces to ODD DEG.
2. R is PPA-hard if ODD DEG reduces to R.
3. R is PPA-complete if R is both in PPA and PPA-hard.

Exercise 22.43. Show that PPAD $\subseteq$ PPA.

The following theorem is due to Smith (unpublished) and is presented in a paper by Thomason [Tho78]; see also a paper by Krawczyk [Kra99], which contains the proof.

Theorem 22.44. *If a 3-regular graph has a Hamiltonian cycle C, then, for all edges e in C, there is a second Hamiltonian cycle that uses e.*

The proof uses theorem 22.41 and hence is nonconstructive. Theorem 22.41 inspires the following problem.

> HAMILTONIAN CYCLES IN LARGE 3-REGULAR GRAPHS (HAM-3REG)
> *Instance:* A 3-regular graph G and a Hamiltonian cycle H in G.
> *Output:* A Hamiltonian cycle that is not H.

The proof of theorem 22.44 easily yields the following theorem.

Theorem 22.45. *HAM-3REG is in PPA.*

Open Problem 22.46. *Is HAM-3REG PPA-complete? (This is believed to be true.)*

Here is a non-graph example.

Definition 22.47. Let D be any integral domain (for our purposes $\mathbb{Z}$ or $\mathbb{Z}_p$).

1. Let m be a monomial in $D[x_1, \ldots, x_n]$. The *degree* of m is the sum of the degrees of its terms. For example, the degree of $x_1^2 x_2^3 x_4^4$ is $2+3+4=9$.
2. Let $p \in D[x_1, \ldots, x_n]$. The *degree* of p is the max of the degrees of the monomials in p.
3. If $q_1, \ldots, q_L \in D[x_1, \ldots, x_n]$, then we refer to that set of polynomials as *a system*. A *solution to the system* is $(a_1, \ldots, a_n) \in D^n$ such that, for all $1 \le i \le L$, $q_i(a_1, \ldots, a_n) = 0$.

Chevalley proved the following.

Theorem 22.48. *Let p be a prime. Let $q_1(x_1, \ldots, x_n), \ldots, q_L(x_1, \ldots, x_n) \in \mathbb{Z}_p[x_1, \ldots, x_n]$. Assume q_i is of degree d_i.*

1. *If $\sum_{i=1}^{L} d_i < n$, then the number of solutions to this system is divisible by p.*
2. *(This is an easy corollary of interest to us.) If $p = 2$ and there is a solution, then there is another solution.*

> CHEVALLEY
> *Instance:* A system of polynomial equations with n variables over $\mathbb{Z}_2$ such that the sum of the degrees is $<n$, and one solution.
> *Output:* A solution that is not the given one.

Theorem 22.49.

1. *(Papadimitriou [Pap94]) CHEVALLEY is in PPA.*
2. *(Goos et al. [GKSZ20]) A variant of CHEVALLEY is PPA_q-complete (you will define PPA_q in exercise 22.50). However, they do not think the original CHEVALLEY is PPA-complete (see their note on page 6).*

Exercise 22.50.

1. Let $q \in \mathbb{N}$ and let G be a bipartite graph. Show that if there is some vertex of degree $\not\equiv 0 \pmod{q}$, then there must be another one.
2. Define PPA_q and PPA_q-complete using part 1 as motivation.
3. Read Goos et al. [GKSZ20], which shows that several problems are PPA_q-complete. Rewrite their proofs in your own words.

Open Problem 22.51. *Is* CHEVALLEY *PPA-complete?*

What about natural problems and completeness? The following are known.

1. Filos-Ratsikas and Goldberg [FG18] showed that the consensus-halving problem, a computational version of the Hobby-Rice theorem, is PPA-complete.
2. Jerábek [Jer16] showed that the following problem randomly reduces to PPA: Given an integer, either declare that it is prime or find a factor. Under the generalized Riemann hypothesis, there is a deterministic reduction. Note that this reduction, randomized or deterministic, is not a hardness result.

22.9.2 Every Directed Acyclic Graph Has a Sink

The following is well known.

Theorem 22.52. *Every directed acyclic graph has a vertex v such that outdeg(v) = 0. Such a vertex is called a* sink.

The proof is nonconstructive. Theorem 22.52 inspires the following problem.

FIND SINK
Instance: A circuit C on $\{0,1\}^n$ that, on input x, outputs a set of elements of $\{0,1\}^n$. We interpret this as a graph on $\{0,1\}^n$ where the circuit outputs a *potential* set of neighbors. The neighbors of v are the elements $u \in C(v)$ such that $u > v$ (interpreted as numbers in binary). This will ensure that the graph is acyclic.
Output: A vertex of out-degree 0. (These are called ***sinks***.)

Johnson et al. [JPY88] defined the following.

Definition 22.53. Let $R \in FNP$.

1. R is in PLS (polynomial local search) if R reduces to FIND SINK. (The name "polynomial local search" comes from using this class to classify certain search problems that have local max [or min], making it difficult to find the global max [or min].)

2. R is PLS-hard if FIND SINK reduces to R.
3. R is PLS-complete if R is both in PLS and PLS-hard.

> LOCAL MAX CUT
> *Instance:* A weighted graph $G = (V, E, w)$.
> *Output:* A partition $V = V_1 \cup V_2$ that is locally optimal (i.e., cannot move any single vertex to the other side to increase the cut size).

Johnson et al. [JPY88] showed the following.

Theorem 22.54. *LOCAL MAX CUT is PLS-complete.*

22.9.3 The Pigeonhole Principle

The following is the well-known pigeonhole principle.

Theorem 22.55. *Let A have n elements and B have $n-1$ elements. For all functions $f: A \to B$, there exists $x_1 \neq x_2 \in A$ such that $f(x_1) = f(x_2)$.*

The proof of theorem 22.55 is nonconstructive.

The following theorem is an easy direct consequence of theorem 22.55, and hence, its proof is nonconstructive.

Theorem 22.56. *If $A \subseteq \{1, \ldots, 2^n - 1\}$ of n elements whose sum is $< 2^n - 1$, then there exist two distinct subsets of A that have the same sum.*

We use theorem 22.55 and 22.56 as the inspiration for problems.

> COLLISON
> *Instance:* A circuit C with input and output both $\{0, 1\}^n$. (We are not requiring that C has range $< 2^n$.)
> *Output:* Either an x such that $C(x) = 0^n$ or an x, y such that $x \neq y$ but $C(x) = C(y)$.

> DISTINCT SUBSETS (DIST SUBSET)
> *Instance:* $A \subseteq \{1, \ldots, 2^n - 1\}$ of n elements whose sum is $< 2^n - 1$.
> *Output:* Two distinct subsets of A with the same sum.

It is believed that both COLLISON and DIST SUBSET are hard and are equivalent. Papadimitriou [Pap94] defined the following classes to make this statement rigorous.

Definition 22.57. Let $R \in FNP$.

1. R is in PPP (polynomial pigeonhole principle) if R reduces to COLLISON.
2. R is PPP-hard if COLLISON reduces to R.
3. R is PPP-complete if R is both in PPP and PPP-hard.

Exercise 22.58. Show that PPAD $\subseteq$ PPP.

The proof of theorem 22.56 easily yields the following theorem.

Theorem 22.59. *DIST SUBSET is in PPP.*

Open Problem 22.60. *Is DIST SUBSET PPP-complete?*

What about natural problems? The following is known.

1. Sotiraki et al. [SZZ18] showed that a variant of the shortest lattice problem is PPP-complete.
2. Jerábek [Jer16] showed that there is a randomized reduction from integer factorization (finding a nontrivial factor) to a weaker version of COLLISON where the domain is 2^n and the range is 2^{n-1}. Under the generalized Riemann hypothesis, the reduction can be derandomized. Note that this reduction, randomized or deterministic, is not a hardness-result for factoring. Nor does it imply that factoring is in PPP.

22.10 How Do the Classes Relate?

We summarize what is known about how the classes relate and what is open.

Exercise 22.61.

1. Show that FP $\subseteq$ PPAD $\subseteq$ PPA $\subseteq$ FNP.
2. Show that FP $\subseteq$ PPAD $\subseteq$ PPP $\subseteq$ FNP.

Open Problem 22.62.

1. *For each subset inclusion in part 1 and 2, resolve if the inclusion is equal or proper. (It is widely believed that all of the inclusions are proper.)*
2. *For each subset inclusion in part 1 and 2, determine whether an equality implies P = NP or some other unlikely conclusion.*
3. *Resolve how PPA and PPP compare.*

23

Quantum Computing

23.1 Introduction

This book is about classical computing. The algorithms and reductions in this book can be carried out by a modern computer running on a single CPU or multiple CPUs. We will call such computers ***classical*** in that they are based on classical physics (at the bit level, computers are based on electricity) and not quantum physics. We use the term ***classical algorithm*** for an algorithm that can be run on a classical computer. What we call a ***classical algorithm*** in this chapter was just an ***algorithm*** in all of the other chapters.

There are theoretical devices called ***quantum computers***. An algorithm that can be run on such a device is a ***quantum algorithm***. There are some problems for which, theoretically, there is a quantum algorithm that is faster than any (known) classical algorithm. Hence quantum algorithms are of interest (we later list other reasons they are of interest).

Quantum computing is a vast topic that we will, for reasons of space, only be able to discuss briefly. We will have very few definitions, proofs, algorithms, or reductions. For basic definitions and more information on quantum computing see (1) the references in this chapter or (2) the books by Aaronson [Aar13], Mermin [Mer07], or Nielson and Chuang [NC16].

Chapter Summary

1. *We discuss results about quantum algorithms.*
2. *We discuss results about quantum streaming algorithms.*
3. *We discuss results about quantum games.*
4. *We discuss the question* "Will quantum computers ever outperform classical computers?"*—the* ***quantum supremecy debate****.*

The topics and results chosen highlight when the classical world and the quantum world differ or seem to differ.

23.2 Since the Hype Is False, Why Study Quantum Computation?

A cautionary note: From the popular press, one might think that quantum computers, if they are built, can solve world hunger, predict the stock market, and solve NP-complete problems. Most experts agree that this is not the case in reality. This misunderstanding may come from a misinterpretation of the *many-worlds interpretation of quantum mechanics*, which gives the false impression that quantum algorithms are massively parallel. They are not. In reality there seem to be only a few problems of interest that quantum computers (if they are built) can do much faster than classical computers. We also note that most experts in quantum do not think that NP-complete problems can be solved quickly by a quantum computer.

Given that the usefulness of quantum algorithms seems limited, why study them?

1. There are two problems, *factoring* and *discrete log*, (1) which are very important and (2) quantum algorithms for them are much faster (polynomial time versus exponential time) than any known classical algorithm.
2. There are many problems that have quantum algorithms that are faster than any known classical algorithm. Childs and Dam [CvD10] survey algebraic problems that have quantum algorithms that seem faster than any known classical algorithm. Jordan [Jor11] maintains a website of problems that have quantum algorithms that seem faster than any known classical algorithm. The speedups are usually not that large. Even so, they are a proof-of-concept for quantum algorithms being useful.
3. Richard Feynman first conceived of quantum computing as a way to potentially simulate quantum mechanics. This is another problem where quantum computers may outperform classical ones.
4. There have been cases where a classical algorithm was inspired by research on quantum algorithms. We give an example. Kerenidis and Prakash [KP17b] had a quantum algorithm for a recommendation system. Tang, while trying to show that no classical algorithm could do as well as the quantum algorithm, found a classical one that did [Tan19]. We stress that this classical algorithm was found because of research in quantum algorithms. For other examples, do a web search for `Quantum Inspired Classical Algorithms`.
5. The study of quantum algorithms has led to results in classical computing. See the survey of Drucker and de Wolf [DdW11] for examples.
6. The attempt to build quantum computers may lead to interesting insights into quantum physics. See next point.
7. The most exciting development that could happen would be if the attempt to build quantum computers leads to a discovery that the current theories of quantum mechanics are wrong or incomplete.

23.3 Factoring

Factoring was an important problem for cryptography. Many cryptosystems would be broken if factoring was easy. Hence, in contrast to work in algorithms, cryptographers hope that factoring is hard.

We will define factoring slightly differently than how it was defined in chapter 0.

> FACTORING (FACTORING)
> *Instance:* A number N
> *Output:* If N is prime, then output **PRIME**. If N is not prime, then output a nontrivial factor of N.

As noted in chapter 0:

1. There are no polynomial-time algorithms for FACTORING, nor is there a proof that it is NP-complete.
2. There are reasons to think it is not NP-complete (See exercise 0.24).
3. There is no clear consensus on whether FACTORING is in P

For a bit more on classical factoring, see section 0.9. For a lot more about classical factoring, see Wagstaff's book [Wag13].

What about quantum polynomial time?

Theorem 23.1. *Let the input to a factoring algorithm be N. Let $L = \lg N$, which is the length of N.*

1. *(Shor [Sho94, Sho99]) There is a polynomial quantum algorithm for FACTORING.*
2. *(Beckman et al. [BCDP96]) There is quantum algorithm for FACTORING that takes time $O(L^2 \log(L) \log(\log(L)))$. (This paper used Shor's algorithm as a starting point.)*

We note the following:

1. The key quantum component of Shor's algorithm for FACTORING is the quantum Fourier transform.
2. The constants in Beckman et al.'s version of Shor's algorithm are small. The biggest obstacle to running the algorithm is building a quantum computer that can handle many qubits.
3. Martin-Lopez et al. [MLLL$^+$12] have factored 21 on a quantum computer using Shor's algorithm. This can be considered a proof-of-concept. Other bigger numbers have been reported to have been factored by *a quantum computer*, but they really used a lot of

classical computing to set the problem up, and hence, we do not count those. Smolin et al. [SSV13] discuss this issue.

Upshot If FACTORING $\notin$ P, then FACTORING will be an example of a problem that quantum computers can do faster than classical computers. Proving FACTORING $\notin$ P is hard since it implies P $\neq$ NP.

23.4 Discrete Log

Discrete log is an important problem for cryptography. Many cryptosystems would be broken if discrete log was easy. Hence, in contrast to work in algorithms, cryptographers hope that discrete log is hard.

We will define discrete log slightly differently than how it was defined in chapter 0.

> DISCRETE LOG
> *Instance:* A prime p, a generator g of $\mathbb{Z}_p$, and $a \in \mathbb{Z}_p$. (g is a generator if $\{g, g^2, \ldots, g^{p-1}\} = \{1, \ldots, p-1\}$.)
> *Output:* x such that $g^x \equiv a \pmod{p}$.

As noted in chapter 0, there are no polynomial-time algorithms for DISCRETE LOG, nor is there a proof that its NP-complete. There are reasons to think it is not NP-complete (See the discussion of DISCRETE LOG in chapter 0.)

Algorithms for DISCRETE LOG are hard to analyze and depend on (widely believed) conjectures in number theory. The fastest known algorithm, the function field sieve, is believed to have running time roughly $2^{1.53 L^{1/3} (\lg L)^{2/3}}$, where $L = \lg N$ is the length of the input number N. This bound is small enough that the algorithm is practical for moderately large inputs. The naïve algorithm for DISCRETE LOG takes time 2^L. Hence reducing the time to roughly $2^{L^{1/3}}$ is a real improvement. Adleman [Adl94] developed the function field sieve and then elaborated the ideas with Huang (see [AH99]).

There is no formal connection between DISCRETE LOG and FACTORING; however, the techniques for one seem to apply to the other. Hence the following two points made about FACTORING are true of DISCRETE LOG also: (1) there is no consensus about if DISCRETE LOG $\in$ P, and (2) if DISCRETE LOG $\in$ P then this will require new techniques.

What about quantum polynomial time?

Theorem 23.2. *Let the input to a DISCRETE LOG algorithm be N. Let $L = \lg N$, which is the length of N.*

1. *(Shor [Sho94, Sho99]) There is a polynomial quantum algorithm for DISCRETE LOG.*
2. *(Folklore though can be obtained from Beckman et al. [BCDP96].) There is a quantum algorithm for DISCRETE LOG that takes time $O(L^2 \log(L) \log(\log(L)))$.*

We note the following:

1. The key quantum component of Shor's algorithm for DISCRETE LOG is the quantum Fourier transform. While discussing classical algorithms for DISCRETE LOG, we noted that while there is no formal connection between DISCRETE LOG and FACTORING, improvements in one tend to lead to improvements in the other. We were referring to classical algorithms. However, the same seems to be true for quantum algorithms: Both the quantum algorithm for FACTORING and that for DISCRETE LOG use the quantum Fourier transform.
2. There do not seem to be any attempts to execute Shor's DISCRETE LOG algorithm on a quantum computer. Since the quantum algorithms for FACTORING and DISCRETE LOG are similar, the same techniques that were used for FACTORING will work on DISCRETE LOG. Hence it is likely that a quantum computer could be used to find DISCRETE LOG when p, g, a are all ≤ 21.

Upshot If DISCRETE LOG $\notin$ P, then DISCRETE LOG will be an example of a problem that quantum computers can do much faster than classical computers. Proving DISCRETE LOG $\notin$ P is hard since it implies P $\neq$ NP.

23.5 The Search Problem

SEARCH
Instance: Access to a function $f\colon \{0, \ldots, N-1\} \rightarrow \{0, 1\}$. We are promised that there is only one x such that $f(x) = 1$. We think of this function as representing a 1-element subset of $\{0, \ldots, N-1\}$.
Output: The x such that $f(x) = 1$.
Note: The basic unit of computation is an evaluation of f that we call a query.

Theorem 23.3. *Let $N \in \mathbb{N}$ and $f\colon \{0, \ldots, N-1\} \rightarrow \{0, 1\}$.*

1. *(Easy) There is a deterministic algorithm for SEARCH that takes N queries in the worst case. There is a randomized algorithm for SEARCH that has expected complexity $\frac{N}{2}$ queries. Both of these are optimal.*
2. *(Grover [Gro96]) There is a quantum algorithm for SEARCH that uses $O(\sqrt{N})$ queries.*
3. *(Bennett et al. [BBBV97]) Any quantum algorithm for SEARCH requires $\Omega(\sqrt{N})$ queries.*
4. *(easy) Assume that instead of having only 1 x with $f(x) = 1$, there are M and the goal is to find one of them. There is a deterministic algorithm that takes $O(N - M)$ queries in the worst case. There is a randomized algorithm that has expected complexity $\frac{N}{M}$ queries. Both of these are optimal.*

5. *For the problem in the last item, there is a quantum algorithm that takes $O(\sqrt{N/M})$ queries.*

Upshot SEARCH, with the complexity measure number-of-queries, is a problem where quantum computers are **provably** faster than classical computers.

23.6 The Traversal Problem

TRAVERSAL

Instance: A graph G with $\Theta(2^n)$ vertices represented by $\Theta(n)$-bit strings. There are two distinguished vertices ENTRANCE and EXIT. The label of ENTRANCE (e.g., Vertex 0110) is given. The label of EXIT is not given.

Output: The label of EXIT.

Note: The graph is large. We think of the graph as being like a database that you ask questions about. The algorithm can ask question of the following type: Given a string w of $\Theta(n)$ bits, return the following information:

- Is w a vertex?
- If w is a vertex, then output all of its neighbors.
- If one of the neighbors of w is EXIT, then indicate this.

The number of queries is the complexity of the algorithm.

Note: TRAVERSAL is only asking to find EXIT. It is not asking to find the path from ENTRANCE to EXIT. We will later comment on this perhaps harder problem of having to find the path.

Note: The basic unit of computation is a query of one of the types above.

This problem looks like it requires $\Omega(2^n)$ queries for either classical or quantum algorithms. And indeed, for the case of general graphs, that is the case. But there is a sequence of graphs where there is a large difference between classical algorithms and quantum algorithms.

One technique that a classical algorithm can use is a CLASSICAL RANDOM WALK: The algorithm picks a random neighbor of ENTRANCE, then a random neighbor of that neighbor, and so on, until it finds EXIT. There is also a notion of a QUANTUM RANDOM WALK, which we will not define.

Theorem 23.4.

1. *(Childs et al. [CFG02]) There is a sequence of graphs $\{G_n\}_{n=1}^{\infty}$ such that the following hold: (a) G_n has $\Theta(2^n)$ vertices, (b) there is a QUANTUM RANDOM WALK algorithm*

that solves TRAVERSAL *on* G_n *using* $\le p(n)$ *queries for some polynomial* p*, and (c) any* CLASSICAL RANDOM WALK *algorithm requires* $2^{\Omega(n)}$ *queries.*

2. *(Childs et al. [CCD+03]) There is a sequence of graphs* $\{G_n\}_{n=1}^{\infty}$ *such that the following hold: (a)* G_n *has* $\Theta(2^n)$ *vertices, (b) there is a* QUANTUM RANDOM WALK *algorithm that solves* TRAVERSAL *on* G_n *with* $p(n)$ *queries for some polynomial* p*, and (c) any classical algorithm (whether or not it uses* CLASSICAL RANDOM WALK*) requires* $2^{\Omega(n)}$ *queries.*
3. *(Jeffery and Zur [JZ22]) In items 1 and 2,* $p(n)$ *can be replaced by* $O(n)$*.*
4. *(Childs et al. [CCG23]) (Informal) Consider the problem of actually finding the path from* ENTRANCE *to* EXIT*. Under reasonable assumptions, any* CLASSICAL RANDOM WALK *or* QUANTUM RANDOM WALK *algorithm requires an exponential number of queries.*

Upshot TRAVERSAL, with the complexity measure number-of-queries, is a problem where quantum computers are **provably** faster than classical computers.

23.7 Subquadratic Approximate Edit Distance

Definition 23.5. Let Σ be a finite alphabet and let $x, y \in \Sigma^*$. The ***edit distance between x and y*** is the number of insertions/deletions/substitutions needed to transform x into y.

> EDIT DISTANCE
> *Instance:* Two strings x, y over some alphabet Σ. We think of Σ as being fixed.
> *Output:* The edit distance between x and y?

Theorem 23.6.

1. *(Easy)* EDIT DISTANCE *can be computed in time* $O(n^2)$*, where* $n = \max\{|x|, |y|\}$.
2. *(Backurs and Indyk [BI15]) Assuming SETH,* EDIT DISTANCE *requires* $\Omega(n^2)$ *time.*
3. *(Abboud et al. [AHWW16]) With an assumption weaker than SETH,* EDIT DISTANCE *requires* $\Omega(n^2)$ *time.*

Theorem 23.6 settles the question for ***exact*** EDIT DISTANCE: Quadratic time is both the upper and lower bound. Is there a subquadratic algorithm for approximating EDIT DISTANCE?

Can a quantum algorithm give a subquadratic approximation algorithm? Yes. Boroujeni et al. [BEG+21] proved the following:

Theorem 23.7.

1. *(Theorem 4.5 of their paper) For all* $\varepsilon > 0$*, there is a quantum algorithm that (a) runs in time* $O(n^{2-(4/21)} \log(\frac{1}{\varepsilon}))$ *and (b) returns a number that is* $\le (3+\varepsilon)OPT(x, y)$*. Note that* $2 - (4/21) \approx 1.8095$.

2. *(Theorem 5.1 of their paper) For all* $\varepsilon > 0$*, there is a quantum algorithm that (a) runs in time* $\tilde{O}(n^{\alpha})$ *where* $\alpha = 2 - (5 - \sqrt{17}/3) \approx 1.7077$ *and (b) returns a number that is* $\leq O(1/\varepsilon)^{O(1/\varepsilon)OPT(x,y)}$.

Open Problem 23.8. *In this open problem, the problem is, of course, approximate* EDIT DISTANCE.

1. *Find a constant* $D > 3$ *such that no classical algorithm can, in subquadratic time, obtain a* $DOPT(x, y)$ *approximation. This would show that a subquadratic* $(3+\varepsilon)$*-approximation for* EDIT DISTANCE *is a problem that a quantum algorithm can do but a classical one cannot.*
2. *Improve the runtime of the quantum algorithm in theorem 23.7.1.*

TWO UPSHOTS: The problem at hand is EDIT DISTANCE.

1. The following can be done by a quantum algorithm but (at least for now) not by a classical algorithm: a subquadratic algorithm that, on input x, y, returns $(3+\varepsilon)\mathrm{OPT}(x, y)$.
2. For this problem a quantum approximation algorithm inspired a classical one.

23.8 Quantum Streaming Algorithms

23.8.1 Classical Streaming for Triangle Counting and Distinguishing

> TRIANGLE COUNTING TC
> *Instance:* Graph $G = (V, E)$
> *Output:* An approximation to the number of triangles in G.

A related problem that is usually considered in the literature is that of TRIANGLE DISTINGUISHING, which is defined as follows.

> TRIANGLE DISTINGUISHING TREE DIAMETER (TD)
> *Instance:* Graph $G = (V, E)$, a number T, and the promise that G has either 0 triangles or T triangles.
> *Question:* Does G have 0 triangles?

Clearly TREE DIAMETER $\leq$ TC. Hence, a lower bound on TD implies a lower bound on TC. We will state lower bounds on TREE DIAMETER (and hence TC). Recall that in chapter 20 lower bounds on streaming algorithms were obtained via lower bounds in communication complexity. In this chapter we will talk about (though not prove) lower bounds on quantum streaming algorithms via lower bounds on quantum communication complexity. We first need a problem with large quantum communication complexity.

Definition 23.9. Let $n \in \mathbb{N}$.

1. A ***perfect matching M over* [2*n*]** is a set of n ordered pairs (i,j), where i and j are distinct elements of $[2n]$, such that every $\ell \in [2n]$ is in exactly one ordered pair.
2. Let M be a perfect matching over $[2n]$. We identify M with the following $n \times 2n$ matrix: For every ordered pair (i,j) in the matching, there is a row with 1's in the ith and jth spot, and 0's everywhere else. Note that a perfect matching can be associated to many different matrices. We will turn this around: We will give Bob a perfect matching by giving him a matrix.

> BOOLEAN HIDDEN MATCHING (BHM)
> *Instance:* Alice gets a string $x \in \{0,1\}^{2n}$. Bob gets (a) a perfect matching M over $[2n]$ via a matrix as described in definition 23.9, and (b) a string $w \in \{0,1\}^n$ where w is promised to satisfy either $Mx = w$ or $Mx = \overline{w}$ (where $\overline{w}$ is w with every bit flipped).
> *Question:* Does $Mx = w$?

Theorem 23.10. *(Gavinsky et al. [GKK$^+$08]) The randomized one-way communication complexity of BHM, with Alice sending, is $\Omega(\sqrt{n})$.*

Notation 23.11. Let n denote the number of vertices, m denote the number of edges, and T is as in the problem statement. Δ_V (respectively Δ_E) is the maximum number of triangles in G that share a vertex (respectively an edge).

The following are known.

Theorem 23.12.

1. *(Jayaram and Kallaugher [JK21]) There is a one-pass streaming algorithm for TC that uses space $\tilde{O}\left(\frac{m\Delta_E}{T} + \frac{m\sqrt{\Delta_V}}{T}\right)$.*
2. *(Braverman et al. [BOV13]) Any one-pass streaming algorithm for TREE DIAMETER (and hence for TC) uses space $\Omega\left(\frac{m\Delta_E}{T}\right)$. This proof uses a reduction from INDEX to TREE DIAMETER.*
3. *(Kallaugher and Price [KP17a]) Any one-pass streaming algorithm for TREE DIAMETER (and hence for TC) uses space $\Omega\left(\frac{m\sqrt{\Delta_V}}{T}\right)$. This proof uses a reduction from BHM to TREE DIAMETER.*

4. *Any one-pass streaming algorithm for* TREE DIAMETER *(and hence for TC) requires space* $\Omega\left(\frac{m\Delta_E}{T}+\frac{m\sqrt{\Delta_V}}{T}\right)$. *This follows from parts 2 and 3. Note that we now have matching bounds for one-pass streaming algorithms for TC.*

23.8.2 Quantum Streaming for Triangle Counting and Distinguishing

Quantum streaming algorithms were first defined by Khadiev et al. [KKM18] (see also Ablayev et al. [AAKV18]). We will discuss modifying the proofs of the lower bounds for streaming on TREE DIAMETER and TC from theorem 23.12 to obtain lower bounds for quantum streaming for these problems.

Theorem 23.12.2 used that INDEX has communication complexity $\Omega(n)$. Fortunately, Ambainis et al. [ANTV02] showed that INDEX also has quantum communication complexity $\Omega(n)$. Hence we have the following analog to theorem 23.12.2 by the same proof:

Theorem 23.13. *Any one-pass quantum streaming algorithm for* TREE DIAMETER *(and hence for TC) requires space* $\Omega\left(\frac{m\Delta_E}{T}\right)$. *This proof uses a reduction from* INDEX *to* TREE DIAMETER. *This follows from theorem 23.12.2 and the work of Ambainis et al. [ANTV02].*

Can we do the same for theorem 23.12.3? No. Gavinsky et al. [GKK+08] showed that the quantum communication complexity of BHM is $O(\log n)$. Hence we do not have a nontrivial lower bound for TC or TREE DIAMETER in the region where $\Delta_E = O(1)$ and $T = \Omega(n)$. Indeed, there is a quantum streaming algorithm that works well in that region. Kallaugher [Kal21] showed the following.

Theorem 23.14. *Restrict TC to the graphs where* $\Delta_E = O(1)$, $\Delta_V = \Omega(T)$, *and* $T = \Omega(m)$. *There is a one-pass quantum streaming algorithm for TC that uses space* $\tilde{O}(m^{2/5})$.

Open Problem 23.15. *Find a lower bound of the form* $\Omega(m^c)$ *for TC in the case where* $\Delta_E = O(1)$, $\Delta_V = \Omega(T)$, *and* $T = \Omega(m)$.

23.8.3 Classical Streaming for *k*-Clique Counting and Distinguishing

In this section, we define two problems for k-clique finding that are analogous to TRIANGLE COUNTING and TRIANGLE DISTINGUISHING.

> k-CLIQUE COUNTING (KCC)
> *Instance:* Graph $G = (V, E)$ and $k \in \mathbb{N}$.
> *Output:* An approximation to the number of cliques of size k in G.

k-CLIQUE DISTINGUISHING (KCD) *Instance:* Graph $G=(V,E)$, $C\in\mathbb{N}$, and the promise that G has either 0 k-cliques or $\geq C$ k-cliques. *Question:* Does G have 0 k-cliques?

Clearly KCD $\leq$ KCC. Hence a lower bound on KCD implies a lower bound on KCC.

Theorem 23.12.2 stated a $\Omega\left(\frac{m\Delta_E}{T}\right)$ space lower bound for one-pass streaming algorithms for TRIANGLE DISTINGUISHING. A similar proof gives the same lower bound for k-CLIQUE DISTINGUISHING (with T being the number of k-cliques); however this gives a trivial lower bound on most graphs, since Δ_E is usually small. We want a stronger lower bound for more general graphs. Additionally, since the quantum streaming complexity of triangle counting in the parameter setting $\Delta_E=O(1)$ and $T=\Omega(m)$ is an open problem, it might be instructive to look for lower bounds on k-CLIQUE COUNTING for $k\geq 4$ in this parameter setting to understand if the difficulty of this problem is unique for triangle counting.

For the next exercise, you need the following definition and theorem.

Definition 23.16. Let $k,n\in\mathbb{N}$.

1. A ***perfect hypermatching M over* [*kn*]** is a set of n ordered k-tuples $(i_1,\ldots,i_k)$, where $i_1,\ldots,i_k$ are distinct elements of $[kn]$, such that every $\ell\in[kn]$ is in exactly one ordered k-tuple.
2. Let M be a perfect hypermatching M over $[kn]$. We identify M with the following $n\times kn$ matrix: For every ordered k-tuple $(i_1,\ldots,i_k)$, in the hypermatching, there is a row with 1's in the i_1th, i_2th, ..., i_kth spot, and 0's everywhere else. Note that a perfect hypermatching can be associated to many different matrices. We will turn this around: We will give Bob a perfect hypermatching by giving him a matrix.

BOOLEAN HIDDEN HYPERMATCHING BHHM *Instance:* Alice gets a string $x\in\{0,1\}^{kn}$. Bob gets (a) a perfect hypermatching M over $[kn]$ via a matrix as described in definition 23.16, and (b) a string $w\in\{0,1\}^n$, where w is promised to satisfy either $Mx=w$ or $Mx=\overline{w}$ (where $\overline{w}$ is w with every bit flipped). *Question:* Does $Mx=w$?

Theorem 23.17.

1. *(Verbin and Yu [VY11]) The randomized one-way communication complexity for BHHM, with Alice sending, is* $\Omega(n^{1-(1/k)})$.
2. *(Shi et al. [SWY15]) the quantum one-way communication complexity for BHHM, with Alice sending, is* $\Omega(n^{1-(2/k)})$.

Exercise 23.18.

1. Prove that any classical one-pass streaming algorithm for κCD requires space $\Omega\left(m^{1-1/k}\right)$.
   (**Hint:** Use the lower bound on BHHM from theorem 23.17.1).
2. Prove that any quantum one-pass streaming algorithm for κCD requires space $\Omega\left(m^{1-2/k}\right)$ (space is measured in qubits).
   (**Hint:** Use the lower bound on BHHM from theorem 23.17.2).

Open Problem 23.19.

1. *We have looked at counting and detecting triangles and k-cliques. Look at the problems of counting and detecting other subgraphs such as k-cycles.*
2. *Obtain classical and quantum upper and lower bounds on p-pass streaming algorithms.*

23.8.4 Separations

So far we have not presented a large separations between classical and quantum streaming algorithms. Note that we have been looking at *natural* streaming problems. There are results for contrived problems.

Theorem 23.20.

1. *(Le Gall [Gal09]) There exists a streaming problem that (a) any classical algorithm requires* $\Omega(n^{1/3})$ *space and (b) there is a quantum algorithm that uses* $O(\log n)$ *space.*
2. *(Gavinsky et al. [GKK⁺08]) There exists a streaming problem that (a) any classical algorithm requires* $\Omega(n^{1/2})$ *space and (b) there is a quantum algorithm that uses* $O(\log n)$ *space.*

(There are reasons why the first result is not quite comparable to the second result.)

Open Problem 23.21. *Find natural streaming problems for which there is a large separation between classical and quantum algorithms.*

Upshot Lower bounds on classical or quantum streaming algorithms are obtained by lower bounds on classical or quantum communication complexity. Hence the difficulty in obtaining a separation for streaming algorithms is to find a separation for communication complexity problems. This has been done for some contrived streaming problems; however, we would like to have a separation for natural problems.

23.9 MIP* = RE

Let's consider 3-Coloring ∈ NP as a game involving two people: an all powerful prover and a poly time verifier. The prover wants to convince the verifier that a given graph is 3-colorable.

- The prover sends the verifier a string y that he hopes will convince the verifier that $x \in A$. The obvious thing to send is a 3-coloring of G.
- The verifier then determines if y really is a 3-coloring of G. If so, then he accepts that G is 3-colorable. If not, then he now believes G is not 3-colorable.

Note that (a) there is only one prover, (b) the conversation is only one direction (prover sends a string to verifier), (c) the verifier can implement any deterministic polynomial time, and (d) the verifier is convinced that $G \in$ 3-Coloring if and only if $G \in$ 3-Coloring.

We can modify the game by (1) allowing more rounds, (2) allowing the verifier to flip coins, or (3) allowing the verifier a small probability of error. Such games are called "interactive proof systems." Note that an interactive proof system has one prover and one verifier. A multiprover interactive proof system allows many provers, who cannot talk to each other.

Multiprover interactive proof systems have been used to get some lower bounds on how well a problem can be approximated in poly time, and can be considered a precursor to PCP (which you saw in chapter 8) and the unique games conjecture (which you saw in chapter 10).

Definition 23.22.

1. A set A is in ***MIP*** if there is a multiprover interactive system such that (a) if $x \in A$, then the verifier accepts, and (b) if $x \notin A$, then the verifier rejects with probability ≥ 0.9.
2. If we allow the provers to share entangled quantum states (the verifier is still classical), then this is ***MIP****.

MIP and MIP* differ a lot:

Theorem 23.23.

1. *(Babai et al. [BFL91]) MIP = NEXPTIME.*
2. *(Ji et al. [JNV+21]) MIP* = RE, where RE is the first level of the arithmetic hierarchy. Since RE contains the Halting set, MIP* contains sets that are undecidable.*

Upshot MIP and MIP* give an example where, in a classical setting, problems are in NEXPTIME and, in a quantum setting, problems can be undecidable.

23.10 Quantum Games

In the previous sections we measured how well an algorithm did by how much time or space it used (queries can be considered time). In this section we look at games and measure how well the players do by looking at their probability of winning.

We discuss two games such that if the players play the game with quantum resources, they can provably do better than if they play the game with classical resources. For further discussion of these games, and other games with this property, see the survey of Brunner et al. [BCP+14].

23.10.1 The CHSH Game

Clauser, Horne, Shimony, and Holt [CHSH69] invented the CHSH Game as a realizable experiment that can differentiate quantum from classical computing. (They did not give it that name; however, it was named using their initials.) We note that Clauser won the 2022 Nobel Prize in Physics for this and other work [Rel22].

> The CHSH Game
> *Instance:* Alice gets bit x, Bob gets bit y. Before they get their bits, they can discuss strategy.
> *Output:* Alice outputs bit a, Bob outputs bit b. Alice and Bob win if and only if $x \wedge y = a \oplus b$.

Clauser et al. [CHSH69] proved the following (see also Aaronson [Aar16, chapter 13] for an exposition).

Theorem 23.24.

1. *If Alice and Bob play the CHSH Game with classical resources, then (a) there is a deterministic strategy where they win with probability* 0.75 *(both always output 0), and (b) there is no strategy, deterministic or randomized, that does better.*
2. *If Alice and Bob play the CHSH Game with quantum resources (they prepare entangled qubits before the game begins), then (a) there is a strategy where they win with probability* $0.5 + \sqrt{2}/4 \approx 0.85$ *(this is complicated), and (b) there is no strategy that does better.*

This game is of interest since it is a case where the quantum world is provably different from the classical world. Note that the gap between the classical and quantum is $0.85 - .0.75 = 0.1$.

23.10.2 Magic Square Game

Cabello [Cab01] defined the magic square game, though he did not call it that. For more information on it, see the survey of Brassard et al. on Quantum pseudo-telepathy [BBT05, section 5] or the Wikipedia entry on Quantum pseudo-telepathy [Wikf].

> The MAGIC SQUARE GAME (MS GAME)
>
> *Instance:* Alice gets $i \in \{1, 2, 3\}$, Bob gets $j \in \{1, 2, 3\}$. They interpret i as the row of a 3×3 matrix and j as the column of a 3×3 matrix. Alice and Bob get to discuss strategy ahead of time.
>
> *Output:* Alice and Bob both output a three-bit sequence. Alice's sequence is used as the ith row of a matrix. Bob's sequence is used as the jth column of a matrix. If the following three conditions hold, then Alice and Bob win; else they lose. (a) The values in row i add to an even number, (b) The values in column j add to an odd number, and (c) Alice and Bob's values are consistent (they agree at (i,j)).

Theorem 23.25.

1. *If Alice and Bob play the MS GAME with classical resources, then (a) there is a deterministic strategy where they win with probability* $\frac{8}{9} = 0.88\ldots$, *and (b) there is no strategy, deterministic or randomized, that does better.*
2. *If Alice and Bob play the MS GAME with quantum resources (they prepare entangled qubits before the game begins), then there is a strategy that wins with probability 1 (so always wins).*

This game is of interest since it is a case where the quantum world is provably different from the classical world. Note that the gap between the classical and quantum is $1.0 - 0.88\ldots = 0.11\ldots$.

Is there an interesting version of the MS GAME on $k \times k$ matrices for $k \geq 4$? The following exercise shows that the answer is no.

Exercise 23.26.

1. Give a 4×4 matrix M of 0's and 1's such that every row sums to an even number and every column sums to an odd number.
2. Show that there is a classical strategy for the 4×4 MS GAME that wins with probability 1. (**Hint:** Use part 1.)
3. Show that, for all $k \geq 4$, there is a classical strategy for the $k \times k$ MS GAME that wins with probability 1.

23.10.3 Comparing the CHSH GAME with The MS GAME

We give two reasons why the MS GAME game is better for distinguishing classical and quantum computation, and one reason why the CHSH GAME game is better.

Two reasons why the MS GAME is better

1. In the CHSH GAME, the gap between the classical and quantum players is 0.1. In the MS GAME, the gap between the classical and quantum players is 0.11, which is bigger!
2. For the MS GAME, the quantum players *always* win. This is better for repeated experiments. Assume the game is played n times.
   a. For the MS GAME:

      If Alice and Bob are classical, then the expected number of wins is $8n/9$.

      If Alice and Bob are quantum, then the expected number of wins is n.

      So if Alice and Bob lose just once, then they must be classical.

   b. For the CHSH GAME

      If Alice and Bob are classical, then the expected number of wins is $0.75n$.

      If Alice and Bob are classical, then the expected number of wins is $0.85n$.

      These two cases are harder to distinguish since a lose by Alice and Bob could happen in either case.

One reason why the CHSH GAME is better. Quantum computers (in 2023) are noisy. The computations are not that reliable. Hence many trials must be run. There is a lot of work on quantum error correction to try to alleviate this.

The CHSH GAME game is simpler and uses fewer operations, and hence less noise. This is not just theoretical. The CHSH GAME is currently used in quantum systems in order to calibrate the quantum computers. The calculations done above for the MS GAME were assuming an error-free quantum computer, which is not a reality yet.

In 2023, the CHSH GAME is better and, as noted above, is actually used. However, if quantum hardware and our level of control on it improve, there may be a time in the future where the MS GAME is better.

Upshot There are games where if the players can use quantum entanglement, then their probability of winning is provably higher than if they cannot.

23.11 Quantum Supremacy

The original point of quantum computing is that it should be better than classical computing on some problems. This goal was crystallized by John Preskill [Pre12] (see also [Pre19]) who defined ***boolean hidden hypermatching***.

Definition 23.27. ***Quantum supremacy*** is the goal of finding a problem where a quantum computer can outperform a classical computer. Note that this involves (a) formulating a problem, (b) coming up with a quantum algorithm for it, (c) coding up that algorithm on a real quantum computer, and (d) having an argument (it need not be a proof) that any classical algorithm, when coded up, will do much worse than the quantum algorithm. Note that this is not theoretical. The goal involves real quantum computers versus real classical computers. The goal is that the quantum computer is ***better*** than the classical one; however, ***better*** may be speed or energy or some other parameter.

In this chapter we discussed many problems where quantum algorithms seem to do better than any known classical algorithm. However, putting these quantum algorithms on a real quantum computer is difficult. Hence other problems have been suggested as candidates for achieving quantum supremacy. We discuss two problems for which quantum supremacy may have been achieved.

23.11.1 Quantum Sampling

RANDOM CIRCUIT SAMPLING (RCS)

Instance: A quantum circuit C. C takes n qubits as input and outputs n qubits that are then measured to get an element of $\{0, 1\}^n$. Note that if you run C on the same input twice, you may not get the same answer. Hence $C(0^n)$ can be viewed as a distribution rather than an element of $\{0, 1\}^n$. We call this distribution D. If the quantum circuit is chosen at random, then the D is far from uniform.

Output: (Informally) Take enough samples of the circuit to show statistically that D is unlikely to be uniform.

The following are known.

1. Bouland et al. [BFNV18] and Aaronson and Chen [AC17] give evidence that if RCS is classically easy, then certain reasonable complexity hypotheses are false. Hence RCS is probably hard classically. We note that this is a theoretical asymptotic result, so it may not be as helpful for quantum supremacy as it appears at first glance.
2. On October 23, 2019, Google published a paper [AA19] that claims to have have solved the 53-bit RCS problem in 200 seconds. They also claim that any classical supercomputer would have taken 10,000 years. The quantum processor used is named ***Sycamore***.
3. IBM [PGA19] (see also Scott Aaronson's blog post [Aar19]) showed how the computation would have taken only 2.5 days on a supercomputer.
4. We will stop here; however, the claims and counter-claims about if Google really did achieve quantum supremacy continue. For intelligent discussions of the issue, go to Scott Aaronson's blog and search for `Quantum Supremacy`.

23.11.2 Gaussian Boson Sampling

Aaronson and Arkhipov [AA13] proposed BOSON SAMPLING as a candidate problem for quantum supremacy. We omit the description as it is somewhat technical, They give evidence that if BOSON SAMPLING is easy in one way, then the polynomial hierarchy collapses, and if it is easy in another way then (with some assumptions) computing the permanent of a matrix is easy, which also implies that the polynomial hierarchy collapses.

Hamilton et al. [HKS$^+$17] introduced a related problem, GAUSSIAN BOSON SAMPLING. They give evidence that if GAUSSIAN BOSON SAMPLING is easy, then computing the permanent of a matrix is easy, so the polynomial hierarchy collapses.

The following are known.

1. In December of 2020, a group based in the University of Science and Technology of China (USTC) [Poa20] achieved quantum supremacy by implementing GAUSSIAN BOSON SAMPLING on 76 photons with their photonic quantum computer Jiuzhang. The paper states that to generate the number of samples the quantum computer generates in two seconds, a classical supercomputer would require 600 million years of computation. For more information on these results, see the articles by Conover [Con21] and Garisto [Gar20].
2. They later increased the number of photons to 113.
3. Martinez-Cifuentes et al. [MCFRQ22] give reasons why the USTC group may not have actually achieved quantum supremacy.

23.11.3 Has Quantum Supremacy Happened?

We close with a quote from a blog post of Scott Aaronson [Aar22]. We insert comments in brackets for clarity.

> The experiments by Google and USTC and now Xanadu [they worked on a graph problem that has a classical $O(n^3)$ algorithm so the quantum advantage cannot be too large] represent a big step forward for the field, but since they started being done, the classical spoofing attacks have also steadily improved, to the point that whether "quantum computational supremacy" still exists depends on exactly how you define it.
>
> Briefly: If you measure by total operations, energy use, or CO2 footprint, then probably yes, quantum supremacy remains. But if you measure by number of seconds, then it doesn't remain, not if you're willing to shell out for enough cores on AWS [Amazon Web Services] or your favorite supercomputer. And even the quantum supremacy that does remain might eventually fall to, e.g., further improvements due to Gao et al. [GKC$^+$21]. For more details, see e.g., the now-published work of Pan, Chen and Zhang [PCZ22] or this good popular summary by Adrian Cho [Cho22] for Science.

Bibliography

[AA13] Scott Aaronson and Alex Arkhipov. The computational complexity of linear optics. *Theory of Computation*, 9:143–252, 2013. https://arxiv.org/pdf/1011.3245.pdf. 486

[AA19] Frank Arute, Kunal Arya, Ryan Babbush, Dave Bacon, Joseph C. Bardin, Rami Barends, Rupak Biswas, et al. Quantum supremacy using a programmable supercomuter processor. *Nature*, 574:505–501, 2019. 485

[AAA$^+$09] Noga Alon, Baruch Awerbuch, Yossi Azar, Niv Buchbinder, and Joseph Naor. The online set cover problem. *SIAM Journal on Computing*, 39(2):361–370, 2009. https://doi.org/10.1137/060661946. 404

[AAB$^+$20] Hugo A. Akitaya, Cordelia Avery, Joseph Bergeron, Erik D. Demaine, Justin Kopinsky, and Jason S. Ku. Infinite all-layers simple foldability. *Graphs and Combinatorics*, 36:231–244, 2020. 158

[AAGH21] Pankaj K. Agarwal, Boris Aronov, Tzvika Geft, and Dan Halperin. On two-handed planar assembly partitioning with connectivity constraints. In Dániel Marx, editor, *Proceedings of the 2021 ACM-SIAM Symposium on Discrete Algorithms, SODA 2021, Virtual Conference, January 10–13, 2021*, pages 1740–1756. SIAM, 2021. https://doi.org/10.1137/1.9781611976465.105. 97

[AAKV18] Farid M. Ablayev, Marat Ablayev, Kamil Khadiev, and Alexander Vasiliev. Classical and quantum computations with restricted memory. In Hans-Joachim Böckenhauer, Dennis Komm, and Walter Unger, editors, *Adventures Between Lower Bounds and Higher Altitudes—Essays Dedicated to Juraj Hromkovič on the Occasion of His 60th Birthday*, volume 11011 of *Lecture Notes in Computer Science*, pages 129–155. Springer, 2018. 478

[AAM21] Mikkel Abrahamsen, Anna Adamaszek, and Tillmann Miltzow. The art gallery problem is $(\exists R)$-complete. *Journal of the ACM*, 69(1):4:1–4:70, 2021. https://doi.org/10.1145/3486220. 298

[Aar11] Scott Aaronson. A linear-optical proof that the permanent is #P-hard. *Electronic Colloquium on Computational Complexity*, page 43, 2011. https://eccc.weizmann.ac.il/report/2011/043. 282

[Aar13] Scott Aaronson. *Quantum Computing Since Democritus*. Cambridge University Press, 2013. 469

[Aar16] Scott Aaronson. Introduction to quantum information science, 2016. https://www.scottaaronson.com/qclec.pdf. 482

[Aar19] Scott Aaronson. Quantum supremacy: The gloves are off, 2019. https://scottaaronson.blog/?p=4372. 485

[Aar22] Scott Aaronson. Summer 2022 quantum supremacy updates, 2022. https://scottaaronson.blog/?p=6645. 486

[ABC$^+$20] Zachary Abel, Jeffrey Bosboom, Michael J. Coulombe, Erik D. Demaine, Linus Hamilton, Adam Hesterberg, Justin Kopinsky, Jayson Lynch, Mikhail Rudoy, and Clemens Thielen. Who witnesses the witness? Finding witnesses in the witness is hard and sometimes impossible. *Theoretical Computer Science*, 839:41–102, 2020. https://doi.org/10.1016/j.tcs.2020.05.031. 21

[ABD$^+$04] Esther M. Arkin, Michael A. Bender, Erik D. Demaine, Martin L. Demaine, Joseph S. B. Mitchell, Saurabh Sethia, and Steven Skiena. When can you fold a map? *Computational Geometry*, 29(1):23–46, 2004. https://doi.org/10.1016/j.comgeo.2004.03.012. 158

[ABD+21] Hayashi Ani, Jeffrey Bosboom, Erik D. Demaine, Jenny Diomidova, Della H. Hendrickson, and Jayson Lynch. Walking through doors is hard, even without staircases: Proving PSPACE-hardness via planar assemblies of door gadgets. In Martin Farach-Colton, Giuseppe Prencipe, and Ryuhei Uehara, editors, *10th International Conference on Fun with Algorithms, FUN 2021, May 30 to June 1, 2021, Favignana Island, Sicily, Italy*, volume 157 of *LIPIcs*, pages 3:1–3:23. Schloss Dagstuhl - Leibniz-Zentrum für Informatik, 2021. https://doi.org/10.4230/LIPIcs.FUN.2021.3. 308

[ABD+23] Hugo Akitaya, Josh Brunner, Erik D. Demaine, Della Hendrickson, Victor Luo, and Andy Tockman. Complexity of simple folding of mixed orthogonal crease patterns. *Thai Journal of Mathematics*, 21(4):1025–1046, December 2023. 158

[ABF+02] Jochen Alber, Hans L. Bodlaender, Henning Fernau, Ton Kloks, and Rolf Niedermeier. Fixed parameter algorithms for DOMINATING SET and related problems on planar graphs. *Algorithmica*, 33(4):461–493, 2002. https://doi.org/10.1007/s00453-001-0116-5. 183, 200

[ABGS21] Divesh Aggarwal, Huck Bennett, Alexander Golovnev, and Noah Stephens-Davidowitz. Fine-grained hardness of CVP(P)—everything that we can prove (and nothing else). In Dániel Marx, editor, *Proceedings of the 2021 ACM-SIAM Symposium on Discrete Algorithms, SODA 2021, Virtual Conference, January 10–13, 2021*, pages 1816–1835. SIAM, 2021. https://doi.org/10.1137/1.9781611976465.109. 167, 168

[ABI+09] Eric Allender, Michael Bauland, Neil Immerman, Henning Schnoor, and Heribert Vollmer. The complexity of satisfiability problems: Refining Schaefer's theorem. *Journal of Computing and System Science*, 75(4):245–254, 2009. https://www.cs.rutgers.edu/~allender/papers/csp.pdf. 46

[ABKM06] Eric Allender, Peter Bürgisser, Johan Kjeldgaard-Pedersen, and Peter Bro Miltersen. On the complexity of numerical analysis. In *21st Annual IEEE Conference on Computational Complexity (CCC 2006), 16–20 July 2006, Prague, Czech Republic*, pages 331–339. IEEE Computer Society, 2006. https://doi.org/10.1109/CCC.2006.30. 301

[Abr21] Mikkel Abrahamsen. Covering polygons is even harder, 2021. https://arxiv.org/abs/2106.02335. 299

[ABS15] Sanjeev Arora, Boaz Barak, and David Steurer. Subexponential algorithms for unique games and related problems. *Journal of the Association of Computing Machinery (JACM)*, 62(5):42:1–42:25, 2015. https://doi.org/10.1145/2775105. 268

[ABSS97] Sanjeev Arora, László Babai, Jacques Stern, and Z. Sweedyk. The hardness of approximate optima in lattices, codes, and systems of linear equations. *Journal of Computing and System Sciences*, 54(2):317–331, 1997. https://doi.org/10.1006/jcss.1997.1472. 255

[ABV15] Amir Abboud, Arturs Backurs, and Virginia Vassilevska Williams. Tight hardness results for LCS and other sequence similarity measures. In Venkatesan Guruswami, editor, *IEEE 56th Annual Symposium on Foundations of Computer Science, FOCS 2015*, Berkeley, CA, USA, 17–20 October, 2015, pages 59–78. IEEE Computer Society, 2015. https://doi.org/10.1109/FOCS.2015.14. 374

[AC06] Sanjeev Arora and Eden Chlamtac. New approximation guarantee for chromatic number. In Jon M. Kleinberg, editor, *Proceedings of the 38th Annual ACM Symposium on Theory of Computing*, Seattle, WA, USA, May 21–23, 2006, pages 215–224. ACM, 2006. https://doi.org/10.1145/1132516.1132548. 263

[AC17] Scott Aaronson and Lijie Chen. Complexity-theoretic foundations of quantum supremacy experiments. In Ryan O'Donnell, editor, *32nd Computational Complexity Conference, CCC 2017, July 6–9, 2017, Riga, Latvia*, volume 79 of *LIPIcs*, pages 22:1–22:67. Schloss Dagstuhl - Leibniz-Zentrum für Informatik, 2017. 485

[ACD+20] Sualeh Asif, Michael Coulombe, Erik D. Demaine, Martin L. Demaine, Adam Hesterberg, Jayson Lynch, and Mihir Singhal. Tetris is NP-hard even with $O(1)$ rows or columns. *Journal of Information Processing*, 28:942–958, 2020. 155

[ACD+22] Hayashi Ani, Lily Chung, Erik D. Demaine, Jenny Diomidova, Della Hendrickson, and Jayson Lynch. Pushing blocks via checkable gadgets: PSPACE-completeness of Push-1F and Block/Box Dude. In *Proceedings of the 11th International Conference on Fun with Algorithms*, pages 2:1–2:30, 2022. 56, 489

[ACD+24] Hayashi Ani, Lily Chung, Erik D. Demaine, Jenny Diomidova, Della Hendrickson, and Jayson Lynch. Pushing blocks via checkable gadgets: PSPACE-completeness of Push-1F and Block/Box Dude. arXiv:2412.20079, 2024. https://arxiv.org/abs/2412.20079. Full version of [ACD+22]. 56

[ACMM05] Amit Agarwal, Moses Charikar, Konstantin Makarychev, and Yury Makarychev. $O(\sqrt{\log n})$ approximation algorithms for min uncut, min 2CNF deletion, and directed cut problems. In *Proceedings of the 37th Annual ACM Symposium on Theory of Computing, Baltimore, MD, USA, May 22–24, 2005*, pages 573–581. ACM, 2005. https://doi.org/10.1145/1060590.1060675. 262

[AD11] Zachary Abel and Erik D. Demaine. Edge-unfolding orthogonal polyhedra is strongly NP-complete. In *Proceedings of the 23rd Annual Canadian Conference on Computational Geometry*, Toronto, Ontario, Canada, August 10–12, 2011. http://www.cccg.ca/proceedings/2011/papers/paper43.pdf. 157

[ADGV15] Greg Aloupis, Erik D. Demaine, Alan Guo, and Giovanni Viglietta. Classic Nintendo games are (computationally) hard. *Theoretical Computer Science*, 586:135–160, 2015. https://doi.org/10.1016/j.tcs.2015.02.037. 59, 60, 61, 126, 308

[ADH+17] Aviv Adler, Erik D. Demaine, Adam Hesterberg, Quanquan Liu, and Mikhail Rudoy. Clickomania is hard even with two colors and columns. In *The Mathematics of Various Entertaining Subjects*, volume 2, pages 325–363. Princeton University Press, 2017. 155

[ADHL22] Hayashi Ani, Erik D. Demaine, Della H. Hendrickson, and Jayson Lynch. Trains, games, and complexity: 0/1/2-player motion planning through input/output gadgets. In Petra Mutzel, Md. Saidur Rahman, and Slamin, editors, *WALCOM: Algorithms and Computation—16th International Conference and Workshops, WALCOM 2022*, Jember, Indonesia, March 24–26, 2022, *Proceedings*, volume 13174 of *Lecture Notes in Computer Science*, pages 187–198. Springer, 2022. https://arxiv.org/abs/2005.03192. 350

[Adi55] S. I. Adian. Algorithmic unsolvability of problems of recognition of certain properties in groups (in Russian). *Doklady Akademii Nauk SSSR*, 103:533–535, 1955. 337

[ADK17] Hugo A. Akitaya, Erik D. Demaine, and Jason S. Ku. Simple folding is really hard. *Journal of Information Processing*, 25:580–589, 2017. 158

[Adl94] Leonard M. Adleman. The function field sieve. In Leonard M. Adleman and Ming-Deh A. Huang, editors, *Algorithmic Number Theory, First International Symposium, ANTS-I, Ithaca, NY, USA, May 6–9, 1994, Proceedings*, volume 877 of *Lecture Notes in Computer Science*, pages 108–121. Springer, 1994. 472

[AFI+09] Esther M. Arkin, Sándor P. Fekete, Kamrul Islam, Henk Meijer, Joseph S. B. Mitchell, Yurai Núñez-Rodrıguez, Valentin Polishchuk, David Rappaport, and Henry Xiao. Not being (super)thin or solid is hard: A study of grid Hamiltonicity. *Computational Geometry: Theory and Applications*, 42(6):582–605, 2009. https://www.sciencedirect.com/science/article/pii/S092577210800117X. 120, 122

[AFM00] Esther M. Arkin, Sándor P. Fekete, and Josephy S. B. Mitchell. Approximation algorithms for lawn mowing and milling. *Computational Geometry*, 17(1):25–50, 2000. http://www.ams.sunysb.edu/~jsbm/papers/lawn.pdf. 124

[AGI+19] Amir Abboud, Loukas Georgiadis, Giuseppe F. Italiano, Robert Krauthgamer, Nikos Parotsidis, Ohad Trabelsi, Przemyslaw Uznanski, and Daniel Wolleb-Graf. Faster algorithms for All-Pairs bounded Min-Cuts. In Christel Baier, Ioannis Chatzigiannakis, Paola Flocchini, and Stefano Leonardi, editors, *46th International Colloquium on Automata, Languages, and Programming, ICALP 2019, July 9–12, 2019, Patras, Greece*, volume 132 of *LIPIcs*, pages 7:1–7:15. Schloss Dagstuhl - Leibniz-Zentrum für Informatik, 2019. https://doi.org/10.4230/LIPIcs.ICALP.2019.7. 389

[AGM+21] Bharat Adsul, Jugal Garg, Ruta Mehta, Milind A. Sohoni, and Bernhard von Stengel. Fast algorithms for rank-1 bimatrix games. *Operations Research*, 69(2):613–631, 2021. https://arxiv.org/abs/1812.04611. 462

[AGMS11] Bharat Adsul, Jugal Garg, Ruta Mehta, and Milind A. Sohoni. Rank-1 bimatrix games: A homeomorphism and a polynomial time algorithm. In Lance Fortnow and Salil P. Vadhan, editors, *Proceedings of the 43rd ACM Symposium on Theory of Computing, STOC 2011, San Jose, CA, USA, 6–8 June 2011*, pages 195–204. ACM, 2011. https://arxiv.org/abs/1010.3083. 462

[AGV15] Amir Abboud, Fabrizio Grandoni, and Virginia Vassilevska Williams. Subcubic equivalences between graph centrality problems, APSP and diameter. In Piotr Indyk, editor, *Proceedings of the Twenty-Sixth Annual ACM-SIAM Symposium on Discrete Algorithms, SODA 2015, San Diego, CA, USA, January 4–6, 2015*, pages 1681–1697. SIAM, 2015. https://doi.org/10.1137/1.9781611973730.112. 382, 383

[AH77a] Kenneth Appel and Wolfgang Haken. Every planar map is four colorable I: Discharging. *Illinois Journal of Mathematics*, 21:429–490, 1977. https://projecteuclid.org/euclid.ijm/1256049011. 69, 94, 170

[AH77b] Kenneth Appel and Wolfgang Haken. Solution of the four color map problem. *Scientific American*, 237:108–121, 1977. 69

[AH99] Leonard M. Adleman and Ming-Deh A. Huang. Function field sieve method for discrete logarithms over finite fields. *Information and Computation*, 151(1–2):5–16, 1999. 472

[AH08] Boris Aronov and Sariel Har-Peled. On approximating the depth and related problems. *SIAM Journal on Computing*, 38(3):899–921, 2008. https://doi.org/10.1137/060669474. 376

[AH24] Zachary Abel and Della Hendrickson. Baba is universal. In *Proceedings of the 12th International Conference on Fun with Algorithms (FUN 2024)*, pages 1:1–1:15, La Maddalena, Italy, June 2024. 335

[AHK77] Kenneth Appel, Wolfgang Haken, and John Koch. Every planar map is four colorable II: Reducibility. *Illinois Journal of Mathematics*, 21:491–567, 1977. https://projecteuclid.org/euclid.ijm/1256049012. 69, 94, 170

[AHWW16] Amir Abboud, Thomas Dueholm Hansen, Virginia Vassilevska Williams, and Ryan Williams. Simulating branching programs with edit distance and friends: or: A polylog shaved is a lower bound made. In Daniel Wichs and Yishay Mansour, editors, *Proceedings of the 48th Annual ACM SIGACT Symposium on Theory of Computing, STOC 2016*, Cambridge, MA, USA, June 18–21, 2016, pages 375–388. ACM, 2016. https://arxiv.org/abs/1511.06022. 475

[Ajt98] Miklós Ajtai. The shortest vector problem in L_2 is NP-hard for randomized reductions (extended abstract). In Jeffrey Scott Vitter, editor, *Proceedings of the Thirtieth Annual ACM Symposium on the Theory of Computing*, Dallas, Texas, USA, May 23–26, 1998, pages 10–19. ACM, 1998. https://doi.org/10.1145/276698.276705. 249

[AK95] Edoardo Amaldi and Viggo Kann. The complexity and approximability of finding maximum feasible subsystems of linear relations. *Theoretical Computer Science*, 147(1&2):181–210, 1995. https://doi.org/10.1016/0304-3975(94)00254-G. 248

[AK00] Paola Alimonti and Viggo Kann. Some APX-completeness results for cubic graphs. *Theoretical Computer Science*, 237(1):123–134, 2000. https://www.sciencedirect.com/science/article/pii/S0304397598001583. 239

[AKI87] H. Adachi, H. Kamekawa, and S. Iwata. Shogi on $n \times n$ board is complete in exponential time. *Transactions IEICE. J70-D*, 88-A:1843–1852, 1987. 329

[AKS87] Miklós Ajtai, János Komlós, and Endre Szemerédi. Deterministic simulation in LOGSPACE. In *Proceedings of the 19th Annual ACM Symposium on Theory of Computing*, 1987, New York, USA, pages 132–140. ACM, 1987. https://doi.org/10.1145/28395.28410. 214

[AKS01] Miklós Ajtai, Ravi Kumar, and D. Sivakumar. A sieve algorithm for the shortest lattice vector problem. In Jeffrey Scott Vitter, Paul G. Spirakis, and Mihalis Yannakakis, editors, *Proceedings on 33rd Annual ACM Symposium on Theory of Computing*, July 6–8, 2001, Heraklion, Crete, Greece, pages 601–610. ACM, 2001. https://doi.org/10.1145/380752.380857. 249

[AKS04] Manindra Agrawal, Neeraj Kayal, and Nitin Saxena. PRIMES is in P. *Annals of Mathematics*, 160:781–793, 2004. 15, 17

[Alf98] Jorge L. Ramírez Alfonsín. On variations of the subset sum problem. *Discrete Applied Mathematics*, 81:1–7, 1998. https://core.ac.uk/download/pdf/82533986.pdf. 134

[All21] Eric Allender. Vaughan Jones, Kolmogorov complexity, and the new complexity landscape around circuit minimization. *New Zealand Journal of Mathematics*, 52, 2021. https://people.cs.rutgers.edu/~allender/papers/jones.pdf. 15

[ALM+98] Sanjeev Arora, Carsten Lund, Rajeev Motwani, Madhu Sudan, and Mario Szegedy. Proof verification and the hardness of approximation problems. *Journal of the ACM*, 45(3):501–555, 1998. https://doi.org/10.1145/278298.278306. 214, 218, 254

[ALN05] Sanjeev Arora, James R. Lee, and Assaf Naor. Euclidean distortion and the sparsest cut. In Harold N. Gabow and Ronald Fagin, editors, *Proceedings of the 37th Annual ACM Symposium on Theory of Computing*, Baltimore, MD, USA, May 22–24, 2005, pages 553–562. ACM, 2005. https://doi.org/10.1145/1060590.1060673. 262

[AM17] Josh Alman and Dylan McKay. Theoretical foundations of team matchmaking. In Kate Larson, Michael Winikoff, Sanmay Das, and Edmund H. Durfee, editors, *Proceedings of the 16th Conference on Autonomous Agents and MultiAgent Systems, AAMAS 2017*, São Paulo, Brazil, May 8–12, 2017, pages 1073–1081. ACM, 2017. http://dl.acm.org/citation.cfm?id=3091277. 404

[AMS99] Noga Alon, Yossi Matias, and Mario Szegedy. The space complexity of approximating the frequency moments. *Journal of Computer and System Science*, 58(1):137–147, 1999. https://doi.org/10.1006/jcss.1997.1545. 405

[AMS20] Mikkel Abrahamsen, Tillmann Miltzow, and Nadja Seiferth. Framework for ER-completeness of two-dimensional packing problems. In Sandy Irani, editor, *61st IEEE Annual Symposium on Foundations of Computer Science, FOCS 2020*, Durham, NC, USA, November 16–19, 2020, pages 1014–1021. IEEE, 2020. https://doi.org/10.1109/FOCS46700.2020.00098. 298

[AN21] Sepehr Assadi and Vishvajeet N. Graph streaming lower bounds for parameter estimation and property testing via a streaming XOR lemma. In Samir Khuller and Virginia Vassilevska Williams, editors, *STOC '21: 53rd Annual ACM SIGACT Symposium on Theory of Computing, Virtual Event, Italy, June 21–25, 2021*, pages 612–625. ACM, 2021. https://doi.org/10.1145/3406325.3451110. 418

[ANTV02] Andris Ambainis, Ashwin Nayak, Amnon Ta-Shma, and Umesh V. Vazirani. Dense quantum coding and quantum finite automata. *Journal of the ACM*, 49(4):496–511, 2002. 478

[APT79] Bengt Aspvall, Michael F. Plass, and Robert Endre Tarjan. A linear-time algorithm for testing the truth of certain quantified boolean formulas. *Information Processing Letters*, 8(3):121–123, 1979. https://doi.org/10.1016/0020-0190(79)90002-4. 33

[AR98] Yonatan Aumann and Yuval Rabani. An $O(\log k)$ approximate min-cut max-flow theorem and approximation algorithm. *SIAM Journal on Computing*, 27(1):291–301, 1998. https://doi.org/10.1137/S0097539794285983. 267

[AR20] Sepehr Assadi and Ran Raz. Near-quadratic lower bounds for two-pass graph streaming algorithms. In Sandy Irani, editor, *61st IEEE Annual Symposium on Foundations of Computer Science, FOCS 2020*, Durham, NC, USA, November 16–19, 2020, pages 342–353. IEEE, 2020. https://doi.org/10.1109/FOCS46700.2020.00040. 418

[Aro98] Sanjeev Arora. Polynomial time approximation schemes for Euclidean traveling salesman and other geometric problems. *Journal of the Association of Computing Machinery (JACM)*, 45(3):501–555, 1998. https://dl.acm.org/doi/10.1145/290179.290180. 207

[ARV09] Sanjeev Arora, Satish Rao, and Umesh V. Vazirani. Expander flows, geometric embeddings and graph partitioning. *Journal of the Association of Computing Machinery (JACM)*, 56(2):5:1–5:37, 2009. https://doi.org/10.1145/1502793.1502794. 262, 267

[AS98] Sanjeev Arora and Shmuel Safra. Probabilistic checking of proofs: A new characterization of NP. *Journal of the ACM*, 45(1):70–122, 1998. https://doi.org/10.1145/273865.273901. 214

[AS18] Divesh Aggarwal and Noah Stephens-Davidowitz. (Gap/S)ETH hardness of SVP. In Ilias Diakonikolas, David Kempe, and Monika Henzinger, editors, *Proceedings of the 50th Annual ACM SIGACT Symposium on Theory of Computing, STOC 2018*, Los Angeles, CA, USA, June 25–29, 2018, pages 228–238. ACM, 2018. https://doi.org/10.1145/3188745.3188840. 168

[ASS+18] Alexandr Andoni, Zhao Song, Clifford Stein, Zhengyu Wang, and Peilin Zhong. Parallel graph connectivity in log diameter rounds. In Mikkel Thorup, editor, *59th IEEE Annual Symposium on Foundations of Computer Science, FOCS 2018*, Paris, France, October 7–9, 2018, pages 674–685. IEEE Computer Society, 2018. https://doi.org/10.1109/FOCS.2018.00070. 425

[Ass22] Sepehr Assadi. A two-pass (conditional) lower bound for semi-streaming maximum matching. In Joseph (Seffi) Naor and Niv Buchbinder, editors, *Proceedings of the 2022 ACM-SIAM Symposium on Discrete Algorithms, SODA 2022*, Virtual Conference / Alexandria, VA, USA, January 9–12, 2022, pages 708–742. SIAM, 2022. https://arxiv.org/abs/2108.07187. 418

[Aus07] Per Austrin. Balanced MAX-2-SAT might not be the hardest. In David S. Johnson and Uriel Feige, editors, *Proceedings of the 39th Annual ACM Symposium on Theory of Computing*, San Diego, California, USA, June 11–13, 2007, pages 189–197. ACM, 2007. https://doi.org/10.1145/1250790.1250818. 260

[AVW14] Amir Abboud, Virginia Vassilevska Williams, and Oren Weimann. Consequences of faster alignment of sequences. In Javier Esparza, Pierre Fraigniaud, Thore Husfeldt, and Elias Koutsoupias, editors, *Automata, Languages, and Programming - 41st International Colloquium, ICALP 2014*, Copenhagen, Denmark, July 8–11, 2014, *Proceedings, Part I*, volume 8572 of *Lecture Notes in Computer Science*, pages 39–51. Springer, 2014. https://people.csail.mit.edu/virgi/hardstrings.pdf. 374, 376

[AVY18] Amir Abboud, Virginia Vassilevska Williams, and Huacheng Yu. Matching triangles and basing hardness on an extremely popular conjecture. *SIAM Journal on Computing*, 47(3):1098–1122, 2018. https://doi.org/10.1137/15M1050987. 390

[AW14] Amir Abboud and Virginia Vassilevska Williams. Popular conjectures imply strong lower bounds for dynamic problems. In *55th IEEE Annual Symposium on Foundations of Computer Science, FOCS 2014*, Philadelphia, PA, USA, October 18–21, 2014, pages 434–443. IEEE Computer Society, 2014. https://arxiv.org/abs/1402.0054. 373, 379

[AWY15] Amir Abboud, Ryan Williams, and Huacheng Yu. More applications of the polynomial method to algorithm design. In Piotr Indyk, editor, *Proceedings of the Twenty-Sixth Annual ACM-SIAM Symposium on Discrete Algorithms, SODA 2015, San Diego, CA, USA, January 4–6, 2015*, pages 218–230. SIAM, 2015. https://doi.org/10.1137/1.9781611973730.17. 374

[BA18] Moti Ben-Ari. Mastermind is NP-complete, 2018. http://arxiv.org/abs/cs/0512049. 49

[Bab16] László Babai. Graph isomorphism in quasipolynomial time [extended abstract]. In Daniel Wichs and Yishay Mansour, editors, *Proceedings of the 48th Annual ACM SIGACT Symposium on Theory of Computing, STOC 2016, Cambridge, MA, USA, June 18–21, 2016*, pages 684–697. ACM, 2016. https://dl.acm.org/doi/10.1145/2897518.2897542. 14

[Bar04] Régis Barbanchon. On unique graph 3-colorability and parsimonious reductions in the plane. *Theoretical Computer Science*, 319(1):455–482, 2004. 275

[BB02] Olivier Bournez and Michael S. Branicky. The mortality problem for matrices of low dimensions. *Theory of Computing Systems*, 35(4):433–448, 2002. https://doi.org/10.1007/s00224-002-1010-5. 336

[BBBV97] Charles H. Bennett, Ethan Bernstein, Gilles Brassard, and Umesh V. Vazirani. Strengths and weaknesses of quantum computing. *SIAM Journal on Computing*, 26(5):1510–1523, 1997. https://arxiv.org/abs/quant-ph/9701001. 473

[BBD+17] MohammadHossein Bateni, Soheil Behnezhad, Mahsa Derakhshan, MohammadTaghi Hajiaghayi, Raimondas Kiveris, Silvio Lattanzi, and Vahab S. Mirrokni. Affinity clustering: Hierarchical clustering at scale. In Isabelle Guyon, Ulrike von Luxburg, Samy Bengio, Hanna M. Wallach, Rob Fergus, S. V. N. Vishwanathan, and Roman Garnett, editors, *Advances in Neural Information Processing Systems 30: Annual Conference on Neural Information Processing Systems 2017, December 4–9, 2017, Long Beach, CA, USA*, pages 6864–6874, 2017. https://proceedings.neurips.cc/paper/2017/hash/2e1b24a664f5e9c18f407b2f9c73e821-Abstract.html. 434

[BBT05] Gilles Brassard, Anne Broadbent, and Alain Tapp. Quantum pseudo-telepathy. *Foundations of Physics (Special Asher Peres Memorial Issue)*, 35(11):1877–1907, 2005. https://arxiv.org/abs/quant-ph/0407221. 483

[BCC+20] Jeffrey Bosboom, Charlotte Chen, Lily Chung, Spencer Compton, Michael Coulombe, Erik D. Demaine, Martin L. Demaine, Ivan Tadeu Ferreira Antunes Filho, Della Hendrickson, Adam Hesterberg, Calvin Hsu, William Hu, Oliver Korten, Zhezheng Luo, and Lillian Zhang. Edge matching with inequalities, triangles, unknown shape, and two players. *Journal of Information Processing*, 28:987–1007, 2020. 153, 273, 285

[BCD$^+$19] Jeffrey Bosboom, Spencer Congero, Erik D. Demaine, Martin L. Demaine, and Jayson Lynch. Losing at checkers is hard. In Jennifer Beineke and Jason Rosenhouse, editors, *The Mathematics of Various Entertaining Subjects: The Magic of Mathematics*. Princeton University Press, 2019. http://arxiv.org/abs/1806.05657. 325

[BCD$^+$21] Josh Brunner, Lily Chung, Erik D. Demaine, Della H. Hendrickson, Adam Hesterberg, Adam Suhl, and Avi Zeff. 1×1 Rush Hour with fixed blocks is PSPACE-complete. In Martin Farach-Colton, Giuseppe Prencipe, and Ryuhei Uehara, editors, *10th International Conference on Fun with Algorithms, FUN 2021, May 30 to June 1, 2021, Favignana Island, Sicily, Italy*, volume 157 of *LIPIcs*, pages 7:1–7:14. Schloss Dagstuhl - Leibniz-Zentrum für Informatik, 2021. https://doi.org/10.4230/LIPIcs.FUN.2021.7. 343

[BCDP96] David Beckman, Amalavoyal Chari, Srikrisha Devabhaktuni, and John Preskill. Efficient networks for quantum factoring. *Physical Review Letters*, 54(2):1034–1063, 1996. https://arxiv.org/abs/quant-ph/9602016. 471, 472

[BCG82] Elwyn R. Berlekamp, John H. Conway, and Richard K. Guy. *Winning Ways for Your Mathematical Plays*, volume 2. Academic Press, New York, 1982. 62

[BCH16] Michele Borassi, Pierluigi Crescenzi, and Michel Habib. Into the square: On the complexity of some quadratic-time solvable problems. *Electronic Notes Theoretical Computer Science*, 322: 51–67, 2016. https://doi.org/10.1016/j.entcs.2016.03.005. 372

[BCP$^+$14] Nicolas Brunner, Daniel Cavalcanti, Stefano Pironio, Valerio Scarani, and Stephanie Wehner. Bell nonlocality. *Review of Modern Physics*, 86(2):419–478, 2014. https://arxiv.org/abs/1303.2849. 482

[BDD$^+$02] Therese C. Biedl, Erik D. Demaine, Martin L. Demaine, Rudolf Fleischer, Lars Jacobsen, and J. Ian Munro. The complexity of Clickomania. In R. J. Nowakowski, editor, *More Games of No Chance*, pages 389–404. Cambridge University Press, 2002. Collection of papers from the MSRI Combinatorial Game Theory Research Workshop, Berkeley, California, July 2000. https://arxiv.org/abs/cs/0107031. 154

[BDD$^+$17] Jeffrey Bosboom, Erik D. Demaine, Martin L. Demaine, Adam Hesterberg, Pasin Manurangsi, and Anak Yodpinyanee. Even $1 \times n$ edge-matching and jigsaw puzzles are really hard. *Journal of Information Processing*, 25:682–694, 2017. https://doi.org/10.2197/ipsjjip.25.682. 153, 248

[BDE$^+$19a] Soheil Behnezhad, Laxman Dhulipala, Hossein Esfandiari, Jakub Lacki, and Vahab S. Mirrokni. Near-optimal massively parallel graph connectivity. In David Zuckerman, editor, *60th IEEE Annual Symposium on Foundations of Computer Science, FOCS 2019, Baltimore, Maryland, USA, November 9–12, 2019*, pages 1615–1636. IEEE Computer Society, 2019. https://arxiv.org/abs/1910.05385. 425, 426, 434

[BDE$^+$19b] Mahdi Boroujeni, Sina Dehghani, Soheil Ehsani, Mohammad Taghi Hajiaghayi, and Saeed Seddighin. Subcubic equivalences between graph centrality measures and complementary problems, 2019. http://arxiv.org/abs/1905.08127. 382, 388, 389

[BDE$^+$21] Soheil Behnezhad, Laxman Dhulipala, Hossein Esfandiari, Jakub Lacki, Vahab S. Mirrokni, and Warren Schudy. Massively parallel computation via remote memory access. *ACM Transactions on Parallel Computing*, 8(3):13:1–13:25, 2021. https://doi.org/10.1145/3470631. 434, 435

[BDH$^+$04] Ron Breukelaar, Erik D. Demaine, Susan Hohenberger, Hendrik Jan Hoogeboom, Walter A. Kosters, and David Liben-Nowell. Tetris is hard, even to approximate. *International Journal of Computational Geometry and Applications*, 14(1–2):41–68, 2004. https://doi.org/10.1142/S0218195904001354. 155

[BDHW20] Josh Brunner, Erik D. Demaine, Della Hendrickson, and Julian Wellman. Complexity of retrograde and helpmate chess problems: Even cooperative chess is hard. In *Proceedings of the 31st International Symposium on Algorithms and Computation (ISAAC 2020)*, pages 33:1–33:14, Hong Kong, December 2020. 328

[BDP08] Ilya Baran, Erik D. Demaine, and Mihai Pătraşcu. Subquadratic algorithms for 3SUM. *Algorithmica*, 50(4):584–596, 2008. https://doi.org/10.1007/s00453-007-9036-3. 356

[BdW02] Harry Buhrman and Ronald de Wolf. Complexity measures and decision tree complexity: A survey. *Theoretical Computer Science*, 288(1):21–43, 2002. https://www.sciencedirect.com/science/article/pii/S030439750100144X. 431

[BEG+21] Mahdi Boroujeni, Soheil Ehsani, Mohammad Ghodsi, MohammadTaghi Hajiaghayi, and Saeed Seddighin. Approximating edit distance in truly subquadratic time: Quantum and mapreduce. *Journal of the ACM*, 68(3):19:1–19:41, 2021. 475

[Ben23] Huck Bennett. The complexity of the shortest vector problem. *SIGACT News*, To Appear, 2023. https://eccc.weizmann.ac.il/report/2022/170/. 168, 249

[BEP05] Cristina Bazgan, Bruno Escoffier, and Vangelis Th. Paschos. Completeness in standard and differential approximation classes: Poly-(D)APX- and (D)PTAS-completeness. *Theoretical Computer Science*, 339(2–3):272–292, 2005. http://dx.doi.org/10.1016/j.tcs.2005.03.007. 246

[Ber66] Robert Berger. *The Undecidability of the Domino Problem*. Memoirs of the American Math Society. American Math Society, 1966. 337

[BFGS20] Guy E. Blelloch, Jeremy T. Fineman, Yan Gu, and Yihan Sun. Optimal parallel algorithms in the binary-forking model. In Christian Scheideler and Michael Spear, editors, *SPAA '20: 32nd ACM Symposium on Parallelism in Algorithms and Architectures, Virtual Event, USA, July 15–17, 2020*, pages 89–102. ACM, 2020. https://doi.org/10.1145/3350755.3400227. 422

[BFL91] László Babai, Lance Fortnow, and Carsten Lund. Non-deterministic exponential time has two-prover interactive protocols. *Compututational Complexity*, 1:3–40, 1991. 481

[BFNV18] Adam Bouland, Bill Fefferman, Chinmay Nirkhe, and Umesh Vazarani. On the complexity and verifiation of quantum random circuit sampling. *Nature Physics*, 15:159–163, 2018. https://arxiv.org/abs/1803.04402. 485

[BFNW93] László Babai, Lance Fortnow, Noam Nisan, and Avi Wigderson. BPP has subexponential time simulations unless EXPTIME has publishable proofs. *Computational Complexity*, 3:307–318, 1993. https://doi.org/10.1007/BF01275486. 18

[BFS99] Jonathan F. Buss, Gudmund Skovbjerg Frandsen, and Jeffrey O. Shallit. The computational complexity of some problems of linear algebra. *Journal of Computing and System Sciences*, 58(3):572–596, 1999. https://doi.org/10.1006/jcss.1998.1608. 292, 294

[BGG97] Egon Börger, Erich Grädel, and Yuri Gurevich. *The Classical Decision Problem*. Perspectives in Mathematical Logic. Springer, 1997. 46

[BGL+18] Davide Bilò, Luciano Gualà, Stefano Leucci, Guido Proietti, and Mirko Rossi. On the PSPACE-completeness of Peg Duotaire and other peg-jumping games. In Hiro Ito, Stefano Leonardi, Linda Pagli, and Giuseppe Prencipe, editors, *9th International Conference on Fun with Algorithms, FUN 2018, June 13–15, 2018, La Maddalena, Italy*, volume 100 of *LIPIcs*, pages 8:1–8:15. Schloss Dagstuhl - Leibniz-Zentrum für Informatik, 2018. https://doi.org/10.4230/LIPIcs.FUN.2018.8. 350

[BGLR93] Mihir Bellare, Shafi Goldwasser, Carsten Lund, and A. Russeli. Efficient probabilistically checkable proofs and applications to approximations. In S. Rao Kosaraju, David S. Johnson, and Alok Aggarwal, editors, *Proceedings of the Twenty-Fifth Annual ACM Symposium on Theory of Computing, May 16–18, 1993, San Diego, CA, USA*, pages 294–304. ACM, 1993. https://doi.org/10.1145/167088.167174.

[BGMW20] Karl Bringmann, Pawel Gawrychowski, Shay Mozes, and Oren Weimann. Tree edit distance cannot be computed in strongly subcubic time (unless APSP can). *ACM Transactions on Algorithms*, 16(4):48:1–48:22, 2020. https://doi.org/10.1145/3381878. 388

[BGRS13] Jaroslaw Byrka, Fabrizio Grandoni, Thomas Rothvoß, and Laura Sanità. Steiner tree approximation via iterative randomized rounding. *Journal of the ACM*, 60(1):6:1–6:33, 2013. https://doi.org/10.1145/2432622.2432628. 244

[BGS98] Mihir Bellare, Oded Goldreich, and Madhu Sudan. Free bits, PCPs, and nonapproximability-towards tight results. *SIAM Journal on Computing*, 27(3):804–915, 1998. https://www.cs.umd.edu/~gasarch/TOPICS/pcp/bgsfreebits.pdf. 256

[BGS17] Huck Bennett, Alexander Golovnev, and Noah Stephens-Davidowitz. On the quantitative hardness of CVP. In Chris Umans, editor, *58th IEEE Annual Symposium on Foundations of Computer Science, FOCS 2017, Berkeley, CA, USA, October 15–17, 2017*, pages 13–24. IEEE Computer Society, 2017. https://doi.org/10.1109/FOCS.2017.11. 168

[BGS20] Huck Bennett, Sasha Golovnev, and Noah Stephens-Davidowitz. Fine-grained hardness of lattice problems: Open questions, 2020. https://blog.simons.berkeley.edu/2020/05/fine-grained-hardness-of-lattice-problems-open-questions/. 168

[BH93] Amir Ben-Dor and Shai Halevi. Zero-one permanent is #P-complete, A simpler proof. In *Second Israel Symposium on Theory of Computing Systems, ISTCS 1993, Natanya, Israel, June 7–9, 1993, Proceedings*, pages 108–117. IEEE Computer Society, 1993. https://shaih.github.io/pubs/01perm.pdf. 282

[BH01] Gill Barequet and Sariel Har-Peled. Polygon containment and translational min-Hausdorff-distance between segment sets are 3SUM-hard. *International Journal of Computational Geometry*, 11(4):465–474, 2001. https://doi.org/10.1142/S0218195901000596. 376

[BH22] Marie Louisa Tølbøll Berthelsen and Kristoffer Arnsfelt Hansen. On the computational complexity of decision problems about multi-player Nash equilibria. *Theory Computing Systems*, 66(3):519–545, 2022. https://doi.org/10.1007/s00224-022-10080-1. 300

[BHP12] Paul C. Bell, Mika Hirvensalo, and Igor Potapov. Mortality for 2×2 matrices is NP-hard. In Branislav Rovan, Vladimiro Sassone, and Peter Widmayer, editors, *Mathematical Foundations of Computer Science 2012 - 37th International Symposium, MFCS 2012, Bratislava, Slovakia, August 27–31, 2012. Proceedings*, volume 7464 of *Lecture Notes in Computer Science*, pages 148–159. Springer, 2012. 336

[BHZ87] Ravi Boppana, Johan Hastad, and Stathis Zachos. Does co-NP have short interactive proofs? *Information Processing Letters*, 25, 1987. 14

[BI06] Jonathan F. Buss and Tarique Islam. Simplifying the weft hierarchy. *Theoretical Computer Science*, 351(3):303–313, 2006. https://doi.org/10.1016/j.tcs.2005.10.002. 186

[BI15] Arturs Backurs and Piotr Indyk. Edit distance cannot be computed in strongly subquadratic time (unless SETH is false). In Rocco A. Servedio and Ronitt Rubinfeld, editors, *Proceedings of the Forty-Seventh Annual ACM on Symposium on Theory of Computing, STOC 2015, Portland, OR, USA, June 14–17, 2015*, pages 51–58. ACM, 2015. https://doi.org/10.1145/2746539.2746612. 374, 375, 475

[BI16] Arturs Backurs and Piotr Indyk. Which regular expression patterns are hard to match? In Irit Dinur, editor, *IEEE 57th Annual Symposium on Foundations of Computer Science, FOCS 2016, 9–11 October 2016, Hyatt Regency, New Brunswick, New Jersey, USA*, pages 457–466. IEEE Computer Society, 2016. https://doi.org/10.1109/FOCS.2016.56. 375

[Bia11] Marzio De Biasi. Boulder Dash is NP-hard, 2011. http://www.nearly42.org/boulderdash/bdnp00.pdf. 126

[Bid20] Stella Biderman. Magic: The Gathering is as hard as arithmetic, 2020. https://arxiv.org/abs/2003.05119. 317, 335

[Bie91] Daniel Bienstock. Some provably hard crossing number problems. *Discrete and Computional Geometry*, 6:443–459, 1991. https://doi.org/10.1007/BF02574701. 298

[Bie20] Nicholas Bieker. Complexity of graph drawing problems in relation to the existential theory of the reals. Master's thesis, Karlsruhe Institute of Technology, 2020. https://i11www.iti.kit.edu/_media/teaching/theses/ba-bieker-20.pdf. 290, 291

[BK98a] Therese C. Biedl and Goos Kant. A better heuristic for orthogonal graph drawings. *Computational Geometry: Theory and Applications*, 9(3):159–180, 1998. https://www.sciencedirect.com/science/article/pii/S0925772197000266. 82

[BK98b] Heinz Breu and David G. Kirkpatrick. Unit disk graph recognition is NP-hard. *Computational Geometry*, 9(1–2):3–24, 1998. https://doi.org/10.1016/S0925-7721(97)00014-X. 297

[BK99] Piotr Berman and Marek Karpinski. On some tighter inapproximability results. In *Automata, Languages and Programming, 26th International Colloquium, ICALP'99, Prague, Czech Republic, July 11–15, 1999, Proceedings*, pages 200–209, 1999. https://eccc.weizmann.ac.il/report/1998/065/. 239

[BK09] Nikhil Bansal and Subhash Khot. Optimal long code test with one free bit. In *50th Annual IEEE Symposium on Foundations of Computer Science, FOCS 2009, October 25–27, 2009, Atlanta,*

Georgia, USA, pages 453–462. IEEE Computer Society, 2009. https://doi.org/10.1109/FOCS.2009.23. 260, 263

[BK10] Nikhil Bansal and Subhash Khot. Inapproximability of hypergraph vertex cover and applications to scheduling problems. In Samson Abramsky, Cyril Gavoille, Claude Kirchner, Friedhelm Meyer auf der Heide, and Paul G. Spirakis, editors, *Automata, Languages and Programming, 37th International Colloquium, ICALP 2010, Bordeaux, France, July 6–10, 2010, Proceedings, Part I*, volume 6198 of *Lecture Notes in Computer Science*, pages 250–261. Springer, 2010. https://cs.nyu.edu/~khot/papers/hypergraph4.pdf. 260

[BKM21] Mark Braverman, Subhash Khot, and Dor Minzer. On rich 2-to-1 games. In James R. Lee, editor, *12th Innovations in Theoretical Computer Science Conference, ITCS 2021, January 6–8, 2021, Virtual Conference*, volume 185 of *LIPIcs*, pages 27:1–27:20. Schloss Dagstuhl - Leibniz-Zentrum für Informatik, 2021. https://doi.org/10.4230/LIPIcs.ITCS.2021.27. 268

[BKP+19] Marthe Bonamy, Lukasz Kowalik, Michal Pilipczuk, Arkadiusz Socala, and Marcin Wrochna. Tight lower bounds for the complexity of multicoloring. *ACM Transactions on Computation Theory*, 11(3):13:1–13:19, 2019. https://doi.org/10.1145/3313906. 170

[BKS17] Paul Beame, Paraschos Koutris, and Dan Suciu. Communication steps for parallel query processing. *Journal of the Assocation of Computing Machinery (JACM)*, 64(6):40:1–40:58, 2017. https://doi.org/10.1145/3125644. 423

[BKW15] Mark Braverman, Young Kun-Ko, and Omri Weinstein. Approximating the best Nash equilibrium in $n^{o(\log n)}$-time breaks the exponential time hypothesis. In Piotr Indyk, editor, *Proceedings of the Twenty-Sixth Annual ACM-SIAM Symposium on Discrete Algorithms, SODA 2015, San Diego, CA, USA, January 4–6, 2015*, pages 970–982. SIAM, 2015. https://doi.org/10.1137/1.9781611973730.66. 462

[BM08] Benjamin E. Birnbaum and Claire Mathieu. On-line bipartite matching made simple. *SIGACT News*, 39(1):80–87, 2008. https://doi.org/10.1145/1360443.1360462. 403

[BM21] Vittorio Bilò and Marios Mavronicolas. $\exists R$-complete decision problems about (symmetric) Nash equilibria in (symmetric) multi-player games. *ACM Transactions Economics and Computation*, 9(3):14:1–14:25, 2021. https://doi.org/10.1145/3456758 300

[Boo58] William Boone. The word problem. *Proceedings of the National Academy of Sciences*, 44: 1061–1065, 1958. https://www.jstor.org/stable/pdf/1970103.pdf. 337

[Bou85] Jean Bourgain. On Lipshitz embeddings of finite metric spaces in Hilbert space. *Israel Journal of Mathematics*, 52:46–52, 1985. https://link.springer.com/article/10.1007/BF02776078. 267

[BOV13] Vladimir Braverman, Rafail Ostrovsky, and Dan Vilenchik. How hard is counting triangles in the streaming model? In Fedor V. Fomin, Rusins Freivalds, Marta Z. Kwiatkowska, and David Peleg, editors, *Automata, Languages, and Programming - 40th International Colloquium, ICALP 2013, Riga, Latvia, July 8–12, 2013, Proceedings, Part I*, volume 7965 of *Lecture Notes in Computer Science*, pages 244–254. Springer, 2013. 477

[BP02] Vincent Blondel and Natacha Portier. The presense of a zero in an integer linear recurrent sequence is NP-hard to decide. *Linear Algebra and Its Applications*, 351–352:91–98, 2002. 330

[BP15] Manuel Bodirsky and Michael Pinsker. Schaefer's theorem for graphs. *Journal of the Association of Computing Machinery (JACM)*, 62(3):19:1–19:52, 2015. https://arxiv.org/pdf/1011.2894.pdf. 46

[BP22] Huck Bennett and Chris Peikert. Hardness of the (approximate) shortest vector problem: A simple proof via Reed-Solomon codes, 2022. https://arxiv.org/abs/2202.07736. 250

[BPT92] Richard B. Borie, R. Gary Parker, and Craig A. Tovey. Automatic generation of linear-time algorithms from predicate calculus descriptions of problems on recursively constructed graph families. *Algorithmica*, 7(5&6):555–581, 1992. https://doi.org/10.1007/BF01758777. 199

[BRG89] David Bernstein, Michael Rodeh, and Izidor Gertner. On the complexity of scheduling problems for parallel/pipelined machines. *IEEE Transactions on Computers*, 38(9):1308–1313, 1989. https://doi.org/10.1109/12.29469. 157

[Bri14] Karl Bringmann. Why walking the dog takes time: Frechet distance has no strongly subquadratic algorithms unless SETH fails. In *55th IEEE Annual Symposium on Foundations of Computer*

Science, FOCS 2014, Philadelphia, PA, USA, October 18–21, 2014, pages 661–670. IEEE Computer Society, 2014. https://doi.org/10.1109/FOCS.2014.76. 375

[Bro41] Rowland Leonard Brooks. On colouring the nodes of a network. *Mathematical Proceedings of the Cambridge Philosophical Society*, 37:194–197, 4 1941. https://www.cambridge.org/core/services/aop-cambridge-core/content/view/546AD533E0FDCFD02755AC34B0972D0E/S030500410002168Xa.pdf/on-colouring-the-nodes-of-a-network.pdf. 69

[BSS88] Lenore Blum, Mike Shub, and Steve Smale. On a theory of computation over the real numbers; NP completeness, recursive functions and universal machines (extended abstract). In *29th Annual Symposium on Foundations of Computer Science, White Plains, New York, USA, 24–26 October 1988*, pages 387–397. IEEE Computer Society, 1988. https://doi.org/10.1109/SFCS.1988.21955. 294

[Bul17] Andrei A. Bulatov. A dichotomy theorem for nonuniform CSPs. In Chris Umans, editor, *58th IEEE Annual Symposium on Foundations of Computer Science, FOCS 2017, Berkeley, CA, USA, October 15–17, 2017*, pages 319–330. IEEE Computer Society, 2017. https://doi.org/10.1109/FOCS.2017.37. 71

[Bur] Burke. I want an easy gadget to prove Planar Hamiltonian Cycle NP-complete (from Hamiltonian Cycle). https://cstheory.stackexchange.com/questions/9587/i-want-an-easy-gadget-to-prove-planar-hamiltonian-cycle-np-complete-from-hamilt. 65

[BW91] Graham Brightwell and Peter Winkler. Counting linear extensions. *Order*, 8:225–242, 1991. https://link.springer.com/content/pdf/10.1007/BF00383444.pdf. 286

[BW05] Graham R. Brightwell and Peter Winkler. Counting Eulerian circuits is #P-complete. In Camil Demetrescu, Robert Sedgewick, and Roberto Tamassia, editors, *Proceedings of the Seventh Workshop on Algorithm Engineering and Experiments and the Second Workshop on Analytic Algorithmics and Combinatorics, ALENEX /ANALCO 2005, Vancouver, BC, Canada, 22 January 2005*, pages 259–262. SIAM, 2005. http://www.siam.org/meetings/analco05/papers/09grbrightwell.pdf. 274

[Cab01] Adan Cabello. Bell's theorem without inequalities and without probabilities for two observers. *Physical Review Letters*, 2001. https://arxiv.org/abs/quant-ph/0008085. 483

[Cam01] Kathie Cameron. Thomason's algorithm for finding a second hamiltonian circuit through a given edge in a cubic graph is exponential on Krawczyk's graphs. *Discrete Mathematics*, 235(1):69–77, 2001. https://doi.org/10.1016/S0012-365X(00)00260-0. 275

[Can88] John F. Canny. Some algebraic and geometric computations in PSPACE. In Janos Simon, editor, *Proceedings of the 20th Annual ACM Symposium on Theory of Computing, May 2–4, 1988, Chicago, Illinois, USA*, pages 460–467. ACM, 1988. https://doi.org/10.1145/62212.62257. 290

[Car15] Jean Cardinal. Computational geometry column 62. *SIGACT News*, 46(4):69–78, 2015. https://doi.org/10.1145/2852040.2852053. 290

[CBH21] Alex Churchill, Stella Biderman, and Austin Herrick. Magic: The Gathering is Turing complete. In Martin Farach-Colton, Giuseppe Prencipe, and Ryuhei Uehara, editors, *Proceedings of the 10th International Conference on Fun with Algorithms*, volume 157 of *LIPIcs*, pages 9:1–9:19, Favignana Island, Italy, 2021. https://doi.org/10.4230/LIPIcs.FUN.2021.9. 335

[CC08] Miroslav Chlebík and Janka Chlebíková. Approximation hardness of dominating set problems in bounded degree graphs. *Information and Computation*, 206(11):1264–1275, 2008. https://doi.org/10.1016/j.ic.2008.07.003. 240

[CCD+03] Andrew M. Childs, Richard Cleve, Enrico Deotto, Edward Farhi, Sam Gutmann, and Daniel A. Spielman. Exponential algorithmic speedup by a quantum walk. In Lawrence L. Larmore and Michel X. Goemans, editors, *Proceedings of the 35th Annual ACM Symposium on Theory of Computing, June 9–11, 2003, San Diego, CA, USA*, pages 59–68. ACM, 2003. https://arxiv.org/abs/quant-ph/0209131. 475

[CCG23] Andrew M. Childs, Matthew Coudron, and Amin Shiraz Gilani. Quantum algorithms and the power of forgetting. In Yael Tauman Kalai, editor, *14th Innovations in Theoretical Computer Science Conference, ITCS 2023, January 10–13, 2023, MIT, Cambridge, Massachusetts, USA*, volume 251 of *LIPIcs*, pages 37:1–37:22. Schloss Dagstuhl - Leibniz-Zentrum für Informatik, 2023. 475

[CCHM15] Rajesh Hemant Chitnis, Graham Cormode, Mohammad Taghi Hajiaghayi, and Morteza Monemizadeh. Parameterized streaming: Maximal matching and vertex cover. In Piotr Indyk, editor, *Proceedings of the Twenty-Sixth Annual ACM-SIAM Symposium on Discrete Algorithms, SODA 2015, San Diego, CA, USA, January 4–6, 2015*, pages 1234–1251. SIAM, 2015. https://doi.org/10.1137/1.9781611973730.82. 418

[CD68] Richard W. Cottle and George B. Dantzig. Complentary pivot theory of mathematical programming. *Linear Algebra and Its Applications*, 1:103–125, 1968. https://www.sciencedirect.com/science/article/pii/0024379568900529. 447

[CD06] Xi Chen and Xiaotie Deng. Settling the complexity of two-player Nash equilibrium. In *47th Annual IEEE Symposium on Foundations of Computer Science (FOCS 2006), 21–24 October 2006, Berkeley, California, USA, Proceedings*, pages 261–272. IEEE Computer Society, 2006. https://doi.org/10.1109/FOCS.2006.69. 456

[CDE+20] Jean Cardinal, Erik D. Demaine, David Eppstein, Robert A. Hearn, and Andrew Winslow. Reconfiguration of satisfying assignments and subset sums: Easy to find, hard to connect. *Theoretical Computer Science*, 806(2):332–343, February 2020. https://arxiv.org/abs/1805.04055. 345

[CDG+08] Fay Chang, Jeffrey Dean, Sanjay Ghemawat, Wilson C. Hsieh, Deborah A. Wallach, Michael Burrows, Tushar Chandra, Andrew Fikes, and Robert E. Gruber. Bigtable: A distributed storage system for structured data. *ACM Transactions on Computer Systems*, 26(2):4:1–4:26, 2008. https://doi.org/10.1145/1365815.1365816. 434

[CDP08] Gruia Călinescu, Adrian Dumitrescu, and János Pach. Reconfigurations in graphs and grids. *SIAM Journal of Discrete Mathematics*, 22(1):124–138, 2008. http://dx.doi.org/10.1137/060652063. 241

[CDP21] Artur Czumaj, Peter Davies, and Merav Parter. Component stability in low-space massively parallel computation. In Avery Miller, Keren Censor-Hillel, and Janne H. Korhonen, editors, *PODC '21: ACM Symposium on Principles of Distributed Computing, Virtual Event, Italy, July 26–30, 2021*, pages 481–491. ACM, 2021. https://arxiv.org/pdf/2106.01880.pdf. 432

[CDR86] Stephen A. Cook, Cynthia Dwork, and Rüdiger Reischuk. Upper and lower time bounds for parallel random access machines without simultaneous writes. *SIAM Journal on Computing*, 15(1):87–97, 1986. https://doi.org/10.1137/0215006. 422

[CE04] Mark Cieliebak and Stephan J. Eidenbenz. Measurement errors make the partial digest problem NP-hard. In Martin Farach-Colton, editor, *LATIN 2004: Theoretical Informatics, 6th Latin American Symposium, Buenos Aires, Argentina, April 5–8, 2004, Proceedings*, volume 2976 of *Lecture Notes in Computer Science*, pages 379–390. Springer, 2004. 157

[Ces03] Marco Cesati. The Turing way to parameterized complexity. *Journal of Computing and System Science*, 67(4):654–685, 2003. https://doi.org/10.1016/S0022-0000(03)00073-4. 199

[CFG02] Andrew M. Childs, Edward Farhi, and Sam Gutmann. An example of the difference between quantum and classical random walks. *Quantum Inf. Processing*, 1(1–2):35–43, 2002. https://arxiv.org/abs/quant-ph/0103020. 474

[CFG+17] Marek Cygan, Fedor V. Fomin, Alexander Golovnev, Alexander S. Kulikov, Ivan Mihajlin, Jakub Pachocki, and Arkadiusz Socala. Tight lower bounds on graph embedding problems. *Journal of the Association of Computing Machinery (JACM)*, 64(3):18:1–18:22, 2017. https://doi.org/10.1145/3051094. 169

[CFK+15] Marek Cygan, Fedor V. Fomin, Lukasz Kowalik, Daniel Lokshtanov, Dániel Marx, Marcin Pilipczuk, Michal Pilipczuk, and Saket Saurabh. *Parameterized Algorithms*. Springer, 2015. https://doi.org/10.1007/978-3-319-21275-3. 171, 192

[CH99] Nadia Creignou and Miki Hermann. On #P-completeness of some counting problems, 1999. 286

[Cha20] Timothy M. Chan. More logarithmic-factor speedups for 3SUM, (median, +)-convolution, and some geometric 3SUM-hard problems. *ACM Transactions on Algorithms*, 16(1):7:1–7:23, 2020. https://doi.org/10.1145/3363541. 357

[Che06] Hubie Chen. A rendezvous of logic, complexity, and algebra. *SIGACT News*, 37(4):85–114, 2006. https://arxiv.org/abs/cs/0611018. 45

[Che09] Hubie Chen. A rendezvous of logic, complexity, and algebra. *ACM Computing Surveys*, 42(1):2:1–2:32, 2009. https://dl.acm.org/doi/pdf/10.1145/1592451.1592453. 45

[CHHN14] Julien Cassaigne, Vesa Halava, Tero Harju, and Francois Nicolas. Tighter undecidability bounds for matrix mortality, zero-in-the-corner problems, and more, 2014. http://arxiv.org/abs/1404.0644. 336

[CHK+21] Jianer Chen, Qin Huang, Iyad Kanj, Qian Li, and Ge Xia. Streaming algorithms for graph k-matching with optimal or near-optimal update time. In Hee-Kap Ahn and Kunihiko Sadakane, editors, *32nd International Symposium on Algorithms and Computation, ISAAC 2021, December 6–8, 2021, Fukuoka, Japan*, volume 212 of *LIPIcs*, pages 48:1–48:17. Schloss Dagstuhl - Leibniz-Zentrum für Informatik, 2021. https://doi.org/10.4230/LIPIcs.ISAAC.2021.48. 418

[CHKX06] Jianer Chen, Xiuzhen Huang, Iyad A. Kanj, and Ge Xia. Strong computational lower bounds via parameterized complexity. *Journal of Computing and System Science*, 72(8):1346–1367, 2006. https://doi.org/10.1016/j.jcss.2006.04.007. 177

[Cho16] Amy Chou. NP-hard triangle packing problems, 2016. https://math.mit.edu/research/highschool/rsi/documents/2015Chou.pdf. 144

[Cho22] Adrian Cho. Ordinary computers can beat Google's quantum computers after all. *Science*, 377(6606):563–564, 2022. 486

[Chr22] Nicos Christofides. Worst-case analysis of a new heuristic for the travelling salesman problem. *Operations Research Forum*, 3(1), 2022. Written in 1976https://doi.org/10.1007/s43069-021-00101-z. 207

[CHSH69] John Clauser, Michael Horne, Abner Shimony, and Richard Holt. Proposed experiment to test local hidden-variable theories. *Physical Review Letters*, 23(15):880–884, 1969. 482

[Chu12] Alex Churchill. Magic: The Gathering is Turing complete, 2012. http://www.toothycat.net/hologram/Turing/. 335

[Chv79] V. Chvatal. A greedy heuristic for the set-covering problem. *Mathematics of Operations Research*, 4:233–235, 1979. https://www.jstor.org/stable/pdf/3689577.pdf. 12, 216, 265

[CJ03] Liming Cai and David W. Juedes. On the existence of subexponential parameterized algorithms. *Journal of Computer and System Sciences*, 67(4):789–807, 2003. https://www.sciencedirect.com/science/article/pii/S0022000003000746. 173, 200

[CKK+06] Shuchi Chawla, Robert Krauthgamer, Ravi Kumar, Yuval Rabani, and D. Sivakumar. On the hardness of approximating multicut and sparsest-cut. *Computational ComplexityBlog*, 15(2):94–114, 2006. https://doi.org/10.1007/s00037-006-0210-9. 262

[CKS01] Nadia Creignou, Sanjeev Khanna, and Madhu Sudan. *Complexity Classifications of Boolean Constraint Satisfaction Problems*, volume 7 of *SIAM Monographs on Discrete Mathematics and Applications*. SIAM, 2001. 304

[CKST99] Pierluigi Crescenzi, Viggo Kann, Riccardo Silvestri, and Luca Trevisan. Structure in approximation classes. *SIAM Journal on Computing*, 28(5):1759–1782, 1999. http://dx.doi.org/10.1137/S0097539796304220. 246, 251

[CKX10] Jianer Chen, Iyad A. Kanj, and Ge Xia. Improved upper bounds for vertex cover. *Theoretical Computer Science*, 411(40–42):3736–3756, 2010. https://doi.org/10.1016/j.tcs.2010.06.026. 173

[CLRS03] Thomas Cormen, Charles Leiserson, Ronald Rivest, and Clifford Stein. *Introduction to Algorithms*. MIT Press, Cambridge, 2003. 422

[CMT20] Moses Charikar, Weiyun Ma, and Li-Yang Tan. Unconditional lower bounds for adaptive massively parallel computation. In Christian Scheideler and Michael Spear, editors, *SPAA '20: 32nd ACM Symposium on Parallelism in Algorithms and Architectures, Virtual Event, USA, July 15–17, 2020*, pages 141–151. ACM, 2020. https://doi.org/10.1145/3350755.3400230. 436

[Cob64] Alan Cobham. The intrinsic computational difficulty of functions. In Bar-Hillel Yehoshua, editor, *Logic, Methodology, and Philosophy of Science*, pages 186–192. North Holland, 1964. xiv

[Col75] George Collins. Quantifier elimination for real closed fields by cylindrical algebraic decompostion. In H. Brakhage, editor, *Automata Theory and Formal Languages*, pages 134–183, Berlin,

Heidelberg, 1975. Springer Berlin Heidelberg. https://link.springer.com/chapter/10.1007/3-540-07407-4_17. 290, 293

[Con21] Emily Conover. The new light-based quantum computer Jiuzhang has achieved quantum supremacy. *ScienceNews*, January 16, 2021, 2021. 486

[Coo71] Stephen A. Cook. The complexity of theorem-proving procedures. In *Proceedings of the Third IEEE Symposium on the Foundations of Computer Science*, pages 151–158, 1971. https://dl.acm.org/doi/10.1145/800157.805047. xiii, xiv, 9, 33, 35

[Cou90] Bruno Courcelle. Graph rewriting: An algebraic and logic approach. In Jan van Leeuwen, editor, *Handbook of Theoretical Computer Science, Volume B: Formal Models and Semantics*, pages 193–242. Elsevier and MIT Press, 1990. https://doi.org/10.1016/b978-0-444-88074-1.50010-x. 199

[CR18] Daniel W. Cranston and Landon Rabern. Planar graphs are 9/2-colorable. *Journal of Combinatorial Theory, Series B*, 133:32–45, 2018. https://arxiv.org/abs/1410.7233. 170

[CS14] Michael S. Crouch and Daniel M. Stubbs. Improved streaming algorithms for weighted matching, via unweighted matching. In Klaus Jansen, José D. P. Rolim, Nikhil R. Devanur, and Cristopher Moore, editors, *Approximation, Randomization, and Combinatorial Optimization. Algorithms and Techniques, APPROX/RANDOM 2014, September 4–6, 2014, Barcelona, Spain*, volume 28 of *LIPIcs*, pages 96–104. Schloss Dagstuhl - Leibniz-Zentrum für Informatik, 2014. https://doi.org/10.4230/LIPIcs.APPROX-RANDOM.2014.96. 418

[CT97] Marco Cesati and Luca Trevisan. On the efficiency of polynomial time approximation schemes. *Information Processing Letters*, 64(4):165–171, 1997. www.sciencedirect.com/science/article/pii/S0020019097001646. 191, 192

[Cul98] J. C. Culberson. Sokoban is PSPACE-complete. In *Proceedings International Conference on Fun with Algorithms (FUN98)*, pages 65–76, Waterloo, Ontario, Canada, June 1998. Carleton Scientific. https://webdocs.cs.ualberta.ca/~joe/Preprints/Sokoban/paper.html. 56

[CvD10] Andrew Childs and Wim van Dam. Quantum algorithms for algebraic problems. *Review of Modern Physics*, 82:1–52, 2010. https://arxiv.org/abs/0812.0380. 470

[CW19] Lijie Chen and Ryan Williams. An equivalence class for orthogonal vectors. In Timothy M. Chan, editor, *Proceedings of the Thirtieth Annual ACM-SIAM Symposium on Discrete Algorithms, SODA 2019, San Diego, California, USA, January 6–9, 2019*, pages 21–40. SIAM, 2019. https://doi.org/10.1137/1.9781611975482.2. 375

[CW21] Timothy M. Chan and R. Ryan Williams. Deterministic APSP, orthogonal vectors, and more: Quickly derandomizing Razborov-Smolensky. *ACM Transactions on Algorithms*, 17(1):2:1–2:14, 2021. https://doi.org/10.1145/3402926. 379

[Dal99] Víctor Dalmau. A dichotomy theorem for learning quantified Boolean formulas. *Machine Learning*, 35(3):207–224, 1999. https://doi.org/10.1023/A:1007582729656. 304

[Dav73] Martin Davis. Hilbert's tenth problem is unsolvable. *American Mathematical Monthly*, 80(3): 233–269, 1973. https://www.math.umd.edu/~laskow/Pubs/713/Diophantine.pdf. 336

[dBK12] Mark de Berg and Amirali Khosravi. Optimal binary space partitions for segments in the plane. *International Journal of Computational Geometry*, 22(3):187–206, 2012. https://www.win.tue.nl/~mdberg/Papers/2010/bk-obspp-10.pdf. 85, 86

[DD07] Erik D. Demaine and Martin L. Demaine. Jigsaw puzzles, edge matching, and polyomino packing: Connections and complexity. *Graphs and Combinatorics*, 23(Supplement-1):195–208, 2007. https://doi.org/10.1007/s00373-007-0713-4. 140, 146

[DD21] Andreas Darmann and Janosch Döcker. On simplified NP-complete variants of Monotone 3-Sat. *Discrete Applied Mathematics*, 292:45–58, 2021. 40

[DDD18] Andreas Darmann, Janosch Döcker, and Britta Dorn. The monotone satisfiability problem with bounded variable appearances. *International Journal of Foundations of Computer Science*, 29(6):979–993, 2018. 40

[DDE02] Erik D. Demaine, Martin L. Demaine, and David Eppstein. Phutball endgames are NP-hard. In R. J. Nowakowski, editor, *More Games of No Chance*, pages 351–360. Cambridge University Press,

2002. Collection of papers from the MSRI Combinatorial Game Theory Research Workshop, Berkeley, California, July 2000. http://www.arXiv.org/abs/cs.CC/0008025. 62

[DDE+17] Erik D. Demaine, Martin L. Demaine, Sarah Eisenstat, Adam Hesterberg, Andrea Lincoln, Jayson Lynch, and Y. William Yu. Total Tetris: Tetris with monominoes, dominoes, trominoes, pentominoes, ... *Journal of Information Processing*, 25:515–527, 2017. 157

[DDO00] Erik D. Demaine, Martin L. Demaine, and Joseph O'Rourke. PushPush and Push-1 are NP-hard in 2D. In Christian Sohler and Martin Ziegler, editors, *Proceedings of the 12th Canadian Conference on Computational Geometry*, Fredericton, New Brunswick, Canada, August 2000. 54, 58

[DDST16] Benjamin Doerr, Carola Doerr, Reto Spöhel, and Henning Thomas. Playing Mastermind with many colors. *Journal of the Association of Computing Machinery (JACM)*, 63(5):42:1–42:23, 2016. https://doi.org/10.1145/2987372. 49

[DdW11] Andrew Drucker and Ronald de Wolf. Quantum proofs for classical theorems. *Theory of Computing*, 2:1–54, 2011. 470

[DE11] Erik D. Demaine and Sarah Eisenstat. Flattening fixed-angle chains is strongly NP-hard. In Frank Dehne, John Iacono, and Jörg-Rüdiger Sack, editors, *Proceedings of the 12th International Workshop on Algorithms and Data Structures (WADS)*, volume 6844 of *Lecture Notes in Computer Science*, pages 314–325, New York, NY, August 2011. Springer. 92

[Der10] Dariusz Dereniowski. Phutball is PSPACE-hard. *Theoretical Computer Science*, 411(44–46): 3971–3978, 2010. https://doi.org/10.1016/j.tcs.2010.08.019. 62

[Der14] Franck Dernoncourt. TrackMania is NP-complete, 2014. https://arxiv.org/abs/1411.5765. 64

[DF86] Martin E. Dyer and Alan M. Frieze. Planar 3DM is NP-complete. *Journal of Algorithms*, 7(2):174–184, 1986. https://www.math.cmu.edu/~af1p/Texfiles/3DM.pdf. 86, 87

[DF99] Rodney G. Downey and Michael R. Fellows. *Parameterized Complexity*. Monographs in Computer Science. Springer, 1999. 171, 186, 187, 188, 199

[DFHT05] Erik D. Demaine, Fedor V. Fomin, Mohammad Taghi Hajiaghayi, and Dimitrios M. Thilikos. Subexponential parameterized algorithms on bounded-genus graphs and H-minor-free graphs. *Journal of the Association of Computing Machinery (JACM)*, 52(6):866–893, 2005. https://doi.org/10.1145/1101821.1101823. 166, 173, 200

[DFL10] Erik D. Demaine, Sándor P. Fekete, and Robert J. Lang. Circle packing for origami design is hard. In Patsy Wang-Iverson, Robert Lang, and Mark Yim, *Origami^5: Proceedings of the 5th International Conference on Origami in Science, Mathematics and Education*, pages 609–626. A K Peters, Singapore, July 2010. http://arxiv.org/abs/1008.1224. 144

[DFR98] Rodney G. Downey, Michael R. Fellows, and Kenneth W. Regan. Parameterized circuit complexity and the W hierarchy. *Theoretical Computer Science*, 191(1–2):97–115, 1998. https://doi.org/10.1016/S0304-3975(96)00317-9. 188

[DGH+02] Evgeny Dantsin, Andreas Goerdt, Edward A. Hirsch, Ravi Kannan, Jon M. Kleinberg, Christos H. Papadimitriou, Prabhakar Raghavan, and Uwe Schöning. A deterministic $(2 - \frac{2}{k+1})^n$ algorithm for k-SAT based on local search. *Theoretical Computer Science*, 289(1):69–83, 2002. https://doi.org/10.1016/S0304-3975(01)00174-8. 160

[DGIM02] Mayur Datar, Aristides Gionis, Piotr Indyk, and Rajeev Motwani. Maintaining stream statistics over sliding windows. *SIAM Journal on Computing*, 31(6):1794–1813, 2002. https://doi.org/10.1137/S0097539701398363. 419

[DGKR05] Irit Dinur, Venkatesan Guruswami, Subhash Khot, and Oded Regev. A new multilayered PCP and the hardness of hypergraph vertex cover. *SIAM Journal on Computing*, 34(5) 1129–1146, 2005. https://doi.org/10.1137/S0097539704443057. 256, 260

[DGP09] Constantinos Daskalakis, Paul W. Goldberg, and Christos H. Papadimitriou. The complexity of computing a Nash equilibrium. *SIAM Journal of Computing*, 39(1):195–259, 2009. https://doi.org/10.1137/070699652. 456

[DH01] Erik D. Demaine and Michael Hoffmann. Pushing blocks is NP-complete for noncrossing solution paths. In Therese Biedl, editor, *Proceedings of the 13th Canadian Conference on*

Computational Geometry, University of Waterloo, Ontario, Canada, August 13–15, 2001, pages 65–68, 2001. http://www.cccg.ca/proceedings/2001/eddemaine-24711.ps.gz. 56

[DH08] Erik D. Demaine and MohammadTaghi Hajiaghayi. The bidimensionality theory and its algorithmic applications. *The Computer Journal*, 51(3):292–302, 2008. 166

[DHH02] Erik D. Demaine, Robert A. Hearn, and Michael Hoffmann. Push-2-F is PSPACE-complete. In Stephen Wismath, editor, *Proceedings of the 14th Canadian Conference on Computational Geometry*, pages 31–35, Lethbridge, Alberta, Canada, August 2002. http://www.cccg.ca/proceedings/2002/31.ps. 54, 56

[DHH04] Erik D. Demaine, Michael Hoffmann, and Markus Holzer. PushPush-k is PSPACE-complete. In *Proceedings of the 3rd International Conference on Fun with Algorithms*, pages 159–170, 2004. 56

[DHHL22] Erik D. Demaine, Robert A. Hearn, Della Hendrickson, and Jayson Lynch. PSPACE-completeness of reversible deterministic systems, August–September 2022. 346, 347

[DHL20] Erik D. Demaine, Della H. Hendrickson, and Jayson Lynch. Toward a general complexity theory of motion planning: Characterizing which gadgets make games hard. In Thomas Vidick, editor, *11th Innovations in Theoretical Computer Science Conference, ITCS 2020*, January 12–14, 2020, Seattle, Washington, USA, volume 151 of *LIPIcs*, pages 62:1–62:42. Schloss Dagstuhl - Leibniz-Zentrum für Informatik, 2020. https://doi.org/10.4230/LIPIcs.ITCS.2020.62. 350

[DHT04] Erik D. Demaine, Mohammad Taghi Hajiaghayi, and Dimitrios M. Thilikos. Bidimensional theory of bounded-genus graphs. *SiAM Journal of Discrete Mathematics*, 20(2):357–371, 2006. https://doi.org/10.1137/040616929. 166

[DK15] Zdeněk Dvořák and Martin Kupec. On planar Boolean CSP. In Magnús M. Halldórsson, Kazuo Iwama, Naoki Kobayashi, and Bettina Speckmann, editors, *Proceedings of the 42nd International Colloquium on Automata, Languages, and Programming*, pages 432–443, Kyoto, Japan, July 2015. 96, 97

[DKK+18a] Irit Dinur, Subhash Khot, Guy Kindler, Dor Minzer, and Muli Safra. On non-optimally expanding sets in Grassmann graphs. In Ilias Diakonikolas, David Kempe, and Monika Henzinger, editors, *Proceedings of the 50th Annual ACM SIGACT Symposium on Theory of Computing, STOC 2018*, Los Angeles, CA, USA, June 25–29, 2018, pages 940–951. ACM, 2018. https://eccc.weizmann.ac.il/report/2017/094/. 223, 254

[DKK+18b] Irit Dinur, Subhash Khot, Guy Kindler, Dor Minzer, and Muli Safra. Towards a proof of the 2-to-1 games conjecture? In Ilias Diakonikolas, David Kempe, and Monika Henzinger, editors, *Proceedings of the 50th Annual ACM SIGACT Symposium on Theory of Computing, STOC 2018*, Los Angeles, CA, USA, June 25–29, 2018, pages 376–389. ACM, 2018. https://doi.org/10.1145/3188745.3188804. 223, 254

[DKL20] Erik D. Demaine, Justin Kopinsky, and Jayson Lynch. Recursed is not recursive: A jarring result. In Yixin Cao, Siu-Wing Cheng, and Minming Li, editors, *Proceedings of the 31st International Symposium on Algorithms and Computation (ISAAC 2020)*, volume 181 of *LIPIcs*, pages 50:1–50:15, Hong Kong, China (Virtual Conference), December 2020. Schloss Dagstuhl - Leibniz-Zentrum für Informatik. https://doi.org/10.4230/LIPIcs.ISAAC.2020.50. 334

[DKP25] Erik D. Demaine, Kritkorn Karntikoon, and Nipun Pitimanaaree. 2-colorable perfect matching is NP-complete in 2-connected 3-regular planar graphs. *Theory of Computing Systems*, 69:Article 22, April 2025. https://arxiv.org/abs/2309.09786. 50

[DKS18] Debarati Das, Michal Koucký, and Michael E. Saks. Lower bounds for combinatorial algorithms for boolean matrix multiplication. In Rolf Niedermeier and Brigitte Vallée, editors, *35th Symposium on Theoretical Aspects of Computer Science, STACS 2018*, February 28 to March 3, 2018, Caen, France, volume 96 of *LIPIcs*, pages 23:1–23:14. Schloss Dagstuhl - Leibniz-Zentrum für Informatik, 2018. https://doi.org/10.4230/LIPIcs.STACS.2018.23. 389

[DKS19] Yuval Dagan, Gil Kur, and Ohad Shamir. Space lower bounds for linear prediction in the streaming model. In Alina Beygelzimer and Daniel Hsu, editors, *Conference on Learning Theory, COLT 2019*, 25–28 June 2019, Phoenix, AZ, USA, volume 99 of *Proceedings of Machine Learning Research*, pages 929–954. PMLR, 2019. http://proceedings.mlr.press/v99/dagan19b.html. 419

[DL25] Erik D. Demaine and Stefan Langerman. Tiling with three polygons is undecidable. In *Proceedings of the 41st International Symposium on Computational Geometry (SoCG 2025)*, pages 39:1–39:16, Kanazawa, Japan, June 2025. https://arxiv.org/abs/2409.11582. 337

[DLL+02] Xiaotie Deng, Guojun Li, Zimao Li, Bin Ma, and Lusheng Wang. A PTAS for distinguishing (sub)string selection. In Peter Widmayer, Francisco Triguero Ruiz, Rafael Morales Bueno, Matthew Hennessy, Stephan J. Eidenbenz, and Ricardo Conejo, editors, *Automata, Languages and Programming, 29th International Colloquium, ICALP 2002, Malaga, Spain, July 8–13, 2002, Proceedings*, volume 2380 of *Lecture Notes in Computer Science*, pages 740–751. Springer, 2002. 189, 190

[DM99] Mauro Dell'Amico and Silvano Martello. Reduction from the three-partition problem. *Journal of Combinatorial Optimization*, 3(1):17–30, 1999. https://doi.org/10.1023/A:1009856820553. 157

[dM21] Carl de Marcken. Computational complexity of air travel planning (slides), 2021. http://www.ai.mit.edu/courses/6.034f/psets/ps1/airtravel.pdf. 337

[dMP13] Leonardo Mendonça de Moura and Grant Olney Passmore. Computation in real closed infinitesimal and transcendental extensions of the rationals. In Maria Paola Bonacina, editor, *Proceedings of the 24th International Conference on Automated Deduction (CADE-24)*, volume 7898 of *Lecture Notes in Computer Science*, pages 178–192, Lake Placid, NY, June 2013. Springer. https://www.cl.cam.ac.uk/~gp351/infinitesimals.pdf. 293

[DMR09] Irit Dinur, Elchanan Mossel, and Oded Regev. Conditional hardness for approximate coloring. *SIAM Journal on Computing*, 39(3):843–873, 2009. https://doi.org/10.1137/07068062X. 263

[DMS+18] Erik D. Demaine, Fermi Ma, Ariel Schvartzman, Erik Waingarten, and Scott Aaronson. The fewest clues problem. *Theoretical Computer Science*, 748:28–39, 2018. https://doi.org/10.1016/j.tcs.2018.01.020. 286, 287

[DOUU14] Erik D. Demaine, Yoshio Okamoto, Ryuhei Uehara, and Yushi Uno. Computational complexity and an integer programming model of Shakashaka. *IEICE Transactions*, 97-A(6):1213–1219, 2014. https://dspace.mit.edu/handle/1721.1/99994. 83

[DPR61] Martin Davis, Hillary Putnam, and Julia Robinson. The decision problem for exponential Diophantine equations. *Annal of Mathematics*, 74:425–436, 1961. 335

[DQS12] Xiaotie Deng, Qi Qi, and Amin Saberi. Algorithmic solutions for envy-free cake cutting. *Operations Research*, 60(6):1461–1476, 2012. https://doi.org/10.1287/opre.1120.1116. 455

[DR17] Erik D. Demaine and Mikhail Rudoy. Hamiltonicity is hard in thin or polygonal grid graphs, but easy in thin polygonal grid graphs. arXiv:1706.10046, 2017. https://arxiv.org/abs/1706.10046. 120, 122

[DS02] Irit Dinur and Shmuel Safra. The importance of being biased. In John H. Reif, editor, *Proceedings on 34th Annual ACM Symposium on Theory of Computing*, May 19–21, 2002, Montréal, Québec, Canada, pages 33–42. ACM, 2002. https://doi.org/10.1145/509907.509915. 256

[DS05] Irit Dinur and Samuel Safra. On the hardness of approximating minimum vertex cover. *Annals of Mathematics*, pages 439–485, 2005. http://www.wisdom.weizmann.ac.il/~dinuri/mypapers/vc.pdf. 223

[DS13] Irit Dinur and David Steurer. Analytical approach to parallel repetition. In David Shmoys, editor, *Proceedings of the 46th Annual ACM Sympsosium on Theory of Computing*, pages 624–633, 2013. https://dl.acm.org/doi/pdf/10.1145/2591796.2591884. 12, 216, 256

[Dub90] Olivier Dubois. On the r, s-SAT satisfiability problem and a conjecture of Tovey. *Discrete Applied Mathematics*, 26(1):51–60, 1990. 39

[Dud24] H. E. Dudeney. SEND + MORE = MONEY. *Strand Magazine*, 68:97 and 214, July 1924. 50

[DvM14] Holger Dell and Dieter van Melkebeek. Satisfiability allows no nontrivial sparsification unless the polynomial-time hierarchy collapses. *Journal of the Association of Computing Machinery (JACM)*, 61(4):23:1–23:27, 2014. https://doi.org/10.1145/2629620. 175

[DVW16] Erik D. Demaine, Giovanni Viglietta, and Aaron Williams. Super Mario Bros. is harder/easier than we thought. In Erik Demaine and Fabrizio Grandoni, editors, *Proceedings of the 8th International Conference on Fun with Algorithms*, pages 13:1–13:14, 2016. 59, 60, 61

[Edm65] Jack Edmonds. Paths, trees, and flowers. *Canadian Journal of Mathematics*, 17:449–467, 1965. xiv

[EG04] Friedrich Eisenbrand and Fabrizio Grandoni. On the complexity of fixed parameter clique and dominating set. *Theoretical Computer Science*, 326(1–3):57–67, 2004. https://doi.org/10.1016/j.tcs.2004.05.009. 198, 389

[EHL+18] Hossein Esfandiari, MohammadTaghi Hajiaghayi, Vahid Liaghat, Morteza Monemizadeh, and Krzysztof Onak. Streaming algorithms for estimating the matching size in planar graphs and beyond. *ACM Transactions on Algorithms*, 14(4):48:1–48:23, 2018. https://doi.org/10.1145/3230819. 417

[EHLM17] Hossein Esfandiari, MohammadTaghi Hajiaghayi, Vahid Liaghat, and Morteza Monemizadeh. Prophet secretary. *SIAM Journal on Discrete Mathematics*, 31(3):1685–1701, 2017. https://arxiv.org/abs/1507.01155. 394, 396

[EJ64] Calrton E. Lemke and J. T. Howson. Equilibrium points of bimatrix games. *SIAM Journal on Applied Mathematics*, 12:412–423, 1964. https://epubs.siam.org/doi/pdf/10.1137/0112033. 447

[ENSS98] Guy Even, Joseph Naor, Baruch Schieber, and Madhu Sudan. Approximating minimum feedback sets and multicuts in directed graphs. *Algorithmica*, 20(2):151–174, 1998. https://doi.org/10.1007/PL00009191. 262

[EP06] Bruno Escoffier and Vangelis Th. Paschos. Completeness in approximation classes beyond APX. *Theoretical Computer Science*, 359(1–3):369–377, 2006. http://dx.doi.org/10.1016/j.tcs.2006.05.023. 240, 246

[Epp87] David Eppstein. On the NP-completeness of cryptarithms. *SIGACT News*, 18(3):38–40, 1987. http://www.ics.uci.edu/~eppstein/pubs/p-cryptarithm.html. 51

[EvdHM24] Jeff Erickson, Ivor van der Hoog, and Tillmann Miltzow. Smoothing the gap between NP and ER. *SIAM Journal on Computing*, 53(6):FOCS20:102–FOCS20:138, 2024. 294

[EY09] Kousha Etessami and Mihalis Yannakakis. Recursive Markov chains, stochastic grammars, and monotone systems of nonlinear equations. *Journal of the ACM*, 56(1):1:1–1:66, 2009. https://doi.org/10.1145/1462153.1462154. 301

[FB02] Gary William Flake and Eric B. Baum. *Rush Hour* is PSPACE-complete, or "Why you should generously tip parking lot attendants." *Theoretical Computer Science*, 270(1–2):895–911, 2002. https://doi.org/10.1016/S0304-3975(01)00173-6. 342

[Fei98] Uriel Feige. A threshold of $\ln n$ for approximating set cover. *Journal of the Association of Computing Machinery (JACM)*, 45:634–652, 1998. https://dl.acm.org/doi/10.1145/285055.285059.

[Fei04] Uriel Feige. Approximating maximum clique by removing subgraphs. *SIAM Journal on Discrete Mathematics*, 18(2):219–225, 2004. https://epubs.siam.org/doi/pdf/10.1137/S089548010240415X. 215

[FG06] Jörg Flum and Martin Grohe. *Parameterized Complexity Theory*. Texts in Theoretical Computer Science. An EATCS Series. Springer, 2006. 171, 175, 199

[FG18] Aris Filos-Ratsikas and Paul W. Goldberg. Consensus halving is PPA-complete. In Ilias Diakonikolas, David Kempe, and Monika Henzinger, editors, *Proceedings of the 50th Annual ACM SIGACT Symposium on Theory of Computing, STOC 2018*, Los Angeles, CA, USA, June 25–29, 2018, pages 51–64. ACM, 2018. https://arxiv.org/abs/1711.04503. 465

[FGJ+78] A. S. Fraenkel, M. R. Garey, D. S. Johnson, T. Schaefer, and Y. Yesha. The complexity of checkers on an $n \times n$ board. In *19th Annual Symposium on Foundations of Computer Science, 1978*, pages 55–64. IEEE Computer Society, 1978. 325

[FGL+96] Uriel Feige, Shafi Goldwasser, László Lovász, Shmuel Safra, and Mario Szegedy. Interactive proofs and the hardness of approximating cliques. *Journal of the ACM*, 43(2):268–292, 1996. https://doi.org/10.1145/226643.226652. 214

[FGLS10] Fedor V. Fomin, Petr A. Golovach, Daniel Lokshtanov, and Saket Saurabh. Algorithmic lower bounds for problems parameterized with clique-width In Moses Charikar, editor, *Proceedings of the Twenty-First Annual ACM-SIAM Symposium on Discrete Algorithms, SODA 2010*, Austin, Texas, USA, January 17–19, 2010, pages 493–502. SIAM, 2010. https://doi.org/10.1137/1.9781611973075.42. 199

[FHM+20] Alireza Farhadi, Mohammad Taghi Hajiaghayi, Tung Mai, Anup Rao, and Ryan A. Rossi. Approximate maximum matching in random streams. In Shuchi Chawla, editor, *Proceedings of the 2020 ACM-SIAM Symposium on Discrete Algorithms, SODA 2020*, Salt Lake City, UT, USA, January 5–8, 2020, pages 1773–1785. SIAM, 2020. https://arxiv.org/abs/1912.10497. 408

[FHRV09] Michael R. Fellows, Danny Hermelin, Frances A. Rosamond, and Stéphane Vialette. On the parameterized complexity of multiple-interval graph problems. *Theoretical Computer Science*, 410(1):53–61, 2009. https://doi.org/10.1016/j.tcs.2008.09.065. 181

[Fil19] Ivan Tadeu Ferreira Antunes Filho. Characterizing Boolean satisfiability variants. Master's thesis, Massachusetts Institute of Technology, 2019. http://erikdemaine.org/theses/ifilho.pdf. 36, 97

[Fin18] Jeremy T. Fineman. Nearly work-efficient parallel algorithm for digraph reachability. In Ilias Diakonikolas, David Kempe, and Monika Henzinger, editors, *Proceedings of the 50th Annual ACM SIGACT Symposium on Theory of Computing, STOC 2018*, Los Angeles, CA, USA, June 25–29, 2018, pages 457–470. ACM, 2018. https://arxiv.org/abs/1711.01700. 440

[FKL+91] Amos Fiat, Richard M. Karp, Michael Luby, Lyle A. McGeoch, Daniel Dominic Sleator, and Neal E. Young. Competitive paging algorithms. *Journal of Algorithms*, 12(4):685–699, 1991. www.cs.cmu.edu/~sleator/papers/competitive-paging.pdf. 399

[FKMP95] M. R. Fellows, J. Kratochvíl, M. Middendorf, and F. Pfeiffer. The complexity of induced minors and related problems. *Algorithmica*, 13(3):266–282, 1995. 73, 74

[FL81] Avierzi S. Fraenkel and David Lichtenstein. Computing a perfect strategy for $n \times n$ chess requires time exponential in n. *Journal of Combinatorial Theory (Series A)*, 31:199–214, 1981. http://www.ms.mff.cuni.cz/~truno7am/slozitostHer/chessExptime.pdf. 328

[Flo62] Robert Floyd. Algorithm 97: Shortest path. *Communications of the ACM*, 5:345, 1962. 378

[For10] Michal Forišek. Computational complexity of two-dimensional platform games. In Paolo Boldi and Luisa Gargano, editors, *Fun with Algorithms*, pages 214–227. Springer, 2010. https://doi.org/10.1007/978-3-642-13122-6_22. 124

[Fre76] Michael Fredmann. New bounds on the complexity of the shortest path problem. *SIAM Journal on Computing*, 5(1):83–89, 1976. https://epubs.siam.org/doi/pdf/10.1137/0205006. 379

[Fri02] Erich Friedman. Spiral galaxies puzzles are NP-complete, 2002. https://erich-friedman.github.io/papers/spiral.pdf. 102

[FSTZ98] Charles M. Fiduccia, Edward R. Scheinerman, Ann N. Trenk, and Jennifer S. Zito. Dot product representations of graphs. *Discrete Mathematics*, 181(1–3):113–138, 1998. https://doi.org/10.1016/S0012-365X(97)00049-6. 297

[FT82] Edward Fredkin and Tommaso Toffoli. Conservative logic. *International Journal of Theoretical Physics*, pages 219–253, 1982. https://link.springer.com/article/10.1007/BF01857727. 347

[FW90] Michael Formann and Frank Wagner. The VLSI layout in various embedding models. In Rolf H. Möhring, editor, *Graph-Theoretic Concepts in Computer Science, 16th International Workshop, WG '90*, Berlin, Germany, June 20–22, 1990, Proceedings, volume 484 of *Lecture Notes in Computer Science*, pages 130–139. Springer, 1990. 157

[Gal09] François Le Gall. Exponential separation of quantum and classical online space complexity. *Theory of Computing Systems*, 45(2):188–202, 2009. https://arxiv.org/pdf/quant-ph/0606066.pdf. 480

[Gar20] Daniel Garisto. Light-based quantum computer exceeds fastest classical supercomputers. *Scientific American*, December 3, 2020. 486

[Gas09a] William Gasarch. Whats your game Mr. Bond? Nim?, 2009. https://blog.computationalcomplexity.org/2009/12/whats-your-game-mr-bond-nim.html. 255

[Gas09b] William Gasarch. Whats your game Mr. Bond?-The sequel!, 2009. https://blog.computationalcomplexity.org/2010/06/whats-your-game-mr-bond-sequel.html. 255

[Gas19] William Gasarch. Complexity theory column 100: The P=NP poll. *SIGACT News*, 50(1):28–34, 2019. https://www.cs.umd.edu/users/gasarch/papers/poll3.pdf. 11

[Gas21] William Gasarch. Hilbert's tenth problem: Refinements and variations, 2021. https://arxiv.org/abs/2104.07220. 336

[GB07] Alexander Grigoriev and Hans L. Bodlaender. Algorithms for graphs embeddable with few crossings per edge. *Algorithmica*, 49(1):1–11, 2007. https://doi.org/10.1007/s00453-007-0010-x. 138

[GGJ76] M. R. Garey, Ronald L. Graham, and David S. Johnson. Some NP-complete geometric problems. In Ashok K. Chandra, Detlef Wotschke, Emily P. Friedman, and Michael A. Harrison, editors, *Proceedings of the 8th Annual ACM Symposium on Theory of Computing*, May 3–5, 1976, Hershey, Pennsylvania, USA, pages 10–22. ACM, 1976. https://doi.org/10.1145/800113.803626. 301

[GGJ20] Mohsen Ghaffari, Christoph Grunau, and Ce Jin. Improved MPC algorithms for MIS, matching, and coloring on trees and beyond. In Hagit Attiya, editor, *34th International Symposium on Distributed Computing, DISC 2020*, October 12–16, 2020, Virtual Conference, volume 179 of *LIPIcs*, pages 34:1–34:18. Schloss Dagstuhl - Leibniz-Zentrum für Informatik, 2020. https://doi.org/10.4230/LIPIcs.DISC.2020.34. 427

[GGN06] Jens Gramm, Jiong Guo, and Rolf Niedermeier. Parameterized intractability of distinguishing substring selection. *Theory of Computing Systems*, 39(4):545–560, 2006. https://link.springer.com/content/pdf/10.1007/s00224-004-1185-z.pdf. 191

[GHM+11] Venkatesan Guruswami, Johan Håstad, Rajsekar Manokaran, Prasad Raghavendra, and Moses Charikar. Beating the random ordering is hard: Every ordering CSP is approximation resistant. *SIAM Journal on Computing*, 40(3):878–914, 2011. https://doi.org/10.1137/090756144. 260, 262

[GHOP22] William Gasarch, Nathan Hayes, Anthony Ostuni, and Davin Park. Open Problems Column 100: The complexit of chromatic number when restricted to graphs with bounded genus or bounded crossing number. *SIGACT News*, 53(2):27–38, 2022. https://www.cs.umd.edu/~gasarch/open/cross.pdf. 70

[GHR95] Raymond Greenlaw, James Hoover, and Walter Ruzzo. *Limits to Parallel Computation: P-Completeness Theory*. Oxford University Press, New York, 1995. https://homes.cs.washington.edu/~ruzzo/papers/limits.pdf. 422

[GHS02] Venkatesan Guruswami, Johan Håstad, and Madhu Sudan. Hardness of approximate hypergraph coloring. *SIAM Journal on Computing*, 31(6):1663–1686, 2002. https://doi.org/10.1137/S0097539700377165. 256

[GJ79] Michael R. Garey and David S. Johnson. *Computers and Intractability: A Guide to the Theory of NP-Completeness*. W. H. Freeman, New York, 1979. xiv, 79, 91, 131, 134, 135

[GJS76] M. R. Garey, David S. Johnson, and Larry J. Stockmeyer. Some simplified *NP*-complete graph problems. *Theoretical Computer Science*, 1(3):237–267, 1976. https://www.sciencedirect.com/science/article/pii/0304397576900591. 66, 68, 94

[GJS21] Michael T. Goodrich, Riko Jacob, and Nodari Sitchinava. Atomic power in forks: A super-logarithmic lower bound for implementing butterfly networks in the nonatomic binary fork-join model. In Dániel Marx, editor, *Proceedings of the 2021 ACM-SIAM Symposium on Discrete Algorithms, SODA 2021, Virtual Conference, January 10–13, 2021*, pages 2141–2153. SIAM, 2021. https://epubs.siam.org/doi/pdf/10.1137/1.9781611976465.128. 422

[GJT76] Michael R. Garey, David S. Johnson, and Robert E. Tarjan. The planar Hamiltonian circuit problem is NP-complete. *SIAM Journal on Computing*, 5(4):704–714, 1976. http://dx.doi.org/10.1137/0205049. 112

[GKC+21] Xun Gao, Marcin Kalinowski, Chi-Ning Chou, Mikhail Lukin, Boaz Barak, and Soonwon Choi. Limitations of linear cross-entropy as a measure for quantum advantage, 2021. https://arxiv.org/abs/2112.01657. 486

[GKK+08] Dmitry Gavinsky, Julia Kempe, Iordanis Kerenidis, Ran Raz, and Ronald de Wolf. Exponential separation for one-way quantum communication complexity, with applications to cryptography. *SIAM Journal of Computing*, 38(5):1695–1708, 2008. https://doi.org/10.1137/070706550. 477, 478, 480

[GKK12] Ashish Goel, Michael Kapralov, and Sanjeev Khanna. On the communication and streaming complexity of maximum bipartite matching. In Yuval Rabani, editor, *Proceedings of the Twenty-Third Annual ACM-SIAM Symposium on Discrete Algorithms, SODA 2012*, Kyoto, Japan, January 17–19, 2012, pages 468–485. SIAM, 2012. https://doi.org/10.1137/1.9781611973099.41. 418

[GKM+12] Rohit Gurjar, Arpita Korwar, Jochen Messner, Simon Straub, and Thomas Thierauf. Planarizing gadgets for perfect matching do not exist. In Branislav Rovan, Vladimiro Sassone, and Peter Widmayer, editors, *Mathematical Foundations of Computer Science 2012 - 37th International Symposium, MFCS 2012*, Bratislava, Slovakia, August 27–31, 2012. Proceedings, volume 7464 of *Lecture Notes in Computer Science*, pages 478–490. Springer, 2012. https://dl.acm.org/doi/10.1145/2934310. 65

[GKM+19] Buddhima Gamlath, Michael Kapralov, Andreas Maggiori, Ola Svensson, and David Wajc. Online matching with general arrivals. In David Zuckerman, editor, *60th IEEE Annual Symposium on Foundations of Computer Science, FOCS 2019*, Baltimore, Maryland, USA, November 9–12, 2019, pages 26–37. IEEE Computer Society, 2019. https://doi.org/10.1109/FOCS.2019.00011. 404

[GKMP09] Parikshit Gopalan, Phokion G. Kolaitis, Elitza N. Maneva, and Christos H. Papadimitriou. The connectivity of Boolean satisfiability: Computational and structural dichotomies. *SIAM Journal on Computing*, 38(6):2330–2355, 2009. https://doi.org/10.1137/07070440X. 345

[GKR00] Naveen Garg, Goran Konjevod, and R. Ravi. A polylogarithmic approximation algorithm for the group Steiner tree problem. *Journal of Algorithms*, 37(1):66–84, 2000. https://doi.org/10.1006/jagm.2000.1096. 244

[GKSZ20] Mika Göös, Pritish Kamath, Katerina Sotiraki, and Manolis Zampetakis. On the complexity of modulo-q arguments and the Chevalley-Warning theorem. In Shubhangi Saraf, editor, *35th Computational Complexity Conference, CCC 2020*, July 28–31, 2020, Saarbrücken, Germany (Virtual Conference), volume 169 of *LIPIcs*, pages 19:1–19:42. Schloss Dagstuhl - Leibniz-Zentrum für Informatik, 2020. https://doi.org/10.4230/LIPIcs.CCC.2020.19. 464, 465

[GKU19] Mohsen Ghaffari, Fabian Kuhn, and Jara Uitto. Conditional hardness results for massively parallel computation from distributed lower bounds. In David Zuckerman, editor, *60th IEEE Annual Symposium on Foundations of Computer Science, FOCS 2019*, Baltimore, Maryland, USA, November 9–12, 2019, pages 1650–1663. IEEE Computer Society, 2019. https://doi.org/10.1109/FOCS.2019.00097. 427, 431

[GLS81] Martin Grötschel, László Lovász, and Alexander Schrijver. The ellipsoid method and its consequences in combinatorial optimization. *Combinatorica*, 1(2):169–197, 1981. https://doi.org/10.1007/BF02579273. 170

[GMVY18] Jugal Garg, Ruta Mehta, Vijay V. Vazirani, and Sadra Yazdanbod. $\exists R$-completeness for decision versions of multi-player (symmetric) nash equilibria. *ACM Transactions on Economics and Computation*, 6(1):1:1–1:23, 2018. https://doi.org/10.1145/3175494. 300

[GO12] Anka Gajentaan and Mark H. Overmars. On a class of $O(n^2)$ problems in computational geometry. *Computational Geometry*, 45(4):140–152, 2012. https://doi.org/10.1016/j.comgeo.2011.11.006. 355, 357, 360, 370

[Goe97] Michel X. Goemans. Semidefinite programming in combinatorial optimization. *Mathematical Programming*, 79:143–161, 1997. https://doi.org/10.1007/BF02614315. 267

[Gol70] Solomon W. Golomb. Tiling with sets of polyominoes. *Journal of Combinatorial Theory*, 9(1): 60–71, 1970. 151, 337

[Gol78] E. Mark Gold. Complexity of automaton identification from given data. *Information and Computation*, 37:302–320, 1978. https://www.sciencedirect.com/science/article/pii/S0019995878905624. 40

[Goo09] Michael T. Goodrich. On the algorithmic complexity of the Mastermind game with black-peg results. *Information Processing Letters*, 109(13):675–678, 2009. https://doi.org/10.1016/j.ipl.2009.02.021. 49

[GP18] Allan Grønlund and Seth Pettie. Threesomes, degenerates, and love triangles. *Journal of the Association of Computing Machinery (JACM)*, 65(4):22:1–22:25, 2018. https://doi.org/10.1145/3185378. 357

[GPS89] Jacob E. Goodman, Richard Pollack, and Bernd Sturmfels. Coordinate representation of order types requires exponential storage. In David S. Johnson, editor, *Proceedings of the 21st Annual*

ACM Symposium on Theory of Computing, May 14–17, 1989, Seattle, Washington, USA, pages 405–410. ACM, 1989. https://doi.org/10.1145/73007.73046. 298

[Gre68] Sheila A. Greibach. A note on undecidable properties of formal languages. *Mathematical Systems Theory*, 2(1):1–6, 1968. https://doi.org/10.1007/BF01691341. 337

[Gro96] Lov K. Grover. A fast quantum mechanical algorithm for database search. In Gary L. Miller, editor, *Proceedings of the Twenty-Eighth Annual ACM Symposium on the Theory of Computing*, Philadelphia, Pennsylvania, USA, May 22–24, 1996, pages 212–219. ACM, 1996. https://doi.org/10.1145/237814.237866. 473

[GST16] Heidi Gebauer, Tibor Szabó, and Gábor Tardos. The local lemma is asymptotically tight for SAT. *Journal of the ACM*, 63(5), December 2016. 39

[GSZ11] Michael T. Goodrich, Nodari Sitchinava, and Qin Zhang. Sorting, searching, and simulation in the mapreduce framework. In Takao Asano, Shin-Ichi Nakano, Yoshio Okamoto, and Osamu Watanabe, editors, *Proceedings of the 22nd International Symposium on Algorithms and Computation (ISAAC 2011)*, volume 7074 of *Lecture Notes in Computer Science*, pages 374–383, Yokohama, Japan, December 2011. Springer. https://arxiv.org/abs/1101.1902. 424

[GU19] Mohsen Ghaffari and Jara Uitto. Sparsifying distributed algorithms with ramifications in massively parallel computation and centralized local computation. In Timothy M. Chan, editor, *Proceedings of the Thirtieth Annual ACM-SIAM Symposium on Discrete Algorithms, SODA 2019*, San Diego, California, USA, January 6–9, 2019, pages 1636–1653. SIAM, 2019. https://doi.org/10.1137/1.9781611975482.99. 427

[GW95] Michel X. Goemans and David P. Williamson. Improved approximation algorithms for maximum cut and satisfiability problems using semidefinite programming. *Journal of the ACM*, 42(6):1115–1145, 1995. https://dl.acm.org/doi/pdf/10.1145/227683.227684. 260, 267

[Had75] F. Hadlock. Finding a maximum cut of a planar graph in polynomial time. *SIAM Journal on Computing*, 4(3):221–225, 1975. https://web.eecs.umich.edu/~pettie/matching/Hadlock-maxcut-planar-graph-reduction-to-T-join.pdf. 94

[Hal35] P. Hall. On representatives of subsets. *Journal of the London Mathematical Society*, 10(1):26–30, 1935. 38

[Ham14] Linus Hamilton. Braid is undecidable, 2014. https://arxiv.org/pdf/1412.0784.pdf. 335

[Has99] Johan Hastad. Clique is hard to approximate within $n^{1-\varepsilon}$. *Acta Mathematica*, 1999. https://www.csc.kth.se/~johanh/cliqueinap.pdf. 216, 256

[Has01] Johan Hastad. Some optimal inapproximability results. *Journal of the Association of Computing Machinery (JACM)*, 48(4):798–859, 2001. https://dl.acm.org/doi/10.1145/502090.502098. 220, 223, 239, 254, 257, 260

[HD09] Robert A. Hearn and Erik D. Demaine. *Game, Puzzles, and Computation*. A K Peters/CRC Press, 1st edition, July 2009. 320, 339, 340, 341, 342, 343, 344, 345, 346, 348, 350

[Hea06] Robert A. Hearn. *Games, Puzzles, and Computation*. PhD thesis, Massachusetts Institute of Technology, Cambridge, MA, USA, 2006. http://hdl.handle.net/1721.1/37913. 321, 339, 345

[Hea09] Robert A. Hearn. Amazons, Konane, and Cross Purposes are PSPACE-complete. *Games of No Chance 3*, 56:287–306, 2009. http://library.msri.org/books/Book56/files/31hearn.pdf. 345, 348

[Hem04] Edith Hemaspaandra. Dichotomy theorems for alternation-bounded quantified Boolean formulas, 2004. http://arxiv.org/abs/cs/0406006. 304

[Her14] Timon Hertli. 3-SAT faster and simpler - Unique-SAT bounds for PPSZ hold in general. *SIAM Journal on Computing*, 43(2):718–729, 2014. https://arxiv.org/abs/1103.2165. 160

[HHHK05] Vesa Halava, Tero Harju, Mika Hirvensalo, and Juhani Karhumaki. Skolem's problem—on the border between decidable and undecidable. Technical Report 683, Turku Centre for Computer Science, 2005. http://oldtucs.abo.fi/publications/view/?pub_id=tHaHaHiKa05a. 330

[HIMRS98] Harry B. Hunt III, Madhav V. Marathe, Venkatesh Radhakrishnan, and Richard E. Stearns. The complexity of planar counting problems. *SIAM Journal on Computing*, 27(4):1142–1167, 1998. https://doi.org/10.1137/S0097539793304601. 278

[Hir22] Shuichi Hirahara. NP-hardness of learning programs and partial MCSP. In *63rd IEEE Annual Symposium on Foundations of Computer Science (FOCS)*. IEEE, 2022. https://eccc.weizmann.ac.il/report/2022/119/. 15

[HJ12] Markus Holzer and Sebastian Jakobi. On the complexity of rolling block and Alice mazes. In Evangelos Kranakis, Danny Krizanc, and Flaminia L. Luccio, editors, *Proceedings of the 6th International Conference on Fun with Algorithms (FUN 2012)*, volume 7288 of *Lecture Notes in Computer Science*, pages 210–222, Venice, Italy, June 2012. Springer. 345

[HKS+17] Craig Hamilton, Regine Kruse, Linda Sansoni, Sonja Barkhofer, Christine Silverhorn, and Igor Jex. Gaussian boson sampling. *Physical Review Letters*, 119(17):170501, 2017. https://arxiv.org/pdf/1612.01199.pdf. 486

[HN90] Pavol Hell and Jaroslav Nesetril. On the complexity of H-coloring. *Journal of Combinatorial Theory, Series B*, 48(1):92–110, 1990. https://doi.org/10.1016/0095-8956(90)90132-J. 71

[Hof00] Michael Hoffmann. Motion planning admidst movable square blocks: Push-$*$ is NP-hard. In *Proceedings of the 12th Canadian Conference on Computational Geometry*, Fredericton, New Brunswick, Canada, August 16–19, 2000. https://people.inf.ethz.ch/hoffmann/pub/h-psnh-00.pdf. 56

[Hop69] John E. Hopcroft. On the equivalence and containment problems for context-free languages. *Mathematical Systems Theory*, 3(2):119–124, 1969. https://doi.org/10.1007/BF01746517. 337

[Hor51] Alfred Horn. On sentences which are true of direct unions of algebras. *The Journal of Symbolic Logic*, 16(1):14–21, 1951. 40

[HR97] Magnús M. Halldórsson and Jaikumar Radhakrishnan. Greed is good: Approximating independent sets in sparse and bounded-degree graphs. *Algorithmica*, 18(1):145–163, 1997. https://doi.org/10.1007/BF02523693. 235

[HR12] Ishay Haviv and Oded Regev. Tensor-based hardness of the shortest vector problem to within almost polynomial factors. *Theory of Computing*, 8(1):513–531, 2012. https://doi.org/10.4086/toc.2012.v008a023. 249

[HSSS21] MohammadTaghi Hajiaghayi, Hamed Saleh, Saeed Seddighin, and Xiaorui Sun. String matching with wildcards in the massively parallel computation model. In Kunal Agrawal and Yossi Azar, editors, *SPAA '21: 33rd ACM Symposium on Parallelism in Algorithms and Architectures*, Virtual Event, USA, 6–8 July, 2021, pages 275–284. ACM, 2021. https://doi.org/10.1145/3409964.3461793. 427

[HT74] John E. Hopcroft and Robert Endre Tarjan. Efficient planarity testing. *Journal of the ACM*, 21(4):549–568, 1974. https://doi.org/10.1145/321850.321852. 72

[Imp03] Russell Impagliazzo. Hardness as randomness: A survey of universal derandomization, 2003. http://arxiv.org/abs/cs/0304040. 18

[INKT21] Fumitaka Ito, Masahiko Naito, Naoyuki Katabami, and Tatsuie Tsukiji. EXPTIME hardness of an n by n custodian capture game. *Algorithms*, 14(3):70, 2021. https://doi.org/10.3390/a14030070. 329

[IP01] Russell Impagliazzo and Ramamohan Paturi. On the complexity of k-SAT. *Journal of Computer and System Science*, 62(2):367–375, 2001. https://doi.org/10.1006/jcss.2000.1727. 160

[IP22] Christian Ikenmeyer and Igor Pak. What is in #P and what is not? In Jelani Nelson, editor, *Proceedings of the 2022 IEEE 63rd Annual Symposium on Foundations of Computer Science*, pages 860–871, Institute of Electrical and Electronics Engineering, October 2022. 276

[IPS82] Alon Itai, Christos H. Papadimitriou, and Jayme Luiz Szwarcfiter. Hamilton paths in grid graphs. *SIAM Journal on Computing*, 11(4):676–686, 1982. https://epubs.siam.org/doi/10.1137/0211056. 113, 116

[IPZ01] Russell Impagliazzo, Ramamohan Paturi, and Francis Zane. Which problems have strongly exponential complexity? *Journal of Computing and System Science*, 63(4):512–530, 2001. https://cseweb.ucsd.edu/~russell/ipz.pdf. 160, 161

[IW97] Russell Impagliazzo and Avi Wigderson. P=BPP if E requires exponential circuits: Derandomizing the XOR lemma. In Frank Thomson Leighton and Peter W. Shor, editors, *Proceedings of the*

Twenty-Ninth Annual ACM Symposium on the Theory of Computing, El Paso, Texas, USA, May 4–6, 1997, pages 220–229. ACM, 1997. https://doi.org/10.1145/258533.258590. 18

[IW05] Piotr Indyk and David P. Woodruff. Optimal approximations of the frequency moments of data streams. In Harold N. Gabow and Ronald Fagin, editors, *Proceedings of the 37th Annual ACM Symposium on Theory of Computing (STOC)*, pages 202–208. ACM, 2005. https://doi.org/10.1145/1060590.1060621. 419

[IZ89] Russell Impagliazzo and David Zuckerman. How to recycle random bits. In *30th Annual Symposium on Foundations of Computer Science*, Research Triangle Park, North Carolina, USA, 30 October - 1 November 1989, pages 248–253. IEEE Computer Society, 1989. https://doi.org/10.1109/SFCS.1989.63486. 214

[Jan68] E. Janovkaya. Equilibrium points in polymatrix games (in Russian). *Litovskii Matematicheskii Sbornik*, 8:381–384, 1968. 458

[JdM12] Dejan Jovanovic and Leonardo de Moura. Solving non-linear arithmetic. *ACM Communications in Computer Algebra*, 46(3–4):104–105, 2012. https://doi.org/10.1145/2429135.2429155. 293

[JDU+74] David S. Johnson, Alan J. Demers, Jeffrey D. Ullman, Michael R. Garey, and Ronald L. Graham. Worst-case performance bounds for simple one-dimensional packing algorithms. *SIAM Journal of Computing*, 3(4):299–325, 1974. https://doi.org/10.1137/0203025. 394

[Jea98] Peter Jeavons. On the algebraic structure of combinatorial problems. *Theoretical Computer Science*, 200(1):185–204, 1998. 45

[Jer94] Mark Jerrum. Counting trees in a graph is #P-complete. *Information Processing Letters*, 51(3):111–116, 1994. https://doi.org/10.1016/0020-0190(94)00085-9. 274

[Jer16] Emil Jeřábek. Integer factoring and modular square roots. *Journal of Computing and System Sciences*, 82(2):380–394, 2016. https://arxiv.org/pdf/1207.5220.pdf. 465, 467

[JK21] Rajesh Jayaram and John Kallaugher. An optimal algorithm for triangle counting in the stream. In Mary Wootters and Laura Sanità, editors, *Approximation, Randomization, and Combinatorial Optimization. Algorithms and Techniques, APPROX/RANDOM 2021*, August 16–18, 2021, University of Washington, Seattle, Washington, USA (Virtual Conference), volume 207 of *LIPIcs*, pages 11:1–11:11. Schloss Dagstuhl - Leibniz-Zentrum für Informatik, 2021. 477

[JLS19] Arun Jambulapati, Yang P. Liu, and Aaron Sidford. Parallel reachability in almost linear work and square root depth. In David Zuckerman, editor, *60th IEEE Annual Symposium on Foundations of Computer Science, FOCS 2019*, Baltimore, Maryland, USA, November 9–12, 2019, pages 1664–1686. IEEE Computer Society, 2019. https://doi.org/10.1109/FOCS.2019.00098. 440

[JNV+21] Zhengfeng Ji, Anand Natarajan, Thomas Vidick, John Wright, and Henry Yuen. Mip* = RE. *Communications of the ACM*, 64(11):131–138, 2021. https://arxiv.org/abs/2001.04383. 481

[Jon98] Peter Jonsson. Near-optimal nonapproximability results for some NPO PB-complete problems. *Information Processing Letters*, 68(5):249–253, 1998. http://dx.doi.org/10.1016/S0020-0190(98)00170-7. 247

[Jor11] Stephen Jordan. Quantum algorithm zoo, 2011. https://quantumalgorithmzoo.org/. 470

[JPY88] David S. Johnson, Christos H. Papadimitriou, and Mihalis Yannakakis. How easy is local search? *Journal of Computer and System Sciences*, 37(1):79–100, 1988. https://doi.org/10.1016/0022-0000(88)90046-3. 465, 466

[JT07] Marcin Jurdzinski and Ashutosh Trivedi. Reachability-time games on timed automata. In Lars Arge, Christian Cachin, Tomasz Jurdzinski, and Andrzej Tarlecki, editors, *Automata, Languages and Programming, 34th International Colloquium, ICALP 2007*, Wroclaw, Poland, July 9–13, 2007, Proceedings, volume 4596 of *Lecture Notes in Computer Science*, pages 838–849. Springer, 2007. http://arxiv.org/abs/0907.3414. 330

[JZ22] Stacey Jeffery and Sebastian Zur. Multidimensional quantum walks, with applications to k-distinctness, 2022. https://arxiv.org/abs/2208.13492. 475

[Kab02] Valentine Kabanets. Derandomization: A brief overview. *Bulletin of the EATCS*, 76:88–103, 2002. https://eccc.weizmann.ac.il/report/2002/008/. 18

[KAI79] Takumi Kasai, Akeo Adachi, and Shigeki Iwata. Classes of pebble games and complete problems. *SIAM Journal of Computing*, 8(4):574–586, 1979. https://epubs.siam.org/doi/abs/10.1137/0208046?journalCode=smjcat. 329

[Kal21] John Kallaugher. A quantum advantage for a natural streaming problem. In Nisheeth Vishnoi, editor, *62nd IEEE Annual Symposium on Foundations of Computer Science, FOCS 2021*, Denver, CO, USA, February 7–10, 2021, pages 897–908. IEEE, 2021. 478

[Kap21] Michael Kapralov. Space lower bounds for approximating maximum matching in the edge arrival model. In Dániel Marx, editor, *Proceedings of the 2021 ACM-SIAM Symposium on Discrete Algorithms, SODA 2021*, Virtual Conference, January 10–13, 2021, pages 1874–1893. SIAM, 2021. https://doi.org/10.1137/1.9781611976465.112. 418

[Kar68] Jerome J Karaganis. On the cube of a graph. *Canadian Mathematical Bulletin*, 11:295–296, 1968. 113

[Kar72] Richard Karp. Reducibilities among combinatorial problems. In Raymond E. Miller and James W. Thatcher, editors, *Complexity of Computer Computations*, pages 85–103. Plenum Press, 1972. https://cs6505.files.wordpress.com/2015/08/karp.pdf. xiv, 9, 47, 56, 112, 133, 134

[Kay00] Richard Kaye. Minesweeper is NP-complete. *Mathematical Intelligencer*, 22(2):9–15, 2000. https://academic.timwylie.com/17CSCI4341/minesweeper_kay.pdf. 105

[KC00] Valentine Kabanets and Jin-yi Cai. Circuit minimization problem. In F. Frances Yao and Eugene M. Luks, editors, *Proceedings of the Thirty-Second Annual ACM Symposium on Theory of Computing*, May 21–23, 2000, Portland, OR, USA, pages 73–79. ACM, 2000. https://doi.org/10.1145/335305.335314. 14

[Kho01] Subhash Khot. Improved inaproximability results for MaxClique, chromatic number and approximate graph coloring. In *42nd Annual Symposium on Foundations of Computer Science, FOCS 2001*, 14–17 October 2001, Las Vegas, Nevada, USA, pages 600–609. IEEE Computer Society, 2001. https://doi.org/10.1109/SFCS.2001.959936. 257, 263

[Kho02] Subhash Khot. On the power of unique 2-prover 1-round games. In *Proceedings of the 17th Annual IEEE Conference on Computational Complexity*, Montréal, Québec, Canada, May 21–24, 2002, page 25. IEEE Computer Society, 2002. https://dl.acm.org/doi/pdf/10.1145/509907.510017. 262

[Kho05] Subhash Khot. Hardness of approximating the shortest vector problem in lattices. *Journal of the Assocation of Computing Machinery (JACM)*, 52(5):789–808, 2005. https://doi.org/10.1145/1089023.1089027. 249

[Kho10] Subhash Khot. On the Unique Games Conjecture (invited survey). In Eric Blais and Ryan O'Donnell, editors, *Proceedings of the 25th Annual IEEE Conference on Computational Complexity, CCC 2010*, Cambridge, Massachusetts, USA, June 9–12, 2010, pages 99–121. IEEE Computer Society, 2010. https://doi.org/10.1109/CCC.2010.19. 254, 257, 262

[Kho19] Subhash Khot. On the proof of the 2-to-2 games conjecture, 2019. https://cs.nyu.edu/~khot/PCP-Spring-20/2-to-2-Exposition.pdf. 223, 254, 256

[Kin15] William B. Kinnersley. Cops and robbers is EXPTIME-complete. *Journal of Combimatorial Theory, Series B*, 111:201–220, 2015. https://doi.org/10.1016/j.jctb.2014.11.002. 329

[KKG21] Anna R. Karlin, Nathan Klein, and Shayan Oveis Gharan. A (slightly) improved approximation algorithm for metric TSP. In Samir Khuller and Virginia Vassilevska Williams, editors, *STOC '21: 53rd Annual ACM SIGACT Symposium on Theory of Computing*, Virtual Event, Italy, June 21–25, 2021, pages 32–45. ACM, 2021. https://doi.org/10.1145/3406325.3451009. 207

[KKM18] Kamil Khadiev, Aliya Khadieva, and Ilanz Mannapo. Quantum online algorithms with respect to space and advice complexity. *Lobachevskii Journal of Mathematics*, 39(9):1377–1387, 2018. 478

[KKMO07] Subhash Khot, Guy Kindler, Elchanan Mossel, and Ryan O'Donnell. Optimal inapproximability results for MAX-CUT and other 2-variable CSPs? *SIAM Journal on Computing*, 37(1):319–357, 2007. https://www.cs.cmu.edu/~odonnell/papers/maxcut.pdf. 260

[KKR18] Alexandr Kazda, Vladimir Kolmogorov, and Michal Rolínek. Even delta-matroids and the complexity of planar boolean CSPs. *ACM Transactions on Algorithms*, 15(2):22:1–22:33, December 2018. 96, 97

[Kle43] Stephen Kleene. Recursive predicates and quantifiers. *Transactions of the American Math Society*, 53(3):41–73, 1943. 23

[KLM$^+$14] Raimondas Kiveris, Silvio Lattanzi, Vahab S. Mirrokni, Vibhor Rastogi, and Sergei Vassilvitskii. Connected components in MapReduce and beyond. In Ed Lazowska, Doug Terry, Remzi H. Arpaci-Dusseau, and Johannes Gehrke, editors, *Proceedings of the ACM Symposium on Cloud Computing*, Seattle, WA, USA, November 3–5, 2014, pages 18:1–18:13. ACM, 2014. https://doi.org/10.1145/2670979.2670997. 434

[KLN91] Jan Kratochvíl, Anna Lubiw, and Jaroslav Nešetřil. Noncrossing subgraphs in topological layouts. *SIAM Journal on Discrete Mathematics*, 4(2):223–244, 1991. 74

[KLS00] Sanjeev Khanna, Nathan Linial, and Shmuel Safra. On the hardness of approximating the chromatic number. *Combinatorica*, 20(3):393–415, 2000. https://doi.org/10.1007/s004930070013. 257, 263

[KLS01] Michael Kearns, Michael L. Littman, and Satinder P. Singh. Graphical models for game theory. In Jack Breese and Daphne Koller, editors, *Proceedings of the 17th Conference on Uncertainty in Artificial Intelligence*, Association of Computing Machinery: 2001, 253–260. http://arxiv.org/abs/1301.2281. 458

[KM94] Jan Kratochvíl and Jirí Matoušek. Intersection graphs of segments. *Journal of Combinatorial Theory, Series B*, 62(2):289–315, 1994. https://www.sciencedirect.com/science/article/pii/S0095895684710719. 296

[KM12] Ross J. Kang and Tobias Müller. Sphere and dot product representations of graphs. *Discrete and Compututational Geometry*, 47(3):548–568, 2012. https://doi.org/10.1007/s00454-012-9394-8. 297

[KMN11] Anna R. Karlin, Claire Mathieu, and C. Thach Nguyen. Integrality gaps of linear and semi-definite programming relaxations for knapsack. In Oktay Günlük and Gerhard J. Woeginger, editors, *Proceedings of the 15th International Conference on Integer Programming and Combinatoral Optimization (IPCO 2011)*, volume 6655 of *Lecture Notes in Computer Science*, pages 301–314, New York, NY, June 2011. Springer. https://homes.cs.washington.edu/~karlin/papers/knapsack.pdf. 267

[KMS98] David R. Karger, Rajeev Motwani, and Madhu Sudan. Approximate graph coloring by semidefinite programming. *Journal of the Assocation of Computing Machinery (JACM)*, 45(2):246–265, 1998. https://doi.org/10.1145/274787.274791. 263

[KMS17] Subhash Khot, Dor Minzer, and Muli Safra. On independent sets, 2-to-2 games, and Grassmann graphs. In Hamed Hatami, Pierre McKenzie, and Valerie King, editors, *Proceedings of the 49th Annual ACM SIGACT Symposium on Theory of Computing, STOC 2017*, Montreal, QC, Canada, June 19–23, 2017, pages 576–589. ACM, 2017. https://doi.org/10.1145/3055399.3055432. 223, 254

[KMS18] Subhash Khot, Dor Minzer, and Muli Safra. Pseudorandom sets in Grassmann graph have near-perfect expansion. In Mikkel Thorup, editor, *59th IEEE Annual Symposium on Foundations of Computer Science, FOCS 2018*, Paris, France, October 7–9, 2018, pages 592–601. IEEE Computer Society, 2018. https://doi.org/10.1109/FOCS.2018.00062. 223, 254

[KN97] Eyal Kushilevitz and Noam Nisan. *Communication Complexity*. Cambridge University Press, Cambridge, England, 1997. 408

[KNR99] Ilan Kremer, Noam Nisan, and Dana Ron. On randomized one-round communication complexity. *Computational Complexity*, 8(1):21–49, 1999. https://doi.org/10.1007/s000370050018. 410

[KO87] Yan Ke and Joseph O'Rourke. Moving a ladder in three dimensions: Upper and lower bounds. In D. Soule, editor, *Proceedings of the Third Annual Symposium on Computational Geometry*, Waterloo, Ontario, Canada, June 8–10, 1987, pages 136–146. ACM, 1987. https://doi.org/10.1145/41958.41972. 370

[KO09] Subhash Khot and Ryan O'Donnell. SDP gaps and UGC-hardness for Max-Cut-Gain. *Theory of Computing*, 5(1):83–117, 2009. https://doi.org/10.4086/toc.2009.v005a004. 260

[Kor96] Ivan Korec. Small universal register machines. *Theoretical Computer Science*, 168(2):267–301, 1996. 332

[Kor04] Simon Korman. On the use of randomization in the online set cover problem. Master's thesis, Weizmann Institute of Science, Rehovot, Israel, 2004. https://www.cs.umd.edu/~gasarch/BLOGPAPERS/korman.pdf. 404

[KP03] Phokion G. Kolaitis and Jonathan Panttaja. On the complexity of existential pebble games. In Matthias Baaz and Johann A. Makowsky, editors, *Computer Science Logic, 17th International Workshop, CSL 2003, 12th Annual Conference of the EACSL, and 8th Kurt Gödel Colloquium, KGC 2003, Vienna, Austria, August 25–30, 2003, Proceedings*, volume 2803 of *Lecture Notes in Computer Science*, pages 314–329. Springer, 2003. https://users.soe.ucsc.edu/~kolaitis/bio11/papers11/csl03.pdf. 329

[KP17a] John Kallaugher and Eric Price. A hybrid sampling scheme for triangle counting. In Philip N. Klein, editor, *Proceedings of the Twenty-Eighth Annual ACM-SIAM Symposium on Discrete Algorithms, SODA 2017*, Barcelona, Spain, Hotel Porta Fira, January 16–19, pages 1778–1797. SIAM, 2017. https://doi.org/10.1137/1.9781611974782.116. 477

[KP17b] Iordanis Kerenidis and Anupam Prakash. Quantum recommendation systems. In Christos H. Papadimitriou, editor, *8th Innovations in Theoretical Computer Science Conference, ITCS 2017*, January 9–11, 2017, Berkeley, CA, USA, volume 67 of *LIPIcs*, pages 49:1–49:21. Schloss Dagstuhl - Leibniz-Zentrum für Informatik, 2017. 470

[KPP16] Tsvi Kopelowitz, Seth Pettie, and Ely Porat. Higher lower bounds from the 3SUM conjecture. In Robert Krauthgamer, editor, *Proceedings of the Twenty-Seventh Annual ACM-SIAM Symposium on Discrete Algorithms, SODA 2016*, Arlington, VA, USA, January 10–12, 2016, pages 1272–1287. SIAM, 2016. https://doi.org/10.1137/1.9781611974331.ch89. 374

[KPS10] Subhash Khot, Preyas Popat, and Rishi Saket. Approximate Lasserre integrality gap for unique games. In Maria J. Serna, Ronen Shaltiel, Klaus Jansen, and José D. P. Rolim, editors, *Approximation, Randomization, and Combinatorial Optimization. Algorithms and Techniques, 13th International Workshop (APPROX 2010) and 14th International Workshop (RANDOM 2010). Proceedings*, volume 6302 of *Lecture Notes in Computer Science*, pages 298–311, Barcelona, Spain, September 2010. Springer. 267

[KR88] Richard Karp and Vijaya Ramachandran. A survey of parallel algorithms for shared-memory machines, 1988. http://www2.eecs.berkeley.edu/Pubs/TechRpts/1988/CSD-88-408.pdf. 422, 440

[KR92] Donald E. Knuth and Arvind Raghunathan. The problem of compatible representatives. *SIAM Journal on Discrete Mathematics*, 5(3):422–427, 1992. https://doi.org/10.1137/0405033. 85, 86

[KR08] Subhash Khot and Oded Regev. Vertex cover might be hard to approximate to within $2-\varepsilon$. *Journal of Computer and System Sciences*, 74(3):335–349, 2008. https://www.sciencedirect.com/science/article/pii/S0022000007000864. 260

[Kra99] Adam Krawczyk. The complexity of finding a second Hamiltonian cycle in cubic graphs. *Journal of Computing and System Science*, 58(3):641–647, 1999. https://doi.org/10.1006/jcss.1998.1611. 275, 463

[Kri75] Mukkai S. Krishnamoorthy. An NP-hard problem in bipartite graphs. *ACM SIGACT News*, 7(1):26–26, 1975. https://dl.acm.org/doi/10.1145/990518.990521. 112

[Kro67] Melven Krom. The decision problem for a class of first-order formulas in which all disjunctions are binary. *Zeitschrift für logik and Grundlagen d. Math (Has changed its name to Math Logic Quarterly)*, 13:15–20, 1967. https://onlinelibrary.wiley.com/doi/10.1002/malq.19670130104. 33

[KS46] Ulrich Krengel and Louis Sucheston. Semimarts and finite values. *Bulletin of the American Mathematical Society*, 83(4):745–747, 1946. 396

[KS92] Bala Kalyanasundaram and Georg Schintger. The probabilistic communication complexity of set intersection. *SIAM Journal on Discrete Mathematics*, 5:545–557, 1992. https://epubs.siam.org/doi/pdf/10.1137/0405044. 410

[KS20] Young Kun Ko and Min Jae Song. Hardness of approximate nearest neighbor search under L-infinity, 2020. https://arxiv.org/abs/2011.06135. 251

[KSS84] J. Kahn, M. Saks, and D. Sturtevant. A topological approach to evasiveness. *Combinatorica*, 4(4):297–306, 1984. https://link.springer.com/content/pdf/10.1007/BF02579140.pdf. 431

[KST93] Jan Kratochvíl, Petr Savický, and Zsolt Tuza. One more occurrence of variables makes satisfiability jump from trivial to NP-complete. *SIAM Journal on Computing*, 22(1):203–210, 1993. 39

[KV15] Subhash Khot and Nisheeth K. Vishnoi. The Unique Games Conjecture, integrality gap for cut problems and embeddability of negative-type metrics into ℓ_1. *Journal of the ACM*, 62(1):8:1–8:39, 2015. https://doi.org/10.1145/2629614. 267

[KVV90] Richard M. Karp, Umesh V. Vazirani, and Vijay V. Vazirani. An optimal algorithm for on-line bipartite matching. In Harriet Ortiz, editor, *Proceedings of the 22nd Annual ACM Symposium on Theory of Computing*, May 13–17, 1990, Baltimore, Maryland, USA, pages 352–358. ACM, 1990. https://doi.org/10.1145/100216.100262. 403

[KW19] Daniel M. Kane and Richard Ryan Williams. The orthogonal vectors conjecture for branching programs and formulas. In Avrim Blum, editor, *10th Innovations in Theoretical Computer Science Conference, ITCS 2019*, January 10–12, 2019, San Diego, California, USA, volume 124 of *LIPIcs*, pages 48:1–48:15. Schloss Dagstuhl - Leibniz-Zentrum für Informatik, 2019. https://doi.org/10.4230/LIPIcs.ITCS.2019.48. 375

[KZ97] Howard Karloff and Uzi Zwick. A 7/8-approximation algorithm for MAX 3-SAT? In *38th Annual IEEE Symposium on Foundations of Computer Science (FOCS)*, pages 406–415. IEEE Computer Society, 1997. https://ieeexplore.ieee.org/document/646129. 218

[Lad75] Richard Ladner. On the structure of polynomial time reducibilities. *Journal of the Association of Computing Machinery (JACM)*, 22:155–171, 1975. https://dl.acm.org/doi/10.1145/321864.321877. 13

[Lam11] Michael Lampis. A kernel of order $2k - c \log k$ for vertex cover. *Information Processing Letters*, 111(23–24):1089–1091, 2011. https://doi.org/10.1016/j.ipl.2011.09.003. 175

[Lar92] P. Laroche. Planar 1-in-3 satisfiability is NP-complete. *Comptes Rendus de l'Académie des Sciences, Série 1, Mathématique*. 316(4):389–392, 1993. 41, 90

[Las01a] Jean B. Lasserre. An explicit exact SDP relaxation for nonlinear 0–1 programs. In Karen Aardal and Bert Gerards, editors, *Integer Programming and Combinatorial Optimization, 8th International IPCO Conference*, Utrecht, The Netherlands, June 13–15, 2001, Proceedings, volume 2081 of *Lecture Notes in Computer Science*, pages 293–303. Springer, 2001. 267

[Las01b] Jean B. Lasserre. Global optimization with polynomials and the problem of moments. *SIAM Journal on Optimzation*, 11(3):796–817, 2001. https://doi.org/10.1137/S1052623400366802. 267

[Lau83] Clemens Lautemann. BPP and the polynomial hierarchy. *Information Processing Letters*, 17(4):215–217, 1983. https://doi.org/10.1016/0020-0190(83)90044-3. 18

[LC89] K. Li and K. H. Cheng. Complexity of resource allocation and job scheduling problems on partitionable mesh connected systems. In Peter Kogge, editor, *Proceedings of the 1st Annual IEEE Symposium on Parallel and Distributed Processing*, pages 358–365, IEEE, May 1989. 141

[Lev73] Leonid Levin. Universal search problems. *Problems of Information Transmission*, 9:115–116, 1973. xiii, xiv, 9, 33

[Lew78] Harry Lewis. Renaming a set of clauses as a Horn set. *Journal of the Association for Computing Machinery (JACM)*, 25(1):134–135, 1978. https://dl.acm.org/doi/pdf/10.1145/322047.322059. 40

[Lew79] Harry R. Lewis. Satisfiability problems for propositional calculi. *Mathematical Systems Theory*, 13:45–53, 1979. https://doi.org/10.1007/BF01744287. 46, 100

[Lic82] David Lichtenstein. Planar formulae and their uses. *SIAM Journal on Computing*, 11(2):329–343, 1982. https://epubs.siam.org/doi/pdf/10.1137/0211025. 72, 73, 74, 77, 80, 305

[Lin02] Nathan Linial. Finite metric spaces: Combinatorics, geometry and algorithms. In Ferran Hurtado, Vera Sacristán, Chandrajit Bajaj, and Subhash Suri, editors, *Proceedings of the 18th Annual Symposium on Computational Geometry*, Barcelona, Spain, June 5–7, 2002, page 63. ACM, 2002. https://doi.org/10.1145/513400.513441. 267

[LLL82] A. Lenstra, H. Lenstra, and L. Lovasz. Factoring polynomials with rational coefficients. *Mathematische Annalen*, 261:513–534, 1982. https://link.springer.com/content/pdf/10.1007/BF01457454.pdf. 249

[LLM+03] J. Kevin Lanctôt, Ming Li, Bin Ma, Shaojiu Wang, and Louxin Zhang. Distinguishing string selection problems. *Information and Computation*, 185(1):41–55, 2003. www.sciencedirect.com/science/article/pii/S0890540103000579. 190

[LLR95] Nathan Linial, Eran London, and Yuri Rabinovich. The geometry of graphs and some of its algorithmic applications. *Combinatorica*, 15(2):215–245, 1995. https://doi.org/10.1007/BF01200757. 267

[LLZ02] Michael Lewin, Dror Livnat, and Uri Zwick. Improved rounding techniques for the MAX 2-SAT and MAX DI-CUT problems. In William J. Cook and Andreas S. Schulz, editors, *Integer Programming and Combinatorial Optimization, 9th International IPCO Conference*, Cambridge, MA, USA, May 27–29, 2002, Proceedings, volume 2337 of *Lecture Notes in Computer Science*, pages 67–82. Springer, 2002. https://dl.acm.org/doi/10.5555/645591.660076. 260

[LMM03] Richard J. Lipton, Evangelos Markakis, and Aranyak Mehta. Playing large games using simple strategies. In Daniel A. Menascé and Noam Nisan, editors, *Proceedings 4th ACM Conference on Electronic Commerce (EC-2003)*, San Diego, California, USA, June 9–12, 2003, pages 36–41. ACM, 2003. https://doi.org/10.1145/779928.779933. 462

[LO02] Peter Looges and Stephan Olariu. Optimal greedy algorithms for indifference graphs. *Computers and Mathematics with Applications*, 7:15–25, 2002. 297

[Lov75] László Lovász. On the ratio of optimal integral and fractional covers. *Discrete Applied Mathematics*, 13(4):383–390, 1975. https://doi.org/10.1016/0012-365X(75)90058-8. 265

[LPV20] Andrea Lincoln, Adam Polak, and Virginia Vassilevska Williams. Monochromatic triangles, intermediate matrix products, and convolutions. In Thomas Vidick, editor, *11th Innovations in Theoretical Computer Science Conference, ITCS 2020*, January 12–14, 2020, Seattle, Washington, USA, volume 151 of *LIPIcs*, pages 53:1–53:18. Schloss Dagstuhl - Leibniz-Zentrum für Informatik, 2020. https://doi.org/10.4230/LIPIcs.ITCS.2020.53. 338

[LR99] Tom Leighton and Satish Rao. Multicommodity max-flow min-cut theorems and their use in designing approximation algorithms. *Journal of the Assocation of Computing Machinery (JACM)*, 46(6):787–832, 1999. https://doi.org/10.1145/331524.331526. 267

[LTW+90] Joseph Y.-T. Leung, Tommy W. Tam, C. S. Wong, Gilbert H. Young, and Francis Y. L. Chin. Packing squares into a square. *Journal of Parallel and Distributed Computing*, 10(3):271–275, 1990. https://doi.org/10.1016/0743-7315(90)90019-L. 142

[Lub86] Michael Luby. A simple parallel algorithm for the maximal independent set problem. *SIAM Journal on Computing*, 15(4):1036–1053, 1986. https://doi.org/10.1137/0215074. 427

[LY89] Ming Li and Yaacov Yesha. New lower bounds for parallel computation. *Journal of the Assocation of Computing Machinery (JACM)*, 36(3):671–680, 1989. https://doi.org/10.1145/65950.65959. 422

[LY94] Carsten Lund and Mihalis Yannakakis. On the hardness of approximating minimization problems. *Journal of the ACM*, 41(5):960–981, 1994. https://dl.acm.org/doi/pdf/10.1145/185675.306789. 216

[Mar13] Daniel Marx. The square root phenomenon in planar graphs (slides), 2013. http://www.cs.bme.hu/~dmarx/papers/marx-faw-2013-planar.pdf. 166

[Mar14] Daniel Marx. W[1]-hardness (slides), 2014. https://fptschool.mimuw.edu.pl/slides/lec4.pdf. 178

[Mat70] Yuri Matijasevic. Enumerable sets are Diophantine (in Russian). *Doklady Academy Nauk, SSSR*, 191:279–282, 1970. Translation in Soviet Math Doklady, Vol 11, 1970. 336

[Mat93] Yuri Matijasevic. *Hilbert's Tenth Problem*. MIT Press, Cambridge, MA 1993. 336

[Mat14] Jirí Matoušek. Intersection graphs of segments and $\exists R$, 2014. http://arxiv.org/abs/1406.2636. 290, 296

[McC81] William F. McColl. Planar crossovers. *IEEE Transactions on Computers*, C-30(3):223–225, 1981. 100

[MCFRQ22] Javier Martinez-Chfuentes, K. M. Fonseca-Romero, and Nicolas Quesada. Classical models may be a better explanation of the Jiuzhang 1.0 Gaussian Boson Sampler than its targeted squeezed light model, 2022. https://arxiv.org/pdf/2207.10058.pdf. 486

[McK87] Michael McKenna. Worst-case optimal hidden-surface removal. *ACM Transactions on Graphics*, 6(1):19–28, 1987. https://doi.org/10.1145/27625.27627. 368

[McP05] Brandon McPhail. Light up is NP-complete, 2005. http://www.mountainvistasoft.com/docs/lightup-is-np-complete.pdf. 102

[Meh84] Kurt Mehlhorn. *Data Structures and Algorithms 2: Graph Algorithms and NP-Completeness*, volume 2 of *EATCS Monographs on Theoretical Computer Science*. Springer, 1984. https://doi.org/10.1007/978-3-642-69897-2. 172

[Meh14] Ruta Mehta. Constant rank bimatrix games are PPAD-hard. In David B. Shmoys, editor, *Symposium on Theory of Computing, STOC 2014*, pages 545–554. ACM, 2014. https://doi.org/10.1145/2591796.2591835. 462

[Mel21] Erika Melder. A chronology of Set Cover inapproximability results, 2021. https://arxiv.org/abs/2111.08100. 216

[Mer07] N. David Mermin, editor. *Quantum Computer Science: An Introduction*. Cambridge Press, 2007. 469

[MG92] Jayadev Misra and David Gries. A constructive proof of Vizing's theorem. *Information Processing Letters*, 41(3):131–133, 1992. https://doi.org/10.1016/0020-0190(92)90041-S. 251

[Mic00] Daniele Micciancio. The shortest vector in a lattice is hard to approximate to within some constant. *SIAM Journal on Computing*, 30(6):2008–2035, 2000. https://doi.org/10.1137/S0097539700373039. 249

[Mic12] Daniele Micciancio. Inapproximability of the shortest vector problem: Toward a deterministic reduction. *Theory of Computing*, 8(1):487–512, 2012. https://doi.org/10.4086/toc.2012.v008a022. 249

[Mil76] Gary L. Miller. Riemann's hypothesis and tests for primality. *Journal of Computing and System Science*, 13(3):300–317, 1976. https://doi.org/10.1016/S0022-0000(76)80043-8. 17

[Mil13] Carl Miller. Evasiveness of graph properties and topological fixed-point theorems (expository). *Foundations and Trends in Theoretical Computer Science*, 7(4):337–415, 2013. https://arxiv.org/pdf/1306.0110.pdf. 431

[Min61] Marvin L. Minsky. Recursive unsolvability of Post's problem of "Tag" and other topics in theory of Turing machines. *Annals of Mathematics*, 74(3):437–455, 1961. 332

[Min67] Marvin Minsky. *Computation: Finite and Infinite Machines*, pages 255–258. Prentice-Hall, Englewood Cliffs, NJ, 1st ed., 1967. 332

[Mit99] Joseph Mitchell. Guillotine subdivisions approximate polygonal subdivisions: A simple polynomial-time approximation scheme for geometric TSP, k-MST, and related problems. *SIAM Journal of Computing*, 28(4):1298–1309, 1999. https://epubs.siam.org/doi/abs/10.1137/S0097539796309764. 207

[MIT24a] MIT Hardness Group, Josh Brunner, Lily Chung, Erik D. Demaine, Della Hendrickson, and Andy Tockman. ASP-completeness of Hamiltonicity in grid graphs, with applications to loop puzzles. In Andrei Z. Broder and Tami Tamir, editors, *Proceedings of the 12th International Conference on Fun with Algorithms*, volume 291 of *LIPIcs*, pages 23:1–23:20, La Maddalena, Italy, June 4–8 2024. 122, 279, 281

[MIT24b] MIT Hardness Group, Della Hendrickson, and Andy Tockman. Complexity of planar graph orientation consistency, promise-inference, and uniqueness, with applications to minesweeper variants. In Andrei Z. Broder and Tami Tamir, editors, *Proceedings of the 12th International Conference on Fun with Algorithms*, volume 291 of *LIPIcs*, pages 25:1–25:20, 2024. 109

[MIT24c] MIT Hardness Group, Josh Brunner, Lily Chung, Erik D. Demaine, Jenny Diomidova, Della Hendrickson, and Jayson Lynch. Pushing blocks without fixed blocks via checkable gizmos: Push-1 is PSPACE-complete. Manuscript, 2024. 56

[MIT24d] MIT Hardness Group, Erik D. Demaine, Holden Hall, and Jeffery Li. Tetris with few piece types. In Andrei Z. Broder and Tami Tamir, editors, *Proceedings of the 12th International Conference on Fun with Algorithms*, volume 291 of *LIPIcs*, pages 24:1–24:18, La Maddalena, Italy, June 2024. 157

[MIT24e] MIT Hardness Group, Hayashi Ani, Erik D. Demaine, Holden Hall, Ricardo Ruiz, and Naveen Venkat. You can't solve these Super Mario Bros. levels: Undecidable Mario games. In Andrei Z. Broder and Tami Tamir, editors, *Proceedings of the 12th International Conference on Fun with Algorithms*, volume 291 of *LIPIcs*, pages 22:1–22:20, La Maddalena, Italy, June 2024. 335

[MLLL+12] Enrique Martin-Lopez, Anthony Laing, Thomas Lawson, Roberto Alvarez, Xiao-Qi Zhou, and Jeremy O'Brian. Experimental realization of Shor's quantum factoring algorithm using qubit recycling. *Nature Photonics*, 6, 2012. https://arxiv.org/abs/1111.4147. 471

[Mnë88] Nicolai Mnëv. The universality theorems on the classication problem of configuration varieties and convex polytopes varieties. In O.Y. Viro, editor, *Topology and Geometry–Rohlin Seminar*, volume 1346 of *Lecture Notes in Mathematics*, pages 527–544. Springer, 1988. 294, 299

[MNV12] Meena Mahajan, Prajakta Nimbhorkar, and Kasturi R. Varadarajan. The planar k-means problem is NP-hard. *Theoretical Computer Science*, 442:13–21, 2012. https://doi.org/10.1016/j.tcs.2010.05.034. 97

[Mor88] Bernard Moret. Planar NAE3SAT is in P. *SIGACT News*, 19(2):51–54, 1988. https://dl.acm.org/doi/pdf/10.1145/49097.49099. 94, 345

[Mos15] Dana Moshkovitz. The projection games conjecture and the NP-hardness of ln n-approximating set-cover. *Theory of Computing*, 11:221–235, 2015. https://doi.org/10.4086/toc.2015.v011a007.

[MR07] Anna Moss and Yuval Rabani. Approximation algorithms for constrained node weighted Steiner tree problems. *SIAM Journal of Computing*, 37(2):460–481, 2007. https://doi.org/10.1137/S0097539702420474. 243

[MR08] Wolfgang Mulzer and Günter Rote. Minimum-weight triangulation is NP-hard. *Journal of the Association of Computing Machinery (JACM)*, 55(2):11:1–11:29, 2008. https://dl.acm.org/doi/10.5555/1379361.1379376. 90, 91

[MS08] Luke Mathieson and Stefan Szeider. The parameterized complexity of regular subgraph problems and generalizations. In James Harland and Prabhu Manyem, editors, *Proceedings of the 14th Symposium on Computing: The Australasian Theory Symposium (CATS 2008)*, volume 77 of *CRPIT*, pages 79–86, Wollongong, Australia, January 2008. http://crpit.scem.westernsydney.edu.au/abstracts/CRPITV77Mathieson.html. 180

[MT20] Shohei Mishiba and Yasuhiko Takenaga. QUIXO is EXPTIME-complete. *Information Processing Letters*, 162:105995, 2020. https://doi.org/10.1016/j.ipl.2020.105995. 329

[MTY13] Kazuhisa Makino, Suguru Tamaki, and Masaki Yamamoto. Derandomizing the HSSW algorithm for 3-SAT. *Algorithmica*, 67(2):112–124, 2013. https://doi.org/10.1007/s00453-012-9741-4. 159

[MV19] Andrew McGregor and Hoa T. Vu. Better streaming algorithms for the maximum coverage problem. *Theory of Computing Systems*, 63(7):1595–1619, 2019. https://doi.org/10.1007/s00224-018-9878-x. 418

[MW17] Cody Murray and Ryan Williams. On the (non) NP-hardness of computing circuit complexity. *Theory of Computing*, 13(4):1–22, 2017. https://doi.org/10.4086/toc.2017.v013a004. 14

[Nas50] John Nash. Equilibrium points in n-person games. *Proceedings of the National Academy of Sciences*, 36:48–49, 1950. https://www.pnas.org/doi/10.1073/pnas.36.1.48. 447

[NC16] Michael A. Nielson and Isaac L. Chuang. *Quantum Computation and Quantum Information (10th Anniversary edition)*. Cambridge University Press, 2016. 469

[Neu28] John Von Neumann. Zur theorie der gesellschaftsspiele. *Mathematische Annalen*, 100:295–320, 1928. 447

[New00] Atlantha Newman. Approximating the maximum acyclic subgraph. Master's thesis, MIT, 2000. 260

[Nie02] Rolf Niedermeier. *Invitation to Fixed-Parameter Algorithms*. PhD thesis, University of Tubingen, 2002. This is the author's Habilitationsschrift. 175

[Nie06] Rolf Niedermeier. *Invitation to Fixed-Parameter Algorithms*. Oxford University Press, 2006. 171, 178, 187

[Nik] Nikoli. Nikoli puzzles, 2001. https://www.nikoli.co.jp/en/puzzles/. 82, 102, 122, 276, 287

[Nik08] Nikoli. *Akari & Suoku*. Sterling, 2008. 82, 102, 122, 276, 287

[Nov55] Pytor Novikov. On the algorithmic unsolvability of the word problem in group theory (in Russian). *Proceedings of the Steklov Institute of Mathematics*, 44:1–143, 1955. 337

[NS22] Danupon Nanongkai and Michele Scquizzato. Equivalence classes and conditional hardness in massively parallel computations. *Distributed Computing*, 35(2):165–183, 2022. https://arxiv.org/abs/2001.02191. 440

[NSS95] Moni Naor, Leonard J. Schulman, and Aravind Srinivasan. Splitters and near-optimal derandomization. In *36th Annual Symposium on Foundations of Computer Science*, Milwaukee, Wisconsin, USA, 23–25 October 1995, pages 182–191. IEEE Computer Society, 1995. https://doi.org/10.1109/SFCS.1995.492475.

[NW94] Noam Nisan and Avi Wigderson. Hardness vs randomness. *Journal of Computer and Systems Science*, 49:149–167, 1994. http://www.math.ias.edu/~avi/PUBLICATIONS/. 18

[OW08] Ryan O'Donnell and Yi Wu. An optimal sdp algorithm for max-cut, and equally optimal long code tests. In Cynthia Dwork, editor, *Proceedings of the 40th Annual ACM Symposium on Theory of Computing, Victoria, British Columbia, Canada, May 17–20, 2008*, pages 335–344. ACM, 2008. https://doi.org/10.1145/1374376.1374425. 260

[OW12] Joël Ouaknine and James Worrell. Decision problems for linear recurrence sequences. In Alain Finkel, Jérôme Leroux, and Igor Potapov, editors, *Proceedings of the 6th International Workshop on Reachability Problems (RP 2012)*, volume 7550 of *Lecture Notes in Computer Science*, pages 21–28, Bordeaux, France, September 2012. Springer. https://doi.org/10.1007/978-3-642-33512-9_3 and https://people.mpi-sws.org/~joel/publications/positivity12.pdf. 330

[Pap85] Christos H. Papadimitriou. Games against nature. *Journal of Computer and System Sciences*, 31(2):288–301, 1985. https://www.sciencedirect.com/science/article/pii/0022000085900455. 315, 316

[Pap94] Christos H. Papadimitriou. On the complexity of the parity argument and other inefficient proofs of existence. *Journal of Computer and Systems Sciences*, 48(3):498–532, 1994. https://doi.org/10.1016/S0022-0000(05)80063-7. 28, 442, 448, 452, 454, 463, 464, 467

[Pas11] Grant Olney Passmore. *Combined Decision Procedures for Nonlinear Arithmetics, Real and Complex*. PhD thesis, University of Edinburgh, 2011. https://www.cl.cam.ac.uk/~gp351/passmore-phd-thesis.pdf. 293

[Pat92] Ramamohan Paturi. On the degree of polynomials that approximate symmetric Boolean functions (preliminary version). In S. Rao Kosaraju, Mike Fellows, Avi Wigderson, and John A. Ellis, editors, *Proceedings of the 24th Annual ACM Symposium on Theory of Computing*, May 4–6, 1992, Victoria, British Columbia, Canada, pages 468–474. ACM, 1992. https://doi.org/10.1145/129712.129758. 437

[Păt10] Mihai Pătraşcu. Towards polynomial lower bounds for dynamic problems. In Leonard J. Schulman, editor, *Proceedings of the 42nd ACM Symposium on Theory of Computing, STOC 2010*, Cambridge, Massachusetts, USA, 5–8 June 2010, pages 603–610. ACM, 2010. https://doi.org/10.1145/1806689.1806772. 359, 371, 373

[PB83] J. Scott Provan and Michael O. Ball. The complexity of counting cuts and of computing the probability that a graph is connected. *SIAM Journal on Computing*, 12(4):777–788, 1983. https://doi.org/10.1137/0212053. 274, 285, 286

[PCZ22] Feng Pan, Keyang Chen, and Pan Zhang. Solving the sampling problem of the Sycamore quantum circuits. *Physical Review Letters*, 129:090502, August 2022. https://arxiv.org/abs/2111.03011. 486

[PGA19] Edwin Pednault, John Gunnels, and Giacomo Scott Aaronson. Levarging secondary storage to simulate deep 54-qubit Sycamore circuits, 2019. https://arxiv.org/abs/1910.09534. 485

[Pie03] Krzysztof Pietrzak. On the parameterized complexity of the fixed alphabet shortest common supersequence and longest common subsequence problems. *Journal of Computer and Systems Science*, 67(4):757–771, 2003. https://doi.org/10.1016/S0022-0000(03)00078-3. 181

[Pil18] Alexander Pilz. Planar 3-SAT with a clause/variable cycle. In David Eppstein, editor, *Proceedings of the 16th Scandinavian Symposium and Workshops on Algorithm Theory*, volume 101 of *LIPIcs*, pages 31:1–31:13, Malmö, Sweden, June 2018. https://doi.org/10.4230/LIPIcs.SWAT.2018.31. 74, 77, 93

[PJ09] Grant Olney Passmore and Paul B. Jackson. Combined decision techniques for the existential theory of the reals. In Jacques Carette, Lucas Dixon, Claudio Sacerdoti Coen, and Stephen M. Watt, editors, *Intelligent Computer Mathematics, 16th Symposium, Calculemus 2009, 8th International Conference, MKM 2009, Held as Part of CICM 2009. Proceedings*, volume 5625 of *Lecture Notes in Computer Science*, pages 122–137, Grand Bend, Canada, July 2009. Springer. 293

[Ple79] Ján Plesník. The NP-completeness of the Hamiltonian cycle problem in planar digraphs with degree bound two. *Information Processing Letters*, 8(4):199–201, 1979. 113

[Plo13] Serge Plotkin. Online algorithms, 2013. http://web.stanford.edu/class/cs369/files/cs369-notes. 403

[PM90] Francis Pelletier and Norman Martin. Post's functional completeness theorem. *Notre Dame Journal of Formal Logic*, 31(2):462–475, 1990. https://www.sfu.ca/~jeffpell/papers/PostPellMartin.pdf. 100

[Poa20] Jian-Wei Pan and 18 other authors. Quantum computational advantage using photons. *Science*, 370, 2020. https://arxiv.org/abs/2012.01625. 486

[Poc10] Henry Pocklington. The determination of the exponent to which a number belongs, the practical solution of certain congruences, and the law of quadratic reciprocity. *Proceedings of the Cambridge Philosophy Society*, 16:1–16, 1910. xiv

[Poo14] Bjorn Poonen. Undecidable problems: A sampler, 2014. https://arxiv.org/pdf/1204.0299.pdf. 336

[Pos41] Emil Post. *The Two-Valued Iterative System of Mathematical Logic*. Annals of Mathematics Studies. Princeton University Press, 1941. 100

[Pos46] Emil Post. A variant of a recursively unsolvable problem. *Bulletin of the American Mathematical Society*, 52(4):264–268, 1946. https://www.ams.org/journals/bull/1946-52-04/S0002-9904-1946-08555-9/S0002-9904-1946-08555-9.pdf. 332

[PPdM12] Grant Olney Passmore, Lawrence C. Paulson, and Leonardo Mendonça de Moura. Real algebraic strategies for MetiTarski proofs. In Johan Jeuring, John A. Campbell, Jacques Carette, Gabriel Dos Reis, Petr Sojka, Makarius Wenzel, and Volker Sorge, editors, *Intelligent Computer Mathematics - 11th International Conference, AISC 2012, 19th Symposium, Calculemus 2012, 5th International Workshop, DML 2012, 11th International Conference, MKM 2012, Systems and Projects, Held as Part of CICM 2012. Proceedings*, volume 7362 of *Lecture Notes in Computer Science*, pages 358–370, Bremen, Germany, July 2012. Springer. 294

[PPZ99] Ramamohan Paturi, Pavel Pudlák, and Francis Zane. Satisfiability coding lemma. *Chicago Journal of Theoretical Computer Science*, 1999. http://cjtcs.cs.uchicago.edu/articles/1999/11/contents.html. 160

[Pre12] John Preskill. Quantum computing and the entanglement frontier, 2012. https://arxiv.org/pdf/1203.5813.pdf. 484

[Pre19] John Preskill. Why I called it quantum supremacy. *Quanta*, 2019. 484

[Pri77] W. L. Price. A topological transformation algorithm which relates the hamiltonian circuits of a cubic planar map. *Journal of the London Mathematical Society*, s2-15(2):193–196, 1977. 275

[PV84] Christos Papadimitriou and Umesh V. Vazirani. On two geometric problems related to the travelling salesman problem. *Journal of Algorithms*, 5(2):231–246, 1984. 118, 119

[PW10] Mihai Pătraşcu and Ryan Williams. On the possibility of faster SAT algorithms. In Moses Charikar, editor, *Proceedings of the Twenty-First Annual ACM-SIAM Symposium on Discrete Algorithms, SODA 2010, Austin, Texas, USA, January 17–19, 2010*, pages 1065–1075. SIAM, 2010. https://doi.org/10.1137/1.9781611973075.86. 372

[PY91] Christos H. Papadimitriou and Mihalis Yannakakis. Optimization, approximation, and complexity classes. *Journal of Computing and System Sciences*, 43(3):425–440, 1991. https://doi.org/10.1016/0022-0000(91)90023-X. 226, 230, 232, 260

[Rab58] Michael Rabin. Recursive unsolvability of group theoretic problems. *Annals of Mathematics*, 67:172–194, 1958. 337

[Rab80] Michael Rabin. A probabilistic algorithm for testing primality. *Journal of Number Theory*, 12(1):128–138, 1980. https://www.sciencedirect.com/science/article/pii/0022314X80900840. 17

[Raz92] Alexander Razborov. On the distributional complexity of disjointness. *Theoretical Computer Science*, 106:385–390, 1992. https://www.sciencedirect.com/science/article/pii/0304397592 90260M. 410

[Raz98] Ran Raz. A parallel repetition theorem. *SIAM Journal on Computing*, 27(3):763–803, 1998. https://dl.acm.org/doi/10.1145/225058.225181. 256

[Rei84] John H. Reif. The complexity of two-player games of incomplete information. *Journal of Computer and System Sciences*, 29(2):274–301, 1984. http://www.sciencedirect.com/science/article /pii/0022000084900345. 329

[Rel22] Press Release. The Nobel prize in physics in 2022, 2022. https://www.nobelprize.org/prizes /physics/2022/summary/. 482

[Ren00b] Paul Rendell. A Turing machine in Conway's Game of Life: Implementation, 2000. http://rendell -attic.org/gol/tm.htm. 334

[Ren11] Paul Rendell. A universal Turing machine in Conway's Game of Life. In Waleed Smari and John Giunta, editors, *Proceedings of the 9th Annual IEEE Conference on High Performance Computing and Simulation (HPCS)*. IEEE, 2011. https://ieeexplore.ieee.org/document/5999906. 334

[RHM+22] Václav Rozhon, Bernhard Haeupler, Anders Martinsson, Christoph Grunau, and Goran Zuzic. Parallel breadth-first search and exact shortest paths and stronger notions for approximate distances. *CoRR*, abs/2210.16351, 2022. https://doi.org/10.48550/arXiv.2210.16351. 440

[Ric53] H. G. Rice. Classes of recursively enumerable sets and their decision problems. *Transactions of the American Mathematics Society*, 74:358–366, 1953. https://www.ams.org/journals/tran/1953 -074-02/S0002-9947-1953-0053041-6/S0002-9947-1953-0053041-6.pdf. 331

[Ric08] Elaine Rich. *Automata, Computability, and Complexity: Theory and Application*. Prentice Hall, 2008. 453

[RK04] Oded Regev and Eyal Kaplan. Lecture 2: LLL algorithm, 2004. 249

[Rob83] John M. Robson. The complexity of Go. *Proceedings of the International Federation for Information Processing*, pages 413–417, 1983. 328

[Rob84a] John M. Robson. Combinatorial games with exponential space complete decision problems. *Proceedings of Mathematical Foundations of Computer Science*, 176, 1984. http://dl.acm.org/doi/10 .5555/645716.665284. 328

[Rob84b] John M. Robson. *N* by *N* checkers is Exptime complete. *SIAM Journal on Computing*, 13: 252–267, 1984. https://epubs.siam.org/doi/10.1137/0213018. 325

[Ros73] Arnold L. Rosenberg. On the time required to recognize properties of graphs: A problem. *SIGACT News*, 5(4):15–16, 1973. https://dl.acm.org/doi/pdf/10.1145/1008299.1008302. 431

[Rot13a] Thomas Rothvoß. Approximating bin packing within o(log OPT * log log OPT) bins. In Ken-ichi Kawarbayashi and Yusuke Kobayashi, editors, *54th Annual IEEE Symposium on Foundations of Computer Science, FOCS 2013*, 26–29 October, 2013, Berkeley, CA, USA, pages 20–29. IEEE Computer Society, 2013. https://doi.org/10.1109/FOCS.2013.11. 251

[Rot13b] Thomas Rothvoß. The Lasserre hierachy in approximation algorithms, 2013. https://sites.math .washington.edu//~rothvoss/lecturenotes/lasserresurvey.pdf. 267

[Roy59] Bernard Roy. Transitive et connexite (in French). *C.R. Acad. Sci Paris*, 249:216–218, 1959. 378

[RR06] Oded Regev and Ricky Rosen. Lattice problems and norm embeddings. In Jon M. Kleinberg, editor, *Proceedings of the 38th Annual ACM Symposium on Theory of Computing*, Seattle, WA, USA, May 21–23, 2006, pages 447–456. ACM, 2006. https://doi.org/10.1145/1132516.1132581. 249

[RSST97] Neil Robertson, Daniel Sanders, Paul Seymour, and Robin Thomas. The four colour theorem. *Journal of Combinatorial Theory (Series B)*, 70:2–44, 1997. https://www.sciencedirect.com /science/article/pii/S0095895697917500. 69, 94, 170

[Rub18] Aviad Rubinstein. Hardness of approximate nearest neighbor search. In Ilias Diakonikolas, David Kempe, and Monika Henzinger, editors, *Proceedings of the 50th Annual ACM SIGACT Symposium on Theory of Computing, STOC 2018*, Los Angeles, CA, USA, June 25–29, 2018, pages 1260–1268. ACM, 2018. https://doi.org/10.1145/3188745.3188916. 250

[RV78] Ronald Rivest and Jean Vuillemin. On recognizing graph properties from adjacency matrices. *Theoretical Computer Science*, 3:371–384, 1978. https://www.sciencedirect.com/science/article/pii/0304397576900530. 431

[RV13] Liam Roditty and Virginia Vassilevska Williams. Fast approximation algorithms for the diameter and radius of sparse graphs. In Dan Boneh, Tim Roughgarden, and Joan Feigenbaum, editors, *Symposium on Theory of Computing Conference, STOC'13*, Palo Alto, CA, USA, June 1–4, 2013, pages 515–524. ACM, 2013. https://doi.org/10.1145/2488608.2488673. 387, 390

[RVW18] Tim Roughgarden, Sergei Vassilvitskii, and Joshua R. Wang. Shuffles and circuits (on lower bounds for modern parallel computation). *Journal of the Assocation of Computing Machinery (JACM)*, 65(6):41:1–41:24, 2018. https://dl.acm.org/doi/10.1145/3232536. 429

[RW90] Daniel Ratner and Manfred Warmuth. The (n^2-1)-puzzle and related relocation problems. *Journal of Symbolic Computation*, 10(2):111–137, 1990. 241

[RZ04] Liam Roditty and Uri Zwick. On dynamic shortest paths problems. In Susanne Albers and Tomasz Radzik, editors, *Algorithms - ESA 2004, 12th Annual European Symposium, Bergen, Norway, September 14–17, 2004, Proceedings*, volume 3221 of *Lecture Notes in Computer Science*, pages 580–591. Springer, 2004. Journal version in Algorithmica 2011. 379

[Sav70] Walter J. Savitch. Relationships between nondeterministic and deterministic tape complexities. *Journal of Computer and System Sciences*, 4(2):177–192, 1970. https://www.sciencedirect.com/science/article/pii/S002200007080006X. 22

[SB86] Jerry Slocum and Jack Botermans. *Puzzles Old and New: How to Make and Solve Them*. University of Washington Press, 1986. 146

[SC79] Larry J. Stockmeyer and Ashok K. Chandra. Provably difficult combinatorial games. *SIAM Journal on Computing*, 8(2):151–174, May 1979. http://www.oocities.org/stockmeyer@sbcglobal.net/games.pdf. 321, 323

[Sch78] Thomas J. Schaefer. The complexity of satisfiability problems. In *Proceedings of the 10th Annual ACM Symposium on Theory of Computing*, pages 216–226, San Diego, California, May 1978. https://dl.acm.org/doi/10.1145/800133.804350. 40, 41, 43, 49, 50, 304

[Sch87] Claus-Peter Schnorr. A hierarchy of polynomial time lattice basis reduction algorithms. *Theoretical Computer Science*, 53:201–224, 1987. https://doi.org/10.1016/0304-3975(87)90064-8. 249

[Sch90] Walter Schnyder. Embedding planar graphs on the grid. In David S. Johnson, editor, *Proceedings of the 1st Annual ACM-SIAM Symposium on Discrete Algorithms*, pages 138–148, San Francisco, Association of Computing Machinery (ACM), January 1990. http://dl.acm.org/citation.cfm?id=320176.320191. 81

[Sch09] Marcus Schaefer. Complexity of some geometric and topological problems. In David Eppstein and Emden R. Gansner, editors, *Graph Drawing, 17th International Symposium, GD 2009, Chicago, IL, USA, September 22–25, 2009. Revised Papers*, volume 5849 of *Lecture Notes in Computer Science*, pages 334–344. Springer, 2009. https://ovid.cs.depaul.edu/documents/convex.pdf. 290, 294, 296

[Sch12] Marcus Schaefer. *Thirty Essays on Geometric Graph Theory*. Chapter Realizability of graphs and linkages, pages 461–482. Springer, 1st edition, 2012. 297

[Sch20] Marcus Schaefer. The real logic of drawing graphs (YouTube talk), 2020. https://www.youtube.com/watch?v=4_Rq2M1wIEQ. 290

[Sch21] Marcus Schaefer. On the complexity of some geometric problems with fixed parameters. *Journal of Graph Algorithms and Applications*, 25(1):195–218, 2021. https://doi.org/10.7155/jgaa.00557. 296, 299

[Ser78] Anatoliy Serdyukov. On some extremal walks in graphs (in Russian). *Upravlyaemye Sisetmy*, 17:76–79, 1978. 207

[Set02] Takahiro Seta. The complexities of puzzles, cross sum, and their another solution problems (ASP) (senior honors thesis, University of Tokyo), 2002. https://www.cs.umd.edu/users/gasarch/BLOGPAPERS/takahiro.pdf. 279

[Sey95] Paul D. Seymour. Packing directed circuits fractionally. *Combinatorica*, 15(2):281–288, 1995. https://doi.org/10.1007/BF01200760. 262

[Sho91] Peter W. Shor. Stretchability of pseudolines is NP-hard. In Peter Gritzmann and Bernd Sturmfels, editors, *Applied Geometry and Discrete Mathematics. Proceedings of a DIMACS Workshop*, Providence, Rhode Island, USA, September 18, 1990, volume 4 of *DIMACS Series in Discrete Mathematics and Theoretical Computer Science*, pages 531–554. DIMACS/AMS, 1991. https://math.mit.edu/~shor/papers/Stretchability_NP-hard.pdf. 292, 294

[Sho94] Peter W. Shor. Algorithms for quantum computation: Discrete logarithms and factoring. In Shafi Goldwasser, editor, *35th Annual Symposium on Foundations of Computer Science*, Santa Fe, New Mexico, USA, 20–22 November 1994, pages 124–134. IEEE Computer Society, 1994. https://doi.org/10.1109/SFCS.1994.365700. 13, 14, 471, 472

[Sho99] Peter W. Shor. Polynomial-time algorithms for prime factorization and discrete logarithms on a quantum computer. In Arto Salomaa and Magnus Steinby, editors, *SIAM Review*, 41(2):303–332, 1999. https://doi.org/10.1137/S0036144598347011. 13, 14, 471, 472

[Sim77] Janos Simon. On the difference between one and many. In *Automata, Languages, and Programming (ICALP 1977)*, volume 52 of *Lecture Notes in Computer Science*, pages 480–491. Springer-Verlag, 1977. 286

[Sip83] Michael Sipser. A complexity theoretic approach to randomness. In David Johnson, editor, *Proceedings of the 15th Annual ACM Symposium on Theory of Computing*, 25–27 April, 1983, Boston, Massachusetts, USA, pages 330–335. ACM, 1983. https://doi.org/10.1145/800061.808762. 18

[Sla97] Peter Slavík. A tight analysis of the greedy algorithm for set cover. *Journal of Algorithms*, 25(2):237–254, 1997. https://doi.org/10.1006/jagm.1997.0887. 216

[Soa96] Robert Soare. Computability and recursion. *Bulletin of Symbolic Logic*, 2(4):284–321, 1996. https://www.cs.umd.edu/~gasarch/BLOGPAPERS/comprec.pdf. 23, 24

[Soa13] Robert Soare. Why Turing's thesis is not a thesis. In *Turing centenary volume*. Cambridge University Press, 2013. 5

[SRG20] Matthew Stephenson, Jochen Renz, and Xiaoyu Ge. The computational complexity of Angry Birds. *Artificial Intelligence*, 280:103232, 2020. https://doi.org/10.1016/j.artint.2019.103232. 330

[SS13] Michael E. Saks and C. Seshadhri. Space efficient streaming algorithms for the distance to monotonicity and asymmetric edit distance. In Sanjeev Khanna, editor, *Proceedings of the Twenty-Fourth Annual ACM-SIAM Symposium on Discrete Algorithms, SODA 2013*, New Orleans, Louisiana, USA, January 6–8, 2013, pages 1698–1709. SIAM, 2013. https://doi.org/10.1137/1.9781611973105.122. 418

[SS17] Marcus Schaefer and Daniel Štefankovič. Fixed points, Nash equilibria, and the existential theory of the reals. *Theory of Computing Systems*, 60(2):172–193, 2017. https://doi.org/10.1007/s00224-015-9662-0. 290, 292, 294, 300

[SS20] Michael Saks and Rahul Santhanam. Circuit lower bounds from NP-hardness of MCSP under Turing reductions. In Shubhangi Saraf, editor, *35th Computational Complexity Conference, CCC 2020*, July 28–31, 2020, Saarbrücken, Germany (Virtual Conference), volume 169 of *LIPIcs*, pages 26:1–26:13. Schloss Dagstuhl - Leibniz-Zentrum für Informatik, 2020. https://doi.org/10.4230/LIPIcs.CCC.2020.26. 14

[SS22] Marcus Schaefer and Daniel Stefankovic. Beyond the existential theory of the reals, 2022. https://arxiv.org/abs/2210.00571. 300

[SSV13] John Smolin, Graeme Smith, and Alexander Vargo. Oversimplifying quantum factoring. *Nature*, 499:163–165, 2013. 472

[SSvR11] Allan Scott, Ulrike Stege, and Iris van Rooij. Minesweeper may not be NP-complete but it is hard nonetheless. *The Mathematical Intelligencer*, 33:5–17, 2011. https://link.springer.com/content/pdf/10.1007/s00283-011-9256-x.pdf. 109

[ST13] Uwe Schöning and Jacob Toran. *The Satisfiability Problem*. Lehaman's Media, 2013. 160

[Sta25] Jack Stade. The point-boundary art gallery problem is $\exists\mathbb{R}$-hard. In Oswin Aichholzer and Haitao Wang, editors, *Proceedings of the 41st International Symposium on Computational Geometry (SoCG 2025)*, volume 332 of *Leibniz International Proceedings in Informatics (LIPIcs)*, pages 74:1–74:23, Kanazawa, Japan, 2025. https://doi.org/10.4230/LIPIcs.SoCG.2025.74. 298

[Ste07] Rob Stegmann. Rob's puzzle page: Pattern puzzles, 2007. 146

[Sto76] L. J. Stockmeyer. The polynomial-time hierarchy. *Theoretical Computer Science*, 3(1):1–22, October 1976. https://www.sciencedirect.com/science/article/pii/030439757690061X. 19

[Sto83] James A. Storer. On the complexity of chess. *Journal of Computer and System Sciences*, 27(1):77–100, 1983. 328

[Str80] Walter Stromquist. How to cut a cake fairly. *American Mathematics Monthly*, 87(8):640–644, 1980. 455

[SU96] Ed Scheinerman and Dan Ullman. *Fractional Graph Theory: A Rational Approach to the Theory of Graphs*. Dover, 1996. http://www.ams.jhu.edu/ers/books/fractional-graph-theory-a-rational-approach-to-the-theory-of-graphs/. 170

[SU08] Marcus Schaefer and Christopher Umans. Completeness in the Polynomial-Time hierarchy, 2008. http://ovid.cs.depaul.edu/documents/phcom.pdf. 21

[SWY15] Yaoyun Shi, Xiaodi Wu, and Wei Yu. Limits of quantum one-way communication by matrix Hypercontractive Inequality, 2015. https://www.cs.umd.edu/~xwu/papers/GHM_v4.pdf. 480

[SY11] Arezou Soleimanfallah and Anders Yeo. A kernel of order $2k - c$ for vertex cover. *Discrete Mathematics*, 311(10–11):892–895, 2011. https://doi.org/10.1016/j.disc.2011.02.014. 175

[SZ06] Jeff Stuckman and Guo-Qiang Zhang. Mastermind is NP-complete. *INFOCOMP Journal of Computer Science*, pages 25–28, 2006. http://arxiv.org/abs/cs/0512049. 49

[SZZ18] Katerina Sotiraki, Manolis Zampetakis, and Giorgos Zirdelis. PPP-completeness with connections to cryptography. In Mikkel Thorup, editor, *59th IEEE Annual Symposium on Foundations of Computer Science, FOCS 2018*, Paris, France, October 7–9, 2018, pages 148–158. IEEE Computer Society, 2018. https://arxiv.org/abs/1808.06407. 467

[Tan19] Ewin Tang. A quantum-inspired classical algorithm for recommendation systems. In Moses Charikar and Edith Cohen, editors, *Proceedings of the 51st Annual ACM SIGACT Symposium on Theory of Computing, STOC 2019*, Phoenix, AZ, USA, June 23–26, 2019, pages 217–228. ACM, 2019. https://eccc.weizmann.ac.il/report/2018/128. 470

[Tar48] Alfred Tarski. A decision method for elementary algebra and geometry, 1948. http://www.rand.org/content/dam/rand/pubs/reports/2008/R109.pdf. 290

[TC05] John Tromp and Rudy Cilibrasi. Limits of Rush Hour Logic complexity. Manuscript, February 2005. http://arxiv.org/abs/cs/0502068. 343

[Tho78] A. G. Thomason. Hamiltonian cycles and uniquely edge colourable graphs. *Annals of Discrete Mathematics*, 3:259–268, 1978. 275, 463

[Tip16] Simon Tippenhauer. *On planar 3-SAT and its variants*. PhD thesis, Fachbeheich Mathematik and Informatik der Freien Universitat Berlin, 2016. https://www.mi.fu-berlin.de/inf/groups/ag-ti/theses/download/Tippenhauer16.pdf. 97

[Tis13] Catalin-Stefan Tiseanu. Promised streaming algorithms and finding psudeo-repetitions. Master's thesis, University of Maryland, College Park, MD, 2013. 411

[Tod91] Seinosuke Toda. PP is as hard as the polynomial-time hierarchy. *SIAM Journal of Computing*, 20(5):865–877, 1991. https://doi.org/10.1137/0220053. 272

[Tov84] Craig Tovey. A simplified NP-complete satisfiability problem. *Discrete Applied Mathematics*, 8:85–89, 1984. https://www.sciencedirect.com/science/article/pii/0166218X84900817. 37, 39, 234

[Tra84] Boris Trakhenbrot. A survey of Russian approaches to perebor (brute-force searches) algorithms. *Annals of the History of Computing*, 6:384–400, 1984. https://ieeexplore.ieee.org/document/4640789. xiv, 9, 33

[Tre11] Luca Trevisan. Lecture 8: A linear programming relaxation of set cover, 2011. http://theory.stanford.edu/~trevisan/cs261/lecture08.pdf. 264

[Tre12] Luca Trevisan. Pseudorandomness and derandomization. *XRDS: Crossroads, The ACM Magazine for Students*, 18(3):27–31, 2012. https://doi.org/10.1145/2090276.2090287. 18

[TSSW00] Luca Trevisan, Gregory B. Sorkin, Madhu Sudan, and David P. Williamson. Gadgets, approximation, and linear programming. *SIAM Journal of Computing*, 29(6):2074–2097, 2000. https://doi.org/10.1137/S0097539797328847. 239

[Tur37] A. M. Turing. On computable numbers, with an application to the Entscheidungsproblem. *Proceedings of the London Mathematical Society 2*, 42 with a correction in 43:230–65, 1937. https://www.cs.virginia.edu/~robins/Turing_Paper_1936.pdf. 46

[Tut46] William T. Tutte. On Hamiltonian circuits. *The Journal of the London Mathematical Society*, 1(2):98–101, 1946. 275

[Tut56] William T. Tutte. A theorem on planar graphs. *Transactions of the American Mathematical Society*, 82(1):99–116, 1956. https://www.ams.org/journals/tran/1956-082-01/S0002-9947-1956-0081471-8/S0002-9947-1956-0081471-8.pdf. 112

[UA93] Unknown-Author. Magic the Gathering official website, 1993. https://magic.wizards.com/en. 335

[UL97] Christopher Umans and William Lenhart. Hamiltonian cycles in solid grid graphs. In Anna Karlin, editor, *Proceedings of the 38th Annual Symposium on Foundations of Computer Science*, pages 496–505. IEEE, 1997. https://ieeexplore.ieee.org/document/646138. 116

[UN96] Nobuhisa Ueda and Tadaaki Nagao. NP-completeness results for NONOGRAM via parsimonius reductions. Technical Report TR96-0008, Dept of Computer Science, Tokyo Institute of Technology, 1996. https://citeseerx.ist.psu.edu/document?doi=1bb23460c7f0462d95832bb876ec2ee0e5bc46cf. 274

[Und78] P. Underground. On graphs with Hamiltonian squares. *Discrete Mathematics*, 21(3):323, 1978. https://doi.org/10.1016/0012-365X(78)90164-4. 112

[Unk] Unknown. The MetiTarski theorem prover, 2014. https://www.cl.cam.ac.uk/~lp15/papers/Arith/. 294

[Unk19] Unknown. (strong) Exponential Time Hypothesis, 2019. https://www.mpi-inf.mpg.de/fileadmin/inf/d1/teaching/summer19/finegrained/lec2.pdf. 198

[Urq10] Alasdair Urquhart. Von Neumann, Gödel and complexity theory. *Bulletin of Symbolic Logic*, 16(4):516–530, 2010. xiv

[US15] Akihiro Uejima and Hiroaki Suzuki. Fillmat is NP-complete and APSP-complete. *Journal of Information Processing*, 23(3):310–316, 2015. https://doi.org/10.2197/ipsjjip.23.310. 276

[Vad01] Salil P. Vadhan. The complexity of counting in sparse, regular, and planar graphs. *SIAM Journal on Computing*, 31(2):398–427, 2001. https://doi.org/10.1137/S0097539797321602. 274

[Val79a] Leslie G. Valiant. The complexity of computing the permanent. *Theoretical Computer Science*, 8:189–201, 1979. http://dx.doi.org/10.1016/0304-3975(79)90044-6. 271, 273, 282, 286

[Val79b] Leslie G. Valiant. The complexity of enumeration and reliability problems. *SIAM Journal on Computing*, 8(3):410–421, 1979. http://dx.doi.org/10.1137/0208032. 271, 274, 286

[Val79c] Leslie G. Valiant. Negative results on counting. In Klaus Weihrauch, editor, *Theoretical Computer Science, 4th GI-Conference, Aachen, Germany, March 26–28, 1979, Proceedings*, pages 38–46, 1979. 271, 286

[Vaz01] Vijay Vazirani. *Approximation Algorithms*. Springer, 2001. 214, 219

[vdZ15] Tom C. van der Zanden. Parameterized complexity of graph constraint logic. In Thore Husfeldt and Iyad Kanj, editors, *10th International Symposium on Parameterized and Exact Computation (IPEC 2015)*, volume 43 of *Leibniz International Proceedings in Informatics (LIPIcs)*, pages 282–293, Dagstuhl, Germany, 2015. Schloss Dagstuhl–Leibniz-Zentrum fuer Informatik. 343

[vdZB15] Tom C. van der Zanden and Hans L. Bodlaender. PSPACE-completeness of Bloxorz and of games with 2-buttons. In Vangelis Th. Paschos and Peter Widmayer, editors, *Proceedings of the 9th International Conference on Algorithms and Complexity*, pages 403–415, 2015. 308

[vEB81] Peter van Emde Boas. Another NP-complete problem and the complexity of computing short vectors in a lattice. Tech Report 8104, University of Amsterdam, Dept of Mathematics, Neverthlands, 1981. https://staff.fnwi.uva.nl/p.vanemdeboas/vectors/abstract.html. 249

[Veg90] Gert Vegter. The visibility diagram: A data structure for visibility problems and motion planning. In John R. Gilbert and Rolf G. Karlsson, editors, *SWAT 90, 2nd Scandinavian Workshop on Algorithm Theory*, Bergen, Norway, July 11–14, 1990, Proceedings, volume 447 of *Lecture Notes in*

Computer Science, pages 97–110. Springer, 1990. https://dl.acm.org/doi/10.5555/88723.88746. 369

[Vig12] Giovanni Viglietta. Hardness of Mastermind. In Evangelos Kranakis, Danny Krizanc, and Flaminia L. Luccio, editors, *Proceedings of the 6th International Conference on Fun with Algorithms (FUN 2012)*, volume 7288 of *Lecture Notes in Computer Science*, pages 368–378, Venice, Italy, June 2012. Springer. https://arxiv.org/abs/1111.6922. 49

[Vig14a] Giovanni Viglietta. Gaming is a hard job, but someone has to do it! *Theory of Computing Systems*, 54(4):595–621, 2014. http://dx.doi.org/10.1007/s00224-013-9497-5. 124, 126, 303, 305, 308

[Vig14b] Giovanni Viglietta. Lemmings is PSPACE-complete. In Alfredo Ferro, Fabrizio Luccio, and Peter Widmayer, editors, *Proceedings of the 7th International Conference on Fun with Algorithms*, pages 340–351, 2014. 313

[Viz64] Vadim Vizing. On an estimate of the chromatic class of a p-graph. *Diskret. Analiz.*, 3:25–30, 1964. 251

[VW18] Virginia Vassilevska Williams and R. Ryan Williams. Subcubic equivalences between path, matrix, and triangle problems. *Journal of the Association of Computing Machinery (JACM)*, 65(5):27:1–27:38, 2018. https://doi.org/10.1145/3186893. 382, 385, 388, 389

[VY11] Elad Verbin and Wei Yu. The streaming complexity of cycle counting, sorting by reversals, and other problems. In Dana Randall, editor, *Proceedings of the Twenty-Second Annual ACM-SIAM Symposium on Discrete Algorithms, SODA 2011*, San Francisco, California, USA, January 23–25, 2011, pages 11–25. SIAM, 2011. https://doi.org/10.1137/1.9781611973082.2. 480

[Wag13] Samuel Wagstaff. *The Joy of Factoring*. AMS, Providence, 2013. 14, 471

[Wan60] Hao Wang. Proving theorems by pattern recognition I. *Communications of the ACM*, 3(4): 220–234, 1960. https://doi.org/10.1145/367177.367224. 337

[War62] Stephen Warshall. A theorem on Boolean matrices. *Journal of the Association of Computing Machinery (JACM)*, 9:11–12, 1962. 378

[Wika] Wikipedia. 15 puzzles, 2012. https://en.wikipedia.org/wiki/15_puzzle. 241

[Wikb] Wikipedia. 2-satisfiability, 2004. https://en.wikipedia.org/wiki/2-satisfiability. 33

[Wikc] Wikipedia. Discrete logarithm, 2003. https://en.wikipedia.org/wiki/15_puzzle. 14

[Wikd] Wikipedia. Eternity II puzzle, 2025. https://en.wikipedia.org/wiki/Eternity_II_puzzle. 146

[Wike] Wikipedia. Eternity puzzle, 2004. https://en.wikipedia.org/wiki/15_puzzle. 140

[Wikf] Wikipedia. Quantum psuedo-telepathy, 2008. https://en.wikipedia.org/wiki/15_puzzle. 483

[Wikg] Wikipedia. SameGame, 2001. https://en.wikipedia.org/wiki/15_puzzle. https://en.wikipedia.org/wiki/SameGame. 153

[Wil05] Ryan Williams. A new algorithm for optimal 2-constraint satisfaction and its implications. *Theoretical Computer Science*, 348(2–3):357–365, 2005. https://doi.org/10.1016/j.tcs.2005.09.023. 167, 374

[Wil18] R. Ryan Williams. Faster all-pairs shortest paths via circuit complexity. *SIAM Journal on Computing*, 47(5):1965–1985, 2018. https://doi.org/10.1137/15M1024524. 379

[WS11] David Williamson and David Shmoys. *Design of Approximation Algorithms*. Cambridge University Press, 2011. 219

[WW10] Virginia Vassilevska Williams and Ryan Williams. Subcubic equivalences between path, matrix and triangle problems. In *51th Annual IEEE Symposium on Foundations of Computer Science, FOCS 2010*, October 23–26, 2010, Las Vegas, Nevada, USA, pages 645–654. IEEE Computer Society, 2010. Journal version is in JACM, 2018. 379

[Yao77] Andrew Chi-Chih Yao. Probabilistic computations: Toward a unified measure of complexity (extended abstract). In Paul Young, editor, *18th Annual Symposium on Foundations of Computer Science*, Providence, Rhode Island, USA, 31 October - 1 November 1977, pages 222–227. IEEE Computer Society, 1977. https://doi.org/10.1109/SFCS.1977.24. 399

[Yat00] Takayuki Yato. On the NP-completeness of the Slither Link puzzle. *IPSJ SIGNotes Algorithms*, 74:25–32, 2000. 123, 281

[Yat03] Takayuki Yato. Complexity and completeness of finding another solution and its application to puzzles. Master's thesis, The University of Tokyo, 2003. 281

[YC24] Howe Choong Yin and Alex Churchill. A programming language embedded in *Magic: The Gathering*. In Andrei Z. Broder and Tami Tamir, editors, *Proceedings of the 12th International Conference on Fun with Algorithms*, volume 291 of *LIPIcs*, pages 31:1–31:19, 2024. 335

[YS03] Takayuki Yato and Takahiro Seta. Complexity and completeness of finding another solution and its application to puzzles. *IEICE Transactions on Fundamentals of Electronics, Communications and Computer Sciences*, E86-A(5):1052–1060, 2003. http://mountainvistasoft.com/docs/ASP.pdf. 274, 276

[YSI05] Takayuki Yato, Takahiro Seta, and Tsuyoshi Ito. Finding yozume of generalized Tsume-Shogi is Exptime-complete. *IEICE Trans. Fundamemtals of Electronics, Communication and Computer Sciences*, 88-A(5):1249–1257, 2005. https://doi.org/10.1093/ietfec/e88-a.5.1249. 329

[Yu16] Jingjin Yu. Intractability of optimal multirobot path planning on planar graphs. *IEEE Robotics and Automation Letters*, 1(1):33–40, 2016. https://doi.org/10.1109/LRA.2015.2503143. 97

[Zhu20] Dmitriy Zhuk. A proof of the CSP dichotomy conjecture. *Journal of the ACM*, 67(5), August 2020. 46, 71, 305

[ZM22] Dmitriy Zhuk and Barnaby Martin. QCSP monsters and the demise of the Chen conjecture. *Journal of the ACM*, 69(5):35:1–35:44, 2022. 305

[Zuc07] David Zuckerman. Linear degree extractors and the inapproximability of Max Clique and Chromatic Number. *Theory of Computing*, 3(6):103–128, 2007. http://theoryofcomputing.org/articles/v003a006/index.html. 216

[Zwi98] Uri Zwick. All pairs shortest paths in weighted directed graph $\frac{3}{4}$ exact and almost exact algorithms. In Jeffery Vitter, editor, *39th Annual Symposium on Foundations of Computer Science, FOCS '98*, November 8–11, 1998, Palo Alto, California, USA, pages 310–319. IEEE Computer Society, 1998. https://doi.org/10.1109/SFCS.1998.743464. 379

[Zwi02] Uri Zwick. All pairs shortest paths using bridging sets and rectangular matrix multiplication. *Journal of the ACM*, 49(3):289–317, 2002. https://doi.org/10.1145/567112.567114. 389

Index

Publisher contact:
The MIT Press
Massachusetts Institute of Technology
77 Massachusetts Avenue, Cambridge, MA 02139
mitpress.mit.edu

EU Authorised Representative:
Easy Access System Europe, Mustamäe tee 50,
10621 Tallinn, Estonia
gpsr.requests@easproject.com

Printed by Integrated Books International,
United States of America